177-200

WITHDRAWN

SCOPE 37

Biological Invasions

Scientific Committee on Problems of the Environment
SCOPE
Executive Committee, elected 10 June 1988

Officers

President: Prof. F. di Castri, CEPE/CNRS, Centre L. Emberger, Route de Mende, BP 5051, 34033 Montpellier Cedex, France.

Vice-President: Academician M. V. Ivanov, Institute of Microbiology, USSR Academy of Sciences, GSP-7 Prospekt 60 letija Oktjabrja 117811, Moscow, USSR.

Vice-President: Professor C. R. Krishna Murti, Scientific Commission for Continuing Studies on Effects of Bhopal Gas Leakage on Life Systems, Cabinet Secretariat, 2nd floor, Sardar Patel Bhavan, New Delhi 110 001, India.

Treasurer: Dr T. E. Lovejoy, Smithsonian Institution, Washington, DC 20560, USA.

Secretary-General: Prof. J. W. B. Stewart, Saskatchewan Institute of Pedology, University of Saskatchewan, Saskatoon, S7N 0W0 Saskatchewan, Canada.

Members

Prof. M. O. Andreae (I.U.G.G. representative), Max-Planck-Institut für Chemie, Postfach 3060, D-6500 Mainz, FRG.

Prof. M. A. Ayyad, Faculty of Science, Alexandria University, Moharram Bey, Alexandria, Egypt.

Prof. R. Herrera, (I.U.B.S. representative), Centro de Ecologia y Ciencias Ambientales (IVIC), Carretera Panamericana km. 11, Apartado 21827, Caracas, Venezuela.

Prof. M. Kecskés, Department of Microbiology, University of Agricultural Sciences, Pater K. utca 1, 2103 Gödöllö, Hungary.

Professor R. O. Slatyer, School of Biological Sciences, Australian National University, P. O. Box 475, Canberra, ACT 2601, Australia.

SCOPE 37

Biological Invasions

A Global Perspective

Edited by

J. A. Drake

Department of Zoology and Graduate Program in Ecology, University of Tennessee, Knoxville, USA

H. A. Mooney

Department of Biological Sciences, Stanford University, Stanford, USA

F. di Castri

CNRS-Centre L Emberger, Route de Mende–BP 5051, 34033 Montpellier Cedex, France

R. H. Groves

CSIRO Division of Plant Industry, GPO Box 1600, Canberra ACT 2601, Australia

F. J. Kruger

South African Forestry Research Unit, PO Box 727, Pretona 0001, Republic of South Africa

M. Rejmánek

Department of Botany, University of California, Davis, California 95616, USA

and

M. Williamson

Department of Biology, University of York, York Y01 5DD

Published on behalf of the
Scientific Committee on Problems of the Environment (SCOPE)
of the
International Council of Scientific Unions (ICSU)
by

JOHN WILEY & SONS

Chichester · New York · Brisbane · Toronto · Singapore

Copyright © 1989 by the
Scientific Committee on Problems of the Environment (SCOPE)

All rights reserved

No part of this book may be reproduced by any means
or transmitted, or translated into a machine language
without the written permission of the copyright holder.

Library of Congress Cataloging-in-Publication Data

Biological invasions: a global perspective/edited by J. A. Drake and
H. A. Mooney.
p. cm.—(SCOPE; 37)
Papers from a number of national and international workshops
resulting from a SCOPE program on the ecology of biological
invasions initiated in mid-1982.
Includes bibliographies and index.
ISBN 0 471 92085 1
1. Biological invasions—Congresses. 2. Ecological succession—
Congresses. I. Drake, James A., 1954– . II. Mooney, Harold A.
III. International Council of Scientific Unions. Scientific
Committee on Problems of the Environment. IV. Series: SCOPE
(Series); 37.
QH353.B56 1989
574.5—dc19

88-20765
CIP

British Library Cataloguing in Publication Data available

Phototypesetting by Thomson Press (India) Limited, New Delhi
Printed and bound in Great Britain by Courier International, Tiptree, Essex

Funds to meet SCOPE expenses are provided by contributions from SCOPE National Committee, an annual subvention from ICSU (and through ICSU, from UNESCO), an annual subvention from French Ministère de l'Environment, contracts with UN Bodies, particularly UNEP, and grants from Foundations and industrial enterprises.

International Council of Scientific Unions (ICSU)
Scientific Committee on the Problems of the Environment (SCOPE)

SCOPE is one of a number of committees established by a non-governmental group of scientific organizations, the International Council of Scientific Unions (ICSU). The membership of ICSU includes representatives from 74 National Academies of Science, 20 International Unions and 26 other bodies called Scientific Associates. To cover multidisciplinary activities which include the interest of several unions, ICSU has established 10 scientific committees, of which SCOPE in one. Currently, representatives of 35 member countries and 20 international scientific bodies participate in the work of SCOPE, which directs particular attention to the needs of developing countries. SCOPE was established in 1969 in response to the environmental concerns emerging at that time; ICSU recognized that many of these concerns required scientific inputs spanning several disciplines and ICSU Unions. SCOPE's first task was to prepare a report on Global Environmental Monitoring (SCOPE 1, 1971) for the UN Stockholm Conference on the Human Environment.

The mandate of SCOPE is to assemble, review, and assess the information available on man-made environmental changes and the effects of these changes on man; to assess and evaluate the methodologies of measurement of environmental parameters; to provide an intelligence service on current research; and by the recruitment of the best available scientific information and constructive thinking to establish itself as a corpus of informed advice for the benefit of centres of fundamental research and of organizations and agencies operationally engaged in studies of the environment.

SCOPE is governed by a General Assembly, which meets every three years. Between such meetings its activities are directed by the Executive Committee.

R. E. Munn
Editor-in-Chief
SCOPE Publications

Executive Secretary: V. Plocq

Secretariat: 51 Bld de Montmorency
75016 PARIS

Contents

Contributors

PETER J. ASHTON	National Institute for Water Research, Council for Scientific and Industrial Research, PO Box 395, Pretoria 0001, South Africa.
G. J. BREYTENBACH	Saasveld Forestry Research Centre, Private Bag X6515, George 6530, South Africa.
JAMES H. BROWN	Department of Biology, University of New Mexico, Albuquerque, NM 87131, USA.
MICHAEL J. CRAWLEY	Department of Pure and Applied Biology, Imperial College at Silwood Park, Ascot, Berkshire SL5 7PY, UK.
FRANCESCO DI CASTRI	CNRS-CEPE, Centre L. Emberger, Route de Mende, B. P. 5051, 34033 Montpellier Cedex, France.
J. A. DRAKE	Department of Zoology and the Graduate Program in Ecology, University of Tennessee, Knoxville, TN 37996, USA.
PAUL R. EHRLICH	Department of Biological Sciences, Stanford University, Stanford, California 94305, USA.
RICHARD H. GROVES	CSIRO Division of Plant Industry, GPO Box 1600, Canberra, ACT 2601, Australia.
O. HAMANN	Botansk Laboratorium, Kobenhavns Universitet, Gothersgade 140, DK-1123 Kobenhavn K., Denmark.

VERNON H. HEYWOOD — Plant Science Laboratories, University of Reading, PO Box 221, Reading, Berkshire, UK.

RICHARD J. HOBBS — CSIRO, Division of Wildlife and Rangelands Research, LMB 4, PO Midland, WA 6056, Australia.

FRED J. KRUGER — Department of Environmental Affairs, South African Forestry Research Institute, 4 Ketjen Street, Pretoria West, PO Box 727, Pretoria 0001, South Africa.

SIMON A. LEVIN — Section of Ecology and Systematics, Ecosystems Research Center, Cornell University, Ithaca, NY, USA.

LLOYD A. LOOPE — Haleakala National Park, Box 369, Makawao, Maui, Hawaii 96768, USA.

IAN A. W. MACDONALD — Percy Fitzpatrick Institute of African Ornithology, University of Cape Town, Rondebosch, 7700 Republic of South Africa.

RICHARD N. MACK — Department of Botany, Washington State University, Pullman, WA 99164-4230, USA.

DAVID S. MITCHELL — CSIRO Centre for Irrigation Research, Private Mail Bag, Griffith NSW 2680, Australia.

H. A. MOONEY — Department of Biological Sciences, Stanford University, Stanford, California 94305, USA.

DIETER MUELLER-DOMBOIS — Department of Botany, University of Hawaii at Manoa, Honolulu, Hawaii, 96822, USA.

IAN R. NOBLE — Research School of Biological Sciences, Australian National University, Box 475 Canberra ACT 2601, Australia.

STUART L. PIMM — Department of Zoology and Graduate Program in Ecology, The University of Tennessee, Knoxville, TN 37996, USA.

P. S. RAMAKRISHNAN — School of Environmental Sciences, Jawaharlal Nehru University, New Delhi 110067, India.

MARCEL REJMANEK — Department of Botany, University of California, Davis, CA 95616, USA.

D. M. RICHARDSON — South African Forestry Research Institute, PO Box 727, Pretoria 0001, South Africa.

DANIEL SIMBERLOFF — Department of Biological Science, Florida State University, Tallahassee, Florida 32306, USA.

MICHAEL B. USHER — Department of Biology, University of York, York YO1 5DD, UK.

PETER M. VITOUSEK — Department of Biological Sciences, Stanford University, Stanford, California 94305, USA.

SHARON L. VON BROEMBSEN — Plant Protection Research Institute, Stellenbosch, South Africa.

MARK WILLIAMSON — Department of Biology, University of York, York YO1 5DD, UK.

Preface

This volume represents the culmination of activity resulting from a SCOPE (Scientific Committee on Problems of the Environment) program on the Ecology of Biological Invasions. This program was initiated in mid-1982 and had its origins at the Third International Conference on Mediterranean Ecosystems held in Stellenbosch, South Africa in 1980. At that conference the considerable impact of invading plants on South African natural ecosystems was highlighted. Particularly noteworthy was that these invasions involved woody plants, such as *Hakea, Acacia*, and *Pinus*, and that they were aggressively entering and even transforming natural systems. The scientific consensus at that time was that invasions were only successful into disturbed ecosystems. It seemed therefore that the time had come to take a new look at the nature of invading plants and animals. At first the idea was to focus on the five mediterranean-climate regions of the world, all of which shared common invading species. However, in keeping with the mandate of SCOPE to advance knowledge on environmental problems of global significance the scale of the project became worldwide in perspective. At the General Assembly of SCOPE held in Ottawa in June 1982 the program was approved. The specific questions the program addressed were:

1. What are the factors that determine whether a species will be an invader or not?
2. What are the site properties that determine whether an ecological system will be relatively prone to, or resistant to, invasion?
3. How should management systems be developed using the knowledge gained from answering these questions?

The primary focus of the program was on the those animals, plants, and micro-organisms that have been successful invaders of non-agricultural regions with an emphasis on those that have disrupted ecosystem function. It was felt that the SCOPE program should build on the considerable knowledge base available on invaders of agricultural systems but that it should concentrate its efforts on natural systems where there had been considerably less attention.

In order to carry out this mandate a scientific advisory committee was established consisting of:

H. A. Mooney, USA (Chairman)
F. di Castri, France
R. H. Groves, Australia
F. J. Kruger, South Africa
M. Rejmánek, Czechoslovakia and USA
M. Williamson, Great Britain
J. A. Drake of the USA served as the program coordinator.

The program has consisted of a number of national workshops (Australia, Czechoslovakia, Great Britain, Netherlands, USA, and South Africa) summarizing regional knowledge, and a series of international working groups addressing special problems. The latter included invasions into nature reserves, invasions into mediterranean-climate ecosystems, and modelling the invasion process. Finally a concluding international workshop was held to synthesize the regional and working group results into cross-continental comparisons. This book is the result of this culminating workshop held at, and co-sponsored by, the East-West Center of Honolulu, Hawaii.

The chapters in this book represent in large part extensions and refinements of treatments given to a topic early in the program. The shift in focus is from regional to cross-continental. A number of chapters, however, cover topics that were not addressed in the national workshops yet were important for a global treatment even though the information base was difficult to address for one reason or another. In total then, these chapters represent an assessment of where we are in our evaluation and understanding of the dramatic rearrangement of the earth's biota that has taken place over the last few centuries.

A list of the publications resulting from the entire SCOPE program is given in Appendix I. Appreciation is expressed to the scientists of many nations who participated in this SCOPE endeavor, to the A. W. Mellon Foundation for support of the effort, and to Anne B. Ferrell for continuous support.

Biological Invasions: a Global Perspective
Edited by J. A. Drake *et al.*
© 1989 SCOPE. Published by John Wiley & Sons Ltd

CHAPTER 1

History of Biological Invasions with Special Emphasis on the Old World

FRANCESCO DI CASTRI

1.1 INTRODUCTION

It is admittedly an almost impossible task to collate, in a single short chapter, all the wealth of information on the history of biological invasions. History, indeed, should be considered in terms of both geological-evolutionary events and human-related factors; there is obviously a continuum as well as feedback relations between them. These interactions should be taken into account to understand trends and mechanisms of two processes: the species invasions across different biogeographical areas and the susceptibility to invasion ('invasibility') of the diverse ecosystems.

Proper attention to the historical background has been given in the previous regional or national syntheses on biological invasions, for instance of Australia (Groves and Burdon, 1986) and of North America and Hawaii (Mooney and Drake, 1986). In particular, Deacon (1986) presents a comprehensive historical profile of biological invasions in South Africa, and Coope (1986) refers to the palaeoecology of invasions of North Atlantic islands. As regards specifically the history of invasion of weeds, reference should be made to the syntheses of Foy *et al.* (1983) and Gwynne and Murray (1985), while the history of insect introductions is covered by Sailer (1983) and that of plant pathogen introductions by Yarwood (1983). Many chapters of this volume itself have quite understandably their own historical background. Finally, the symposium sponsored by SCOPE and held in Montpellier (France) from 21 to 23 May 1986 specifically addressed 'History and Patterns of Biological Invasions in Europe and the Mediterranean Basin'; the contributions of Le Floc'h *et al.* (1989), Marcuzzi (1989), Pons *et al.* (1989), Sykora (1989) and Vernet (1989) were among the most historically oriented articles.

Having discarded the possibility of compiling a real 'state of knowledge' report on the history of invasions because of the space limitations, and having considered it inappropriate to present too many unrelated episodic or anecdotal examples of invasions, I am compelled to go rather far towards a context of generalization. This is in line with the objectives of this volume, but it may appear still premature at this stage.

Clearly, an analysis of the chronological steps of biological invasions within the Old World, from the Old World, and to the Old World, provides some of the best examples to propose a tentative framework for historical generalization. It must be stressed, however, that a cautious frame of mind is needed for a progressive conceptualization of a theory of biological invasions.

First of all, a number of concepts that ecologists use, namely those of resilience and disturbance, have very different meanings when applied to such different biota as plants, large animals, invertebrates, parasites and microbes, and terrestrial and aquatic organisms. These meanings are even more difficult to reconcile when different ecosystem types are considered.

Secondly, historical analyses—both the geological and human ones—are viewed at such a large scale of space and time that (in accordance with the hierarchical theory) they are more able to provide information on the relevance and constraints of a given process than explore the mechanisms of testable hypotheses. Experimental research on biological invasions is badly needed at this time.

Thirdly, a general interpretation of the invasion problem implies an inextricable mixture of ecological and genetic attributes of species from one side, and of chance, timing and human-derived opportunities from the other side (di Castri, 1989). If a geologically based analysis can provide insights on the emergence of genetic and ecological attributes, facilitating the invasion potential of species, human historical studies can provide the background to explain chance and opportunities for invasions.

Having already stressed the continuum between the geological and human dimensions of history, this chapter—nevertheless—emphasizes the latter dimension. A human historical approach is important, partly because it is inherent to the peculiarities of human occupation and migrations in the Old World, and partly because few attempts have been made in the previous workshops to explicate human history as a driving force of biological invasions. After all, the breakdown of natural biogeographical realms and their barriers, such is the cornerstone of the overall problem of biological invasions, is immediately imputable to unbounded economic and socio-cultural forces around the year 1500 AD rather than explicable on solely biological grounds.

In accordance with the objectives of this chapter, where generalities on the relevance of historical factors have to be combined with specificities related to the Old World, I will propose a few grossly defined steps as a framework to understand the relation between historical analyses and the invasion of species and the susceptibility of ecosystems to invasion. Later, I will reply in a preliminary way to such questions as: 'Do Old World organisms have a greater invasion potential in newly colonized ecosystems?', 'If so, is it because of intrinsic evolutionary-shaped attributes, or because of more opportunities provided by Old World men?', and 'Are the Old World ecosystems less susceptible to invasion by alien species than New World ecosystems?' With this background, I will conclude by proposing another preliminary framework on human history as a driving force in the Old World in relation to biological invasions.

1.2 THE RELEVANCE OF AN HISTORICAL BACKGROUND TO UNDERSTANDING TRENDS AND PATTERNS OF BIOLOGICAL INVASIONS

The main aspects related to the above point are schematically presented in Table 1.1. The five steps along a time scale cannot be precisely defined, because of a time lag as regards the occurrence of similar events in different regions. This is particularly true in the pre-historical period, since there may be a difference of thousands of years regarding the presence and effects of early man, even within a given biogeographical area. Anyway, the duration of any given period decreases massively from millions of years as regards the geological one, tens of thousands of years for the pre-historical, a few thousands for the first historical period, a few hundreds from 1500 AD up to present. Presumably, a time span of only a few dozen years are considered in so far as the immediate perspective is concerned.

Admittedly, the selection for each period of the main events and of the lessons to be learned as regards invasion processes constitutes an arbitrary choice. Some of the points of Table 1.1 will be discussed in some length in this chapter, others are hopefully self-explanatory.

Geologically speaking, invasions of species from one continent to another are true evolutionary processes, somewhat like speciation and extinction. Classical examples are the rapid intercontinental expansion of the primitive horse, or the waves of migrations between the two Americas. This migration was more massive from North to South, when the broken isthmus of Panama in the Tertiary was bridged in successive times; invasions and extinctions—the latter concerning particularly South American mammals—are well documented processes. These processes can provide geological insights for a better understanding of the present-day phenomena of biological invasions (Elton, 1958).

In general, all large-scale climatic changes and geological crises—even very old ones like the Permian-Triassic crisis—are at the origin of massive exchanges of flora and fauna. At a smaller scale, it is well known that physical barriers such as oceans, mountains or deserts can be overcome by many organisms. These organisms move on rafts of vegetation carried by rivers and marine currents, long-distance wind transport, or phoresis (non-flying species tied up to, and carried by, birds or larger flying arthropods). The importance of these factors has probably been underestimated in the past by a number of biogeographers.

Nevertheless, the main lesson to be learned from studies of invasion processes, at the geological scale, may well be to understand whether or not, or to what extent, a given evolutionary and geological history has facilitated or undermined the potential of a species to become an invader. Additionally, one might ask how history has increased or decreased the susceptibility of an ecosystem and a region

Table 1.1. Historical steps as related to specific invasion processes

Time scale	Main events	Knowledge to be gained as regards the invasion of species and the invasibility of ecosystems
Geological	Climatic changes Tectonic processes	Understand the evolution of genetic and ecological attributes facilitating species invasion potential—as well as of ecosystem resilience—in relation to different disturbance regimes. Follow patterns of invasion of species from one biogeographical realm to another (e.g. from and to North America and South America) in the absence of man.
Pre-historical	Emergence of the human condition Synchronous climatic and tectonic changes	Pinpoint the concomitant effects on species of natural (endogenous) disturbance and of man-made (exogenous) disturbance, and in particular the interactions that may have led to positive or negative feedbacks increasing or decreasing the invasion potential of species.
Historical up to 1500 AD	Stronger and more extended man-made perturbations Migrations and 'shaking up' of human populations within a given biogeographical realm Improved marine transportation systems	Study patterns of success (and failure) of biological invaders closely associated with man's actions, within a given biogeographical realm. Follow the effects—as regards biological invasions and colonizations—of the first break-down by man of biogeographical realms (e.g. human migration from Southeast Asia to Pacific islands). Compare the effects of early biological invasions on the biota of islands of different size and located at different distances from continental masses.
Historical from 1500 AD to present	Opening of new man-made routes across biogeographical realms Intercontinental human migrations and colonizations, driven by strong economic and social pressures, and made possible by new transportation and communications systems	Highlight trends of species introduction and colonization (or their failure to colonize) from one biogeographical realm to another, due to intentional or inadvertent introduction by man. Compare the susceptibility to invasion of ecosystems, mainly those of the same type (tropical, savannas, mediterranean-climate, etc.) but

Table 1.1. (*Contd.*)

Time scale	Main events	Knowledge to be gained as regards the invasion of species and the invasibility of ecosystems
	Progressive globalization of trade and exchanges Enlargement of the scale of space and 'acceleration' of the scale of time	belonging to different biogeographical realms, and with different phylogenetic heritage and land-use history. Defect patterns of biological invasions as related to new large-scale uniform agricultural systems in 'colonized' territories.
Perspectives from present	Global interdependence of market economies and worldwide urbanization, thus implying further landscape uniformity and agricultural simplification, massive deforestation, and dramatically declining biological diversity New types of disturbance (e.g. pollutants, including nuclear fall-out) Genetic bioengineering Man-induced climatic changes	Monitor the changing behaviour of species or varieties that may have acquired an invader potential under new biologically simplified situations or thanks to new dispersal opportunities. Monitor the release, dispersal patterns and behaviour in natural 'free' conditions of genetically engineered organisms. Compare the invasibility of ecosystems submitted to increased human perturbations, from protected areas to strongly stressed ecosystems. Perceive the urgent need of a new generation of experimental predictive research on biological invasion of species and susceptibility of ecosystems to invaders.

to invasion. A subsequent question to be posed is whether the emergence of man-made ('exogenous') disturbance—as adding to or interacting with natural ('endogenous') disturbance—has had a synergistic or an opposite effect as regards the invasion problem of species and ecosystems. The concept of disturbance is used here *sensu* Fox and Fox (1986).

Once the human-history factor is taken into consideration, a major task is to characterize the kinds of disturbance caused by man and man's role in shaping new routes of dispersal, which facilitate colonization by invaders. In a similar way as for the geological crises, some human-history crises or breaking points are of particular importance: (1) for having 'fixed' the patterns of the 'man-biological invader' relationships (a kind of symbiotic relation *sensu lato*), (2) for having unlocked the biogeographic enclosures and triggered successive waves of invaders, and (3) for having changed the rules and rhythm of the natural evolutionary game.

The first of the three human-history crises outlined above can be placed in early

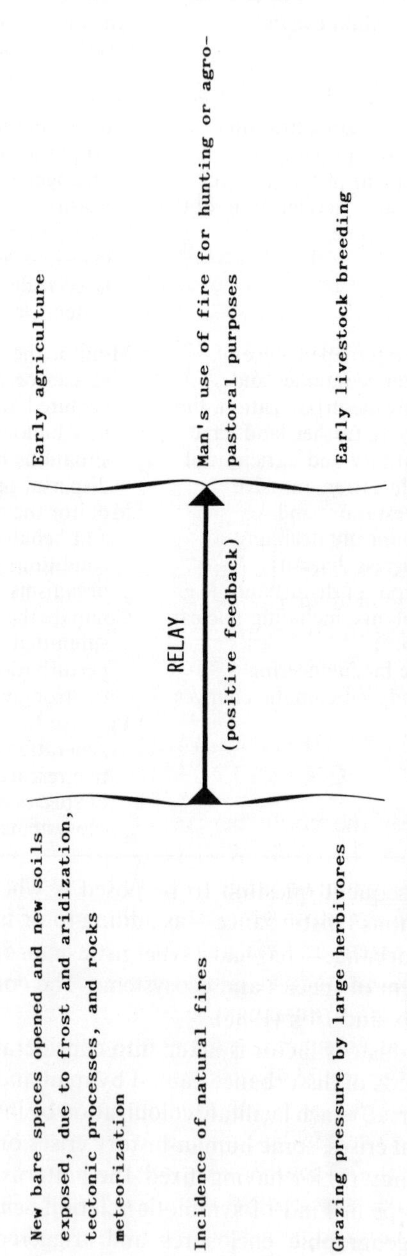

Figure 1.1. The 'relay' between natural and man-made disturbance

Neolithic times, when and wherein a sort of 'relay' took place between natural and man-made disturbance (as sketched in Figure 1.1). This relay may have synergistically strengthened some trends of natural selection, and created close ties between specific human actions and specific *potential* invaders. Under conditions of greater similarity between the effects of natural and man-made disturbance, of more ancient and more concomitant overlap of the two types of disturbance, and of more long-lasting association between man and an invader, one may assume that this relay should have been more efficient. These conditions seem to be particularly applicable to the patterns of early occupation by man of the lands around the Mediterranean Basin (Pons *et al.*, 1989; Vernet, 1989). Fortunately, these hypotheses can be challenged at present through a combination of the flourishing research on pollen analysis and profiles, on charcoal analysis, and on dendrochronology and archaeology.

The second, and more spectacular, historical crisis corresponds to the times of the 'great discoveries' and the ensuing European trade networks and 'colonization' (see Figures 1.8 and 1.9). Special attention to this aspect will be given in Section 1.5.

We are likely to face now the third crisis referred to above. Being able to act on a large scale, as the most impacting 'geological' agent (including inducing climatic changes), man has accelerated the pace of biological changes. In particular, man has genetically simplified many species. Other species have simply become extinct. Many 'surprises' may occur as new invaders and new trends and patterns of biological invasion emerge in a near future.

1.3 PECULIARITIES OF THE OLD WORLD AS RELATED TO THE INVASION POTENTIAL OF ITS SPECIES

The Old World covers the most extensive and continuous continental land mass. At least from East (Siberia) to West (Atlantic coasts) the physical barriers, represented by some mountain ranges (e.g. the Urals) and several large rivers, are not insurmountable for most organisms. Also because of the relative higher land/sea ratio, the extent and effects of the Quaternary glaciations have been greater in the Old World than anywhere else. Climate shows strong continental patterns (more than 20 °C difference between the mean temperature in summer and winter), and the occurrence of 'killing frosts' is frequent. This also happens—to a certain extent—in the Mediterranean Basin, so that this climate is the most 'continental' one (di Castri, 1981) as compared with those of other regions of the world with a mediterranean-type climate (California, Chile, Cape Province in South Africa, Western and southern Australia). The heterogeneity and roughness of the landscape have favoured the existence of several 'massifs de refuge' that have permitted recolonization by several taxa in interglacial periods. In addition, large open spaces exist (e.g. the Gobi desert, the Ethiopian plateaux, the Arabian, Persian and Turkmen deserts) together with large forested areas.

Man's impacts are detectable as long as 40 000 years before the present (BP)

(Verner, 1989), and strong human effects on the environment appear some 8000 years BP. Around the Mediterranean Basin, clearing by fire, pastoralism and primitive agriculture were the primary impacts (Pons *et al.*, 1989). In addition, frequent and sometimes massive migrations of human populations have taken place in Eurasia from the oldest periods up to the present times, more often—but not exclusively—from East to West.

From the practical point of view of biological invasions, I tend to limit the Old World to the zone situated north of the line (Figure 1.2, Braudel, 1979a) separating the areas with a hoe-based agriculture (towards the south) from those where the instruments for earth turn-over were the spade and particularly the plough (often using an animal-labour power). Deeply removing soil by ploughing has far-reaching effects on biological processes in soil, including germination. Tropical and South Africa (not eastern Africa), and part of south and South-east Asia, are therefore excluded from my considerations of the Old World.

Taking into account what has been said up to now, one could postulate that most of the species of the Old World have been and are submitted to frequent endogenous disturbances. Some of these disturbances (aridization, killing frosts) have been concomitant and analogous to some man-made disturbance that have led together to opening new spaces and creation of new habitats. Very frequent and extremely old biological invasions took place within the limits of the Old World itself. Expressed in a slightly caricatural way, the fact of having already been an invader and of still being submitted to high spatial and temporal variability, as well as to varied man's impacts, makes it easier for species to continue to have an 'invader destiny' when transported by man—intentionally or accidentally—to colonize new territories (Figure 1.3).

This statement on the invasion potential of species of the Old World is admittedly a speculation and an overgeneralization. If some species of the Old World have been and are 'good' invaders, others have been less resistant to invasion and prone to be displaced or extinct. For instance, all the original mammal faunas of Corsica have become extinct at the limit of the historical times because of the accidental introduction of new rodents, among them some *Rattus* originating from the Far and Middle East (see Figure 1.4, after Vigne, 1983). This provides a classical example of historical 'island biogeography' as applied within the Old World.

Furthermore, the condition for an invader to behave already like an invader in the region of origin is by no means a general law. For instance, *Pinus radiata* (Monterrey pine) does not exhibit an invasive behaviour in its original stands in California. However, it has become a strongly aggressive invader in South Africa, Australia and New Zealand (not so in the Mediterranean Basin, Atlantic France and Chile).

While keeping the pitfalls of generalization in mind, there are numerous facts in favour of a higher invasion potential—as an average—of Old World organisms

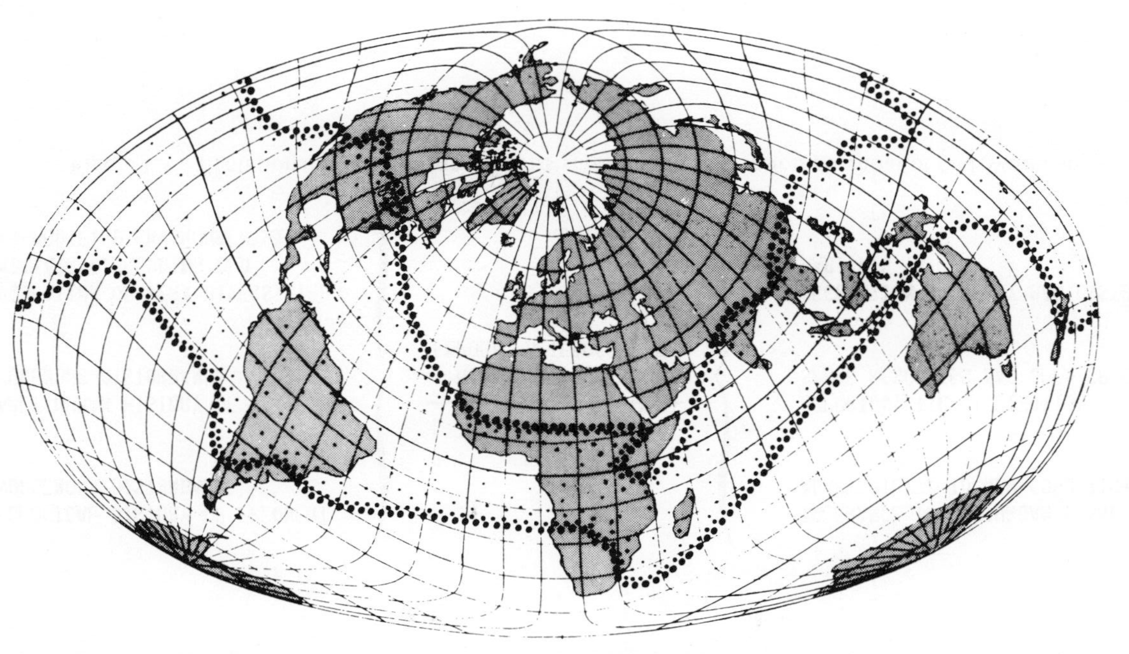

Figure 1.2. The central belt of 'hoe' cultivations (after Braudel 1979a; with permission of Armand Colin, Paris)

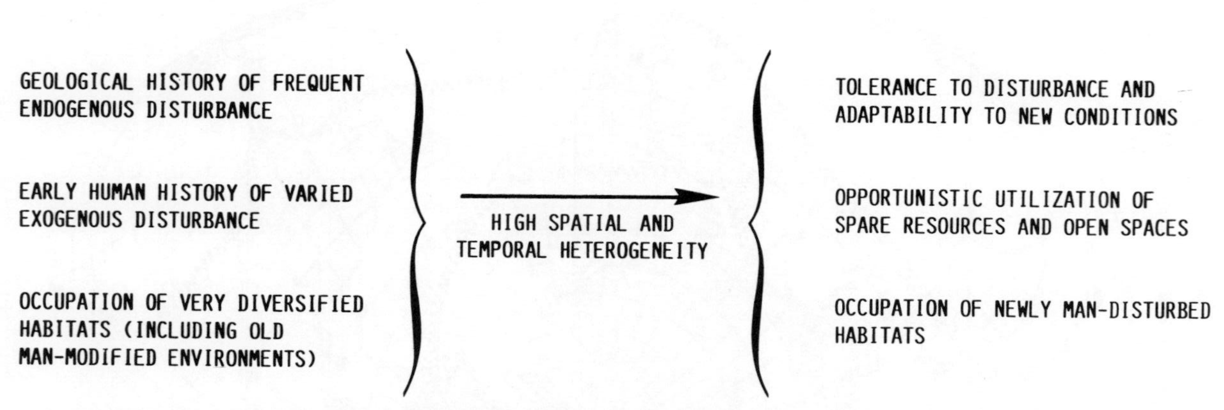

Figure 1.3. Evolutionary and historical factors supporting the invasion potential of Old World species

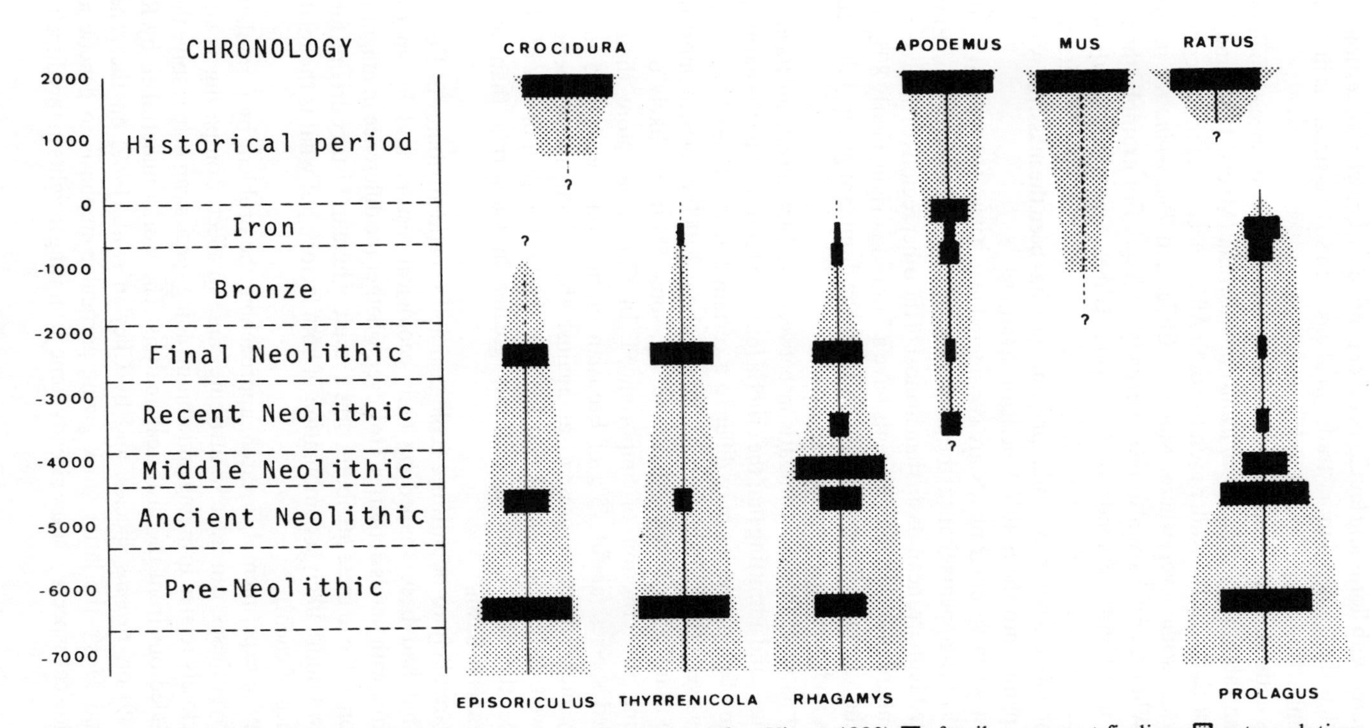

Figure 1.4. Introduction and extinction of mammals in Corsica (after Vigne, 1983) ■ fossil or present findings ▨ extrapolation

as compared with floras and faunas of other continents. Crawley's question (1985) on 'why has the trans-Atlantic "trade" in aliens been so one-sided?' is just one of many similar statements found in literature.

One is understandably impressed by the invasion ability of species such as the European rabbit (*Oryctolagus cuniculus*) in Australia (Myers, 1986) and Chile, *Rattus rattus* (an early invader even to South Africa, according to Deacon, 1986) and *Mus musculus* everywhere, *Rubus* in Chile, and *Pinus pinaster* in South Africa. Whole floras have also been virtually replaced. For example, the native species of the grazing lands of California, Chile, Argentina, South Africa, Australia, have been replaced by Old World species, particularly Mediterranean Basin annuals and herbs. A similar phenomenon has been the massive invasion of African plants into the South American savannas.

At least when vascular plants are considered (e.g. Allan, 1937; Frenkel, 1970; Raven, 1977), it appears that: (a) the proportion of alien species is lower in Europe and in the Mediterranean Basin than in most of the other regions of the world; and (b) most aliens in the other continents have a Eurasian origin, nearly 80% of the total of the adventive flora; annuals constitute the dominant group, followed by biennials and perennials, shrubs, and trees.

Taking as an example a comparison between the Mediterranean Basin and California—and admitting that the criteria followed for recording the nature and the origin of plant species may differ to a certain extent—Quezel *et al.* (1989) recognize over a total Mediterranean flora of about 20 000 species, no more than 400 'important' alien species, that is to say, some 5% (more likely 6–7% not considering the criterion of 'importance'). In California, more than 1000 introduced species (about 75% of Eurasian origin) were recognized in 1968 (certainly many more at present) as against about 5200 native species (see also Howell, 1972, and Spicher and Josselyn, 1985). The proportion of aliens would be therefore about three times greater in California than in the Mediterranean Basin.

There are likely to be intrinsic ecological and genetic attributes of Eurasian species that had been shaped by their geological history and by an early 'association' with man's activities. However, whether the difference in magnitude of invasion is due to these attributes or is a result of human-history driving forces is unknown (more likely a combination of both factors, but what is the relative importance of each?).

Only more experimental research, approaching the problem by formulating testable hypotheses, can hope to attribute cause to effect. Comparing as far as possible closely related invading and non-invading species can help resolve these issues singled out from the historical analysis. The works undertaken by Roy *et al.* (1989) on *Bromus* species, and by Cheylan *et al.* (1989) on the genera *Rattus* and *Mus*—the latter work giving particular emphasis to genetic and eco-ethological aspects—show already some promising insights in this direction.

1.4 BIOLOGICAL INVASIONS TO OLD WORLD ECOSYSTEMS

In spite of the so-called resistance to invasion of Old World ecosystems, even an incomplete list of real invaders is impressive. Obviously because of large oceanic barriers, the first invaders came from one region of the Old World to another. In addition, as far as the oldest invasions are concerned, it is almost impossible to determine whether or to what extent they have been favoured by the activities of primitive human populations. This is the case, for instance, of the slow migrations towards the northern Mediterranean Basin of tenebrionid beetles (a predominantly xerophilous family of Coleoptera). These species proceeded from the central Asian (Turkmenistan) deserts and North Africa northward since or before the Tertiary (Marcuzzi, 1989).

Conversely, species that are really indigenous may have had their distribution and their density increased largely due to different kinds of human impact on the environment. For instance, the Mediterranean species *Pinus halepensis* reached its largest distribution from the times of early man's activities up to those of the Roman Empire (Pons, personal communication).

The most common trend, in any event, is that of invasions from East to West. For example, annual plants moved from the east Mediterranean and the Irano-Turanian region towards the western Mediterranean. Rodents spread from the Far and Middle East to Europe and the Mediterranean. The black rat, *Rattus rattus*, probably indigenous to Indochina, and the house mouse, *Mus musculus domesticus*, from the Middle East reached the Mediterranean in the first or second millennium BC, while invasion by the Norway rat *Rattus norvegicus*, indigenous to southern China, seems to date only from a few centuries ago (but some recent fossil findings may prove the contrary). Also, *Cricetus cricetus* arrived in central Europe from central Asia in the late 19th century—but it was found as a fossil in central Europe (see Cheylan *et al.*, 1989, and Marcuzzi, 1989).

With the emergence of an agricultural civilization in the Mediterranean and later in central Europe, it is not surprising that most of the successful plant invaders came from the Mediterranean Basin. Kornas (1989) illustrates this trend in the Polish flora. The effect of these invasions has been an enrichment rather than a loss for the overall central European flora as exemplified in Figure 1.5 (after Kornas, 1982). An interesting case is that of rye; rye was apparently transported accidentally towards northern Europe as a weed of other cereals, but became a very useful crop under the new colder conditions.

Quite understandably, the discovery of the New World and Australia, and the increased communications with South Africa, opened a wealth of new possibilities of invasion of Old World ecosystems (and even more in the other direction). Interestingly enough, several succulents were introduced which became a most conspicuous and peculiar feature of Mediterranean Basin landscapes, such as *Agave americana, Yucca* and particularly *Opuntia* from

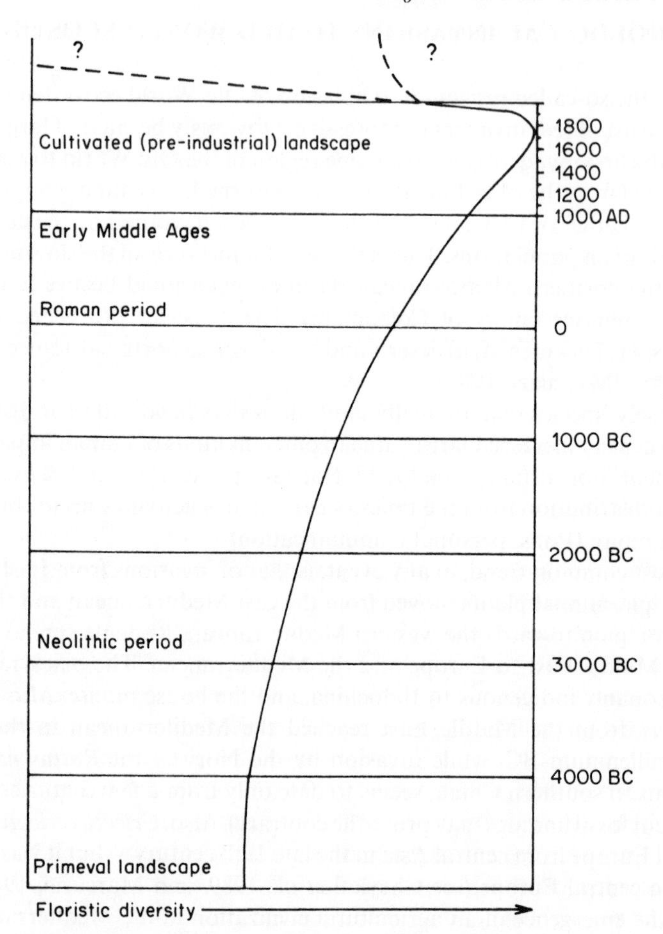

Figure 1.5. Changes in plant diversity in central Europe in prehistoric and historic times (after Kornas, 1982 and 1983, modified from Fukarek 1979)

southern North America, as well as *Mesembryanthemum* from South Africa. In addition, some notable shrubs or trees were successful invaders, including: *Nicotiana glauca* from South America, *Robinia pseudoacacia* from North America, *Acacia* and *Eucalyptus* from Australia. Incidentally, most of these succulents and some of the southern hemisphere trees have suffered so severely (including their total elimination in some regions) because of the 'killing frost' of the winter 1984–1985 in Europe that one can wonder to what extent they are really naturalized in this continental environment. This is particularly true for some species of *Mesembryanthemum* and *Acacia*.

As regards the Mediterranean Basin, it is worthwhile to point out that most of the new invaders were indigenous to eastern North America. Surprisingly, few invaders came from the 'homoclimatic' California, Chile, South Africa and southern Australia (Quezel *et al.*, 1989). This undermines to a certain extent the value of the homoclimatic origin as a key factor for species invasions.

In spite of the relatively low percentage of true invaders, the Mediterranean Basin has undergone several alterations in the 'face' of its landscapes, because of the occurrence of many different waves of invasion of fruit and ornamental trees. As Braudel (1985) cogently said, if Herodotus (5th century BC), the father of history and a remarkable explorer and naturalist *ante litteram*, were able to revisit his eastern Mediterranean region (or even to visit it in Roman times), he would find himself in a very strange 'humanized' environment. Many of the most common and widespread plants introduced later by Mediterranean man such as citrus trees, peach tree, cypress, not to speak of the more recent introductions from the Americas, Australia and South Africa already mentioned, would be completely unrecognizable.

In highlighting the main historical events of biological invasions to the Old World, I would like to pinpoint three more recent cases of invasions. These invasions are well defined from a chronological viewpoint and represent three very distinct patterns of invasion.

The first one refers to the so-called Lessepsian migrations (Por, 1978), that is the migrations that took place from the Red Sea towards the Mediterranean Sea once the Suez Canal was opened in 1869. Most migrations of these marine organisms are one-sided (only recently some anti-Lessepsian migrations towards the Red Sea have been described), and concentrate on the eastern side. Most of the invaders have settled on the coasts rather than in the open sea. Furthermore, biogeographical explanations must take into account the distribution of the Tethys fauna and the characteristics of the Tethys Sea with a Tethyan Gulf having covered part of the present-day northern Red Sea. This is a further example of the significance of geological and evolutionary factors in understanding the present patterns of invasions (for a most comprehensive discussion of these topics see Por, 1978).

The second case points out a rather sudden change in the invasion behaviour of a Mediterranean plant, *Dittrichia viscosa*. This species has only recently begun invading ruderal habitats along roads, railways, and abandoned fields with a variety of different soil conditions as regards salinity and calcareous or siliceous substrate. This represents a testable case of an indigenous plant enlarging its ecological tolerance and poses quite interesting research questions on nutritional and genetic aspects (Wacquant, 1989).

A third classical case study is of the lepidopteran *Hyphantria cunea* (fall webworm). *Hyphantria* was accidentally introduced in central Europe from North America in 1940. A secondary introduction took place in 1978 on the French Atlantic coast. From the primary site of introduction, the trends and speed of this invasion have been followed. By the 1980s, this species had reached the south of Italy and the Black Sea, and is further expanding—apparently more slowly—towards the north and the east (Marcuzzi, 1989).

Having shown—admittedly in a rather superficial and descriptive way—that Old World ecosystems are also prone to invasion by both Old World and exotic species, what is the degree of susceptibility to invasion ('invasibility') of different ecosystems and habitats?

There is no doubt that many new plant invaders initially become established by colonizing ruderal places and arable fields. In the ranking of central European ecosystems susceptible to invasion by plants proposed by Kornas (1983), it could be implied that openness and disturbance are the two main factors required for a successful colonization by a non-native species. Only very few introduced plants such as *Elodea canadensis* seem to be able to 'jump over' undisturbed and close sites (Kornas, 1983). There does not seem to be anything in the Old World comparable to the impressive invasions of apparently undisturbed habitats of South Africa by *Hakea* (from Australia) and *Pinus pinaster* (from Europe).

The problem is more complex when other taxa, life forms, ecological niches and habitats are jointly considered (animals, microbes, aquatic organisms, coastal and marine environments, urbanized environments, etc.). It is unquestionable that urbanized and densely populated sites favour the expansion of commensals and pathogens. In addition, while lacking precise statistics in this respect, I am impressed by the number and importance of invaders (intentionally or accidentally introduced species) in such environments as the coasts (including coastal dunes), and rivers and adjacent riverine habitats of Europe. It is more difficult to find real cases of biological invasions in the open sea, perhaps because there were few obstacles for a very old mixture of compatible species. On European coasts, where large engineering works or intensive culture of aquatic organisms have been developed, there is a wealth of invaders (molluscs and their numerous parasites, worms, sea-weeds, etc.; see Maillard and Raibaut, 1989). Classical invasions have taken place in and along most European rivers: fishes such as *Gambusia affinis* from southeastern North America, *Salmo gairdneri* (rainbow trout) from North America, *Ameiurus nebulosus* from the United States, several molluscs as well as parasites with complex cycles (Combes and Le Brun, 1989). Mammals such as the North American muskrat (*Ondatra zibethica*) and *Myocastor coypus* from South America, and even the water hyacinth (*Eichhornia crassipes*) in some thermophilous sites of south Portugal have also invaded riverine ecosystems. In some of these streams, the degree of disturbance was negligible at the moment of the introduction. However, it may be that the fact of introducing a species, even accidentally, represents in itself an exogenous disturbance.

In fact, I am not convinced that the resistance to invasion of these Eurasian aquatic and semi-aquatic ecosystems is really higher as compared with similar temperate-climate habitats in non-insular conditions. There are several reasons justifying this difference of invasibility between terrestrial and aquatic (or quasi-aquatic) ecosystems. First of all, the latter are very mobile environments (waves on coasts, wind on sand-dunes, running waters, frequent flooding of riverine habitats), submitted therefore to an almost permanent disturbance regime. Secondly, as regards the marine environments, the possible barriers are not quite of the same kind as those commonly described to define biogeographical realms. Thirdly, it is unlikely that the same considerations already made on old 'man-invader associations' and on 'exogenous–endogenous disturbance overlaps' are applicable to these aquatic environments. Finally, as regards mainly Europe in the Old World, it is not yet known what the invasibility danger will be as related to strongly stressed ecosystems (as in forest die-back or in highly eutrophic waters) and the repercussions of invasions on the extinction rate of native species.

1.5 HUMAN-HISTORY DRIVING FORCES IN THE OLD WORLD AS RELATED TO BIOLOGICAL INVASIONS

The third section of this chapter has been rather speculative in trying to answer the question of why so many Old World species seem to be highly invasive. The fourth section has been of a mostly descriptive nature in mentioning some diverse types of invasions—in different times—within or towards the Old World.

I would now like to attempt a typification and categorization of the main human-history driving forces which originated in the Old World. These forces have had an effect in facilitating the introduction of alien species, either in direct relation to the Old World or—indirectly—by promoting exchanges. For instance, between North and South America, among regions of the inter-tropical belt, and between South Africa and Australia many exchanges have occurred, which are attributable to European economic expansion.

I am indebted to Fernand Braudel and his school of history which links and relates the patterns of the daily life with long-term historical trends, thus embracing different scales of space and time (as some modern approaches in ecology are also trying to do). Braudel's conception is free from the too common ethnocentric (Europocentric) distortions, and the different perceptions are presented in a very comprehensive and harmonious design. I have been particularly impressed, in preparing this chapter, by the relevance of his last monumental work in three volumes (Braudel, 1979 a, 1979 b, 1979 c) on 'civilization, economy, and capitalism' from the 15th to the 18th century. From another historical point of view, it is also worth reading Crosby's book (1986) on European 'Ecological Imperialism'.

Table 1.2 summarizes most of my preliminary thoughts. The columns are separated by 'breaking points' of high historical significance, at least as regards the chance and timing of species transportation to new territories. The general trend is certainly that of a progressive globalization of the problem, as well as of a very rapid acceleration of invasion processes.

Of course, an ideal presentation should have been to provide, for each one of

Table 1.2. Human-history driving forces in the Old World as related to biological invasions

Before 1500 AD	After 1500 AD	From last century in a worldwide perspective
Forest clearing	Exploration, discovery and early colonization by Europeans of other territories and continents	Improvement of transportation systems (roads, railways, internal navigation canals)
Primaeval agriculture	Establishment of new market economies and crossroads places (e.g. Amsterdam, London) favouring the 'globalization' of trade exchanges	Large engineering works for irrigation and hydropower
Sheep and cattle-raising	Large 'colonies' under the rule of Europeans, often entailing introduction of European-like agriculture and increasing *inter alia* intertropical exchanges	Opening of inter-oceanic canals (e.g. Suez, Panama, Volga-Don)
Migrations and nomadism	'Revolution' of food customs in wealthy Europe (e.g. increased use of tea, coffee, chocolate, rice, sugar, potatoes, maize, beef and lamb)	Aircraft transportation
Inshore coastal traffic	Increased demand in Europe of products such as cotton, tobacco, wool, etc.	World wars and displacement of human populations
Settlement of islands (e.g. Corsica)	Negro slavery; Indian and Chinese migrations	'Decolonization'; international aid to newly independent countries following 'western' patterns
Intensification of agriculture by ploughing	Missionary establishments	Emergence of multinational companies
Offshore traffic and trade	Occupation by Russians of northern and part of central Asia, up to Siberia	Tropical deforestation and resettlement schemes
Coastal 'colonies' (e.g. Phoenician and Greek colonies)	Intentional introduction into the Old World of exotic species through activities of acclimatization societies, botanical gardens and zoos, and for agricultural, forestry, fishery or ornamental purposes	Afforestation of arid lands with exotic species
Building up of large empires (e.g. Persian, Roman, Arab, Mogul) with considerable expansion of communication and transportation systems	Large-scale emigration from the Old World due to persecution during religious conflicts, civil and 'independence' wars, and to increased demography, unemployment and famine	Environmental impacts decreasing ecosystems' resilience
Long-ranging wars and military expansion		Increased urbanization and creation of ruderal habitats
Invasions of German and Asian people, mainly from east to west		International interdependence of markets
Long-distance shipping trade		Release of genetically engineered organisms
Establishment of 'market economies' (e.g. Venice) covering the 'known world' up to the Far East		

the historical driving forces mentioned in Table 1.2, a precise dating and a definition of a given set of parallel biological invasions. This is not feasible in this introductory general chapter, but more detailed work is underway towards these results. Furthermore, at least for the first part of the left column, a comparative dating is impossible, even within such a restricted region as the Mediterranean Basin. 'Neolithization' progressed slowly from east to west, and there may be a time lag of 4000 to 6000 years for the establishment, for instance, of the Bronze and Iron periods (Le Houerou, 1981; Vernet, 1989).

The left column covers a period of several millennia up to about 1500 AD, and refers exclusively to human historical events which favoured invasions and migrations *within* the Old World. For example, the existence of an extremely extended Persian Empire (500 years BC) with its imperial roads has conceivably promoted the expansion of Irano-Turanian and central Asian elements towards the Mediterranean Basin. The Phoenician and Greek colonies, far away from their eastern Mediterranean homelands, have certainly promoted the expansion of cereals (and their weeds) towards the western Mediterranean and the Black Sea. Later, the continuous movement of human populations in Europe and Asia at the end of the Roman Empire and in the Middle Ages—and in Asia up to very recent times—has produced a real biological 'brassage' ('shaking and mixing up' of species). It should be stressed that the so-called 'barbarian' invasions to Europe, and the migrations and wars in central Asia and the Far East (Figures 1.6 and 1.7) were not 'wars' in a modern meaning, but slow movements of populations with entire families, their domestic animals, seeds, parasites, pathogens and commensals.

During the last centuries of this period, Venice developed a quite modern market and trade economy. While its action in exchanging goods and products was not consistently different from the present-day patterns, its coverage was reduced to the 'known' world (see Figure 1.8, after Braudel, 1979c).

A historical turning point occurred around 1500 AD with the exploration, discovery and colonization of new territories. It also marked the beginning of the globalization of exchanges, while still keeping a very strong European focus. A comparison between Figure 1.8 (1500 AD) and Figure 1.9 (1775 AD) gives a fairly good idea of the enormous increase of communications and trades (and of the 'opportunities' for biological invasions) in a relatively short period. It is not a coincidence that the delimitating date between archaeophytes and neophytes as regards alien plants is also 1500 AD (see Kornas, 1983 and Sykora, 1984).

The second period (central column) has a much shorter duration that the previous one: some 350–400 years. Among the facts quoted here, the change in food and dressing customs, thanks to the unusual wealth of a part of the European people, may seem to be a most trivial one. On the contrary, in addition to prefiguring some of the most peculiar features of modern society, it has had far-reaching consequences as regards biological invasions; for example, through the importation of new products and species from other continents, or from poorer

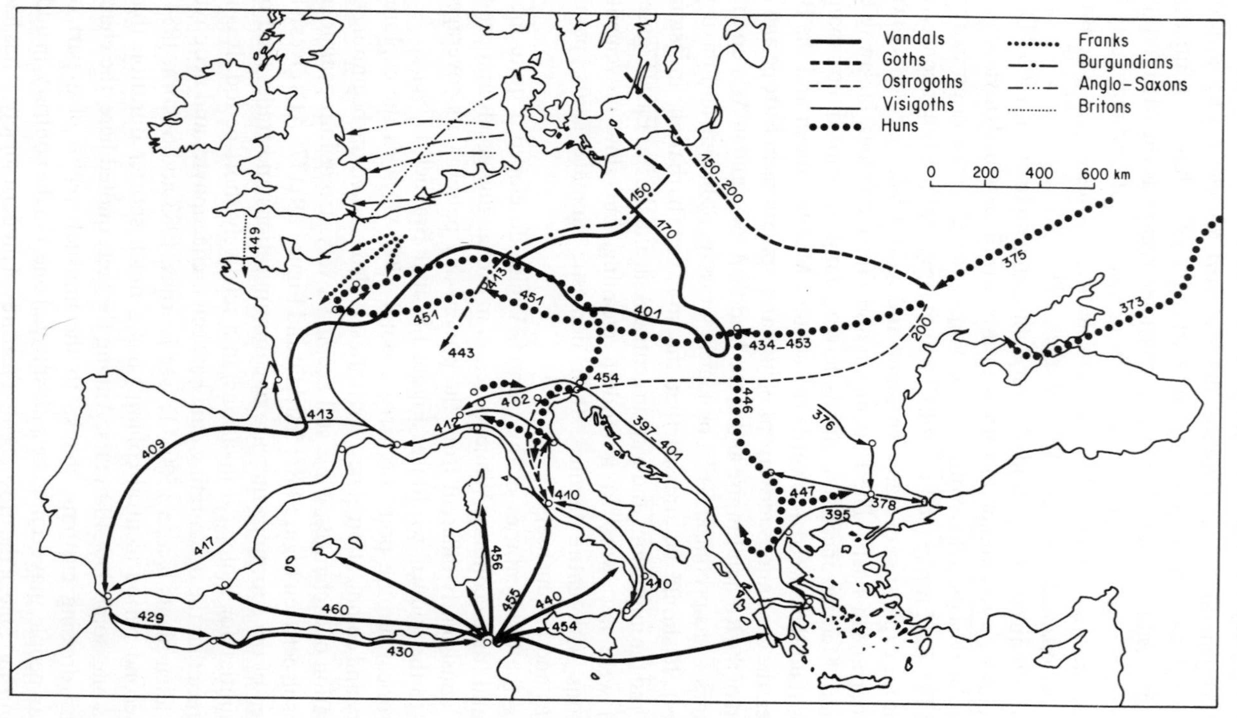

Figure 1.6. The German migrations around 500 AD

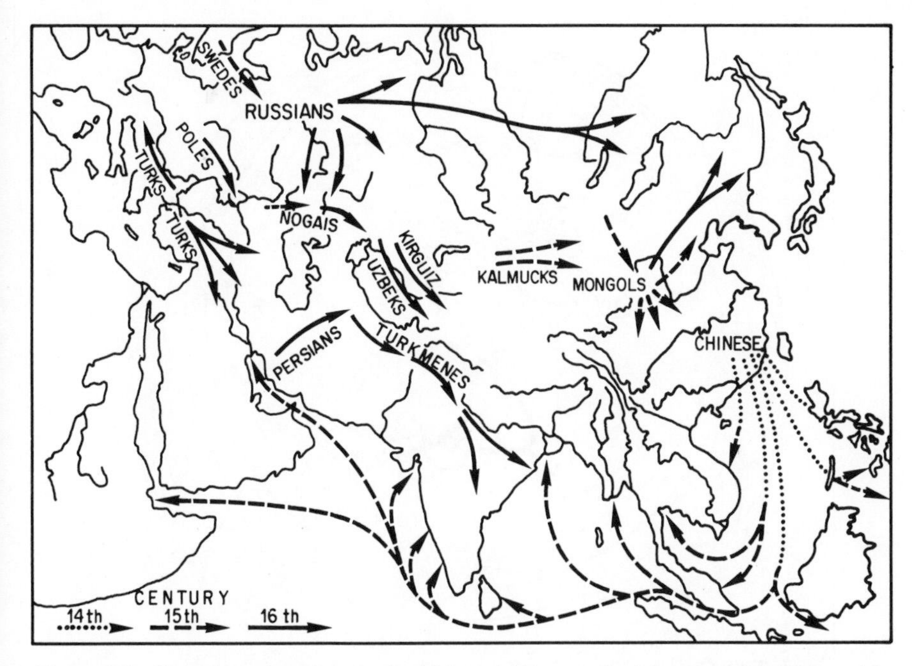

Figure 1.7. Eurasian migrations in the 14th to 16th centuries (after Braudel, 1979c; with permission of Armand Colin, Paris)

parts of Europe (e.g. cattle from eastern Europe), or by contributing to fundamental changes in tropical agriculture (including new human labour needs) and increasing intertropical exchanges. Consider the neotropical (South American) water hyacinth. It began to spread about a century ago, and is now found throughout the tropical world. It is worth pointing out the fate of water hyacinth, because it constitutes the case of a very harmful invader that has become a valuable resource in some invaded areas, such as in southern China, as a staple part of the diet of pigs and other domesticated animals (di Castri and Hadley, 1980).

Some of these historical factors, as for instance the increased demand for wool, have been analysed in a very detailed way from the viewpoint of the accidental introduction of alien plants. The very comprehensive and extremely stimulating monograph of Thellung (1908–1910) on the adventive flora of Montpellier was largely inspired by the expansion of aliens due to the import, hanging out and drying of wool at Port-Juvénal (near Montpellier). Similarly, very precise chronological steps can be established for the introduction of exotics to the botanical gardens of Europe at different times, and their intentional or accidental spreading out (see Sykora, 1984).

In spite of the fact that the last two examples are supportive of invasions towards the Old World, it is unquestionable that the period up to the middle of

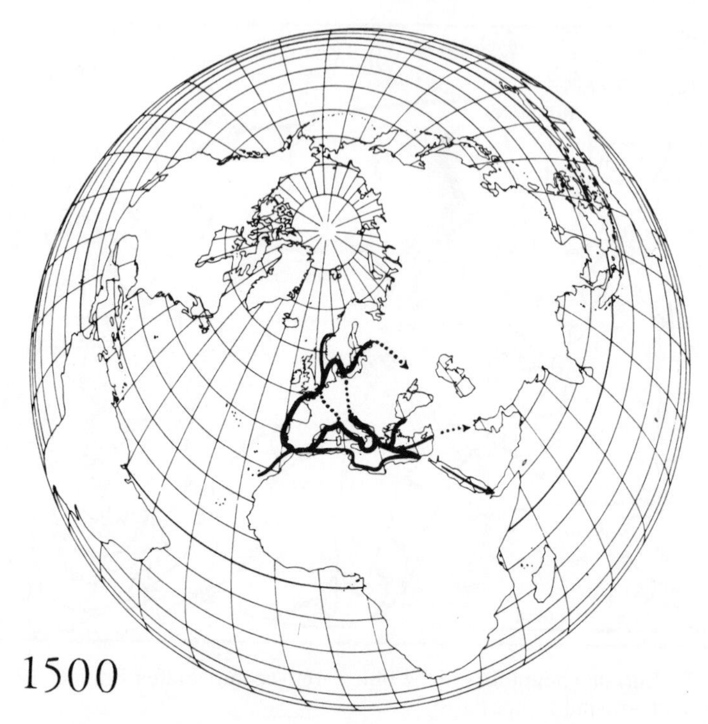

Figure 1.8. The expansion of the European trades and economies, with special emphasis on Venice as its centre, around 1500 AD (after Braudel, 1979c; with permission of Armand Colin, Paris)

the last century, has greatly facilitated opportunities for Old World invaders to colonize the other continents. As a matter of fact, at the turning time of 1500 AD when the main explorations and colonizations began, intensive agriculture was restricted to the Old World (including south and Southeast Asia, but occurring in very few parts of Africa south of the Sahara, and of central Asia). The rest of the 'unknown' world was populated mainly by nomads and hunter-gatherers, and by primitive agriculturists, while some more advanced agricultural practices—of a less perturbing nature as compared with the Eurasian ones—were applied mostly in Central and Andean America and in Madagascar. This does not imply a 'superiority' of European man vis-à-vis other extremely rich and diversified cultures and civilizations existing in the other continents. It means simply the 'implant' of a more aggressive technology.

Even the human emigrations from Europe at the end of this period—that have marked its political vicissitudes and a decline of the absolute economic predominance of Europe—further favoured the 'export' of culturally oriented agricultural practices. The waves of Germans, Greeks, Swedes, Italians, Irish,

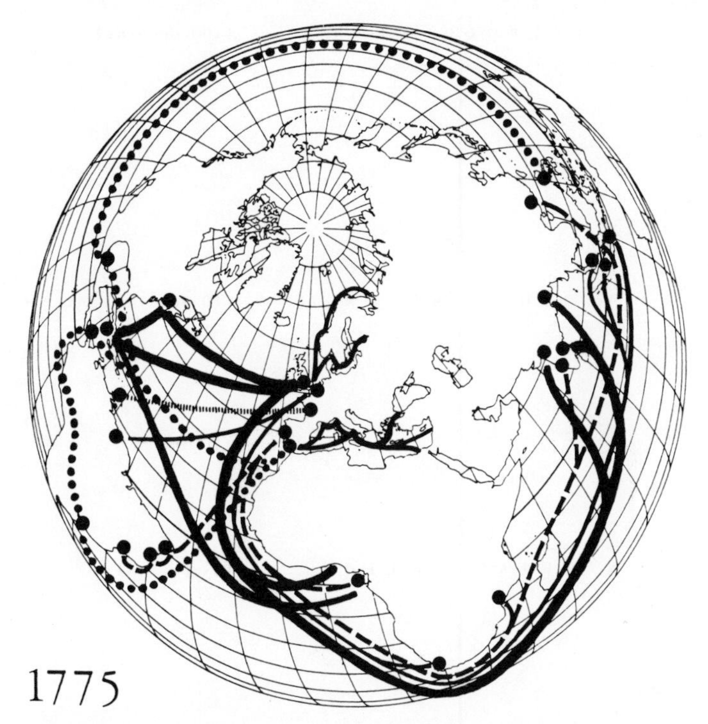

Figure 1.9. The 'globalization' of the European trades (particularly of the British, Dutch, Spanish, Portuguese and French networks) around 1775 AD; London as the main crossroads of economies (after Braudel, 1979c; with permission of Armand Colin, Paris)

Spaniards, Poles, etc., that have shaped the cultivated landscapes of so many regions of the Americas or Australia, have tended to use their most familiar plants and animal breeds. Too often species were imported with insufficient (or badly applied) quarantine regulations.

As a matter of fact, even at present, any development scheme guided by another country (or by an international organization) is not 'aseptic', in the sense that a given technology carries with it a 'cortège' not only of cultural repercussions, but also of biological changes.

Accordingly, European man has not only greatly helped the introduction of Old World species (with their weeds and parasites), but also—by adopting his own original agricultural practices—he is likely to have facilitated colonization and naturalization processes (see Figure 1.10). Probably, the approximate mean rates given in Figure 1.10 for each successive process of invasion should be slightly increased, in the light of these considerations regarding the biological invasions from the Old World.

When one looks at the factors favouring invasion of new territories by Old

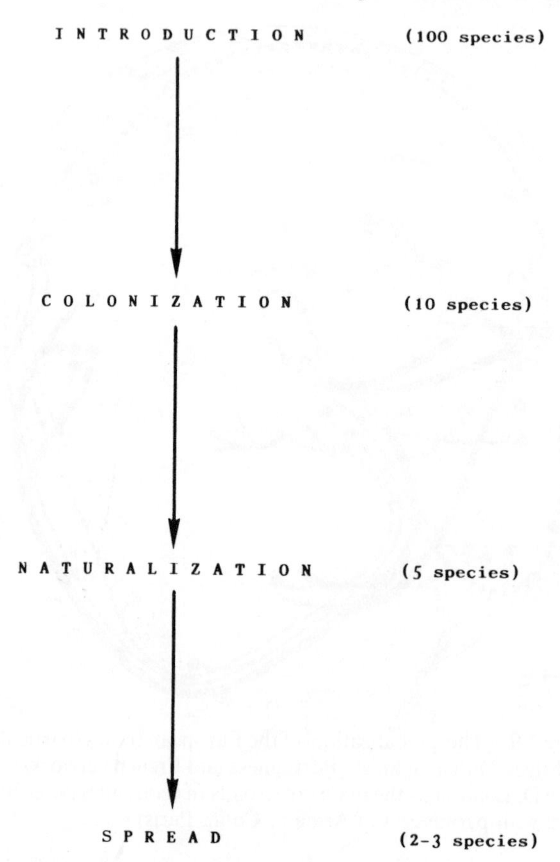

Figure 1.10. Main steps in the biological invasion processes

World species (Figure 1.11),—at least up to the end of the second historical period shown in Table 1.2—the first factors shown in Figure 1.11 and most likely the last one (the transformation of local conditions to resemble those of the original homeland) result in an incontestable advantage for Old World species compared with invaders from other continents. It is also true that several new territories colonized by Europeans had some biogeographical conditions of 'insularity' *sensu lato*, such as the southernmost tips of America and Africa, most of Chile, Australia, etc., being therefore more susceptible to invasion by alien organisms. The other factors mentioned in Figure 1.11 do not seem to be peculiar to invasion from the Old World.

The key place of the Mediterranean Basin should be stressed again. Having occupied a focal position as regards invasions within the Old World, due to its crossroads role in geological, evolutionary, biogeographical and human-cultural

Patterns of man's migration, invasion and colonization.
Trade and market trends, transportation systems.

Leaving behind their own pathogens, parasites, predators, competitors.

Finding open spaces and spare resources in newly man-disturbed environments.

'Insularity' conditions in islands, southernmost tips of continents, western fringes of continents.

Homoclimatic regions.

Homocultural conditions.

Figure 1.11. Factors affecting the chance of invasion of new territories by Old World species

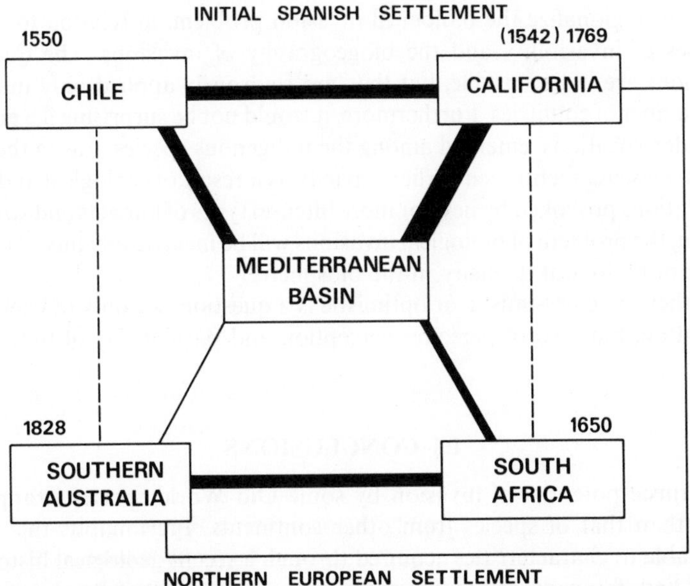

Figure 1.12. The similarities between the five regions with a mediterranean-type climate. The degree of similarity is proportional to the thickness of connecting bars and lines (simplified from Fox, 1989, on the basis of Figure 1.10. from di Castri, 1981)

terms, the Mediterranean Basin has maintained a similar role for invasions on a world-wide scale at least up to the 17th century or later. Figure 1.12, based on Fox (1989) on the basis of di Castri's (1981) typology of inter-mediterranean affinities, including here the affinities resulting from invasion processes, emphasizes the Mediterranean Basin focus. Furthermore, a 'homocultural' dimension is added to the geological and ecological similarities. In fact, the degree of affinities increases between the mediterranean-climate regions settled by Spaniards on the one side, and between those settled by northern European people on the other side.

Finally, the right side column of Table 1.2 only covers the last 100–150 years, and is open towards the future. It is difficult to restrict here to simply Old World-based considerations, because a multifoci globalization of problems is almost completely achieved. This globalization, after vicissitudes like world wars and progressive cultural homogenization, is leading to a close international interdependence of markets and other exchanges. Together with an acceleration of the pace of changes, there is *de facto* a kind of increased 'smallness' of space because of the efficiency of new transportation and communication systems.

Most of the factors presented in this column are self-explanatory from an invasion viewpoint. It would be more interesting to foresee what the continuation of the present period would look like. It will certainly become more and more difficult to regionalize the biological invasion problem, in relation to both the processes of invasibility and the biogeography of invasions. The quarantine regulations are indispensable, but they are rigorously applied only in a rather small number of countries. Furthermore, it would not be surprising if a great deal of 'invader vocations' emerged among the indigenous species—as in the case of *Dittrichia viscosa* mentioned earlier—partly as a result of ecological and genetic modifications provoked by new (or more intense) types of impacts and stresses. In addition, the problem of biological invasions will be inextricably linked to that of the loss of biological diversity in the biosphere.

Whether to be pessimistic or optimistic is a question not only of background information, but also of personal perception and particularly of time scale of concern.

1.6 CONCLUSIONS

The intrinsic potential of invasion by some Old World species is apparently greater than that of species from other continents. Presumably, this skew is attributable to characteristics acquired through a recent geological history with diverse and frequent endogenous disturbance, connected later in positive feedbacks with the emerging exogenous disturbance of early man's impacts. Conversely, there is more evidence that human-historical events have really provided the Old World species with a greater wealth of opportunities for invading new man-colonized territories.

A broad historical overview from the geological past towards a perspective of possible future trends (Table 1.1) outlines three main 'crises' as regards biological invasions: the first, in pre-historical times, when the initial conditions of 'man-invader associations' were settled; the second, around 1500 AD, when the barriers of the biogeographical realms were broken owing to new transportation systems of man (and the tied invaders); the third, at present, because of the breakdown of the previous 'scaling rules', with much more extended space available in a much shorter time.

Taking a stricter human-history viewpoint—and more specific to the Old World—three periods characterized by different human-driven forces as regards biological invasions are defined (Table 1.2). The first one, ranging from the Neolithic times up to 1500 AD only refers to human-history factors and biological invasions having taken place within the Old World. The second one extended almost up to the end of the last century. This period shows the occurrence of flows of invaders from, to and within the Old World. The progressive globalization of biological invasions has had, however, a decided Europocentric focus as regards the human-responsibility driving forces. The third period, covering the last 100–150 years, is leading to a complete multifocal globalization of the human-governed forces that promote biological invasions.

An historical analysis cannot provide testable elements to highlight the genetic and ecological attributes that give a species potential as an invader. Similarly, history cannot give us firm evidence on the reasons why some ecosystems are vulnerable to invasion while others are not. More experimental research is needed on these topics, particularly to understand not only why some species have been so successful as invaders, but also why closely relate species *are not* invaders. Furthermore, the 'invasion' approach can help in finding stimulating 'biological models' to study an 'evolution in march' through processes of genetic differentiation, ecological colonization, and extinction.

From both an historical and a biological viewpoint, the invasion problems may be considered as a play of chance and necessity, of human-derived opportunities and of evolutionary heritages.

ACKNOWLEDGEMENTS

I would like to thank the publisher Armand Colin, Paris, for having graciously granted permission to reprint four figures from Braudel (1979), and Dr James A. Drake for his help in editing the manuscript of this chapter.

REFERENCES

Allan, H. H. (1937). The origin and distribution of the naturalized plants of New Zealand. *Proc. Linn. Soc. London*, **150**, 25–46.

Braudel, F. (1979a). *Civilisation Matérielle, Économie et Capitalisme, XVe-XVIIIe Siècle. Tome 1. Les Structures du Quotidien: le Possible et l'Impossible.* Armand Colin, Paris. 544 pp. (Published in English by Collins with the title *The Structures of Everyday Life.*)

Braudel, F. (1979b). *Civilisation Matérielle, Économie et Capitalisme, XVe-XVIIIe Siècle. Tome 2. Les Jeux de l'échange.* Armand Colin, Paris. 600 pp. (Published in English by Collins with the title *The Wheels of Commerce.*)

Braudel, F. (1979c). *Civilisation Matérielle, Économie et Capitalisme, XVe-XVIIIe Siècle.*

Tome 3. Le temps du monde. Armand Colin, Paris. 607 pp. (Published in English by Collins with the title *The Perspective of the World.*)

Braudel, F. (1985). Mère Méditerranée. *Courrier UNESCO*, **38** (12), 4–12.

Cheylan, G., Michaux, J., and Croset, H. (1989). Of mice and men. In: di Castri, F., Hansen, A. J., and Debussche, M. (Eds), *Biological Invasions in Europe and the Mediterranean Basin*, Kluwer Academic Publications, Dordrecht (in press).

Combes, C., and Le Brun, N. (1989). Invasion by parasites in continental Europe. In: di Castri, F., Hansen, A. J., and Debussche, M. (Eds), *Biological Invasions in Europe and the Mediterranean Basin*, Kluwer Academic Publications, Dordrecht (in press).

Coope, G. R. (1986). The invasion and colonization of the North Atlantic islands: a palaeoecological solution to a biogeographic problem. *Phil. Trans. R. Soc. Lond., B*, **314**, 619–35.

Crawley, M. J. (1985). What makes a community invasible? *Summaries Symposium on Colonisation, Succession and Stability, 15–19 July 1985, Southampton University*, p. 32.

Crosby, A. W. (1986). *Ecological Imperialism. The Biological Expansion of Europe, 900–1900.* Cambridge University Press, Cambridge. 368 pp.

Deacon, J. (1986). Human settlement in South Africa and archaeological evidence for alien plants and animals. In: Macdonald, I. A. W., Kruger, F. J., and Ferrar, A. D. (Eds), *The Ecology and Management of Biological Invasions in Southern Africa*, pp. 3–19, Oxford University Press, Cape Town.

di Castri, F. (1981). Mediterranean-type shrublands of the world. In: di Castri, F., Goodall, D. W., and Specht, R. L. (Eds), *Mediterranean-type Shrublands, Ecosystems of the World 11*, pp. 1–52. Elsevier, Amsterdam.

di Castri, F. (1989). On invading species and invaded ecosystems: a play of historical chance and biological necessity. In: di Castri, F., Hansen, A. J., and Debussche, M. (Eds), *Biological Invasions in Europe and the Mediterranean Basin*, Kluwer Academic Publications, Dordrecht (in press).

di Castri, F., and Hadley, M. (1980). Research and training for ecologically-sound development: problems, challenges and strategies. In: Furtado, J. I. (Ed.), *Tropical Ecology and Development, Part 2*, pp. 1229–1252. The International Society of Tropical Ecology, Kuala Lumpur.

Elton, C. S. (1958). *The Ecology of Invasions by Animals and Plants.* Methuen, London. 181 pp.

Fox, M. D. (1989). Mediterranean weeds. Exchanges of invasive plants between the five mediterranean regions of the world. In: di Castri, F., Hansen, A. J., and Debussche, M. (Eds), *Biological Invasions in Europe and the Mediterranean Basin*, Kluwer Academic Publications, Dordrecht (in press).

Fox, M. D., and Fox, B. J. (1986). The susceptibility of natural communities to invasion. In: Groves, R. H., and Burdon, J. J. (Eds), *Ecology of Biological Invasions: an Australian Perspective*, pp. 57–66. Australian Academy of Science, Canberra.

Foy, C. L., Forney, D. R., and Cooley, W. E. (1983). History of weed introduction. In: Wilson, C. L., and Graham, C. L. (Eds), *Exotic Plant Pests and North American Agriculture*, pp. 65–92. Academic Press, New York.

Frenkel, R. E. (1970). Ruderal vegetation along some California roadsides. *Univ. Calif. Publ. Geogr.*, **20**, 1–163.

Fukarek, F. (1979). Der Mensch beeinflusst die Pflanzenwelt. In: Fukarek, F. (Ed.), *Pflanzenwelt der Erde*, pp. 65–77. Urania Verlag, Leipzig.

Groves, R. H., and Burdon, J. J. (Eds) (1986), *Ecology of Biological Invasions: an Australian Perspective.* Australian Academy of Science, Canberra. 166 pp.

Gwynne, D. C., and Murray, R. B. (1985). Weed origins and distribution. In: Gwynne, D. C., and Murray, R. B. (Eds), *Weed Biology and Control in Agriculture and Horticulture*, pp. 13–21. Batsford Academic and Educational, London.

Howell, J. T. (1972). A statistical estimate of Munz Supplement to California Flora. *Wasmann J. Biol.*, **30**, 93–6.

Kornas, J. (1982). Man's impact upon the flora: processes and effects. *Memorabilia Zool.*, **37**, 11–30.

Kornas, J. (1983). Man's impact upon the flora and vegetation in Central Europe. In: Holzner, W., Werger, M. J. A., and Ikusima, I. (Eds), *Man's Impact on Vegetation*, pp. 277–86. Dr W. Junk Publ., The Hague.

Kornas, J. (1989). Plant invasions in Central Europe: historical and ecological aspects. In: di Castri, F., Hansen, A. J., and Debussche, M. (Eds), *Biological Invasions in Europe and the Mediterranean Basin*, Kluwer Academic Publications, Dordrecht (in press).

Le Floc'h, E., Le Houérou, H. N., and Mathez, J. (1989). History and patterns of plant invasion in Northern Africa. In: di Castri, F., Hansen, A. J., and Debussche, M. (Eds), *Biological Invasions in Europe and the Mediterranean Basin*, Kluwer Academic Publications, Dordrecht (in press).

Le Houérou, H. N. (1981). Impact of man and his animals on Mediterranean vegetation. In: di Castri, F., Goodall, D. W., and Specht, R. L. (Eds). *Mediterranean-type Shrublands, Ecosystems of the World 11*, pp. 479–521. Elsevier, Amsterdam.

Maillard, C., and Raibaut, A. (1989). Human activities and modifications of the Ichtyofauna of the Mediterranean Sea: effect on parasitosis. In: di Castri, F., Hansen, A. J., and Debussche, M. (Eds), *Biological Invasions in Europe and the Mediterranean Basin*, Kluwer Academic Publications, Dordrecht (in press).

Marcuzzi, G. (1989). Biogeographical fluctuations of European animals from the Miocene to the Present. In: di Castri, F., Hansen, A. J., and Debussche, M. (Eds), *Biological Invasions in Europe and the Mediterranean Basin*, Kluwer Academic Publications, Dordrecht (in press).

Mooney, H. A., and Drake, J. A. (Eds) (1986). *Ecology of Biological Invasions of North America and Hawaii. Ecological Studies*, Vol. 58. Springer-Verlag, New York. 321 pp.

Myers, K. (1986). Introduced vertebrates in Australia, with emphasis on the mammals. In: Groves, R. H., and Burdon, J. J. (Eds), *Ecology of Biological Invasions: an Australian Perspective*, pp. 120–36. Australian Academy of Science, Canberra.

Pons, A., Couteaux, M., de Beaulieu, J. L., and Reille, M. (1989). The plant invasions in southern Europe from the paleoecological point of view. In: di Castri, F., Hansen, A. J., and Debussche, M. (Eds), *Biological Invasions in Europe and the Mediterranean Basin*, Kluwer Academic Publications, Dordrecht (in press).

Por, F. D. (1978). *Lessepsian Migration. The Influx of Red Sea Biota into the Mediterranean by way of the Suez Canal. Ecological Studies*, Vol. 23. Springer-Verlag, Berlin. 228 pp.

Quézel, P., Barbero, M., Bonin, G., and Loisel, R. (1989). Recent invasions of plant species in the circum-Mediterranean region. In: di Castri, F., Hansen, A. J., and Debussche, M. (Eds), *Biological Invasions in Europe and the Mediterranean Basin*, Kluwer Academic Publications, Dordrecht (in press).

Raven, P. H. (1977). The California flora. In: Barbour, M. G., and Major, J. (Eds), *Terrestrial Vegetation of California*, pp. 109–37. Wiley, New York.

Roy, J., Navas, M. L., and Sonié, L. (1989). Invasion by annual brome grasses: a case study challenging the homocline approach to invasions. In: Groves, R. H., and di Castri, F. (Eds), *Biogeography of Mediterranean Invasions*, Cambridge University Press, Cambridge (in press).

Sailer, R. I. (1983). History of insect introductions. In: Wilson, C. L., and Graham, C. L., (Eds), *Exotic Plant Pests and North American Agriculture*, pp. 15–38. Academic Press, New York.

Spicher, D., and Josselyn, M. (1985). Spartina (Gramineae) in northern California. *Madrono*, **32**, 158–67.

Sykora, K. V. (1984). Plants in the footsteps of man. *Endeavour, New Series*, **8**(3), 5 pp.
Sykora, K. V. (1989). History of the impact of man on the distribution of plant species, and the behaviour of exotic plant species in the country of introduction. In: di Castri, F., Hansen, A. J., and Debussche, M. (Eds), *Biological Invasions in Europe and the Mediterranean Basin*, Kluwer Academic Publications, Dordrecht (in press).
Thellung, A. (1908–1910). La flore adventice de Montpellier. *Mém. Soc. Nat. Sc. Natur. Math. Cherbourg*, **37**, 57–728.
Vernet, J. L. (1989). Man and vegetation in the Mediterranean area during the last 40 000 years. In: di Castri, F., and Hansen, A. J. (Eds), *Biological invasions in Europe and the Mediterranean Basin*, Kluwer Academic Publishers, Dordrecht (in press).
Vigne, J. D. (1983). Le remplacement des faunes de petits mammifères en Corse lors de l'arrivée de l'homme. *C. R. Soc. Biogéogr.*, **59**, 41–51.
Wacquant, J. P. (1987). Biogeographical and physiological aspects of the invasion of *Dittrichia* (ex: *Inula*) *viscosa* W. Greuter, a ruderal species in the Mediterranean Basin. In: di Castri, F., Hansen, A. J., and Debussche, M. (Eds), *Biological Invasions in Europe and the Mediterranean Basin*, Kluwer Academic Publications, Dordrecht (in press).
Yarwood, C. E. (1983). History of plant pathogen introductions. In: Wilson, C. L., and Graham, C. L. (Eds), *Exotic Plant Pests and North American Agriculture*, pp. 39–63. Academic press, New York.

Biological Invasions: a Global Perspective
Edited by J. A. Drake *et al.*
© 1989 SCOPE. Published by John Wiley & Sons Ltd

CHAPTER 2

Patterns, Extents and Modes of Invasions by Terrestrial Plants

VERNON H. HEYWOOD

2.1 INTRODUCTION

The botanical traveller soon becomes aware of the fact that there is scarcely a region in the world where the vegetation has not been disturbed to some degree by man's activities, usually leading to the introduction of alien species. The extent and pattern of these invasions varies widely from one part of the world to another and it is the aim of this paper to seek any regularities or generalizations that might be made concerning either the geography, origins or taxonomic affiliations of the invader species. As Harper (1977) notes, man has introduced a new order of magnitude into distances of dispersal, and through the transportation, by accident or design of seeds or other propagules, through the disturbance of native plant communities and of the physical habitat, and by the creation of new habitats and niches, the invasion and colonization by adventive species is made possible.

In some cases the invaders make their presence felt only too conspicuously, as in the Mediterranean Basin where much of what until 200 years ago was regarded as native vegetation is in fact man-modified, such as the matorral, garrigue, etc. Ellenberg (1979) observes that the reason that he travelled to Peru and other tropical countries was to study 'real nature' but after several months of field work he could not fail to discover traces of man's impact there too, even in the Amazonian rainforest area. Human influence was apparent too in the remote highlands and lonely dry valleys of the Pacific slopes. The ecosystems of the Andes are far from being untouched nature and Mediterranean people introduced their agropastoral system into tropical mountainous countries where similar landuse systems had been practised for thousands of years previously.

A global survey of the pattern and extent of invasion by terrestrial plant species is bound to be anecdotal to a degree because of the extreme diversity in the sources and in reliability of the available data. Apart from the obvious lack of a fixed geographical starting point for a global overview, the main factor that makes my task virtually impossible is the lack of data and the wide inconsistency of those that can be traced in the literature. For many parts of the world, notably in temperate regions, Floras or catalogues listing native plants exist and often analyses of adventive or weed species. However, the terms used and the criteria employed (when they can be determined) vary widely between these surveys. As discussed below, the situation in many parts of the tropics and subtropics is much more difficult in that few Floras or handbooks exist and those that do are usually highly incomplete.

As Webb (1985) in a perceptive paper on the criteria for presuming native status notes, 'Most Floras make some attempt to distinguish between native and introduced species though some of them do so half-heartedly. But their authors seldom disclose the evidence which has led them to their decision, and all too often it would appear that the assignment has been made by copying from earlier works or on essentially intuitive grounds.' This problem is also discussed by Smith (1986) in connection with annual species of *Bromus* in Europe and South west Asia.

In reviewing literature for this paper, I have become acutely conscious of the difficulties of interpreting the available data because of either failure to distinguish between different categories of introduction—between native and introduced weeds, between casual and established (naturalized) weeds and/or aliens and so on. Invaders have been termed aliens, immigrants, exotics, adventives, neophytes, xenophytes or simply introduced species. Invaders may be native to the region or country but not to the community in question. Naturalization of species may occur within their native country or region but outside their natural geographical range (Robinson *et al.*, 1986; Robin and Carr, 1986). Two useful definitions are: 'Plant invaders are alien plants that invade and oust native vegetation' (Stirton, 1979) and 'Aliens... can be exotic species (plants introduced from overseas) as well as indigenous species (...native species) that have successfully taken up residence in plant communities in which there is good reason to believe that the species are newcomers' (Williams, 1985, citing Michael, 1981). Mack (1985) regards as an invader any taxon entering a territory in which it has never occurred before, regardless of the circumstances (e.g. transoceanic migrations). All entrants even if they fail to establish will be regarded as invaders, including recurring Holocene migrations. The term weed, however, is a value judgement.

There is a long-recognized connection between habitat disturbance or environmental alteration by humans and the invasion and spread of aliens. The invasion process is divided by Groves (1986) into three main stages: introduction, colonization and naturalization. Introduction itself is a function of dispersal and Berg (1983) distinguishes between dispersal and successful (and effective) dispersal, i.e. followed by establishment, since only successful dispersal can bring about distributional and evolutionary change. He quotes Fosberg: 'Transport without establishment has no significance.' Colonization depends in turn not only on successful dispersal but on successful reproduction. And this according to

Stebbins (1971) depends on a compromise between the often conflicting demands of three separate processes—diaspore production, plant dispersal and establishment which in turn depend on a whole series of factors such as pollination biology, seed production, predation, size, vigour, longevity, storage, germination requirements, seedling vigour, establishment and so on.

Invasions can proceed without any apparent disturbance, as in the case of pines from temperate biomes which are repeatedly introduced into tropical and subtropical areas, an example being *Pinus radiata* in Australia which was found to invade native eucalyptus forest from adjacent populations (Burdon and Chilvers, 1977).

Another problem stems from the different criteria used, implicitly or explicitly, in the application of the terms: for example the term neophyte (or neosynanthropic plant) is regarded by some as post-Columbian, as followed, for example, by *MedChecklist* (Greuter *et al.*, 1986) whilst archaeophytes are those that were present in the area concerned before that time. Even this dividing line is disputed—*MedChecklist* uses 'before the end of the fifteenth century' while Webb (1985) proposes 'about A.D. 1550, when as a result of voyages of discovery, plants from America and Asia came flooding in an unprecedented scale'. Neophytes, are on the other hand, regarded by some authors as much more recent introductions (cf. Dafni and Heller, 1982). Also archaeophytes are sometimes regarded as part of the native flora (as in *MedChecklist*) while others consider native plants as those which either evolved *in situ* or which arrived in the area concerned before the beginning of the Neolithic period or which arrived there subsequently by a method entirely independent of human influence.

These, then are some of the factors that have to be taken into account in trying to interpret the literature on invasions on a global scale. It has to be noted, however, that it is only in the case of selected countries or regions such as Australia, Europe, North America, South Africa, New Zealand, that the literature is sufficiently detailed for such problems to be really troublesome. Usually the difficulty is simply lack of information of any depth or quality. A major source of information is the very large weed literature, which I have used extensively. As we have already seen, however, not all weeds are invaders and in any case weed information for many parts of the world is sadly defective.

2.2 TAXONOMIC PATTERNS: NAMES AND NUMBERS

It is evident from what we noted above about the floristic literature that it is virtually impossible to give a reliable estimate as to the numbers of taxa worldwide that can be regarded, however loosely, as invaders. Theoretically this could be done by making a country-by-country survey of the whole world and listing those species that have been recorded as invaders or aliens and then editing the result. Alas this is no more feasible than attempting to calculate the total numbers of native species, let alone aliens, due to the gaps in the floristic

Table 2.1. Distribution of plant species throughout the world (after Heywood, 1985a)

	Flowering plants	Fungi	Ferns	Mosses
Worldwide	250 000	120 000	12 000	14 000
Tropical	160 000	90 000	11 000	9 000
Tropical Africa	30 000	20 000	1 000	1 500
Tropical Asia	35 000	20 000	6 000	2 000
Tropical America	95 000	50 000	5 000	4 000
Europe	11 500	—	150	1 100

literature. We can, however, present some figures and make some extrapolations or approximations from these.

Firstly, some statistics on the numbers of species in the various groups of plants are needed to provide a context. These are summarized in Table 2.1. It should be noted that even these estimates are disputed and the figures for tropical groups are the least reliable. Toledo (1985) in a review of Latin American floristics suggests higher figures for the New World tropics than are normally accepted and the figures for individual countries also vary quite widely. If has been suggested that there are up to 50 000 species of angiosperms in Brazil while other estimates put the figure at no more than 30 000. What is evident, however, is that the greatest floristic richness is found in regions where our knowledge of the flora is least studied.

Perhaps we should note in passing that when it comes to aliens numbers of species are not necessarily an indication of the extent to which the vegetation of the area is invaded or otherwise affected since species vary widely in their invasive capacity, aggressiveness, degree of spread and permanence and their interaction with the natural ecosystem. The literature is full of examples of single species which have shown dramatic spread and major impact on plant communities (see below).

The only global assessments of aliens are a few publications on weeds, notably Holm *et al.*, *The World's Worst Weeds* (1977). In this valuable compendium it is suggested that certainly fewer than 250 plant species have become important weeds of the world (as judged by one particular set of criteria) and they list the 18 most serious weeds in approximate order of their troublesomeness to the world's agriculturalists. These are listed below. Those marked with an asterisk are regarded as a group apart since they are not only cited more often than other world weeds but are ranked as the greatest troublemakers in the largest number of crops.

*1. *Cyperus rotundus*
*2. *Cynodon dactylon*
*3. *Echinochloa crusgalli*
*10. *Chenopodium album*
*11. *Digitaria sanguinalis*
*12. *Convolvulus arvensis*

*4. *Echinochloa colonum*
*5. *Eleusine indica*
*6. *Sorghum halepense*
*7. *Imperata cylindrica*
8. *Eichhornia crassipes*
9. *Portulaca oleracea*
13. *Avena fatua* etc.
14. *Amaranthus hybridus*
15. *Amaranthus spinosus*
16. *Cyperus esculentus*
17. *Paspalum conjugatum*
18. *Rottboellia exaltata*

Looking through this list many will be surprised not only by inclusions but by omissions. Certainly such a list does not closely relate with those species that are considered to be the most important invasive species in terms of their effect on vegetation. It should be noted that 10 of the 18 species are grasses, most of them tropical.

Attention must be drawn to bracken (*Pteridium aquilinum*) which has been described as either the world's worst weed or the most successful pteridophyte, depending on one's interests. It is the most widely distributed of pteridophytes and with the possible exception of some annual weeds it is probably the most widely distributed of vascular plants (Page, 1976). It became established in many open forest communities long before the advent of man or his agriculture although its frequency of occurrence and spread has, of course, been greatly expanded as a result of man's activities. It has been estimated that the rate of spread of dense bracken communities is 2% per annum. Some of the reasons for bracken's success as a 'permanent ecological opportunist' are its high disease resistance, low palatability to herbivores, its allelopathic effect on competing species, the effectiveness of the long distance dispersal by it minute spores, its legendary vegetative lifespan (up to hundreds of years), its tolerance of burning, its wide edaphic and climatic tolerance and its broad cytological and genetical variability (see Smith and Taylor, 1986).

In *A Geographical Atlas of World Weeds* by Holm *et al.* (1979) it is noted that the agricultural literature suggests that up to 5000 species of plants have been recorded as weeds but this is now regarded as an unrealistic estimate. After years of intensive study, Holm *et al.* consider that a reasonable estimate of the number of weeds of agriculture is of the order of 8000. Unfortunately their *Atlas* is not analysed but for the purposes of this paper we have organized the lists into families in order of number of contained species and also into those genera containing 10 species or more in descending order. These analyses are given in Appendixes 2.1 and 2.2. Clearly the statistics cover a mixture of both introduced and indigenous species but as we have noted above, indigenous species are often invasive in plant communities in their native territory. From these analyses it will be seen that the leading two families are the Compositae (Asteraceae) with 224 genera and 830 species and the Gramineae (Poaceae) with 166 genera and 753 species, which is what a quick perusal of the literature would suggest. The Papilionaceae comes a good third with 87 genera and 415 species followed by a cluster of 11 families with over 100 species each, Euphorbiaceae, Lamiaceae,

Brassicaceae, Convolvulaceae, Cyperaceae, Solanaceae, Apiaceae, Rosaceae, Scrophulariaceae, Polygonaceae and Malvaceae. More than half the remaining families contain 10 weed species or fewer, 44 of them with one species only. It is clear that weediness occurs very selectively amongst the angiosperms. In terms of monocotyledons versus dicotyledons the division is 1193:6218 species. In the gymnosperms the leading families in this regard are the Pinaceae with 32 species (20 of them *Pinus* spp.) and Cupressaceae with 15 species.

Looking at the genera, *Cyperus* and *Euphorbia* head the list with 81 and 80 species respectively, followed by *Solanum* (66), *Polygonum* (62), *Panicum* (57), *Ipomoea* (57), and *Acacia* (53).

Considering the high-ranking families, Asteraceae and Poaceae, they represent 13% and 12% of the species total respectively. Similar high percentages have been recorded in the alien flora of individual regions such as California where Raven and Axelrod (1978) report the naturalized flora to contain 674 species of which 137 (20%) are grasses and 112 (16.5%) are Asteraceae. For North America as a whole Rollins and Al-Shehbaz (1986) have analysed the herbaceous weed species and of the 460 species in 253 genera, the Compositae (with 69 species), the Cruciferae (with 52 species) and the Gramineae (with 52 species) far outrank any other family. In an overview of European weeds as exemplified by a comparison of the agrestal floras of Finland, Austria and Italy, Holzner and Immonen (1982) found that the weediest families were the Compositae with 16% of all the weed species of the three countries, followed by the Gramineae (13%), Leguminosae (13%), Cruciferae (7%), Caryophyllaceae (7%), Labiatae (6%) and Polygonaceae (5%). In the west Mediterranean 50% of the weed species belong to four families, the Compositae, Papilionaceae, Gramineae and Cruciferae (Guillerm and Maillet, 1982). The figures for Italy (which are almost certainly an underestimate) given by Franzini (1982) are a total of 466 species belonging to 51 families with the leading families being: Papilionaceae 17%, Gramineae 14%, Compositae 16%. In a review of the weed flora of South Africa, Wells and Stirton (1982) give the following figures (based on the first national weed list of South Africa by Harding *et al.*, 1980):

Exotic weeds	78 families	284 genera	503 species
Indigenous weeds	75 families	211 genera	381 species

The plant families containing most weed species (introduced and indigenous) were: Asteraceae (86 and 70), Poaceae (85 and 40), Fabaceae (50 and 22), Solanaceae (21 and 7), Cactaceae (17 and 3), Brassicaceae (16 and 11), Onagraceae (12 and 3), Rosaceae (11 and 7). Another analysis by Wells *et al.* (1983), confirmed by a further and larger sample (Wells *et al.*, 1986), indicated that 50% of the plants introduced to South Africa belonged to four families: Poaceae, Fabaceae, Asteraceae and Solanaceae (in that order) with the Brassicaceae also supplying many species that are marginally invasive. However, if one considered only the so-called transformer species (i.e. those that transform habitats or landscapes) about 50% belong to three families: Fabaceae, Myrtaceae and Pinaceae (in that order) with a substantial part of the remainder belonging to the Cactaceae, Poaceae, Proteaceae, Salicaceae and Solanaceae.

It is not possible to quote comparable figures for tropical countries since complete analyses have not been made. Some indication can, however, be obtained from partial listings.

It is not perhaps surprising that in general terms the largest angiosperm families supply such a large percentage of the world's aliens or invader species. To a large degree the very features that have been responsible for the evolutionary success and diversity of these families are those that have been responsible for their successful spread and establishment as aliens.

The Compositae (Asteraceae) are regarded as one of the most advanced families from an evolutionary point of view and few families contain such an abundance of weedy species, many of which are extremely successful and have spread especially through temperate areas of the world (see Table 2.2). Their success derives largely from the development of biological features which both ensure survival under adverse conditions and a high reproductive rate. They possess a complex series of integrated reproductive biological features in the pseudanthial capitula. The aggregation of reduced flowers into heads, their geitonogamous breeding system, often with superimposed agamospermy, a series of complicated dispersal mechanisms involving involucral bracts, capitular scales, wings, tubercles, hooks, spines, pappus scales, bristles and parachutes, and the streamlined single-seeded pseudofruits (cypselas) developed from an inferior ovary have all contributed to their success. Their diversity of habit, too, is reflected in those species that occur as invasive species—annual, biennial or perennial herbs, shrubs, trees are all

Table 2.2. Some weedy members of the Compositae (after Heywood *et al.* 1977)

Achillea millefolium—yarrow
Ambrosia artemisifolia—roman ragweed
Anthemis cotula—stinking mayweed
Bellis perennis—daisy
Centaurea nigra—lesser knapweed
Chrysanthemum segetum—corn marigold
Cirsium spp.—thistles
Cotula coronopifolia—brass buttons
Crepis spp.—hawk's beards
Hieracium spp.—hawkweeds
Leontodon spp.—hawkbits
Matricaria matricarioides—pineapple weed
Parthenium hysterophorus—wild feverfew
Senecio jacobaea—ragwort
Senecio vulgaris—groundsel
Sonchus spp.—sow thistles
Taraxacum officinale—dandelion
Xanthium strumarium—cockle bur

represented. Chemical factors are also important in their success in providing protection from overgrazing. The common groundsel (*Senecio vulgaris*) which produces between 50 000 and 60 000 cypselas per plant, with a germination frequency of over 80% is well protected from the majority of potential herbivores by the presence in the leaf tissue of the plant of toxic levels of pyrrolizidine alkaloids (Heywood *et al.*, 1977).

The grasses are a major source of weeds in many parts of the world. Again a streamlined and highly evolved inflorescence containing reduced and aggregated flowers, coupled with a series of dispersal mechanisms in the flowers and associated parts, together with a diversity of habit has been largely responsible for their evolutionary success and diversification in general and their successful role as weeds and aliens in particular. Annual grasses are amongst the most noxious and invasive species in both temperate and tropical regions. Pantropical annual grass weeds include *Eleusine indica*, *Echinochloa* spp., *Roettboellia exaltata*, *Digitaria sanguinalis*, and *Setaria* spp. Perennial grasses are exceedingly difficult to eradicate and are highly competitive, making them amongst the most pernicious weeds in the world (Kasasian, 1971). Four perennial grasses which are especially important in the tropics are *Cynodon dactylon* (also grown as a lawn grass), *Imperata cylindrica*, *Paspalum conjugatum* and *Sorghum halepense*.

The rich representation of Leguminosae in weed floras and as successful invaders is again not unexpected in view of the large size of the family: 650 genera and 18 000 species (Polhill *et al.*, 1981) and its enormous diversity of habit. As Polhill *et al.* point out 'the Leguminosae are notably "generalists", ranging from forest giants to tiny ephemerals, with great diversity in their methods of acquiring the essentials of growth and in their modes of reproduction and defence'. It is significant that nearly a third of the 18 000 species are contained in only six genera—*Acacia*, *Astragalus*, *Cassia sensu lato*, *Crotalaria*, *Indigofera*, and *Mimosa*—all of which are characteristic of open and disturbed habitats. It may be noted that five of these genera rank highly in the list based on Holm *et al.* in Appendix 2.2, with *Acacia*, *Crotalaria* and *Cassia* in the first 25. The Leguminosae possess many unique features which have been responsible for their evolutionary and ecological success, not the least being their frequent ability to fix atmospheric nitrogen through the possession of *Rhizobium* in their root nodules—general in the subfamilies Mimosoideae and Papilionoideae although in only about 30% of species of Caesalpinoideae, especially in the *Dimorphandra* group of the Caesalpinieae and *Chamaecrista* in the Cassieae (see Corby, 1981). Malloch *et al.* (1980) in a review of mycorrhizae in vascular plants point out that in the Caesalpinoideae the Detarieae-Amherstieae, which often occur on infertile soils, probably always have ectotrophic mycorrhizae which may provide an alternative to root nodulation. Other major ectotrophic trees include Dipterocarpaceae, *Nothofagus*, *Eucalyptus* and *Quercus*, which like the Detarieae-Amherstieae are ecologically aggressive and have been successful in invading primarily or exclusively endotrophic forest in their respective ecological regions. Attention should also be focussed on the Leguminosae's remarkably successful pollination mechanisms and reciprocal co-evolution with Hymenoptera (Arroyo, 1981). Legumes have successful dispersal mechanisms and as Raven and Polhill (1981) comment the often pantropical distribution of genera and major groups reflects this. They point out that no fewer than 13 genera of legumes reached the Hawaiian Islands over water barriers as wide as any to be found in the world. Madagascar too has been repeatedly colonized by legumes at various stages of the family's evolution.

Several other so-called 'natural' families provide us with many examples of invader species, such as the Brassicaceae (Cruciferae), Apiaceae (Umbelliferae), Euphorbiaceae, Lamiaceae (Labiatae), Polygonaceae, Amaranthaceae, etc. Similar analyses could be provided to a greater or lesser extent of the adaptive syndromes of these groups in terms of their success as weeds and aliens. At the generic level too, many illustrative examples could be given but there is room here for only a few examples. A tropical example is the weedy amaranths. Of the 60 or so species of *Amaranthus* only a handful are nowadays used as crops (National Research Council, 1984) while 37 have been reported as weeds; several of them such as *A. viridis*, *A. spinosus*, *A. retroflexus*, and *A. hybridus* are serious weeds of pastures, crops, roadsides and even in urban areas. The seeds remain viable for long periods, some germinating after 40 years. Their ability to adapt to any environment and tolerate adverse conditions may be partly explained by their C_4 photosynthetic pathway as well as their abundant pseudocereal-type fruits.

Nearly all the literature is concerned with invasions by higher plants, especially the angiosperms and gymnosperms, although a number of ferns and allies, notably bracken (mentioned above) and species of *Equisetum* are also included. There are few recorded cases of invasion among bryophytes. One of the best documented examples is *Campylopus introflexus*, an American/southern temperate species that has spread rapidly in Britain and western Europe since the first British record in 1941 (Richards and Smith, 1975).

Although it may not appear to be directly connected with considerations of invasions by plants, mention must be made of the special taxonomic problems often posed by invading alien species. Quite simply the fact that the alien belongs to another flora frequently increases the possibilities of misidentification or misunderstanding. Several authors have drawn attention to the difficulties that can arise from these problems. This is especially true of what are termed 'critical' groups where even taxonomists are divided as to their correct classification. The absence of an agreed taxonomy as between countries or continents, for example, exacerbates the difficulties. North American taxonomists frequently adapt, for example, different generic classifications from those widely used in Europe and Raven and Axelrod (1977) sensibly advocate the use by Californian (and by implication other North American) botanists of the taxonomy and nomenclature worked out by Flora Europaea. A particular problem arises when both native and alien variants of the same species occur in the territory concerned as in the case of, for example, *Leucanthemum vulgare* in Britain where both native and introduced tetraploid races occur in addition to the diploid. Burtt (1986) draws attention to the practical problems that can arise from the incorrect identification of aliens as in the case of *Hypericum perforatum* where biological control in New Zealand proved unsuccessful until it was realized that it was the Mediterranean race involved. Recognition of these and similar problems has led us in Britain recently to propose the establishment of a Weed Identification Centre at Kew and Reading. Other related initiatives are the establishment of a European-Mediterranean Weed Flora and Computerized Database in association with the European Economic Community's Agro-Med programme.

2.3 THE EXTENT OF INVASIONS

Only very broad generalizations can be made about the global extent of invasions by terrestrial plants. Some idea can be obtained from looking at the figures for native and alien plants for different parts of the world. Table 2.3 gives such a selection although it has to be repeated that the data vary widely in their accuracy and comparability. Even so it is clear that in terms of species numbers alone there is wide diversity with the percentage of aliens in the flora ranging from a few per cent in some tropical regions to nearly 50% in New Zealand. But number of species is not necessarily a good measure of the extent of invasion since there are numerous cases of the disproportionate effect of single species such as *Casuarina litorea* in the Bahamas which Correl (1982) describes as undoubtedly the most

Table 2.3. Percentages of introduced species in selected floras

Country	Native species	Introduced species	Percentage introduced
Antigua/Barbuda	900	180	10
Australia	15–20 000	1500–2000	10
Sydney	1500	4–500	26–33
Victoria	2750	850	27.5
Austria	3000	300	10
Canada	3160	881	28
Ecuador			
Rio Palenque	1100	175	15
Finland	1250	120	10
France	4400	500	11
Guadeloupe	1668	149	9
Hawaii	12–1300	228	17.5–19
Java	4598	313	7
New Zealand	1790	1570	47
Spain	4900	750	15

successful alien plant, especially in coastal areas, creating dense shade and producing a toxic effect on most plants so that very few native plants are able to reproduce under it. Or *Andropogon pertusus* which in less than 100 years has become the commonest grass in lowland Jamaica (Adams, 1972). And the European shrubs *Cytisus scoparius* and *Ulex europaeus* which have become widely naturalized in the Nilgiri hills in Tamil Nadu province in South India (Nair and Henry, 1983) or bracken, *Pteridium aquilinum*, which is rampant on Horton Hills in the hill country of Sri Lanka. Or *Casuarina* and *Melaleuca* which have established themselves in natural communities in the Cape region of South Africa, and constitute a threat to their ecological balance. The recent explosion of the North American *Parthenium hysterophorus* in Egypt was apparently initiated from a single sowing in 1960 of a large area with impure grass seed imported from Texas (Boulos and el-Hadidi, 1984). And so the list of examples could continue for pages with examples from most parts of the world and from most bioclimatic regions.

The extent of invasion of a particular territory is closely linked with the history and mode of invasions that have occurred, especially those that are man-induced. This is discussed in Section 2.4. What has to be stressed is that the present pattern and extent of invasion is often the culmination of centuries if not thousands of years of change in the vegetation, usually caused directly or indirectly by man's action. This is all too clear when one considers the extent to which the original vegetation has been converted or modified for agriculture.

2.3.1 North temperate regions

In north temperate regions such as Europe the native vegetation has been largely destroyed or modified by deforestation, agriculture, grazing, urbanization and other of man's activities. In these regions non-active species have come to play a major role in our perception of the landscape. In Great Britain, for example, agriculture affects 80% of the land surface and most of the native vegetation had been destroyed or heavily altered already by 1700. Natural forest cover had been reduced to 5.4% in 1924 although the forested area was increased subsequently to 9.4%, largely due to the plantation of alien conifers (Ratcliffe, 1984). What is not perhaps realized is the extent to which the distribution of plant species can change over relatively short periods due to man's activities. The situation in the Netherlands as described by Mennema (1984) is so serious that he considers that plant geography as a subject can no longer be practised there. By comparing the products of the number of Netherland vascular plant species and the number of their localities before and after 1950, he found that 70% of the flora (not the species!) had disappeared. What is in many ways worse is that many wild species have been planted in educational flora parks and other man-made nature areas or along roadsides, and commercial nurseries have invested considerable sums of money in cultivating less common species for planting on roadsides or other nature areas. As a result genotypes and ecotypes have been introduced which are different from those that naturally belong to the areas concerned. Attention needs to be drawn to this dangerous practice which is now becoming widespread due to the actions of well-meaning but ill informed people. The consequences in terms of the invasion of communities and populations by not so much alien species as alien ecotypes is insidious. This is of course also a very serious danger in countless reafforestation projects where non-local if not exotic provenances have been employed.

2.3.2 Mediterranean-climate regions

In the Mediterranean Basin, most of the natural vegetation has been modified by man's activities, a process dating back thousands of years. Many of the plant communities are secondary and exist today in the form of matorral or garrigue as the result of human interference with the climax communities of oaks and pines (di Castri, 1981). These shrublands, which largely consist of invasive subseral species and often other more weedy elements depending on the state of degradation of the soil and vegetation, cover an immense area and constitute a third of the total vegetation. Much of the remaining forest cover in the Mediterranean is highly modified, especially by reafforestation or other forms of management and the oak forests in particular have suffered from the introduction of alien species, usually in the form of conifers, some of which can become invasive so long as man controls the regeneration of the natural oak climax forest. The pine forests themselves, which cover vast areas of the Mediterranean and sub-Mediterranean, are seldom climax communities but a replacement of the broad-leaved forest that man has destroyed over the centuries. With the exception of some montane forests these pinewoods are heavily modified in composition through the planting of species such as *Pinus halepensis, P. pinaster and P. sylvestris.* Indeed, in the west Mediterranean *P. halepensis*, the Aleppo pine, was considered to be introduced until pollen analysis revealed it to be native. Eucalypts have been extensively employed in reafforestation and in Spain cover nearly 80% of the acreage of broadleaves planted and represent nearly 10% of all reafforestation acreage (ICONA, 1984). The vigorous growth and competitive ability of several of the introduced species of *Eucalyptus* pose an increasing threat to the native residual vegetation.

In Australia, as Specht (1981) has observed, in the areas of mediterranean climate and vegetation, man through modern agriculture has managed to achieve in 50 years what has taken over 2000 years in the Mediterranean Basin. Little remains in its original form, having been replaced largely by fields of wheat and grazing land in all but the driest zones.

Likewise in the Cape region of South Africa, which is loosely mediterranean in climatic terms, the impact of man on the mediterranean-type scrublands on the more fertile soils has been such that little of the original vegetation remains. Intensive landuse has reduced the renosterveld from 36% of the area to 1% and today farmland and other non-native vegetation, plus urbanization covers over 50% of the southwest Cape region. Much of the remaining vegetation of the Cape region is threatened by the large numbers of alien species that have been introduced, often invading and replacing native plant communities. Particularly serious are the problems posed by the large number of woody plants, many of them of Australian origin. Counted amongst these are several wattles or acacias, especially *Acacia cyclops*, *A. decurrens*, *A. elata*, *A. longifolia*, *A. saligna*, as well as the stinkbean *Albizzia lophantha*, and *Eucalyptus cladocalyx*, *E. gomphocephala* and *E. lehmannii*. Other Australian tree invaders include several species of *Hakea* and from the Mediterranean Basin two species of pine, *Pinus pinaster* and *P. halepensis*. These invasive species often grow more strongly than the native scrub and have better powers of regeneration after fire with the result that the original communities are being largely replaced. This is especially true of the fynbos which are particularly susceptible to invasion. It is estimated that 60% of the natural vegetation of the fynbos has been replaced in this way. Major problems are caused by other introductions such as *Lantana camara*, one of the world's 10 worst weeds (*sensu* Holm *et al.*, 1977) which is toxic to livestock and has invaded the veld and agricultural or derelict land. Notorious too are the prickly pears (*Opuntia*), especially *Opuntia aurantiaca*, which is the most widespread and which is largely resistant to eradication by chemical or mechanical means.

In the mediterranean vegetation zones of Chile little remains of the natural vegetation today as a result of man's action over thousands of years and most of the remaining matorral communities are heavily modified by invader species.

2.3.3 Grasslands and pastures

The grassland and pastures of the world again owe their origin to a large extent to man's action, often dating back thousands of years and these areas are often rich in alien invaders, some of them deliberately introduced. As Williams (1985) reminds us the principal plants used to create Australia's sown pastures have been neophytes and their derivatives. The long history of planned plant introductions into northern Australia is reviewed by Mott (1986) in a paper on planned invasions of tropical Australian savannas. Fire has frequently played a major role in the creation of these pasture lands. In Australia Williams (1985) believes that the ease by which so many aliens have been able to invade so successfully and become part of the Australian landscape can be explained in part by the fire regimes of the Aboriginal populations for as long as 10 000 years. Fire too has been suggested as one cause of the extensive grassland communities of California; the European settlers upset the equilibrium of the landuse system involving fire established by the native Indians and so increased the scale and frequency of fires as a means of clearing forest and chaparral to open up ranges for grazing, agriculture and mining (Trabaud, 1981).

Whatever the nature of the original grassland in California, whether it was dominated by perennial grasses such as *Aristida*, *Poa*, *Stipa* or other plants, there can be no denying that today it is dominated by introduced annual grass species to such an extent that they have been considered as new and permanent members of the native flora on the grounds that they are unlikely now ever to be eliminated (Heady, 1977). Again in Africa the grassland ecosystems usually result from the destruction of the forest (with the exception of some savannas of edaphic origin) and are often maintained through the use of fire regimes and are frequently infested by alien invader species. Likewise in South America, the region of the great savannas, including the cerrados and campos of Brazil, and the llanos of Colombia and Brazil, covering 3 million square kilometres of barely usable land, are increasingly being seen as areas for exploitation leading to their invasion by weedy and pioneer species. Many of the ranch lands of South America have been produced by forest destruction after the European conquest. Similarly in Madagascar vast areas of the natural vegetation have been burned, since the Palaeo-Indonesians invaded the island, to provide pasture land for the zebu cattle. These pastures are poor in species and characterized by weedy species such as the pantropical *Imperata cylindrica* and *Heteropogon contortus*.

A general review of pasture weeds of the tropics and subtropics is given by Tothill *et al.* (1982). They concur with Moore (1971) in considering that woody weeds are by far the most important in native grazing lands in the semiarid/arid/subhumid tropics and subtropics. Many of these woody weeds are native species but there has also been invasion from exotics such as *Calotropis procera*, which has extended throughout tropical Asia and Africa into South America and is now spreading rapidly in northern Australia. Other invaders in Australia include *Acacia farnesiana*, *Mimosa* spp., *Cryptostegia grandiflora* and *Zizyphus mauritiana* which form dense thickets. The effects of woody invaders on the natural rangelands of south and southwest Africa have already been mentioned, whether by native or exotic species. Herbaceous weeds also affect tropical and subtropical pastures. A grass that is widespread and used for permanent pasture in various parts of the tropics, such as Florida, West Indies, Fiji, Malaysia, Guyana and Hawaii, is *Axonopus affinis*. It invades degenerate pasture of *Paspalum dilatatum*, *Trifolium repens* and *Pennisetum clandestinum*, leading to a decline in animal production (UNESCO, 1979). Even amenity grassed areas in the tropics are subject to severe invasions. In Brasilia, for example, the grassed areas in parks, gardens, etc., mainly comprise two species, *Paspalum notatum* and *Cynodon dactylon* and these are invaded by no fewer than 60 species belonging to 38 genera and 15 families (Gramineae 15 species, Compositae 10 species) and this is a common pattern in other areas of Brazil (Brandao Ferreira and Borges Machado, 1976).

2.3.4 Tropical forests

The widespread destruction or conversion of the primary forest ecosystem in both the dry and humid tropics has led to a whole series of secondary successional communities (when degradation has not proceeded too far) in which numerous invaders play a role—herbaceous weeds, grasses, shrubs and secondary trees. These successional stages can last for hundreds of years and have not been studied in detail. The human impact on the tropical forest differs in the major regions. Large-scale commercial logging has already been responsible for clearing out a large part of the forests of Africa and Southeast Asia. Additionally the introduction of large scale plantation crops such as sugar, rubber, oil palm, coffee, tea, etc., especially in the 19th century, has been the cause of massive deforestation. While few of these crops have themselves become weedy and invasive, the plantations have each attracted their own suite of weeds. The weeds of tea plantations, for example, are reviewed by Ohsawa (1982) and include herbs, lianas, shrubs, trees and even epiphytes. In Central and South America logging has so far been less serious a cause of deforestation or forest conversion than clearing and burning for cattle ranching together with slash and burn cultivation which when practised on a massive scale, especially by unskilled immigrants, can have a serious effect on the forest cover. Although weeds are not unknown in the relatively untouched forests of large areas of the Amazon, such ecosystems are perhaps amongst the least affected by invasion by higher plants.

A perhaps underestimated source of invaders in tropical forest ecosystems are the minor crops which are often cultivated by local peasants on an appreciable scale. A good example is cardamom (*Elettaria cardamomum*) which has become a serious invader of valleys in humid forests in Sri Lanka and South India, even in reserve areas where its cultivation is illegal.

2.3.5 A pattern of islands

Island ecosystems are a special case in terms of plant invaders and a good deal has been written about them. It is now accepted as highly probable that the flora of remote islands have been derived from long distance dispersal, the classic case being the Hawaiian Islands (Wagner *et al.*, 1985). However, by far the most devastating effects have been caused by man's introductions, accidental or deliberate, of plant species that have become serious invaders of the often fragile island ecosystems. The introduction of domestic animals has been another important factor since these not only affected the vegetation directly by grazing or browsing but were responsible indirectly for the introduction of many vigorous weeds and invaders through the imports of fodder plants and seeds.

The extent of invasion by alien species in island floras varies widely. Islands such as New Zealand, Hawaii and Madagascar are extreme examples with greatly modified vegetation and large numbers of introduced species. Patterns for the islands of the Indian Ocean, including Madagascar, the Mascareignes, Seychelles and Sri Lanka, are reveiwed by Renvoize (1979) while those of temperate island floras are considered by Moore (1979, 1983). Other useful contributions will be found in the symposium volume *Plants and Islands* (Bramwell, 1979).

2.4 MODES OF INVASIONS

Viewed globally it might seem impossible to seek out any overall pattern or trends in the invasion process by higher plants. There is a risk too that what appears to be a pattern is simply a reflection of our state of knowledge of different parts of the world. Nevertheless some regularities and generalizations do seem to emerge.

Historically one can often detect several distinct phases in the invasion process—for example, prehistoric man's effect on the vegetation, that of the early settlers, the period of colonial expansion, the phase of modern agriculture and most recently the massive degradation of the vegetation caused by the population explosion, coupled with the demand for increased resources and improved living standards. A detailed example of such a chronology is given by Harris (1965) in his outstanding ecological study on the Outer Leeward Islands (Antigua, Barbuda and Anguilla).

These do not all fit into exactly the same chronological pattern in different parts of the world but I shall consider the main phases briefly in turn, giving what I hope are representative examples. The emphasis will be on invasions which have been facilitated by man, either deliberately or accidentally, since these are by far the main source of recent introductions across the world. Human action has been responsible for the extension of range of native plants through habitat disturbance which, as Baker (1986) points out is invasion in only a technical sense. Clear cases of invasions are the many examples of native species which have been introduced with man's help into other parts of the country and often into communities where they were not known previously. But by far the most numerous invaders are those that have been introduced by some means of transport from foreign lands by man, his domesticated animals or machinery. The actual modes of entry are numerous and varied, ranging from packing material and soil around introduced plants to bird seed and clothing and footwear. As we shall see man has transported thousands of species of plants from one biogeographic region to another through the deliberate introduction of economically important plants such as crop or plantation species, timber trees, forage plants, or those of ornamental or amenity importance, often together with their accompanying weeds. This has been done on a massive scale and the extent is not perhaps appreciated until one takes a global view. There are few areas in the world where the consequences of this process are not strongly reflected.

The geographical source of these invaders is, therefore, often a reflection of the agricultural and economic history of the region concerned and of the mode of invasion. High percentages of invaders in North America and Austalia, for example, have come from Europe, western Asia and the Mediterranean Basin. These sources are understandable when one considers the pattern of colonization and the source of the human and animal introductions. The eastern and southern fringes of the Mediterranean Basin and the adjacent Mediterraneo-Irano-Turanian steppes of the Middle East are probably the largest source of weeds and the cradle of many that are common to temperate and warm-temperate zones of the world (Zohary, 1962). Outstanding members of such weed communities are annual species of Gramineae, Compositae, Leguminosae, Umbelliferae, Cruciferae, etc. This is partly explained by the long history of human modification of the vegetation of this region, dating back to at least 9000 BP and the origins of agriculture. This together with the climatic conditions and the nature of the soils has created a whole series of habitats that favour the evolution and spread of species with the characteristics that today we associate with weeds. Not surprisingly a large part of the synanthropic flora of countries such as Israel and Egypt is made up of species of Mediterranean origin or distribution, with again high percentages of grasses, composites, crucifers, legumes, etc. The number of alien weeds and invaders is, not surprisingly, small: Dafni and Heller (1982) record only 73 adventive species for Israel. In Egypt the occurrence of tropical weeds seems to be of recent date and a considerable number have recently been reported from the Nile basin. The winter and summer weed communities of Egypt were recently studied by Kosinova (1975) who found that summer weeds are predominantly of tropical distribution or origin while winter weeds are represented by species of Mediterranean origin or distribution.

2.4.1 The early historical and aboriginal phases

In a number of cases we have some information on the effects of aboriginal activities on the plant life of various parts of the world. For example the agricultural practices of the Arawaks and Caribs on the Leeward Islands is discussed in detail by Harris (1965). Widespread and naturalized aliens at this period (600–1500 AD) probably included guava, soursop and sweetsop. In Hawaii when the Polynesians arrived some 1600 years ago, they found the vegetation was virtually untouched by man. The Polynesians burned and cleared large parts of the lowlands for their crops; they also introduced pigs and rats with devastating consequences (see account in Wagner *et al.*, 1985). In Australia Groves (1986) notes that the first record of an introduction leading to a plant 'invasion' was tamarind *Tamarindus indica*, brought in by the Macassans on their annual visits to the northern shores of Australia for their own diet from about 1700 AD. In East Africa extensive agriculture was thought to have been introduced by Bantu-speaking peoples in 0–500 AD who caused widespread forest clearance. The main crops were probably sorghum, eleusine millet and bananas (Lind and Morrison, 1974). Possible introductions to southern Africa are considered by Wells *et al.* (1986) who point out that several of the early crop plants are themselves invasive there, such as *Cannabis sativa*, *Cocos nucifera*, *Jatropha curcas*, and *Pisidium guajava*. An elegant review of pre-Columbian plant migrations from lowland South America to Meso-America is given in a series of papers edited by Stone (1984). In pre-Columbian times grain amaranths *Amaranthus hypochondriacus*, was one of the basic food crops in the New World and thousands of hectares were cultivated in Aztec and Inca farmlands. It was the Spanish conquistadores who stopped its use as a staple (National Research Council 1984). Some of the earliest records of synanthropic plants come from the excavations of Egyptian prehistorical and historical settlements, most importantly from ancient tombs, from the Neolithic to the Coptic periods. A composite list of the synanthropic species identified from the various periods and dynasties is given by Kosinova (1974) from which it will be seen that numerous well-known invasive species are represented. What is evident from these and other sources is that many plant introductions took place in many parts of the world long before the effects of the European colonial phase were noted.

2.4.2 The European/colonial phase

The voyages of exploration and gradual settlement of the East and the New World by the European powers led to the most intensive and extensive invasions by higher plants that the world has witnessed. Although there had been some movement of plants by man prior to this along the trade routes, dating back many hundreds of years, it was the opening up of the tropics, in particular by the East India Companies, for the development of the spice trade especially, and the discovery of the New World leading to the great wave of post-Columbian exchanges of crop plants that was responsible for the major impact on the world's ecosystems. This was essentially a European-dominated phenomenon involving the great colonial powers such as France, Spain, Portugal, the Netherlands and most notably Britain, who eventually became the proprietors of some of the most important agricultural lands in the world. This phase has been aptly described in recently published account by Crosby (1986) 'ecological imperialism'.

The European colonization of Australia, to take a major example, has been described as an 'apocalyptic event' for Australian ecosystems. Since the settlement by the British in 1788 the flow of new plant material was almost uncontrollable. The sequence of events has been graphically summarized by Williams (1985):

> The first trickle of plants was represented mainly by species from Britain, few of which grew without massive and continuous disturbance of the native vegetation and soils. By the mid-1800s this trickle had become a torrent and by the 1880s it was a flood...

A similar pattern has been described for other countries. Globally the extent to which the vegetation has been destroyed or modified by the introduction of crop plants is closely related to the history of colonization. In the tropics of Asia the effects have been the most dramatic because of the extent and duration of the colonial period; in tropical Africa the consequences on the native ecosystems, although substantial, were less due to the later timing of the process. In Latin America, the early withdrawal of the colonial powers of Portugal and Spain and the minor involvement of Britain and other colonial countries has meant that the vegetation has suffered very much less than in other areas of the tropics. In the Caribbean, on the other hand, the sustained period of colonization has had devastating effects on the native plant life, especially through the widescale introduction of the plantation crops, notably sugarcane on which the local economies became dependent (see below).

2.4.3 The role of the botanic gardens

The early European botanic gardens, founded from the 16th century onwards in Italy, France, Britain, the Netherlands and so on, were responsible for a considerable amount of plant introduction for medicinal and ornamental and amenity purposes, as well as later for scientific study. However, it was the tropical botanic gardens which were used mainly as an instrument of colonization. They were created in many cases as instruments of colonial expansion and commercial development and played a major role in establishing the patterns of agriculture in several parts of the world, notaly in Southeast Asia (Heywood, 1985b, 1986). Great tropical botanic gardens that have been significant in this process include that of Pamplemousses in Mauritius (the first tropical botanic garden to be established in 1735), Buitenzorg (today the Kebun Raya, Bogor) founded in 1819 and which was responsible for the introduction of the oil palm (*Elaeis guineensis*) as a plantation crop in Southeast Asia from seedlings obtained from Mauritius in 1848; Peradeniya (1821) and its associated gardens at Gampaha and Hakgala in Sri Lanka and Singapore (1822) which along with Gampaha and the Royal Botanic Gardens at Kew were the source of the major rubber plantations in Southeast Asia (Holttum, 1970; Purseglove, 1959). Very many crop plants were introduced through these gardens, often in association with European botanic gardens in Britain, the Netherlands and France—cloves, chocolate, cinchona, tea, coffee, breadfruit and so on. These tropical gardens were often more to be regarded as staging posts or introduction centres than botanic gardens in the modern sense. The celebrated Royal Botanic Gardens of Calcutta were originally created as, in effect, a commercial nursery and indeed their founder Colonel Kyd called the place a botanic garden although no botany was to be practised there and he explicitly wrote that there would be none. Purseglove (1959) comments on a curious aspect of these gardens:

> It has always seemed strange to me that many of the world's major tropical plant products are produced largely in countries far removed from their region of origin, e.g. South American rubber in Malaya and Indonesia, South America cocoa in West Africa, South American quinine in the East Indies, African coffee in Brazil, cloves and nutmeg from the Moluccas in Zanzibar and Grenada respectively, sugar, banana and limes from South-East Asia in the West Indies, and

> vanilla from Central America in Madagascar. Obviously this cannot be attributed to any one particular reason and many factors are involved, including economics, available land and labour supply, technical skill in processing, suitability of the crop for plantation or peasant agriculture, etc. Nevertheless, one would have expected that a crop was more suited ecologically to its country of origin than to its new home. I suspect that one of the major reasons is that when a new crop has been introduced without many of its normal pests and diseases, it has more chance of flourishing and giving high yields. Many of the crops are introduced as seeds, which limits the number [of] pests and diseases they can carry with them, and this is further enhanced by the plant quarantine regulations now enforced in many countries. One can only contemplate what would happen to the Malayan rubber industry if the South American Leaf Blight (Dothidella ulei) were accidently introduced.

Just how important the colonial tropical botanic gardens were considered is indicated by the attitude of Sir Joseph Hooker, the celebrated onetime Director of the Royal Botanic Gardens, Kew, who regarded the economic revival of the West Indies as dependent on the wise application of botanical science. He had seen the emancipation of slaves, the rise and fall of the sugar industry and growing poverty, discontent and demoralization. In a letter in 1897 he wrote 'I am interested greatly in the W. Indian sugar situation... I had so much to do with the vegetable industries of the W. Indies when I was at Kew, that I cannot but feel deeply interested... I see nothing for it but the establishment of cheap Botanical Gardens, confined to economic plants, in the other colonies.' The quite remarkable influence of the network of British botanic gardens that stretched around the world and centred on Kew is described in Brockway's somewhat provocative study entitled *Science and Colonial Expansion. The Role of the British Royal Botanic Gardens* (1979). Another important historical study of the role of a colonial botanic garden, that of the Company's Garden, later the Cape Town Botanic Garden and now the Cape Town Public Gardens, is given by Shaughnessy (1986), and in particular its activities in the 1850s and 1860s in introducing and spreading Australian plants.

While human agencies have been responsible for the spread and maintenance of many of these introduced species, and the disturbance of the vegetation that had to take place before the crops or plantations could be established, only a limited number of the species involved have become invasive independently of man's continuing action in aiding their dispersal. It is often the crop and plantation weeds that create the problems rather than the cultivated plants. Man has provided the conditions suitable for invasion through these large scale agricultural activities and the surprise, perhaps, is that not more species have responded to the opportunities.

2.4.4 Recent changes in the spread and decline of invaders

In recent years, the changes caused by the impact of man on the environment as a result of modern agricultural methods, machinery, weedkillers, crop rotations, etc., has altered the agricultural landscape and has had an effect on the synanthropic flora, causing not only the increase of some weed species but the decline of others. In Sweden, for example, and other Nordic countries two of the most threatened species are *Agrostemma githago* and *Bromus secalinus.* The former is an archaeophyte, known from archaeological sites. It is at risk today because the germination biology of the seeds is not adapted to modern agricultural practices (see Svenson and Wigren, 1986). In Egypt, *Ceruana pratensis* (Compositae) was formerly a common weed along the Nile and in irrigation channels, especially in Upper Egypt, with records dating back to Pharaonic times. It is now extremely rare and on the way to extinction (Boulous and el-Hadidi, 1984).

There are several factors causing an increase in the availability of habitats suitable for successful colonization by exotic species. The accelerating trends in the deforestation of parts of South America, for example, mainly for cattle ranching, expansion of plantation crops such as coffee and cane, the introduction of plantations of exotic timber and firewood species, and the shortening of the fallow cycle in slash and burn cultivation which is itself extending considerably, are almost certain to have a major effect on the patterns of invasion and on the composition of the exotic flora. Already we can predict some of the candidate species for invasions since there is today a strongly developing interest in wider scale introduction of lesser known species of potential use in agriculture and forestry (National Research Council, 1975). Examples include the forgotten crops of the Incas such as oca (*Oxalis tuberosa*) which is second only to potato in the Andes. It is already a commercial crop in New Zealand ('yam') and as Vietmeyer (1986) puts it would like to sweep round the world if given modern attention.

As a plant conservationist I must confess to viewing the present situation with some alarm. So much of the world's vegetation has already been modified or destroyed by invasions, and we have become all to accustomed to accepting secondary or even artificial vegetation and landscapes as though they were today's norms. We must make strenuous efforts to safeguard as far as possible the remaining part of our native vegetation, a natural heritage that is under ever increasing threat.

2.5 SUMMARY AND CONCLUSIONS

There are few ecosystems in the world that have not been affected to a greater or lesser degree by invasions by terrestrial plants, especially flowering plants and conifers. Human intervention has been the major causal factor in these invasions, especially through the clearance of natural vegetation for agriculture and forestry and the subsequent invasions by weedy species. The extent of this habitat modification varies considerably in different bioclimatic regions—ranging from large scale transformation in many mediterranean-climate regions and in many island ecosystems, to relatively minor incursions in remote humid tropical forests. The species involved belong to a wide range of families but with notable concentrations in large 'natural' families such as the Asteraceae, Fabaceae and Poaceae which possess complex reproductive and dispersal mechanisms. Invasion of natural communities, in many parts of the world, by introduced plants, especially woody species, constitutes one of the most serious threats to their survival, although it is one that is not fully acknowledged by conservationists.

ACKNOWLEDGEMENTS

Many colleagues have assisted me with references or suggestions, especially J. Akeroyd, A. H. Bunting, D. Drennan, D. M. Moore, B. Pickersgill, P. H. Raven, W. Wagner. Their help is gratefully acknowledged.

REFERENCES

Adams, C. D. (1972). *Flowering Plants of Jamaica.* University of West Indies, Mona.

Arroyo, M. T. Kalin (1980). Breeding systems and pollination biology in Leguminosae. In: Polhill, R. M., and Raven, P. H. (Eds.) *Advances in Legume Systematics*, pp. 723–69. Royal Botanic Gardens, Kew.

Baker, H. G. (1986). Patterns of plant invasions in North America. In: Mooney, H. A., and J. A. Drake (Eds.), *Ecology of Biological Invasions of North America and Hawaii.* Springer-Verlag, New York.

Berg, R. Y. (1983). Plant distribution as seen from plant dispersal—general principles and basic methods of dispersal. *Sonderb. Naturwiss. Ver. Hamburg*, **7**, 13–46.

Boulos, L., and el-Hadidi, M. Nahil (1984). *The Weed Flora of Egypt.* American University of Cairo Press.

Bramwell, D. (Ed.) (1979). *Plants and Islands.* Academic Press, London.

Brandao Ferreira, M., and Borges Machado, J. (1976). Invasoras dos gramados do Distrito Federal. *An Soc. Bot. Brasil XXV Congreso Nacional de Botanica.* pp. 389–94. Recife.

Brockway, L. H. (1979). *Science and Colonial Expansion. The role of the British Royal Botanic Gardens.* Academic Press, New York. 215 pp.

Burdon, J. J., and Chilvers, G. A. (1977). Preliminary studies on a native eucalypt forest invaded by exotic pines. *Oecologia*, **31**, 1–12.

Burtt, B. L. (1986). Concluding remarks and the way ahead. *Proc. Roy. Soc. Edinburgh*, **89B**, 301–7.

Corby, H. D. L. (1981). The systematic value of leguminous root nodules. In: Polhill, R. M., and Raven, P. H. (Eds.), *Advances in Legume Systematics*, pp. 657–76. Royal Botanic Gardens, Kew.

Correll, D. S. (1982). Flora of the Bahamas Archipelago. Cramer, Vaduz.

Crosby, A. W. (1986). *Ecological Imperialism. The Biological Expansion of Europe, 900–1900.* Cambridge University Press, Cambridge. 368 pp.

Dafni, A., and Heller, D. (1982). Adventive flora of Israel— phytogeographical, ecological

and agricultural aspects. *Pl. Syst. Evol.*, **140**, 1–18.
di Castri, F. (1981). Mediterranean—type shrublands in the world. In: di Castri, F., Goodall, D. W., and Specht, R. L. (Eds.), *Ecosystems of the World 11. Mediterranean-type Shrublands*, pp. 1–52. Junk, The Hague.
Ellenberg, H. (1979). Man's influence on tropical mountain ecosystems in South America. *J. Ecol.*, **67**, 401–16.
Franzini, E. (1982). Italy. In: Holzner, W. and Numata, M. (Eds.), *Biology and Ecology of Weeds*, pp. 245–56. Junk, The Hague.
Greuter, W., Burdet, H., and Long, G. (1986) *MedChecklist*, Vol. 3. OPTIMA, Berlin and Geneva.
Groves, R. H. (1986). Plant invasions of Australia: an overview. In: Groves, R. H., and Burdon, J. J. (Eds.), *Ecology of Biological Invasions*, pp. 137–49. Cambridge University Press, Cambridge.
Guillerm, J. L., and Maillet, J. (1982). West Mediterranean countries of Europe. In: Holzner, W., and Numata, M. (Eds.), *Biology and Ecology of Weeds*, pp. 227–44. Junk, The Hague.
Harding, G. B., Stirton, C. H., Wells, M. J., Balsinhas, A., and van Hoepen, E. (1980). A first national weed list for southern Africa. Cited in Wells M. J., and Stirton, C. H. (1982).
Harper, J. L. (1977). *Population Biology of Plants*, Academic Press, London.
Harris, D. R. (1965). *Plants, Animals, and Man in the Outer Leeward Islands, West Indies.* Univ. Calif. Publ. Geogr. 18.
Heady, H. F. (1977). Valley grassland. In: Barbour, M. G., and Major, J. (Eds.), *Terrestrial Vegetation of California*, pp. 491–514. Wiley, New York.
Heywood, V. H. (1985a). *Times Higher Education Supplement.* 19 March, p. 13.
Heywood, V. H. (1985b). Botanic Gardens and taxonomy—their economic role. *Bull. Bot. Survey India*, **25**, 134–47.
Heywood, V. H. (1986). *Botanic Gardens as Resource and Conservation Centres.* Commonwealth Science Council (in press).
Heywood, V. H., Harborne, J. B., and Turner, B. L. (Eds.). (1977). *The Biology and Chemistry of the Compositae.* London, Academic Press. xiv + 1189 ps.
Holm, L., Pancho, J. V., Herberger, J. P., and Plucknett, D. L. (1979). *A Geographical Atlas of World Weeds*, Wiley, New York.
Holm, L. G., Plucknett, D. L., Pancho, J. V., and Herberger, J. P. (1977). *The World's Worst Weeds*, University Press Hawaii, Honolulu.
Holttum, R. E. (1970). The historical significance of botanic gardens in S.E. Asia. *Taxon*, **19**, 707–14.
Holzner, W., and Immonen, R. (1982). Europe: An overview. In: Holzner, W., and Numata, M. (Eds.), *Biology and Ecology of Weeds*, pp. 203–26. Junk, The Hague.
ICONA (1984). *Inventario Forestal Nacional ICONA. Años 1965 a 1974.* Madrid.
Kasasian, L. (1971). *Weed Control in the Tropics.* Leonard Hill, London.
Kosinova, J. (1974). Studies on the weed flora of cultivated land in Egypt. 4 Mediterranean and tropical elements. *Candollea*, **29**, 281–95.
Kosinova, J. (1975). Weed communities of winter crops in Egypt. *Preslia*, **47**, 58–74.
Lind, E. M. and Morrison, M. E. S. (1974). *West African Vegetation.* Longman, London.
Mack, R. M. (1985). Invading plants: their potential contribution to population biology. In: White, J. (Ed.), *Studies on Plant Demography*, pp. 127–42. Academic Press, London.
Malloch, D. W., Pyrozynski, K. A., and Raven, P. H. (1980). Ecological and evolutionary significance of mycorrhizal symbiosis in vascular plants. *Proc. Nat. Acad. Sci.*, **77**, 2113–18.
Mennema, J. (1984). The end of plant geography in the Netherlands. *Norrlinia*, **2**, 99–106.

Michael, P. W. (1981). Alien plants. In: Groves, R. H. (Ed.), *Australian Vegetation*, pp. 44–64. Cambridge University Press, Cambridge.
Moore, D. M. (1979). Origins of temperate island floras. In: Bramwell, D. (Ed.), *Plants and Islands*, pp. 66–85. Academic Press, London.
Moore, D. M. (1983). Human impact on island vegetation. In: Holzner, W., Werger, M. J. A., and Ikusima, M. (Eds.), *Man's Impact on Vegetation*, pp. 237–46. Junk, The Hague.
Moore, R. M. (1971). Weeds and weed control in Austalia. *J. Aust. Inst. Agric. Sci.*, **37**, 181–91.
Mott, J. J. (1986). Planned invasions of Australian tropical savannas. In: Groves, R. H., and Burdon, J. J. (Eds.), *Ecology of Biological Invasions*, pp. 89–96. Cambridge University Press, Cambridge.
Nair, N. C., and Henry, A. N. (1983). *Flora of Tamil Nadu Ser. 1 Analysis*, Vol. 1. Botanical Survey of India, Southern Circle, Coimbatore.
National Research Council (1975). *Underexploited Tropical Plants*. National Academy of Science, Washington, DC.
National Research Council (1984). *Amaranth. Modern Prospects for an Ancient Crop.* National Academy Press, Washington, DC. 80 pp.
Ohsawa, M. (1982). Weeds of tea plantations. In: Holzner, W., and Numata, M. (Eds.), *Biology and Ecology of Weeds*, 435–48. Junk, The Hague.
Page, C. N. (1976). The taxonomy and phytogeography of bracken—a review. *Bot. J. Linn. Soc.*, **73**, 1–34.
Polhill, R. M., Raven, P. H., and Stirton, C. H. (1981). Evolution and systematics of the Leguminosae. In: Polhill, R. M., and Raven, P. H. (Eds.), *Advances in Legume Systematics*, pp. 1–26. Royal Botanic Gardens, Kew.
Purseglove, J. W. (1959). History and functions of botanic gardens with special reference to Singapore. *Garden's Bulletin Singapore*, **17**, 125–54.
Ratcliffe, D. A. (1984). Post-medieval and recent changes in British Vegetation: the culmination of human influence. *New Phytologist*, 98, 73–100.
Raven, P. H., and Axelrod, D. E (1978). *Origin and Relationships of the Californian Flora.* Univ. Calif. Publ. Bot. 72.
Raven, P. H., and Polhill, R. M. (1981). Biogeography of the Leguminosae. In: Polhill, R. M., and Raven, O. H. (Eds.), *Advances in Legume Systematics*, pp. 27–34. Royal Botanic Gardens, Kew.
Renvoize, S. A. (1979). The origins of Indian Ocean Island Floras. In: Bramwell, D. (Ed.), *Plants and Islands*, pp. 107–29. Academic Press, London.
Richards, P. W., and Smith, D. (1975). A progress report on *Campylopus introflexus* (Hedw.) and *C. polytrichordis* De Not. in Britain and Ireland. *J. Bryol.*, **8**, 293–8.
Robin, J. M., and Carr, G. W. (1986). Hybridization between introduced and native plants in Victoria, Australia. In: Groves, R. H., and Burdon, J. J. (Eds.) *Ecology of Biological Invasions*, p. 56. Cambridge University Press, Cambridge.
Robinson, R. W., Carr, G. W., and Robin, J. M. (1986). Plants naturalized outside their natural geographical range in Victoria, Australia: dire forboding for the flora. In: Groves, R. H., and Burdon, J. J. (Eds.), *Ecology of Biological Invasions*, p. 57. Cambridge University Press, Cambridge.
Rollins, R. C., and Al-Shehbaz, I. A. (1986). Weeds of South-West Asia in North America with special reference to the Cruciferae. *Proc. Roy. Soc. Edinburgh*, **89B**, 289–99.
Shaughnessy, G. L. (1986). A case study of some woody plant introductions to the Cape Town area (in press).
Smith, P. M. (1986) Native or introduced? Problems in the taxonomy of some widely introduced annual brome grasses. *Proc. R. Soc. Edinburgh, Sect. B*, 273–81.

Smith, R. T., and Taylor, J. A. (1986). *Bracken: Ecology, Land Use and Control Technology*, Parthenon Publishing, Carnforth.
Specht, R. L. (1981). Major vegetation formations in Australia. In: Keast, A. (Ed.), *Ecological Biogeography of Australia*, pp. 165–297. Junk, The Hague.
Stebbins, G. L. (1971). Adaptive radiation of reproductive characters of angiosperms, II: Seeds and seedlings. *Ann. Rev. Ecol. Sys.*, **2**, 237–60.
Stirton, C. H. (1979). Taxonomic problems associated with invasive alien trees and shrubs in South Africa. *Proc. 9th Plenary Meeting AETFAT*, pp. 218–19.
Stone, D. (ed.), (1984). Pre-Columbian Plant Migration. Papers Peabody Museum Archaeology and Ethnology, Vol. 26.
Svenson, R., and Wigren, M. (1986). Observations on the decline of some farmland weeds. *Mem. Soc. Fauna Flora Fennica*, **62**, 63–7.
Toledo, V. (1985). *A Critical Evaluation of the Floristic Knowledge in Latin America and the Caribbean.* The Nature Conservancy, Washington DC.
Tothill, J. C. Mott, J. J., and Gillard, P. (1982). Pasture weeds of the tropics and subtropics with special reference to Australia. In: Holzner, W., and Numata, M. (Eds), *Biology and Ecology of Weeds*, pp. 403–28. Junk, The Hague.
Trabaud, L. (1981). Man and fire: impacts on Mediterranean vegetation. In di Castri, F. Goodall, D. W., and Specht, R. L. (Eds.), *Ecosystems of the World. Mediterranean-type Scrublands*, pp. 523–37. Junk, The Hague.
UNESCO (1979). *Tropical Grassland Ecosystems*, UNESCO, Paris.
Vietmeyer, N. D. (1986). Lesser-known plants of potential use in agriculture and forestry. *Science*, **232**, 1379–84.
Wagner, W. L., Herbst, D. R., and Yee, R. S. N. (1985). Status of the native flowering plants of the Hawaiian islands. In: Stone, C. P., and Scott, J. M. (Eds.), *Hawaii's Terrestrial Ecosystems: Preservation and Management*, pp. 23–74. University of Hawaii, Honolulu.
Webb, D. A. (1985). What are the criteria for presuming native status? *Watsonia*, **15**, 231–6.
Wells, M. J., and Stirton, C. H. (1982). South Africa. In: Holzner, W., and Numata, M. (Eds.), *Biology and Ecology of Weeds*, pp. 339–45. Junk, The Hague.
Wells, M. J., Engelbrecht, V. M., Balsinhas, A. A., and Stirton C. H. (1983). Weed flora of South Africa 2. Power shifts in the veld. *Bothalia*, **14**, 961–5.
Wells, M. J., Poynton, R. J., Balsinhas, A. A. van Hoepen, E., and Abbott, S. K. (1986). The history of introduction of invasive alien plants to Southern Africa, (in press).
Williams, O. B. (1985). Population dynamics of Australian plant communities, with special reference to the invasion of neophytes. In: White, J. (Ed.), *The Population Structure of Vegetation*, pp. 623–35. Junk, The Hague.
Zohary, M. (1962). *Plant life of Palestine.* Reynold Press, New York.

Appendix 2.1. Families, genera and species listed in Holm *et al.* (1979), arranged in descending order of species richness.

Family	No. of genera	No. of species
Angiosperms		
Asteraceae	224	830
Poaceae	166	753
Papilionaceae	87	415
Euphorbiaceae	24	183
Lamiaceae	49	178
Brassicaceae	57	175
Convolvulaceae	14	147
Cyperaceae	18	146
Solanaceae	26	143
Apiaceae	54	142
Rosaceae	24	130
Scrophulariaceae	46	123
Polygonaceae	12	115
Malvaceae	21	111
Mimosaceae	15	99
Rubiaceae	32	97
Amaranthaceae	18	88
Chenopodiaceae	21	82
Caryophyllaceae	24	80
Boraginaceae	27	75
Ranunculaceae	12	74
Verbenaceae	14	67
Acanthaceae	24	59
Onagraceae	8	58
Fagaceae	5	55
Caesalpiniaceae	14	48
Asclepiadaceae	19	46
Commelinaceae	8	42
Cucurbitaceae	17	42
Liliaceae	23	39
Oxalidaceae	2	36
Rhamnaceae	10	34
Lythraceae	6	33
Tiliaceae	5	31
Urticaceae	12	31
Cactaceae	4	30
Orobanchaceae	4	30
Capparidaceae	5	29
Ericaceae	11	29
Juncaceae	2	29
Potamogetonaceae	1	28
Salicaceae	2	28
Saxifragaceae	2	28
Apocynaceae	16	26
Melastomataceae	11	26
Araceae	20	25
Papaveraceae	9	25
Caprifoliaceae	5	23
Geraniaceae	2	23
Alismataceae	4	22
Anacardiaceae	8	22
Hydrocharitaceae	11	22
Moraceae	8	21
Heliotropiaceae	1	20
Iridaceae	10	19
Plantaginaceae	1	19
Ulmaceae	4	19
Myrtaceae	7	17
Oleaceae	5	17
Primulaceae	4	17
Vitaceae	6	17
Bignoniaceae	11	16
Dipsacaceae	5	16
Nymphaeaceae	4	16
Aizoaceae	6	15
Campanulaceae	10	15
Eriocaulaceae	1	15
Gentianaceae	10	15
Hydrophyllaceae	6	15
Zygophyllaceae	6	15
Alliaceae	1	14
Hypericaceae	1	14
Sapindaceae	9	14
Fumariaceae	2	13
Piperaceae	3	13
Portulacaceae	5	13
Sterculiaceae	6	13
Juglandiaceae	1	12
Passifloraceae	2	12
Polygalaceae	2	12
Typhaceae	1	12
Amaryllidaceae	7	11
Betulaceae	4	11
Haloragidaceae	1	11
Lentibulariaceae	1	11
Lobeliaceae	1	11
Nyctaginaceae	3	11
Valerianaceae	4	11

Appendix 2.1. (*Contd.*)

Family	No. of genera	No. of species	Family	No. of genera	No. of species
Berberidaceae	3	10	Orchidaceae	3	4
Loranthaceae	6	10	Santalaceae	3	4
Menispermaceae	8	10	Saururaceae	3	4
Pontederiaceae	4	10	Zingiberaceae	4	4
Rutaceae	7	10	Araliaceae	3	3
Violaceae	3	10	Butomaceae	2	3
Aceraceae	1	9	Ceratophyllaceae	1	3
Lemnaceae	3	9	Elaeagnaceae	2	3
Pedaliaceae	6	9	Hamamelidaceae	2	3
Phytolaccaceae	4	9	Juncaginaceae	1	3
Smilacaceae	1	9	Tamaricaceae	1	3
Sparganiaceae	1	9	Xyridaceae	1	3
Agavaceae	4	8	Aponogetonaceae	1	2
Arecaceae	5	8	Asparagaceae	1	2
Aristolochiaceae	1	8	Calyceraceae	2	2
Balsaminaceae	1	8	Dichapetalaceae	1	2
Combretaceae	2	8	Dioscoreaceae	1	2
Crassulaceae	3	8	Epacridaceae	1	2
Najadaceae	1	8	Eryothroxilaceae	1	2
Sapotaceae	3	8	Frankeniaceae	1	2
Vacciniaceae	1	8	Hydrangeaceae	1	2
Annonaceae	1	7	Icacinaceae	2	2
Callitrichaceae	1	6	Loasaceae	1	2
Cornaceae	1	6	Myrsinaceae	2	2
Ebenaceae	3	6	Nelumbonaceae	1	2
Lauraceae	5	6	Nyssaceae	1	2
Linaceae	1	6	Ochnaceae	1	2
Proteaceae	2	6	Parkeriaceae	1	2
Aquifoliaceae	1	5	Pittosporaceae	2	2
Basellaceae	3	5	Plumbaginaceae	1	2
Bromeliaceae	4	5	Zannichelliaceae	2	2
Cannaceae	1	5	Avicenniaceae	1	1
Cannabaceae	2	5	Barringtoniaceae	1	1
Elatinaceae	2	5	Batidaceae	1	1
Martyniaceae	3	5	Begoniaceae	1	1
Musaceae	2	5	Buddlejaceae	1	1
Myricaceae	2	5	Burseraceae	1	1
Polimoniaceae	3	5	Canellaceae	1	1
Resedaceae	1	5	Caryocaraceae	1	1
Simaroubaceae	5	5	Cassythaceae	1	1
Thymelaeaceae	4	5	Celastraceae	1	1
Meliaceae	4	4	Clusiaceae	1	1
Dilleniaceae	2	4	Cochlospermaceae	1	1
Loganiaceae	4	4	Connaraceae	1	1
Malpighiaceae	4	4	Cyrillaceae	1	1
Marantaceae	4	4	Flacourtiaceae	1	1
Molluginaceae	1	4	Garryaceae	1	1
Myoporaceae	2	4	Gesneriaceae	1	1

Appendix 2.1. (*Contd.*)

Family	No. of genera	No. of species
Haemodoraceae	1	1
Hippuridaceae	1	1
Koeberliniaceae	1	1
Magnoliaceae	1	1
Melianthaceae	1	1
Myristicaceae	1	1
Olacaceae	1	1
Philydraceae	1	1
Platanaceae	1	1
Restionaceae	1	1
Rhizophoraceae	1	1
Ruppiaceae	1	1
Sonneratiaceae	1	1
Styracaceae	1	1
Taccaceae	1	1
Tetragoniaceae	1	1
Trapaceae	1	1
Tropaeolaceae	1	1
Turneraceae	1	1
Xanthorrhoeaceae	1	1
Gymnosperms		
Pinaceae	6	32
Cupressaceae	4	15
Cycadaceae	2	3
Taxaceae	2	3
Ephedraceae	1	2
Taxodiaceae	1	1

Family	No. of genera	No. of species
Ferns, Horsetails and Clubmosses		
Salviniaceae	2	14
Equisetaceae	1	11
Marsileaceae	2	7
Sinopteridaceae	4	7
Thelypteridaceae	3	7
Aspidiaceae	3	6
Dennstaedtiaceae	2	5
Davalliaceae	1	5
Schizaeaceae	1	5
Blechnaceae	3	4
Polypodiaceae	4	4
Selaginellaceae	1	3
Aspleniaceae	2	2
Athyriaceae	1	2
Gleicheniaceae	1	2
Lindsaeaceae	2	2
Osmundaceae	1	2
Pteridaceae	1	2
Cyatheaceae	1	1
Dryopteridaceae	1	1
Gymnogrammaceae	1	1
Ophioglossaceae	1	1
Sphenocleaceae	1	1
Lycopodiaceae	1	1

Appendix 2.2. Genera listed in Holm *et al.* (1979), containing 10 or more species.

Genera		No. of species	Genera		No. of species
Angiosperms					
Cyperus	(Cyperaceae)	81	*Astragalus*	(Papilionaceae)	20
Euphorbia	(Euphorbiaceae)	80	*Indigofera*	(Papilionaceae)	20
Solanum	(Solanaceae)	66	*Medicago*	(Papilionaceae)	20
Polygonum	(Polygonaceae)	62	*Bidens*	(Asteraceae)	20
Panicum	(Poaceae)	57	*Cleome*	(Capparidaceae)	20
Ipomoea	(Convolvulaceae)	57	*Commelina*	(Commelinaceae)	20
Acacia	(Mimosaceae)	53	*Croton*	(Euphorbiaceae)	20
Cuscuta	(Convolvulaceae)	49	*Phyllanthus*	(Euphorbiaceae)	20
Eragrostis	(Poaceae)	48	*Fimbristylis*	(Cyperaceae)	20
Quercus	(Fagaceae)	48	*Heliotropium*	(Heliotropiaceae)	20
Senecio	(Asteraceae)	45	*Hibiscus*	(Malvaceae)	20
Chenopodium	(Chenopodiaceae)	39	*Galium*	(Rubiaceae)	19
Amaranthus	(Amaranthaceae)	37	*Lactuca*	(Asteraceae)	19
Centaurea	(Asteraceae)	35	*Plantago*	(Plantaginaceae)	19
Oxalis	(Oxalidaceae)	34	*Acalypha*	(Euphorbiaceae)	18
Eleocharis	(Cyperaceae)	33	*Asclepias*	(Asclepiadaceae)	18
Rumex	(Polygonaceae)	33	*Aster*	(Asteraceae)	18
Crotalaria	(Papilionaceae)	32	*Lepidium*	(Brassicaceae)	18
Desmodium	(Papilionaceae)	31	*Salix*	(Salicaceae)	18
Carex	(Cyperaceae)	30	*Verbena*	(Verbenaceae)	18
Scirpus	(Cyperaceae)	30	*Brachiaria*	(Poaceae)	17
Rubus	(Rosaceae)	30	*Cirsium*	(Asteraceae)	17
Sida	(Malvaceae)	30	*Xanthium*	(Asteraceae)	17
Salvia	(Lamiaceae)	29	*Baccharis*	(Asteraceae)	16
Cassia	(Caesalpiniaceae)	29	*Carduus*	(Asteraceae)	16
Digitaria	(Poaceae)	28	*Delphinium*	(Ranunculaceae)	16
Ranunculus	(Ranunculaceae)	28	*Eryngium*	(Apiaceae)	16
Juncus	(Juncaceae)	28	*Hydrocotyle*	(Apiaceae)	16
Potamogeton	(Potamogetonaceae)	28	*Ludwigia*	(Onagraceae)	16
Ribes	(Saxifragaceae)	27	*Lupinus*	(Papilionaceae)	16
Trifolium	(Papilionaceae)	26	*Sporobolus*	(Poaceae)	16
Vicia	(Papilionaceae)	26	*Stachys*	(Lamiaceae)	16
Vernonia	(Asteraceae)	25	*Anthemis*	(Asteraceae)	15
Setaria	(Poaceae)	25	*Gnaphalium*	(Asteraceae)	15
Opuntia	(Cactaceae)	25	*Aristidia*	(Poaceae)	15
Orobanche	(Orobanchaceae)	25	*Brassica*	(Brassicaceae)	15
Physalis	(Solanaceae)	23	*Borreria*	(Rubiaceae)	15
Artemisia	(Asteraceae)	22	*Eriocaulon*	(Eriocaulaceae)	15
Lathyrus	(Papilionaceae)	22	*Geranium*	(Geraniaceae)	15
Oenothera	(Onagraceae)	22	*Scleria*	(Cyperaceae)	15
Veronica	(Scrophulariaceae)	22	*Atriplex*	(Chenopodiaceae)	14
Bromus	(Poaceae)	21	*Allium*	(Alliaceae)	14
Eupatorium	(Asteraceae)	21	*Ceanothus*	(Rhamnaceae)	14
Silene	(Caryophyllaceae)	21	*Cenchrus*	(Poaceae)	14

Appendix 2.2. (*Contd.*)

Genera		No. of species	Genera		No. of species
Echinochloa	(Poaceae)	14	*Convolvulus*	(Convolvulaceae)	11
Corchorus	(Tiliaceae)	14	*Hyptis*	(Lamiaceae)	11
Hypericum	(Hypericaceae)	14	*Leucas*	(Lamiaceae)	11
Rhus	(Anacardiaceae)	14	*Lindernia*	(Scrophulariaceae)	11
Sagittaria	(Alismataceae)	14	*Striga*	(Scrophulariaceae)	11
Alternanthera	(Amaranthaceae)	13	*Myriophyllum*	(Haloragidaceae)	11
Erigeron	(Asteraceae)	13	*Nymphaea*	(Nymphaeaceae)	11
Hordeum	(Poaceae)	13	*Lobelia*	(Lobeliaceae)	11
Pennisetum	(Poaceae)	13	*Passiflora*	(Passifloraceae)	11
Linaria	(Scrophulariaceae)	13	*Polygala*	(Polygalaceae)	11
Merremia	(Convolvulaceae)	13	*Prosopsis*	(Mimosaceae)	11
Rosa	(Rosaceae)	13	*Rorippa*	(Brassicaceae)	11
Triumfetta	(Tiliaceae)	13	*Utricularia*	(Lentibulariaceae)	11
Agrostis	(Poaceae)	12	*Avena*	(Poaceae)	10
Chloris	(Poaceae)	12	*Chrysanthemum*	(Asteraceae)	10
Sorghum	(Poaceae)	12	*Cerastium*	(Caryophyllaceae)	10
Arctostaphylos	(Ericaceae)	12	*Daucus*	(Apiaceae)	10
Bacopa	(Scrophulariaceae)	12	*Lippia*	(Verbenaceae)	10
Carya	(Junglandaceae)	12	*Justicia*	(Acanthaceae)	10
Hedyotis	(Rubiaceae)	12	*Piper*	(Piperaceae)	10
Fumaria	(Fumariaceae)	12	*Populus*	(Salicaceae)	10
Mentha	(Lamiaceae)	12	*Prunus*	(Rosaceae)	10
Sisymbrium	(Brassiaceae)	12	*Sesbania*	(Papilionaceae)	10
Typha	(Typhaceae)	12	*Tephrosia*	(Papilionaceae)	10
Abutilon	(Malvaceae)	11	*Gymnosperms*		
Malva	(Malvaceae)	11			
Alopecurus	(Poaceae)	11	*Pinus*	(Pinaceae)	20
Andropogon	(Poaceae)	11	*Juniperus*	(Cupressaceae)	11
Ambrosia	(Asteraceae)	11			
Crepis	(Asteraceae)	11	*Ferns, Horsetails and Clubmosses*		
Solidago	(Asteraceae)	11			
Ammannia	(Lythraceae)	11	*Equisetum*	(Equisetaceae)	11

Biological Invasions: a Global Perspective
Edited by J. A. Drake *et al.*
© 1989 SCOPE. Published by John Wiley & Sons Ltd

CHAPTER 3

Which Insect Introductions Succeed and Which Fail?

DANIEL SIMBERLOFF

3.1 INTRODUCTION

Why some insect introductions succeed (survive) and others fail to survive has been a central question since the first systematic studies of introduced insects (e.g. Elton, 1958; Baker and Stebbins, 1965) but no satisfactory answer has emerged for several reasons.

Firstly, the question has been asked at several levels. For example, some studies have focussed at a very high level: why insects from one biogeographic region tend to be more successful at invading another region than vice versa, or why insects of certain kinds of habitats seem to be more successful at invading new regions than insects of other habitats. Other studies have looked more proximately at the causes of success and failure: for example, what life history traits affect success? The nexus between the levels has not been established. For example, if it were true that mainland species are more likely to be successful when they invade islands than vice versa, is the reason that island and mainland species differ in characteristic ways with respect to life history?

Secondly, attempts to understand success or failure of invasions have not always controlled for opportunity. For example, it is well known (e.g. Greathead, 1976) that European insects have successfully invaded many other regions more frequently than species from those regions have invaded Europe. This imbalance is part of a larger pattern—European plants, vertebrates, even disease organisms seem to have been more successful at invading a number of other regions than vice versa (Crosby, 1986). However, before seeking to explain this disparity by invoking superiority of European species on one grounds or another, one must ask if some fraction of the pattern simply results from more European species being transported to other regions than vice versa. Many ships sailing from Europe to North America loaded soil as ballast in southwestern England (Lindroth, 1957). In North America, the soil was unloaded and exchanged for cargo. Small wonder that 90% of the insects known to have been introduced into North America before 1820 were beetles, many of them soil-dwellers found in southwestern England (Sailer, 1983). A similar transport of ballast from South

America to North America about a century later brought, among other species, the imported fire ant, *Solenopsis invicta* (Sailer, 1983). In both cases, because the transport was one-way, one does not try to explain the greater establishment in one site as a consequence of competitive or other superiority of one biota's species over the other's. However, these cases happen to be ones for which we know about the differential opportunity for successful invasion because of differential transport. Usually there are no firm data on rates of transport by humans from one site to another.

This gap is part of a larger problem that besets attempts to explain success or failure: we know about a much larger fraction of successes than of failures. Without knowledge of the failures, it is impossible to answer questions at any of the levels listed above simply by observing numbers of successful invasions in different habitats or regions, or numbers of successful invasions by species with particular suites of life history traits. Of course it is possible to say which source area or which habitat or which sort of life history typifies the most survivors but one cannot determine anything about probabilities: whether invaders of disturbed habitats are more likely to succeed, for example, as opposed simply to being more numerous. For insects, the only data that begin to approach equal coverage of successes and failures are those on biological control.

Thirdly, even without equal data on successes and failures, it would be possible to seek generalities in the proximate reasons why particular species succeed or fail if the trajectories of enough species were studied carefully, particularly trajectories of failures. The best way to do this would usually be experimentally because controlled experiments are best at implicating or eliminating potential factors as reasons for failure. If one suspects predation by ants prevents establishment, one precludes ants. Experiments are costly and difficult, however, and one would not expect strong experimental evidence on most cases. Thus, perhaps the best one could hope for would be detailed observations, especially on naturally occurring mortality in recently initiated populations. But detailed observations are exceptional. Usually the reports, if they suggest any reason at all for a failure, base the suggestion on rather casual observation: ants appeared to be eating the invader, the weather seemed too cold, etc.

Fourthly, often the only information available on an insect is that it reached some site and either succeeded or failed. We lack information on how many propagules were involved and how large those propagules were. Yet in many biological control projects, of several apparently replicated introductions, some succeed and others fail. For example, the parasitic wasp *Hungariella peregrina* was released in five southern California locations to control *Pseudococcus longispinus*, but only one release resulted in establishment (Clausen, 1978). This apparent lack of replicability is an outstanding problem in the ecology of introductions (Simberloff, 1985). There is probably a stochastic component to success, analogous to the stochastic aspects of flour beetle competition in Park's classic experiments (Park, 1962). By 'stochastic,' I do not mean there are no good reasons why one propagule persists and another disappears. Rather, I mean that all the forces that contribute to this difference cannot be specified and measured but the result can be viewed as a random draw from a specified distribution.

A fraction of what passes for stochasticity in introductions may result simply from different propagule sizes. The concept of the 'minimum viable population size' (Shaffer, 1981) is relevant here. An accruing literature in conservation biology (e.g. Soulé and Simberloff, 1986) indicates that, for a variety of reasons, the probability of persistence is characteristically less for small than for large propagules, even if there is no precise threshold population size above which establishment in an appropriate habitat is assured and below which it is precluded. Biological control records sometimes contain accounts of number of attempts for particular species and information on the numbers of individuals released. Greathead (1971), observing that some very small propagules have succeeded and some very large ones have failed in African biological control projects, downplays the importance of propagule size. He suggests that, if a few modest releases fail, this usually means the habitat is unsuitable and larger propagules will not help. However, there are counterexamples. The chrysomelid *Chrysolina hyperici* was introduced as a propagule of 120 adult beetles in Victoria in 1930 to control St John's wort and failed, while an inoculum of 1, 340 adults in 1934 succeeded (Clausen, 1978). One cannot prove that the sizes of these propagules were responsible for the different outcomes, but there are enough similar examples that one suspects large propagule size aids establishment. Williamson (this volume) tabulates data on 159 introductions for biocontrol in Canada and finds that increasing numbers of individuals released and increasing numbers of releases both increase the probability that a species will be successfully established.

In spite of these problems with the available data base, there are a number of interesting syntheses, at one level or another, on which insect introductions succeed and which ones fail. In the remainder of this contribution I will summarize many of these and examine the prospects that they offer for accurate predictions on the fates of introduced insects. Much of the literature on insect introductions treats biological control projects and most of these are conducted in more or less modified habitats such as those of agriculture and silviculture. The degree of modification has not been systematically recorded. Other papers, both theoretical and data-based, treat pristine and modified habitats indiscriminately. One can imagine that conclusions drawn about introductions into pristine habitats may differ from those about introductions into modified habitats, so I will indicate when an author's comments are meant to be restricted to biocontrol introductions.

3.2 OPPORTUNITY FOR COLONIZATION

For North America north of Mexico there is a continually updated list of all insects known to be introduced into any habitat whatever, compiled by the United States. Department of Agriculture (1986) (Sailer, 1983). Of the 1554 species, two-thirds are from the western palearctic region. Only 14% are from South and Central America and the West Indies, which surely have many more insects than the western palearctic and would seem by geographic proximity to be at least as likely to provide introductions into North America (Simberloff, 1986). Sailer argues convincingly that the vast majority of introduced insects reached North America not under their own steam but through anthropogenous transport, either deliberate or inadvertent. The main means have been ship ballast, introduced plants, and deliberate introduction of beneficial insects for biological control (Sailer, 1983). The latter category alone encompasses 227 species (as opposed to only two of 177 successful insect introductions into the British Isles, indicated by data from Brown (1985)). Further, most traffic that would have provided the opportunity for invasion, especially by the first two means, was from the western palearctic.

Greathead (1976) observes that Europe, especially Britain, has played a disproportionate role as source for natural enemies in biological control projects in North America, Australia, and New Zealand. For example, about half of the species brought into the United States for this purpose up to 1950 came from Europe (Clausen, 1956). On the other hand, very few came from Africa (Greathead, 1971). Early classical biological control was mainly an Anglo-American enterprise. It is thus unsurprising that a large fraction of the deliberate introductions (and therefore a large fraction of the successes) were European, especially British. In fact, as Greathead (1976) notes, this is in spite of the fact that Britain would not, on objective grounds, have seemed a promising source for predators and parasites.

The preponderance of European species among successful insect introductions of all sorts in North America cannot automatically be taken to indicate inherent qualities that make European insects better colonists than those of other regions. Yet this hypothesis is tempting. Matthew (1915) argued that Palearctic animal species are stronger than others, having been honed in the 'more efficient workshops of the north,' and that the biogeographic history of the earth was the progressive replacement of species in other regions by occasional immigrants from the Palearctic. Crosby (1986) constructs a similar scenario for the success of animals, plants, and pathogens introduced from Europe to other temperate regions and relative failure of species that moved in the opposite direction. One would like comprehensive data on the number of introduced nearctic species in Europe (and preferably some data on failures as well as successes) in order to assess whether it is true that there is a higher probability of success for a European colonist in the nearctic than vice versa. The opportunity for European species to reach other areas was greater, so the same success rate may simply have produced a larger number of successes. The dearth of African insects introduced into the United States (and, in fact, into most other regions) probably results at least partly from lack of opportunity. There has historically been less traffic from Africa to North America, for example, than from Europe.

3.3 SUITABLE HABITAT

Wilson (1960) detected a correlation between the climate of source areas of insects successfully introduced for biological control into Australia and the climate of the part of Australia where the introduction occurred. The similarity of habitat between much of Europe and much of North America may partially account for the fact that such a large fraction of the successful introductions into North America came from Europe, though it is impossible with available data to separate this factor from opportunity. In addition to climatic similarity between the palearctic and the nearctic, there is broad vegetational similarity in both growth form and taxonomic composition (far more plant taxa held in common than, for example, between Africa and South America). The arguments about the reasons for the taxonomic similarity (e.g. Sailer, 1983) need not concern us, but the fact remains that the vegetation is similar. Lattin and Oman (1983) suggest that vegetational similarity between source and target for a potential introduction is at least as important as climatic similarity for insects that use plants as food or habitat.

Several biological control workers have argued that successful invasions are more likely between sites within one region than between different regions. Few of these reports examine both failures and successes systematically so it is impossible to say whether intra- or inter-region introductions are more likely to succeed, but the impression is that it is the former. For example, Wilson (1960) reports several successful introductions from eastern to western Australia and from one part of New Guinea to another, while Clausen (1956) describes successful transfer of species within the United States. A higher fraction of transfers of insects within Canada resulted in successful establishment than did introduction of insects into Canada from other regions (Commonwealth Institute of Biological Control, 1971). A preliminary hypothesis for the apparently greater success rate of intra-region introductions is that the climatic and vegetational similarity is greater.

Holdgate (1986) argues that invasion depends more on species–habitat interactions once dispersal has occurred than on dispersal itself. For insects it is difficult to see on what basis such a claim can currently be made, especially without much more information on failed invasions. It is, of course, true that those introduced species surviving at any site have found a suitable habitat, and one possible reason for the failure of those that have landed but not survived is that the habitat is unsuitable. On the other hand, all those species that survived at a site also had to reach it, and to assess the importance of dispersal we would have to know how many species dispersed there but failed to survive, and how many have not dispersed there, by their own means or with human help, but could have survived if they had gotten there. No entomological data fill these needs.

3.4 THE BIOTIC RESISTANCE HYPOTHESIS

Several hypotheses about why some introductions succeed and some fail rest on the argument that biotic resistance to an invader is the key (Simberloff, 1986). For example, it is often argued, for insects as for other taxa, that mainland species typically invade island communities more readily than island species invade mainlands (e.g. Elton, 1958; Carlquist, 1965). The underlying reason is often thought to be superior competitive ability of the mainland species. Carlquist (1965) views continental species as 'steeled by competition.' One could also imagine that, since continents have more predator, pathogen, and parasite species than islands do, mainland species would deal more successfully with all sorts of interactions. Greathead (1971) says that, because there are fewer species per unit area on islands, island species occupy broader niches and a wider array of habitats. Thus they are likely to be less well adapted to any particular food or habitat than is a potential competitor from a continent.

However, in spite of the many well-known examples, for insects as for other species, of the devastation of island communities by mainland invaders (e.g. Elton, 1958), it is far from clear that mainland species are really more likely to invade successfully. For one beetle and five wasp genera widely used in biological control efforts around the world, Simberloff (1986) classified the 281 introductions (Clausen, 1978) into four categories: mainland to island, mainland to mainland, island to island, and island to mainland. A multiway contingency test was performed with three dimensions of two levels each: source (mainland or island), target (mainland or island), and outcome (success or failure). The null hypothesis, that success or failure of an introduction does not depend on whether source or target is island or mainland, could not be rejected. Of course almost all these introductions were into agricultural communities that are typically both highly modified and more similar from site to site than other communities. It might be that a different result would be obtained for introductions into pristine habitats, but no comparable data exist for such introductions. As to why the view has arisen that island communities are more easily invaded or that mainland species are better colonists, it may be that mainland species simply have more opportunities to invade islands than vice versa, both because there are more mainland species and because they are more likely to be transported deliberately or by accident. For example, of the 281 biocontrol introductions just discussed, 71 were from mainland to island but only 15 were from island to mainland. The source for 247 of these introductions was mainland, while only 34 originated on islands.

Because many early successes of biological control were on islands, the view arose (e.g. Imms, 1931) that only the simplified communities of islands would allow successful invasion and propagation. However, Doutt and DeBach (1964), DeBach (1965), Huffaker *et al.* (1976), and Laing and Hamai (1976) all present data on mainland successes that contradict the 'island theory.' None of these studies closely examines success rate on islands versus mainland. Greathead (1971) compared 73 projects in Africa to 53 on islands and found the African success rate to be 23% as opposed to 45% on the islands. However, 'success' here means economic success of the project, not survival of the introduction as throughout the rest of this paper. Greathead (1971) gives examples of species that have survived on both Africa and islands but that provide effective control only on islands; he implies such situations are common.

Another pattern often detected, for plants as for insects, is that 'disturbed habitats' are much more readily invaded than 'undisturbed habitats.' Elton (1958) and Sharples (1983) attribute this pattern to biotic resistance, the competitors, predators, parasites, and diseases that an invader encounters in pristine habitat. These authors assume that disturbance decreases numbers of species in all categories. Neither author tabulates success and failure rates in the two kinds of habitats, so it is difficult to know whether this pattern is real or simply an impression. Two possible tendencies may make it appear that invasions are more likely to be successful in disturbed habitats (Simberloff, 1986). First, the disturbed habitats studied are almost all modified habitats associated with humans—primarily agricultural, and secondarily associated with dwellings. These habitats are most important to us and so are studied more carefully than pristine habitats. We are thus more likely to detect successes. Secondly, the opportunity was almost certainly greater for introduction into this kind of disturbed habitat. Biological control agents were imported into these habitats, as were agricultural and ornamental plants and the insects they might have carried. There must have been far fewer deliberate or inadvertent introductions into pristine habitat.

So a tentative hypothesis would be that disturbed habitats *seem* to be more easily invaded because there have been more introductions into some of them (because of greater opportunity) and successes are more likely to be detected. It would be interesting to study systematically and exhaustively naturally rather than anthropogenously disturbed habitats (e.g. fire disclimaxes) to see if disturbance *per se*, rather than a particular kind of disturbance, makes invasion easier. Even if it should turn out that generic disturbance did increase the ease of invasion, one would still have to demonstrate that reduced biotic resistance is the reason. Even the assumption that disturbance leads to a reduced number of species may not be correct. For some sorts of communities, intermediate levels of disturbance are associated with highest species richness (Connell, 1978), so it is not automatically clear why biotic resistance should be lessened as disturbance increases. Furthermore, Howarth (1985) even argues that increasing the number of species in a site *increases*, rather than decreases, the probability that a future propagule will establish, because creation of new niches is facilitated.

Yet another example of the biotic resistance hypothesis is the contention by Tallamy (1983) that, for parasitic insects introduced to control gypsy moth (*Lymantria dispar*) in the United States, later species were competitively precluded by earlier ones. Washburn (1984) concedes that later introductions had a lower success rate but argues that the earlier ones were exactly those species that were *a priori* more likely to succeed. Hall and Ehler (1979) and Ehler and Hall (1982) similarly found, for a veriety of insects introduced for biological control, that species introduced simultaneously with or after other introductions were less likely to succeed than those introduced early and alone. Although Ehler and Hall (1984) were loath to draw strong conclusions about why this pattern was obtained, Keller (1984) was quick to argue that it need *not* imply competitive exclusion of later invaders by earlier ones. He suggested a number of biases in typical control procedures (such as the tendency for early, single releases to be of species *a priori* most likely to succeed) as well as statistical reasons why the biological control literature would make it *appear* that some introductions would exclude others even if this were not happening.

Lawton and Brown (1986) emphasize predators, parasitoids, and diseases rather than competitors as likely determinants of success or failure of all kinds of insect introductions. Though they provide several interesting examples this proposition has yet to be sufficiently closely studied. Many discussions of deliberate release of biological control agents cite predators as frequently preventing establishment (e.g. Commonwealth Institute of Biological Control, 1971; Goeden and Louda, 1976; Crawley, 1986a, 1986b) but much of the evidence consists of casual observation. Crawley (1986b) finds that predators were reported to cause reduced impact of insects introduced for weed control in 22% of cases, parasitoids in 11%, and diseases in 8%. Climate, by contrast, was reported to be important in 44% of cases.

3.5 BIOLOGICAL TRAITS FAVORING SUCCESSFUL INTRODUCTION

There is a long tradition of trying to predict success or failure of an introduced species from its characteristics, particularly its life history traits such as birth rate, death rate, etc. This approach is epitomized by several papers in *The Genetics of Colonizing Species* (Baker and Stebbins, 1965) in which lists of traits that should conduce to successful invasion are proposed. Though such lists invariably make good sense and may, on average, be borne out when comprehensive data become available on opportunities, successes, and failures for given taxa, there is widespread doubt that this approach will be very useful in predicting the outcome of a particular introduction (e.g. Crawley, 1986b). Thompson (1939) believed that introduction of insects for biological control would always be a trial and error process because of the number and complexity of factors acting on an introduction in nature. Sharples (1983) cites many other authors, writing on a variety of taxa, who made the same claim.

Lawton and Brown (1986) point out that the problem is that extinction in population dynamic models based more or less on the models of MacArthur and Wilson (1967) is generated by demographic stochasticity, whereas environmental stochasticity is likely to be a far more important force in the field. This argument may well be true, although much of the conservation literature, focussing on the relation between demographic and genetic stochasticity, emphasizes the importance of demographic stochasticity (National Research Council, 1986). As the conservation modelers realize, environmental stochasticity is extremely difficult to model. There are two components—rare catastrophes that occur randomly in time (such as hurricanes or catastrophic fires) and the average, day-to-day fluctuations that are the composite of many forces that are themselves continually varying and all of which affect a population. By 'random' and 'stochastic,' I do not mean that there are not deterministic reasons why a catastrophe occurs or why a particular force fluctuates from day to day. All that is meant is that, with respect to the forces included in the model, these components are not completely predictable, and their occurrence can at best be drawn from a specified statistical distribution.

It is difficult to imagine how to incorporate catastrophes into a predictive model, but Leigh (1981) suggests that the effects of day-to-day fluctuations can be captured by looking at the coefficient of variation of a population's fluctuations in size among generations. His conclusion is that, the higher the coefficient of variation, the lower the likelihood of persistence (or of initial establishment for an introduction). For most insect species, the coefficients of variation of their population fluctuations are no better known than their intrinsic rates of increase, birth and death rates, and carrying capacities.

Lawton and Brown (1986) find a nearly perfect inverse correlation between mean body length and probability of successful establishment for six insect orders whose successful and failed introductions into England have been recorded, but the reason for the correlation is obscure. The relationship between mean body size for these orders and mean coefficient of variation of population fluctuation or other population parameters is unknown. The importance of size would be more strongly implicated if probabilities of establishment for the individual species within the orders were shown, for different size ranges. Without this information, it seems just as reasonable to say that some traits of the natural history of, say, Hemiptera predispose them to successful invasion while traits of Lepidoptera predispose them to fail. In any event, as interesting as this correlation among the six points is, it is still far from allowing us to predict the fate of an individual invader. For insects introduced to control weeds, Crawley (1986a, 1986b) detected positive correlations between intrinsic rate of increase and probability of success for insects on plants other than *Opuntia*, even within orders. He attributes an inverse correlation between size and probability of success to an inverse relationship between size and intrinsic rate of increase. However, for neither intrinsic rate of increase nor size is the correlation so strong as to allow sound prediction of the fate of a particular introduction.

'Niche breadth' often surfaces as an important feature in discussions of reasons for success or failure of introductions. There have been exactly opposite predictions about how niche breadth should affect invasion success. On the one hand, Mayr (1965), discussing birds, Howarth (1985), discussing invertebrates, and Holdgate (1986), summarizing reports on many taxa, suggest that a broad range of possible foods and/or habitats should facilitate invasion. The underlying reasoning seems to be that the such species are more 'versatile' ecologically, thus more likely to carve out a niche they can persist in. Wilson's 'taxon cycle' (Wilson, 1961), in which generalized ant species from the mainland or large islands continually invade smaller islands and replace their more specialized inhabitants, rests on such reasoning. Dritschilo *et al.* (1985) examined several hundred insects introduced by a variety of means into California and found that typically they could survive in a variety of habitats and on a variety of foodstuffs. On the other hand, Greathead (1971) argues that mainland species are more likely to drive out island species exactly because the island species have broader niches, and so are less suited to any particular habitat and/or food than are the more specialized mainland species. DeBach (1965) finds that, among entomophagous insects introduced for biological control, most successes are very specialized for their hosts, rather than having a broad food niche (though he did not analyze *probability* of success for narrow- versus broad-niched species). Phytophagous insects deliberately introduced for weed-control are chosen to be very host-specific and are often rigorously tested for this trait before release. Thus the fact that most of the successful species in this realm of biological control have a narrow range of host plants tells us little about whether host range is an important predictor of success.

Part of the confusion about how niche breadth bears on invasion success rests on the fact that consistent, measurable, and meaningful measures of 'niche breadth' and 'specialization' have not been used. Another part of the problem is that failures have not been examined as thoroughly as successes, so, even if one *could* measure niche breadth, one would need more data in order to say how it is associated with probability of invasion success, if at all. If consistent definitions and comprehensive data are available, it would be interesting to see what correlations arise, but I doubt if the single compound trait 'niche breadth' will ever be a very strong predictor of invasion success for insects.

The method of sex determination may predispose certain insects to succeed upon introduction of small propagules. Howarth (1985) found that a large fraction of invertebrates successfully introduced into Hawaii are either hermaphroditic or parthenogenetic and argued that such species are favored because the problem early in the invasion process of finding a mate is lessened or eliminated. Simberloff (1986) noted that haplo-diploidy, spanandry, and intense inbreeding might be expected to favor parasitic Hymenoptera by lowering the threat of inbreeding depression, often cited (e.g. Shaffer, 1981) as a threat to very small populations. As a preliminary test of this idea, I checked Clausen (1978) for all biocontrol introductions in which the propagule size was given and did not exceed 1000. I counted only field releases, not projects in which a small propagule was used for mass rearing and subsequent large-scale release. Multiple releases in one region as part of one project were counted only once. Results are given in Table 3.1. The results are generally consistent with the hypothesis, although Hymenoptera, Coleoptera, and Diptera differ in so many other ways that one could hypothesize other reasons for these results. It is interesting that, in total and in the categories '1–20' and '101–1000' the order of probability of success, Hymenoptera > Coleoptera > Diptera, is exactly the opposite of that determined by Lawton and Brown (1986) for all insects introduced, by any means, into the British Isles.

I did not count projects in which a single small propagule was used for mass rearing and subsequent large-scale release, on the grounds that such a procedure can rapidly lessen inbreeding depression (Senner, 1980; Soulé, 1980). However, there are several such examples, most involving parasitic wasps. Simberloff (1986) discusses the braconid *Aphidius smithi.* Here may be added the encyrtid *Hambletonia pseudococcina.* Two females of a unisexual race were used in Puerto Rico to rear about 7000 individuals, which established and spread rapidly upon release. Also the encyrtid *Pauridia peregrina*: stocks were propagated from a single female, and release resulted in permanent establishment in California. And the encyrtid *Anarhopus sydneyensis.* For this wasp, stocks were reared from eight individuals, and several subsequent releases in California were successful. In Hawaii, the chalcidid *Brachymeria agonoxenae* was propagated from one female and several males, after which three males and eight females were released and established. The eulophid *Tetrastichus incertus*, imported into the United States from France, was mass reared and successfully released from eight adults. *T. giffardianus* in Fiji was mass reared from ten individuals and successfully established. The only such non-hymenopteran cited by Clausen (1978) is the coccinellid beetle *Scymnus smithianus.* From Sumatra, 27 living adults reached Cuba, but the stock declined to a single female. However, it then built up sufficiently to allow a successful release.

Table 3.1. Intentional small field releases of biological control agents in three insect orders. Multiple releases as parts of single projects counted only once. S = survival; E = extinction. Data from Clausen (1978)

	Propagule size					
	1–20		21–100		101–1000	
	S	E	S	E	S	E
Hymenoptera	10	11	17	37	43	72
Coleoptera	3	10	9	27	18	52
Diptera	1	4	3	7	3	26

3.6 CONCLUSIONS

Predicting which insect introductions succeed is still not a very precise business and there may be an inherent stochasticity to the process so that predictions will never be better than statements of probability of success given a certain number of propagules of certain sizes. Even the broad generalizations about which sorts of introductions historically have tended to succeed at the highest rate are beclouded by deficiencies in the data base, particularly regarding failed introductions. The reasons for such venerable patterns as the predominance of European successes, the frequency of invasion of islands, and the frequency with which disturbed habitats are invaded cannot be definitively given without much more information on opportunities for various sorts of invasions. There is absolutely no doubt that the habitat in the target area is always important to the potential invader, and there are some instances where other species already there seem to be crucial, but there is no basis for saying that patterns of observed introduction are more a reflection of species–habitat or species–species interaction than of the historical distribution of dispersal opportunities.

Continuing efforts to make more precise predictions about successful introductions based on such gross inherent biological traits of the potential invaders as size or intrinsic rate of increase do not seem to be very promising. There may be interesting patterns relating life history, population dynamic, genetic, and other traits to probability of success, but there will always be exceptions, perhaps partly owing to inherent stochasticity in the invasion process, certainly partly owing to interactions between a potential invader and the habitat and/or species it encounters. It is depressing to be unable to draw striking generalizations about introduced insects but it would serve no worthwhile purpose to generalize prematurely.

However, very detailed study of the natural history of an organism, including its phenology and life history, may allow more precise prediction. For example, DeBach (1974) describes the results of the accidental introduction into Fiji of *Pediculoides ventricosus*, a mite that attacks larvae and pupae (but not eggs and adults) of the coconut leaf-mining beetle *Promecotheca reichei*. The mite destroyed all the larvae and pupae during the dry season in some sites. The adult beetles then oviposited and died, converting the beetle population into one with synchronous, non-overlapping generations. The consequent absence during certain periods of larvae and pupae caused the mite population to crash. Similarly, two native parasitoids that had controlled the beetle were almost eliminated because they did not live long enough to persist during the intervals between occurrences of the host stages that they require for oviposition. The beetle population greatly increased. However, a parasitoid was sought that was not so restricted in the host stages it requires for oviposition and with sufficient longevity that it could survive periods when hosts are rare or absent. The chalcid wasp *Pleurotropis parvulus* in Java satisfied these criteria and controlled the beetle remarkably well after it was imported into Fiji.

There is every reason to think that careful consideration of the biotic and physical habitat into which a species is to be introduced, plus insightful study of its natural history, will yield sounder prediction than the sorts of general models that have been attempted to date. Crawley (1986a) says that ecological theory has contributed little or nothing to the practice of choosing insects for weed control and is quite pessimistic that it ever will. Part of the problem may well be that the models deployed to date are mostly too general and idealized, but that models tailored to particular cases will be more successful in predicting the outcomes not only of biocontrol introductions but of other types as well.

ACKNOWLEDGMENTS

This paper is dedicated to Dr Reece I. Sailer, who had the foresight to see that comprehensive data on introduced insects would aid in many inquiries, both applied and academic.

REFERENCES

Baker, H. G., and Stebbins, G. L. (Eds.) (1965). *The Genetics of Colonizing Species.* Academic Press, New York.

Brown, K. C. (1985). Animals, plants, and micro-organisms introduced into the British Isles. Unpublished report prepared for the Department of the Environment, UK.

Carlquist, S. (1965). *Island Life.* Natural History Press, Garden City, NY.

Clausen, C. P. (1956). *Biological Control of Insect Pests in the Continental United States.* Technical Bulletin No. 1139, US Department of Agriculture, Washington, DC.

Clausen, C. P. (Ed.) (1978). *Introduced Parasites and Predators of Arthropod Pests and Weeds: A World Review.* Agriculture Handbook 480, US Department of Agriculture, Washington, DC.

Commonwealth Institute of Biological Control. (1971). *Biological Control Programmes Against Insects and Weeds in Canada. 1959–1968.* Techn. Comm. No. 4, Commonwealth Agricultural Bureaux, Farnham Royal.

Connell, J. H. (1978). Diversity in tropical rain forests and coral reefs. *Science*, **199**, 1302–10.

Crawley, M. J. (1986a). What makes a community invasible? In: Gray, A. J., Crawley, M. J., and Edwards, P. J. (Eds.), *Colonization, Succession and Stability.* Blackwell Scientific Publications, Oxford.

Crawley, M. J. (1986b). The population biology of invaders. *Phil. Trans. Roy. Soc. Lond. B*, **314**, 711–31.

Crosby, A. W. (1986). *Ecological Imperialism: The Biological Expansion of Europe, 900–1900.* Cambridge University Press, Cambridge.

DeBach, P. (1965). Some biological and ecological phenomena associated with colonizing entomophagous insects. In: Baker, H. G., and Stebbins, G. L. (Eds.), *The Genetics of Colonizing Species*, pp. 287–303. Academic Press, New York.

DeBach, P. (1974). *Biological Control by Natural Enemies.* Cambridge University Press, London.

Doutt, R. L., and DeBach, P. (1964). Some biological control concepts and questions. In: DeBach, P. (Ed.), *Biological Control of Insect Pests and Weeds*, pp. 118–42. Reinhold, New York.

Dritschilo, W., Carpenter, D. E., Meyn, O., Moss, D., and Weinstein, M. N. (1985).

Implications of data on introduced ~cies in California for field releases of recombinant DNA organisms. Unpublished ms.
Ehler, L. E., and Hall, R. W. (1982). Evidence for competitive exclusion of introduced natural enemies in biological control. *Environm. Entomol.*, **11**, 1–4.
Ehler, L. E., and Hall, R. W. (1984). Evidence for competitive exclusion of introduced natural enemies in biological control: An addendum. *Environm. Entomol.*, **13**, v–vii.
Elton, C. S. (1958). *The Ecology of Invasions by Animals and Plants.* Methuen, London.
Goeden, R. D., and Louda, S. M. (1976). Biotic interference with insects imported for weed control. *Annu. Rev. Entomol.*, **21**, 325–42.
Greathead, D. J. (1971). *A Review of Biological Control in the Ethiopian Region.* Techn. Comm. No. 5, Commonwealth Agricultural Bureaux, Farnham Royal.
Greathead, D. J. (ed.) (1976). *A Review of Biological Control in Western and Southern Europe.* Techn. Comm. No. 7, Commonwealth Agricultural Bureaux, Farnham Royal.
Hall, R. W., and Ehler, L. E. (1979). Rate of establishment of natural enemies in classical biological control. *Bull. Entomol. Soc. Amer.*, **25**, 280–2.
Holdgate, M. W. (1986). Summary and conclusions: Characteristics and consequences of biological invasions. *Phil. Trans. Roy. Soc. Lond B*, **314**, 733–42.
Howarth, F. G. (1985). Impacts of alien land arthropods and mollusks on native plants and animals in Hawai'i. In Stone, C. P., and Scott, J. M. (Eds.), *Hawaii's Terrestrial Ecosystems: Preservation and Management*, pp. 149–79. University of Hawaii, Honolulu.
Huffaker, C. B., Simmonds, F. J., and Laing, J. E. (1976). The theoretical and empirical basis of biological control. In: Huffaker, C. B., and Messenger, P. S. (Eds.), *Theory and Practice of Biological Control*, pp. 41–78. Academic Press, New York.
Imms, A. D. (1931). *Recent Advances in Entomology.* Churchill, London.
Keller, M. A. (1984). Reassessing evidence for competitive exclusion of introduced natural enemies. *Environm. Entomol.*, **13**, 192–5.
Laing, J. E., and Hamai, J. (1976). Biological control of insect pests and weeds by imported parasites, predators, and pathogens. In: Huffaker, C. B., and Messenger, P. S. (Eds.), *Theory and Practice of Biological Control*, pp. 685–743. Academic Press, New York.
Lattin, J. D., and Oman, P. (1983). Where are the exotic insect threats? In: Graham, C., and Wilson, C. (Eds.), *Exotic Plant Pests and North American Agriculture*, pp. 93–137. Academic Press, New York.
Lawton, J. H., and Brown, K. C. (1986). The population and community ecology of invading insects. *Phil. Trans. Roy. Soc. Lond. B*, **314**, 607–17.
Leigh, E. G., Jr. (1981). The average lifetime of a population in a varying environment. *J. Theoret. Biol.*, **90**, 213–39.
Lindroth, C. H. (1957). *The Faunal Connections between Europe and North America.* John Wiley, New York.
MacArthur, R. H., and Wilson, E. O. (1967). *The Theory of Island Biogeography.* Princeton University Press, Princeton, NJ.
Matthew, W. D. (1915). Climate and evolution. *Ann. New York Acad. Sci.*, **24**, 171–318.
Mayr, E. (1965). The nature of colonizations in birds. In: Baker, H. G., and Stebbins, G. L. (Eds.), *The Genetics of Colonizing Species*, pp. 29–43. Academic Press, New York.
National Research Council (1986). *Ecological Knowledge and Environmental Problem-Solving.* National Academy Press, Washington, DC.
Park, T. (1962). Beetles, competition, and populations. *Science*, **138**, 1369–75.
Sailer, R. I. (1983). History of insect introductions. In: Graham, C., and Wilson, C. (Eds.), *Exotic Plant Pests and North American Agriculture*, pp. 15–38. Academic Press, New York.
Senner, J. W. (1980). Inbreeding depression and the survival of zoo populations. In: Soulé,

M. E., and Wilcox, B. A. (Eds.), *Conservation Biology: An Evolutionary-Ecological Perspective*, pp. 209–24. Sinauer Associates, Sunderland, Mass.

Shaffer, M. L. (1981). Minimum population sizes for species conservation. *BioScience*, **31**, 131–4.

Sharples, F. E. (1983). Spread of organisms with novel genotypes: Thoughts from an ecological perspective. *Recombinant DNA Techn. Bull.*, **6**, 43–56.

Simberloff, D. (1985). Predicting ecological effects of novel entities: Evidence from higher organisms. In: Halvorson, H. O., Pramer, D., and Rogul, M. (Eds), *Engineered Organisms in the Environment/Scientific Issues*. American Society for Microbiology, Philadelphia.

Simberloff, D. (1986). Introduced insects: A biogeographic and systematic perspective. In: Mooney, H. A., and Drake, J. A. (Eds.), *Ecology of Biological Invasions of North America and Hawaii*. Springer-Verlag, New York.

Soulé, M. E. (1980). Thresholds for survival: Maintaining fitness and evolutionary potential. In: Soulé, M. E., and Wilcox, B. A. (Eds.), *Conservation Biology: An Evolutionary-Ecological Perspective*, pp. 151–69. Sinauer Associates, Sunderland, Mass.

Soulé, M. E., and Simberloff, D. (1986). What do genetics and ecology tell us about the design of nature reserves? *Biol. Conserv.*, **35**, 19–40.

Tallamy, D. W. (1983). Equilibrium biogeography and its application to insect host-parasite systems. *Amer. Nat.*, **121**, 244–54.

Thompson, W. R. (1939). Biological control and the theories of the interactions of populations. *Parasitology*, **31**, 299–388.

United States Department of Agriculture (1986). *Western Hemisphere Immigrant Arthropod Database*. Agricultural Research Service, USDA, Beltsville, Maryland.

Washburn, J. O. (1984). The gypsy moth and its parasites in North America: A community in equilibrium? *Amer. Nat.*, **124**, 288–92.

Wilson, E. O. (1961). The nature of the taxon cycle in the Melanesian ant fauna. *Amer. Nat.*, **95**, 169–93.

Wilson, F. (1960). *A Review of the Biological Control of Insects and Weeds in Australia and Australian New Guinea*. Techn. Comm. No. 1, Commonwealth Agricultural Bureaux, Farnham Royal.

Biological Invasions: a Global Perspective
Edited by J. A. Drake *et al.*
© 1989 SCOPE. Published by John Wiley & Sons Ltd

CHAPTER 4

Invasions of Natural Ecosystems by Plant Pathogens

Sharon L. von Broembsen

4.1 INTRODUCTION

Disease epidemics have had a great influence on the history of mankind. While most disease epidemics have involved humans, crops and domesticated animals, natural ecosystems have not gone unscathed. Invading plant pathogens have led to some of the most serious disruptions of natural ecosystems ever recorded. The devastating impact of chestnut blight on North American hardwood forests and of *Phytophthora* root rot on Western Australian jarrah forests are two striking examples.

This paper examines the extent, modes and patterns of invasions of natural ecosystems by plant pathogens not originating in these ecosystems. Agricultural systems, cultivated forests and forests whose natural vegetation structure has been significantly altered are excluded from the concept of natural ecosystems used here. Pathogens occurring on invasive plants in natural ecosystems are not considered invasive unless they cause epidemics in the indigenous vegetation.

In this chapter I address a number of important questions about invasions of natural ecosystems by plant pathogens on a global basis. How successful have plant pathogens been in invading natural ecosystems? Are there any similarities amongst modes of invasion by plant pathogens? What has been the human role in successful invasions? What can we expect in the future? Are there promising management strategies that can be employed?

4.2 SUCCESSFUL INVASIONS

The extent and effects of invasions of natural ecosystems by plant pathogens can probably best be appraised by examining some successful invasions. The following descriptions of successful invasions also provide the basis for further analysis of the underlying patterns common to such invasions.

4.2.1 Chestnut blight

The destruction of the American chestnut, *Castanea dentata*, is one of the best known and most devastating results of an invasion by a plant pathogen.

Cryphonectria parasitica, the chestnut blight fungus, was introduced into North America on ornamental nursery material from Asia late in the 1890s (Walker, 1957). Quimby (1982) and Horsfall and Cowling (1978) provide much information regarding the impact of the disease. From the original introduction in New York, the pathogen spread throughout 91 million hectares of hardwood forests of the eastern USA in less than 50 years. The chestnut was the dominant tree in large areas of forest and composed up to 25% of some forests. Chestnut blight virtually eliminated the American chestnut throughout its natural range. In the words of Harper (1977), 'This is probably the largest single change in any natural plant population that has ever been recorded by man.'

4.2.2 White pine blister rust

The white pine blister rust fungus, *Cronartium ribicola*, is indigenous to Asia and southern parts of Europe (Bingham *et al.*, 1971). The fungus was introduced from Europe into the New England area of North America in the early 1900s on white pine nursery stock. It is also thought to have been introduced into British Columbia, Canada, at about the same time. Blister rust rapidly became epidemic in northern populations of highly susceptible five needle pines (Walker, 1957). The disease is especially severe on seedlings and young trees. Although the rust requires *Ribes* spp. as an alternate host to complete its life cycle, susceptible currants and gooseberries occur along with five needle pines throughout their natural ranges in both North America and Eurasia. Extensive *Ribes* eradication programs were carried out to try to control the disease, especially in managed forests, but these were not effective.

4.2.3 Dutch elm disease

Dutch elm disease, which first appeared in Europe in the early 1900s, received its name because of its rapid spread through elm populations in Holland. Shortly after that, the fungus causing the disease, *Ophiostoma ulmi*, was introduced to North America on diseased timber from Europe. The pathogen's natural vector, the European elm bark beetle, was also introduced to North America at about the same time (Bingham *et al.*, 1971). The disease spread rapidly throughout American elm populations, aided by both its introduced vector and an American elm bark beetle, (Quimby, 1982). Elms from eastern Asia are resistant and this area is believed to be the origin of the fungus (Bingham *et al.*, 1971; Horsfall and Cowling, 1978). An interesting aspect of this disease is that recent outbreaks of the disease in Europe apparently have resulted from the introduction of more pathogenic strains from North America (Brasier, 1979). Logs used as dunnage in ships have been implicated in these recent introductions.

4.2.4 Pine wilt disease

The pine wood nematode, *Bursaphelenchus xylophilus*, was introduced into Japan in the early 1900s on diseased timber. Mamiya (1983) has recently reviewed the development of the epidemic caused by this insect-vectored nematode. The nematode was spread amongst the Japanese islands by the movement of infested timber and locally disseminated by a native insect vector, the Japanese pine sawyer. Pine wilt disease has affected more than 650 000 hectares or 25% of Japan's 2.6 million hectares of pine forest and killed more than 10 million Japanese pine trees. The pine wood nematode was not recognized as the cause of the epidemic until the early 1970s. Subsequently, the nematode was found in North America, where it is considered to be indigenous (Wingfield *et al.*, 1984).

4.2.5 *Phytophthora* root disease

The most dramatic and well-known invasion of a natural ecosystem by a plant pathogen in the southern hemisphere, is the invasion of the jarrah forests of Western Australia by *Phytophthora cinnamomi.* However, this root fungus has also invaded other natural ecosystems. *Phytophthora cinnamomi* has a host range of nearly 1000 different plants, including many in natural ecosystems (Zentmyer, 1980). This pathogen is therefore different from the other pathogenic invaders considered above.

The background and salient features of the jarrah forest invasion are covered in recent reviews (Shea, 1976; Zentmyer, 1980). *Phytophthora cinnamomi* is presumed to have been introduced to Western Australia on diseased nursery material from eastern Australia. More than 300 000 hectares of jarrah forest have been affected by this disease. Water supplies of the region are threatened by increasing salination due to loss of forest cover. Spread of disease within the forests was facilitated by roadbuilding, logging and mining activities that involved movement of soil or gravel containing the fungus. From roadsides and other points of introduction, the disease moved into forests along drainage lines. The canopy vegetation of the forests is comprised almost entirely of the highly susceptible jarrah (*Eucalyptus marginata*). The situation was further aggravated by altered fire regimes that favored more susceptible understory vegetation. Widescale quarantine has been implemented to protect areas not yet affected by the disease.

Eucalypt forests in Victoria, Australia have also been invaded by *Phytophthora cinnamomi* (Weste and Taylor, 1971). The effects on the canopy vegetation have not been as great as in Western Australia, but a greater number of plant species have been affected. More than half the species in plant communities have been destroyed in some places. The pathogen was presumably introduced to that region on diseased nursery material.

The earliest known invasion of a natural ecosystem by *Phytophthora cinnamomi* occurred sometime in the 19th century in the forests of the southeastern USA (Crandall *et al.*, 1945). Chestnut root disease spread rapidly throughout these forests and eliminated the chestnut in its southern range long before chestnut blight appeared in the northeastern USA. Certain native *Pinus* spp. were also affected and the disease on pines became known as little leaf disease. The means of introduction of the pathogen is not known.

Phytophthora cinnamomi has been involved in tree mortality in other forest ecosystems including the *Nothofagus* forests of Papua New Guinea (Arentz, 1983), the rain forests of northeastern Australia (Brown, 1976), the ohia forests of Hawaii (Kliejunas and Ko, 1976), and the afromontane forests of South Africa (Von Broembsen *et al.*, 1986). However, it is not clear in these cases whether the fungus is an introduced invader or a native pathogen.

The center of origin of *Phytophthora cinnamomi* is believed to be the Australasian region (Zentmyer, 1980). However, recent evidence (Von Broembsen and Kruger, 1984) indicates the fungus is also indigenous to South Africa. This suggests that the fungus may once have had a wide southern distribution before the breakup of Gondwanaland and that the present disjunct natural distribution has arisen through vicariance. Superimposed on this natural distribution is dispersal by movement of the fungus on cultivated plants. The origin of the *Phytophthora cinnamomi* occurring in a particular natural ecosystem thus can be difficult to determine.

4.3 PATTERNS AND MODES OF INVASION

From the information available on successful invasions, a number of patterns are apparent. The most obvious is that there have been relatively few successful invasions of natural ecosystems by plant pathogens. Micro-organisms are relatively inconspicuous and undoubtedly some invasions have not been recognized. However, plant disease epidemics seldom occur in undisturbed natural ecosystems (Harlan, 1976). Moreover, except for certain unspecialized pathogens, most pathogens are able to attack only a very limited number of closely related host species, and the possibility of finding a susceptible host in a new ecosystem is low.

4.3.1 The invaders

The successful invaders represent three of the five kingdoms of living organisms (Margulis and Schwartz, 1982) and include nematodes and both lower and higher fungi. Bacterial and viral phytopathogens are conspicuous by their apparent absence. One of the successful invaders, *Phytophthora cinnamomi*, is a generalist pathogen and has invaded several different natural ecosystems at the expense of a wide range of plant species. Host specificity is not an important constraint for a generalist invader, thus there are potentially more ecosystems that could be invaded by generalist pathogens. The remaining invaders are specialist pathogens that cause epidemics only on groups of closely related species. None of the specialist invaders are important pathogens at thier centers of origin.

Complex life cycles do not seem to deter invasion of natural ecosystems by plant pathogens as has been suggested for invasions by animal and human pathogens (Dobson and May, 1986). Several of these successful plant pathogenic invaders depend on insect vectors or alternate hosts to complete their life cycles. In at least one case (Dutch elm disease), an insect vector was also introduced. Local insect vectors proved suitable for other invaders.

4.3.2 The invaded ecosystems

All of the successfully invaded ecosystems are temperate forests with climates similar to the areas of pathogen origin. All of the devastating invasions were facilitated by the marked susceptibility of fairly uniformly distributed, dominant trees of one or a few closely related species. Harlan (1976) has suggested that forests of the wet tropics are not good candidates for epidemics because of their heterogeneity. However, other important regions such as the taiga, tropical savanna, grassland steppes, and desert shrublands are often characterized by vast uniform stands of one of a few dominant species. It is not clear whether these regions are climatically unsuitable for successful invasions by plant pathogens or merely awaiting the arrival of invasive pathogens.

An obvious but significant characteristic of the successfully invaded ecosystems is that they have been separated from the ecosystems where the invasive pathogens originated for a considerable period of time by geographical barriers. The resulting isolation means that more susceptible genotypes that might have been eliminated or suppressed in the presence of the pathogen would not be constrained by this factor. Hosts that have evolved in the absence of their previous pathogens are often more susceptible than their constantly challenged relatives. When the physical barriers between the isolated regions are bridged, the consequences can be disastrous.

4.3.3 The human role

One of the most important patterns evident is that all of the pathogens that have successfully invaded natural ecosystems have been introduced to these regions by humans. Inter-continental dispersal of viable propagules of plant pathogens by air streams certainly occurs, and is believed to have been responsible for the movement of a few agricultural pathogens across immense ocean barriers (Harlan, 1976). However, humans have been a much more efficient agent of dispersal. Throughout history, cultivated plants and their pathogens have been carried to new regions of human colonization and development. Thus, humans have been responsible for moving many plant pathogens, of which only a few have invaded natural ecosystems. All of the successful invaders discussed above were introduced on diseased nursery material or timber prior to strict implementation of quarantine.

Fox and Fox (1986) have suggested that disturbance is a prerequisite for invasion of natural ecosystems by plants and animals. In the case of *Phytophthora* root disease, invasions have apparently been facilitated by disturbance of natural ecosystems. However, disturbance does not seem to be essential for invasions of natural ecosystems by plant pathogens.

4.4 CONCLUSIONS

It is important to consider whether management strategies, which will act to constrain future invasions, can be implemented. From the preceding analysis, certain strategies would seem to be indicated. The mode of spread of invasive plant pathogens is similar. All the invaders were human introductions before the institution of effective quarantine policies. The introductions have occurred on diseased nursery material or timber. The pathogens were thus introduced with suitable host material that served as temporary repositories from which subsequent invasions took place. The possibility that inoculum from this material could establish on susceptible hosts would be greater than for inoculum not associated with host material.

Quarantine has apparently been effective in limiting invasions. Quarantine efforts should, therefore, be emphasized and given greater support. The movement of high risk materials such as rooted nursery stock and barked timber between and within countries is now generally prohibited. Some problems regarding ornamental nursery stock, timber, dunnage, and packing materials are still evident. Where movement of such materials is not prohibited, this ought to be implemented. Quarantine programs in Third World countries, where facilities may not be adequate to exclude potential invaders, require additional support.

The paucity of successful invasions of natural ecosystems by plant pathogens pays tribute to the powerful natural mechanisms operating to prevent such invasions. The operation of these mechanisms in natural ecosystems needs to be fully investigated and more clearly understood. A general management policy that minimizes the alteration of intact ecosystems so that these protecting mechanisms can remain fully operational seems prudent. This is especially important for temperate forest ecosystems, which are apparently more susceptible to invasion. This protectionist policy would also minimize epidemics that are caused by indigenous pathogens and that frequently result from disturbances.

ACKNOWLEDGEMENT

I thank Michael J. Wingfield for many stimulating discussions on the biogeography of plant pathogens and for useful criticisms of this paper.

REFERENCES

Arentz, F. (1983). *Nothofagus* dieback on Mt. Giluwe, Papua New Guinea. *Pacific Science*, **37**, 453–8.
Bingham, R. T., Hoff, R. J., and McDonald, G. I. (1971) Disease resistance in forest trees. *Ann. Rev. Phytopathol.*, **2**, 433–52.
Brasier, C. M. (1979). Dual origin of recent Dutch elm disease outbreaks in Europe, *Nature*, **281**, 78–80.
Brown, B. N. (1976). *Phytophthora cinnamomi* associated with patch death in tropical rain forests in Queensland. *Aust. Plant Pathol. Newsl.*, **5**, 1–4.
Crandall, B. S., Gravatt, G. F., and Ryan, M. M. (1945). Root diseases of *Castanea* species and some coniferous and broadleaf nursery stocks, caused by *Phytophthora cinnamomi*, *Phytopathology*, **35**, 162–80.
Dobson, A. P., and May, R. M. (1986). Patterns of invasions by pathogens and parasites. In: Mooney, H. A., and Drake, J. A. (Eds.), *Ecology of Biological Invasions of North America and Hawaii*, pp. 58–76. Springer-Verlag, New York.
Fox, M. D., and Fox, B. J. (1986). The susceptibility of natural ecosystems to invasions. In: Groves, R. H., and Burdon, J. J. (Eds.), *Ecology of Biological Invasions: An Australian Perspective*, pp. 57–66. Australian Academy of Sciences, Canberra.
Harlan, J. R. (1976). Diseases as a factor in plant evolution, *Ann. Rev. Phytopathol.*, **14**, 31–51.
Harper, J. L. (1977). *Population Biology of Plants*. Academic Press, New York.
Horsfall, J. G., and Cowling, E. B. (1978). Some epidemics man has known. In: Horsfall, J. G., and Cowling, E. B. (Eds.), *Plant Disease: An Advanced Treatise Volume II–How Disease Develops in Populations*, pp. 17–32. Academic Press, New York.
Kliejunas, J. T., and Ko, W. H. (1976). Association of *Phytophthora cinnamomi* with ohia decline on the island of Hawaii. *Phytopathology*, **66**, 116–21.
Mamiya, Y. (1983). Pathology of the pine wilt disease caused by *Bursaphelenchus xylophilus*. *Ann. Rev. Phytopathol.*, **21**, 201–20.
Margulis, L., and Schwartz, K. V. (1982). *Five Kingdoms*, W. H. Freeman, Sam Francisco.
Quimby, P. C., Jr. (1982). Impact of diseases on plant populations. In: Charudattan, R., and Walker, H. L. (Eds.), *Biological Control of Weeds with Plant Pathogens*, pp. 47–60 John Wiley & Sons, New York.
Shea, S. R. (1976). Jarrah forest could be destroyed. *Aust. For. Ind. J.*, **42**, 16–22.
Von Broembsen, S. L., and Kruger, F. J. (1984). *Phytophthora cinnamomi* associated with mortality of native vegetation in South Africa. *Plant Dis.*, **69**, 715–17.
Von Broembsen, S. L., Lubbe, W. A., and Geldenhuys, C. J. (1986). *Phytophthora cinnamomi* associated with decline of stinkwood (*Ocotea bullata*) in southern Cape indigenous forests. *Phytophylactica*, **18**, 44 (Abstract).
Walker, J. C. (1957). *Plant Pathology*. McGraw-Hill Book Company, New York.
Weste, G. and Taylor, P. (1971). The invasion of native forests by *Phytophthora cinnamomi* I. Brisbane Ranges, Victoria. *Aust. J. Bot.*, **19**, 281–94.
Wingfield, M. J., Blanchette, R. A., and Nicholls, T. H. (1984). Is the pine wood nematode an important pathogen in the United States? *J. For.*, **19**, 281–94.
Zentmyer, G. A. (1980). *Phytophthora cinnamomi and the Diseases It Causes*. American Phytopathological Society, St. Paul, MN.

Biological Invasions: a Global Perspective
Edited by J. A. Drake *et al.*
© 1989 SCOPE. Published by John Wiley & Sons Ltd

CHAPTER 5

Patterns, Modes and Extents of Invasions by Vertebrates

JAMES H. BROWN

5.1 INTRODUCTION

Within just the last 200 years human beings have transformed the globe, drastically altering its land, water, atmosphere, and life. Probably only in the few, poorly understood episodes of mass extinctions has the earth experienced such rates and magnitudes of change, and if present trends continue, the human impacts in the next century threaten to rival the natural catastrophes that occurred at the Permian–Triassic and Cretaceous–Tertiary boundaries (Nitecki, 1984).

Three interrelated activities of modern humans have drastically altered the distribution and abundance of living things and the structure and function of ecosystems on a truly global scale: (1) the transformation of the earth's surface into a landscape of fields, pastures, suburbs, cities, roads, reservoirs, canals, dumps, and mines; (2) the extraction and appropriation of the earth's physical and biotic resources to support the human population; and (3) the loss of biotic diversity owing to the disturbance of ecological communities, extinction of native species, and spread of exotics. Increasingly ecologists and biogeographers are being called upon to assess the rates and magnitudes of these changes, predict the future impact of human activities, and manage ecosystems so as to avoid or mitigate the most deleterious effects. The task of the present conference is to synthesize what is known about invasions of exotic species, and my job is to summarize information on 'patterns, extents and modes of invasions by vertebrates.'

This is a virtually impossible task for several reasons. First, this is not an area in which I can claim any special expertise beyond being a generalized community ecologist and biogeographer who has worked primarily with vertebrates. Second, the literature on vertebrate invasions is so vast that one would have to devote one's entire career in order to become a real expert on this subject. Third, the information that is available is widely scattered, of uneven quality, and of questionable comparability, making quantitative analyses difficult and suspect.

Given these limitations, I shall not attempt to synthesize all of the relevant data, analyses, and interpretations. Instead, I offer a uniquely personal assessment of the state of our understanding of vertebrate invasions. First, I shall use examples from vertebrates to illustrate some of the general patterns and processes that characterize successful and unsuccessful invasions. These will be presented in the context of tentative rules that characterize attributes of species that influence their colonizing ability and attributes of environments that affect their susceptibility to invasion. Second, I shall show that these rules are of limited utility in making specific predictions about the probability of establishment of a particular species in a certain region or habitat. This unpredictability is due to the uniqueness of species and places, which in turn is in large part a consequence of their distinct histories. This influence of history complicates the interchange between basic ecologists, interested in general patterns and processes, and applied scientists, concerned with solving the case specific problems caused by alien species. Finally, I shall argue for a reassessment of scientific attitudes toward exotic species. The current trend toward homogenization of the earth's biota is inevitable. Given that anthropogenic habitat change and elimination of native species are likely to increase, the ability of certain species to tolerate human activities and invade disturbed habitats may provide one of the best hopes for preserving any functional ecosystems.

5.2 GENERAL PATTERNS AND PROCESSES

5.2.1 Patterns of successful invasion

There appear to be general patterns that can be induced from what is known about the differential success and failure of exotic vertebrates (Elton, 1958; Diamond and Case, 1986). Most of these inferences are necessarily qualitative, because the diverse data on different taxónomic groups, habitats, and geographic regions have not been assembled and analyzed quantitatively (but see Long, 1981). Nevertheless, I believe that these provisional generalizations could be developed much more rigorously, and I suspect that they apply equally well to plant and invertebrate groups that have been well studied, and probably to all organisms. I suggest that these might tentatively be called the rules of biological invasions. I develop five of these rules below, illustrate them with examples, and discuss some apparent exceptions.

Rule 1: isolated environments with a low diversity of native species tend to be differentially susceptible to invasion. The claim requires careful qualification. Those environments that are particularly susceptible to invasion by exotic species are generally those that are in some sense insular and have experienced low rates of natural colonization. They are usually characterized by small size, habitats that contrast markedly with the surrounding matrix, and a long history of effective isolation from similar environments that would be the most likely source of suitable colonists. Examples include oceanic islands such as Hawaii and

New Zealand, insular continents such as Australia and Madagascar, insular habitats such as lakes and desert springs, and other isolated distinctive environments such as the subtropical part of the Florida peninsula and the temperate tip of southern South America. When immigration rates have historically been low, increasing the rate of colonization (in this case through human transport) is likely to result in the establishment of new species. Qualified in this way, this rule is simply a restatement of one of the basic tenets of the theory of island biogeography (MacArthur and Wilson, 1967; Williamson, 1981) and the extension of the theory to other kinds of insular habitats (e.g., Brown and Gibson, 1983; Brown, 1986). Some environments have low biotic diversity because they are unproductive and/or physically harsh. They are not isolated by barriers from sources of potential colonists, and they are not particularly susceptible to invasion. Examples include the Sahara Desert and the tundra and taiga regions of northern North America and Eurasia.

Island biogeography theory predicts that, with equal opportunities for increased colonization, smaller and more isolated insular habitats should accumulate relatively larger numbers of alien species than larger and/or less isolated regions. In some circumstances the small, isolated islands will even be expected to acquire absolutely more species of exotics. This is largely a consequence of species–area relationships (e.g., see MacArthur and Wilson, 1967; Williamson, 1981; Brown and Gibson, 1983). Whenever similar environments are compared, large regions typically contain a larger sample of species than small ones. This holds for both insular habitats and nonisolated environments, but the species–area relationship for insular habitats characteristically has a steeper slope and lower intercept than for similar, but nonisolated environments (Figure 5.1). This is because insular habitats, surrounded by inhospitable environments, have high extinction rates; small islands cannot sustain species (such as many birds of prey and large mammalian carnivores and herbivores) that are constrained to occur at low population densities. Human activity usually reduces the effective isolation of insular habitats by increasing the rate of colonization by alien species. Although the quantitative increase in insular species richness will depend on the size of the pool of available colonists, whether or not native species go extinct, and other factors (see Schoener, 1974), given the range of species area relationships reported for habitats isolated to varying degrees, the result will sometimes be a greater increase in the absolute as well as the relative number of species in small, isolated habitats than in otherwise similar but larger and/or less isolated environments (Figure 5.1).

Figure 5.2 presents evidence that long-isolated regions with few species of native vertebrates are differentially susceptible to invasion. Because of their histories of exploitation by human colonists of British descent, Hawaii, New Zealand, Australia, and North America have had roughly comparable opportunities for colonization by exotics. For these four regions there is a strong negative relationship between percent of the fauna that is comprised of established exotics

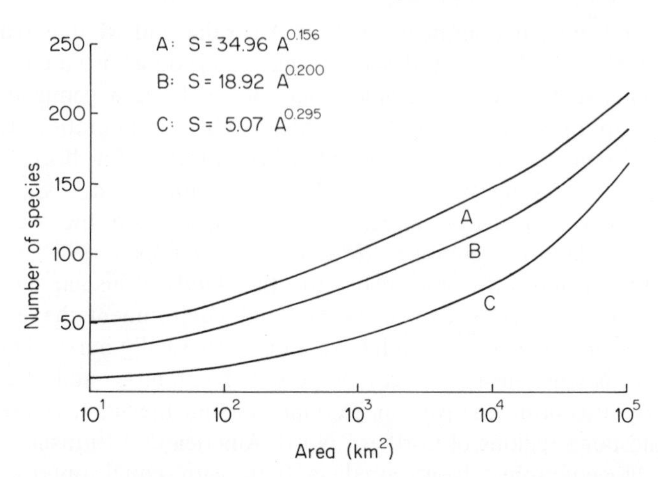

Figure 5.1. Hypothetical species–area relationships showing how previously isolated biotas should acquire additional species when colonization rates are increased as a result of human activity. The three curves show the expected equilibrium number of species (*sensu* MacArthur and Wilson, 1967) for areas with different degrees of isolation. The upper curve with a low slope and high intercept represents nonisolated sample areas of varying sizes within a large area, such as a continent. The lower curve with a steep slope and a low intercept represents historically isolated oceanic islands of varying sizes before invasion by aliens with human assistance. The intermediate curve represents the same islands after equal opportunity for invasion by alien species. Although species–area curves are usually plotted on logarithmic axes, here the ordinate is linear to show that small islands with few native species can sometimes be expected to acquire slightly more species than islands or continents with more diverse native biotas

and the number of native species. Interestingly, data points for birds and mammals seem to suggest approximately the same relationship. Percentages of exotics range from over 90% for mammals in New Zealand and Hawaii to less than 10% for both birds and mammals in Australia and North America.

Data for other vertebrate groups and other insular regions are consistent in suggesting that a large proportion of the species of isolated oceanic islands and other long-isolated environments is comprised of invaders. All of the nonmarine reptiles (14 species), amphibians (four species), and primary freshwater fishes (31 species) of the Hawaiian Islands have been introduced (McKeown, 1978; Maciolek, 1984). Fourteen (34%) of the 41 freshwater fish species of New Zealand are exotic (McDowall, 1978). Ten of 15 species of terrestrial mammals on the Galapagos are aliens (Thornton, 1971; H. Snell, personal communication). It is not only islands with depauperate biotas that are highly susceptible to invasion.

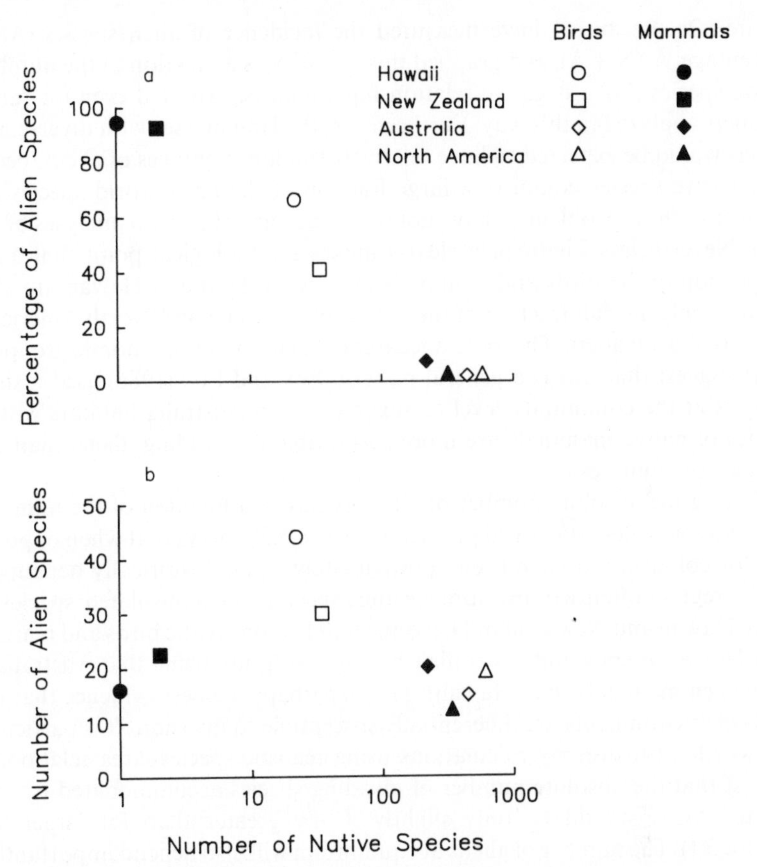

Figure 5.2. (a) The percentage of exotic species of mammals and birds in the total mammalian or avian fauna as a function of the number of native species for the Hawaiian Islands, New Zealand, Australia, and North America north of Mexico. (b) The absolute number of exotic species of mammals and birds as a function of the number of native species for these same four regions. Note that both ways of plotting the data suggest that historically isolated areas with low diversity of native species are differentially susceptible to invasion. Compiled from works cited in the text, de Vos *et al.* (1956), Falla *et al.* (1956), Kramer (1971), Pizzey (1980), van Riper and van Riper (1982), and other sources

Because of the low diversity and small area of aquatic habitats, the native fish fauna of the arid southwestern United States contains few species. The creation of artificial impoundments and introduction of exotics has greatly added to this number. For example, Minkley (1973) lists approximately 95 species that have bred in Arizona, and 67 (71%) of these have been introduced.

There are serious statistical problems with presenting data as I have in

Figure 5.2a. Because I have measured the incidence of alien species (A) as a percentage, A/(N + A), and graphed this quantity as a function of the number of native species (N), a negative relationship would be expected even for random numbers analyzed in this way. Furthermore, the land masses with diverse native faunas would be expected to have relatively small percentages of aliens, because their native species comprise a large fraction of the finite world species pool. Because of these problems, I have not performed any statistical analyses of these data. Nevertheless Figure 5.2a clearly makes the biological point that a large proportion of the birds and mammals of New Zealand and Hawaii are aliens, whereas only a small fraction of the faunas of Australia and North America are comprised of invaders. The more anecdotal data for other taxonomic groups and areas suggest that this is a general pattern. Fox and Fox (1986) used a similar analysis at the community level to suggest that in Australia habitats with few species of native mammals are more susceptible to invading aliens than more diverse communities.

Plotting the absolute number of alien species as a function of the number of natives avoids these statistical problems. Figure 5.2b shows that when opportunities for colonization have been approximately equal, historically depauperate insular regions often acquire more invading species than nonisolated, species-rich areas. Hawaii and New Zealand have absolutely more exotic birds and mammals than North America and Australia (the only exception being that Australia has more alien mammals than Hawaii). This is perhaps the best evidence that these kinds of environments are differentially susceptible to invasion. It is particularly noteworthy, because my calculations using realistic species–area relationships suggest that the absolute number of invading species accommodated by small, isolated areas should be only slightly if any greater than for larger areas (Figure 5.1). The number of aliens at equilibrium will also depend importantly on the number of native species that have gone extinct; it is possible to accommodate more invaders if some of them 'replace' extinct natives. As mentioned below, isolated islands such as Hawaii and New Zealand have lost a substantial fraction of their endemic species, whereas larger land masses such as Australia and North America have suffered proportionately fewer extinctions.

Susceptibility of a region or habitat to invasion can also be assessed in terms of the impact of alien species on native biota and ecosystem function. There is abundant evidence that invading species not only comprise a large proportion of the biota in many small, isolated environments, they have also directly caused or indirectly contributed to the extinction of native species and substantially changed the structure and dynamics of both natural and human-modified ecosystems. The examples are overwhelming, even though only a few have been systematically quantified. The most important effects of exotic vertebrates can be attributed to predation (including herbivory) on native biota, competition with native species, effects of accompanying parasites and diseases, and disturbance of the physical environment. One of the best documented recurring effects is the decimation of native sea and land birds by introduced rats and other predatory mammals on isolated oceanic islands (e.g., Atkinson, 1977, 1985; King, 1980, 1984). Table 5.1 summarizes some of the data compiled by King (1980) documenting the probable causes of extinctions of birds on islands throughout the world. Note that introduced predators are the single most important cause of extinctions, accounting for 42% of the total. Of those extinctions caused by predators, King attributes 54% to rats, 26% to cats, and the remainder to mongeese, weasels, stoats, and other species. Alien competitors and diseases account for an additional 7% and 6%, respectively, of the extinctions of island birds. Introduced mammalian herbivores, especially feral goats and pigs, have had equally dramatic but less thoroughly documented impact on the vegetation of small islands, causing extinctions of native species, pronounced changes in dominance and physiognomy, and indirectly affecting many other organisms (e.g. Holdgate and Nace, 1961; Coblentz, 1978; Diamond and Veitch, 1981; Loope and Scowcroft, 1985; Loope and Mueller-Dombois, this volume). The impact of exotic fishes as predators and competitors on native fishes and invertebrates in many small, isolated lakes and streams with depauperate native ichthyofaunas rivals that of alien mammalian predators on small oceanic islands. Introduction of fish, especially largemouth and smallmouth bases (*Micropterus salmoides* and *M. dolomieui*), mosquitofish (*Gambusia affinis*), and tilapia (*Tilapia mosambica*), has been accompanied by extinction of native species and pronounced changes in dominance and food webs in aquatic ecosystems as different as the Great Lakes of temperate North America (Smith, 1972; Moyle, 1986), small lakes in tropical Panama (Zaret and Paine, 1973), springs, streams, and lakes in the arid southwestern United States (Miller, 1972; Moyle, 1976), and streams in Hawaii (Maciolek, 1984).

Table 5.1. Apparent causes of extinction of birds on islands throughout the world since 1600 AD (from King, 1980; 1984). Note that predation, competition, disease, and genetic swamping, which together account for 57% of all extinctions, are all effects of alien species

	Indian Ocean	Atlantic Ocean	Pacific Ocean	Total
Predation	0	16	96	112
Competition	0	0	18	18
Disease	0	0	15	15
Genetic swamping	0	0	2	2
Hunting	7	13	20	40
Habitat destruction	1	5	46	52
Weather	0	1	0	1
Unknown	13	4	11	28
Total	21	39	208	258

It should be emphasized that impacts of alien species on native biota through indirect interactions can sometimes be as great or greater than the effects of direct competition or predation. Thus many effects of alien mammalian herbivores on the vertebrates and invertebrates of small oceanic islands are indirect. The direct effects of consumption of plants and disturbance of the physical environment cause major changes in vegetation and habitat (e.g. Holdgate and Nace, 1961; Coblentz, 1978; Diamond and Veitch, 1981). Alien species may also have indirect effects by facilitating the invasions of still more exotics. Of course this has been one of the most important effects of human colonization, but other vertebrates provide additional examples. A substantial part of the impact of introduced birds on the native avifauna of Hawaii is due to the alien populations importing and serving as a reservoir for avian malaria, an epizootic to which some native species are highly susceptible (van Riper *et al.*, 1986). Also in Hawaii, soil disturbance caused by the rooting of introduced pigs facilitates the invasion of native forest by exotic plant species (Smith, 1985).

There can be little doubt that the impacts of alien vertebrates in small, insular environments with depauperate native faunas are on average greater than their effects in larger, less isolated areas, such as continental habitats. Although the impact of exotics on continental biotas and ecosystems is sometimes substantial, I know of no case in which the extinction of a native species of terrestrial animal or plant on any of the major continental land masses can be attributed with reasonable certainty to the effect of an alien vertebrate (other than *Homo sapiens*). On the other hand, there are many well-documented cases of extinctions of insular animals and plants that can be attributed directly to predation by introduced mammalian carnivores or herbivores (see references above). Perhaps the most dramatic example is the extinction of the Stephens Island wren, whose demise can be reliably attributed to a single cat belonging to the lighthouse-keeper; in 1894 this animal 'collected' all 22 specimens known to science (King, 1984). But not all of these extinctions occurred long ago before the importance of conservation was widely recognized. King (1984) describes the case of Big South Cape Island, an important predator-free refuge in New Zealand until an irruption of rats in 1964 eliminated five species of birds and one of bats (one of the two land mammals native to New Zealand).

Rule 2: species that are successful invaders tend to be native to continents and to extensive, nonisolated habitats within continents. This rule is also a restatement of a long-held generalization of insular biogeography to the effect that there is a differential immigration from mainland to island. This is the basis of the so-called insular taxon cycle, which occurs if colonization is sufficiently infrequent that successful immigrants evolve into endemic insular species that are eventually replaced by new colonists (Wilson, 1961, MacArthur and Wilson, 1967; Ricklefs and Cox, 1972; Brown and Gibson, 1983). As an empirical generalization, this rule is illustrated in Tables 5.2–5.5. Of the introduced vertebrates of isolated oceanic islands, such as Hawaii (Table 5.2) and New Zealand (Table 5.3), the vast

Table 5.2. Identity and origin of the introduced land bird species of the Hawaiian Archipelago (from Long, 1981; Moulton and Pimm, 1986, 1987; and other sources). Unless specifically designated as being tropical, the native ranges of these species include temperate latitudes on the continent of origin. Although Hawaii is tropical, 17 of the 37 species (46%) are of temperate origin

Species	Native region
California quail (*Lophortyx californicus*)	North America
Chukar (*Alectoris chukar*)	Eurasia
Coturnix (*Coturnix coturnix*)	Eurasia
Jungle fowl (*Gallus gallus*)	Tropical Asia
Ring-necked pheasant (*Phasianus colchicus*)	Eurasia
Peafowl (*Pavo cristatus*)	Tropical Asia
Turkey (*Meleagris gallopavo*)	North America
Rock dove (*Columbia livia*)	Eurasia
Spotted dove (*Streptopelia chinensis*)	Asia
Zebra dove (*Geopelia striata*)	Tropical Asia
Skylark (*Alauda arvensis*)	Eurasia
Red-vented bulbul (*Pycnonotus cafer*)	Tropical Asia
Red-whiskered bulbul (*Pycnonotus jocosus*)	Tropical Asia
Melodious laughingthrush (*Garrulax canorus*)	Tropical Asia
Greater necklaced laughingthrush (*Garrulax pectoralis*)	Tropical Asia
Red-billed leiothrix (*Leiothrix lutea*)	Tropical Asia
Mockingbird (*Mimus polyglottus*)	North America
Shama (*Copsychus malabaricus*)	Tropical Asia
Bush warbler (*Cettia diphone*)	Asia
Hill myna (*Gracula religiosa*)	Tropical Asia
Common myna (*Acridotheres tristis*)	Tropical Asia
Japanese white-eye (*Zosterops japonica*)	Asia
Lavender waxbill (*Estrilda caerulescens*)	Tropical Africa
Orange-cheeked waxbill (*Estrilda melopoda*)	Tropical Africa
Red-cheeked cordon-bleu (*Ureaginthus bengalus*)	Tropical Africa
Strawberry finch (*Amandava amandava*)	Tropical Asia
Common silverbill (*Lonchura malabarica*)	Tropical Asia
Spice finch (*Lonchura punctulata*)	Tropical Asia
Java sparrow (*Padda oryzivora*)	Tropical Asia
House sparrow (*Passer domesticus*)	Eurasia and North Africa
Western meadowlark (*Sturnella neglecta*)	North America
Cardinal (*cardinalis cardinalis*)	North America
Red-crested cardinal (*Paroaria cristata*)	South America
Yellow-faced grassquit (*Tiaris olivacea*)	Tropical America
House finch (*carpodacus mexicanus*)	North America
Yellow-fronted canary (*Serinus mosambicus*)	Tropical Africa
Saffron finch (*Sicalis flaveola*)	Tropical America

majority have come from diverse continental biotas, and the few other invaders have colonized from other islands with more speciose faunas. Vertebrates that have invaded continents, such as Australia (Table 5.4) and North America

Table 5.3. Identity and origin of the introduced mammals of New Zealand (from Gibb and Flux, 1973)

Species	Origin
Brush-tailed possum (*Trichosurus vulpecula*)	Australia
Red-necked wallaby (*Macropus rufogriseus*)	Australia
Hedgehog (*Erinaceus europeus*)	Eurasia
Rabbit (*Oryctolagus cuniculus*)	Eurasia
Hare (*Lepus capensis*)	Eurasia
Polynesian rat (*Rattus exulans*)	Southeast Asia
Black rat (*Rattus rattus*)	Eurasia
Brown rat (*Rattus norvegicus*)	Eurasia
House mouse (*Mus musculus*)	Eurasia
Weasel (*Mustela nivalis*)	Eurasia
Stoat (*Mustela erminea*)	Eurasia
Ferret (*Mustela furo*)	Eurasia
Cat (*Felis cattus*)	Eurasia
Pig (*Sus scrofa*)	Eurasia
Red deer (*Cervus elaphus*)	Eurasia and North America
Sambar (*Cervus unicolor*)	Tropical Asia
Japanese deer (*Cervus nippon*)	Asia
Rusa deer (*Cervus timorensis*)	Tropical Asia
Fallow deer (*Dama dama*)	Eurasia
Moose (*Alces americana*)	North America
Cow (*Bos taurus*)	Eurasia
Sheep (*Ovis aries*)	Eurasia
Goat (*Capra hircus*)	Eurasia
Himalayan thar (*Hemitragus jemlahicus*)	Asia
Chamois (*Rupicapra* rupicapra)	Eurasia

Table 5.4. Identity and origin of the introduced mammals of Australia (from Strahan, 1983; Myers, 1986)

Species	Origin
Rabbit (*Oryctolagus cuniculus*)	Eurasia
Hare (*Lepus capensis*)	Eurasia
Black rat (*Rattus rattus*)	Eurasia
Brown rat (*Rattus norvegicus*)	Eurasia
House mouse (*Mus musculus*)	Eurasia
Dingo (*Canis familiaris*)	Eurasia
Fox (*Vulpes vulpes*)	Eurasia
Cat (*Felis cattus*)	Eurasia
Horse (*Equus caballus*)	Eurasia
Donkey (*Equus asinus*)	Eurasia and North Africa
Pig (*Sus scrofa*)	Eurasia
One-humped camel (*Camelus dromedarius*)	Asia and North Africa

Table 5.4. (*Contd.*)

Species	Origin
Red deer (*Cervus elaphus*)	Eurasia and North America
Sambar (*Cervus unicolor*)	Tropical Asia
Rusa deer (*Cervus timorensis*)	Tropical Asia
Fallow deer (*Dama dama*)	Eurasia
Chital (*Axis axis*)	Tropical Asia
Hog deer (*Axis porcinus*)	Tropical Asia
Water buffalo (*Bubalus bubalus*)	Tropical Asia
Goat (*Capra hircus*)	Eurasia

Table 5.5. Identity and origin of introduced land bird species with established breeding populations in North America north of the United States–Mexican border (from Long, 1981; National Geographic Society, 1983; and other sources). Note that the vast majority of alien species in temperate North America originated in temperate regions of the Old World, whereas most of these exotics restricted to subtropical southern California and southern Florida invaded from tropical regions of the Old or New World

Species	Origin
Temperate North America	
Chukar (*Alectoris chukar*)	Eurasia
Black francolin (*Francolinus francolinus*)	Eurasia
Gray partridge (*Perdix perdix*)	Eurasia
Himalayan snowcock (*Tetraogallus himalayensis*)	Asia
Ring-necked pheasant (*Phasianus colchicus*)	Eurasia
Rock dove (*Columbia livia*)	Eurasia
Ringed turtle-dove (*Streptopelia risoria*)	Eurasia
Monk parakeet (*Myiopsitta monachus*)	Temperate South America
Skylark (*Alauda arvensis*)	Eurasia
Crested myna (*Acridotheres cristitellus*)	Asia
Starling (*Sturnus vulgaris*)	Eurasia
Eurasian tree sparrow (*Passer montanus*)	Eurasia
House sparrow (*Passer domesticus*)	Eurasia and North Africa
Restricted to subtropical southern California or southern Florida	
Spotted dove (*Streptopelia chinensis*)	Asia
Rose-ringed parakeet (*Psittacula krameri*)	Tropical Asia and Africa
Budgerigar (*Melopsittacus undulatus*)	Australia
Canary-winged parakeet (*Brotogeris versicolorus*)	South America
Yellow-headed parrot (*Rhynchopsitta pachyrhyncha*)	Tropical America
Hill myna (*Gracula religiosa*)	Tropical Asia
Java sparrow (*Padda oryzivora*)	Tropical Asia

(Table 5.5) almost without exception have come from other continents, usually from larger ones with more diverse biotas.

This pattern, by itself, is not really sufficient to demonstrate differential invasive ability of species from diverse biotas, because opportunities for colonization may not be equal. Since continents and large habitats tend to have more species and larger populations than islands and restricted habitats, they would be expected to send out more colonists. Thus, even if the success rate of colonists were independent of their origin, we would expect asymmetrical exchange between diverse and depauperate biotas. Furthermore, because the dispersal of alien species is assisted by humans, patterns of commerce and peculiarities of culture affect the direction and probability of dispersal. Perhaps the most glaring examples were the naturalization or acclimatization societies formed by British colonists in North America, Australia, and New Zealand to promote the introduction of European songbirds and other plants and animals in order to provide familiar surroundings in foreign lands.

The fact that there are almost no good examples of successful invaders of continents that have come from small islands and other depauperate faunas (e.g. Tables 5.2–5.5), however, suggests that biotic resistance from diverse native species can be effective in repelling invaders. The susceptibility of southern Florida to many exotic vertebrates, including some from islands in the Caribbean, is an apparent exception, but actually one that supports the rule. Southern Florida is an ecological island of tropical, subtropical, and human-modified habitats, with a highly depauperate and endemic vertebrate fauna. It has proven susceptible to invasion by many alien vertebrates, including 25 species of reptiles and amphibians. Among the latter, six species from the diverse *Anolis* lizard fauna of the Greater Antilles have successfully colonized the small island of mesic tropical habitat provided by suburban Miami, Florida (Wilson and Porras, 1983).

Although it seems logical that the larger the biota, the higher the probability that it will contain competitors or predators capable of inhibiting initial population growth of a new colonist sufficiently to prevent its establishment, it is hard to test this conjecture. Both competition and predation can potentially be concentrated in a few, strongly interacting, specialized species or diffused among many, generalized species. Since abiotic factors and chance events can also affect the fate of invaders, it is usually very difficult to identify the cause(s) for the success or failure of a particular colonization.

These problems can potentially be overcome by statistical analysis of large samples of exotic species and the biotas they have or have not been able to invade. Moulton and Pimm (1986, 1987) have been able to compare attributes of the many successfully and unsuccessfully introduced land bird species in the Hawaiian Archipelago, and their analyses suggest that competition from native and previously established exotic species has played a significant role. Success of exotics was higher on islands and in habitats where the diversity of established species was lower, and successful invaders tended to be more different in morphology from (and hence presumably less likely to compete with) resident species than unsuccessful introduced species. Fox and Fox (1986) reach similar conclusions about the ability of alien mammals to invade local communities in Australia.

It might be expected that absence or low diversity of predators would also facilitate establishment of exotics. I know of no direct data to support this prediction. However, the devastating effects of introduced predators and pathogens on the native vertebrates of many oceanic islands suggests that native enemies are either absent or much less effective than the predators from diverse continental biotas. Although this is somewhat indirect evidence, it strongly suggests that colonists of depauperate islands and habitats should usually face less severe predation than they experienced in the diverse biotas where they originated.

Rule 3: successful invasion is enhanced by similarity in the physical environment between the source and target areas. This pattern probably can be attributed primarily to the direct effects of physical factors in limiting the abundance and distribution of populations of vertebrates. In addition, however, availability of food resources and attributes of coexisting species tend to be correlated with physical variables. Consequently, colonists are more likely to be able to invade environments that provide climatic, geological, limnological, and oceanographic conditions similar to where they originated.

This rule is difficult to quantify accurately without analysis of the entire geographic ranges of species and of physical conditions in both source and target regions. Tables 5.2–5.5 provide much less precise information. Nevertheless it is apparent that the temperate regions of North America and Australia have been colonized almost exclusively by species that are native to temperate regions of Europe, Asia, and North Africa. Similarly, exotics from tropical America and Southeast Asia have been relatively successful in colonizing the small subtropical to tropical regions of southern Florida, southern California, and northern Australia. There are some conspicuous exceptions to this rule. In Australia, for example, species of temperate-zone origin, such as brown rat (*Rattus norvegicus*), house mouse (*Mus musculus*), and donkey (*Equus asinus*) have invaded tropical habitats, whereas the tropical rusa deer (*Cervus timorensis*) and hog deer (*Axis porcinus*) have become established in temperate regions. The reason for this lack of precise correspondence between the physical environments of source and target areas seems fairly straightforward. Many vertebrates can tolerate a much wider range of physical conditions than they encounter within their native geographic and habitat ranges. As long as these broad tolerances are not exceeded, they may invade whenever the biotic environment offers an abundance of food and a relative absence of competitors and predators. For example, largemouth bass (*Micropterus salmoides*) from temperate North America has been successfully introduced into tropical and subtropical lakes and reservoirs around the world (e.g. in central Africa, Central America, the West Indies, and Hawaii).

Islands and other depauperate regions provide many exceptions to this rule. Thus, although the Hawaiian Islands are tropical, a relatively large proportion of the mammalian, avian, and amphibian exotics have come from temperate regions (Table 5.2). Similarly, of the 67 species of exotic freshwater fishes in Arizona reported by Minkley (1973) at least 21 have native ranges confined to more northerly latitudes in North America or Eurasia, and at least 10 species are native to more tropical regions of the Americas or Africa. The fact that similarity of physical environments of source and target seems to have less effect on success of invaders when the diversity of the target biota is low suggests the interesting possibility that exotics can tolerate a greater variety of abiotic conditions when the biotic resistance is low.

Rule 4: invading exotics tend to be more successful when native species do not occupy similar niches. This rule is related to Rules 1 and 2, above, and like them suggests that competition from native species with similar requirements plays a significant role in resisting invaders. Space does not permit a discussion of how to measure niches and whether unfilled niches exist. Suffice to say that many successful invaders exhibit tolerances for abiotic and biotic conditions and use habitats and resources differently from native species. Good examples are provided by the birds and mammals that have successfully colonized relatively undisturbed habitats in either North America or Eurasia from the other landmass. North America has no native ecological equivalents of the horse, donkey, or chukar partridge (although horses inhabited North America in the Pleistocene), and Europe has no truly amphibious rodents equivalent to the muskrat and coypu (which is native to South America, although it has also invaded the southeastern United States). Perhaps the best continental example is provided by comparison of the Australian bird and mammal faunas. Because of the superior ability of birds to colonize across water barriers, the native Australian avifauna is diverse and contains representatives of many orders and families; in contrast, native mammals consist only of monotremes, marsupials, rodents, and bats. The number of established introduced bird and mammal species is about the same (approximately 20). But, only two birds, the European blackbird (*Turdus merula*) and the mallard duck (*Anas platyrhynchos*) have spread into relatively undisturbed habitats (Fox and Adamson, 1980). This is in dramatic contrast to the mammals; at least 12 species (two lagomorphs, three carnivores, and at least seven ungulates) that have no ecological vicars in the native fauna have spread widely to colonize undisturbed habitats (Fox and Adamson, 1980; Strahan, 1983; Fox and Fox, 1986; Myers, 1986).

The same pattern is seen on oceanic islands. Because of their differential abilities to colonize over saltwater barriers, native terrestrial mammals and freshwater fishes are few or absent on isolated islands whereas land birds are much better represented. Introduced fish and mammals have in general been more successful than birds in colonizing such islands and especially in becoming established in undisturbed habitats.

Rule 5: species that inhabit disturbed environments and those with a history of close association with humans tend to be successful in invading man-modified habitats. This is really an important special case of Rule 4, above. There are many examples (see also Orians, 1986). Establishment of exotic fishes in the southwestern United States has been enormously facilitated by construction of reservoirs in a region where natural lakes were almost nonexistent. More than half of the 67 species of exotic fishes in Arizona are confined to artificial impoundments. More species of Eurasian vertebrates have become established in North America than vice versa. Most of the successful Old World invaders are very successful around human habitation in their native region, and they are largely restricted to urban, suburban, and agricultural ecosystems in the New World (e.g. black rat (*Rattus rattus*), brown rat, house mouse, house sparrow (*Passer domesticus*), starling (*Sturnus vulgaris*), rock dove (*Columbia livia*), ring-necked pheasant (*Phasianus colchicus*), and gray partridge (*Perdix perdix*)). If these species are discounted, then the number of exotics that have been able to colonize relatively undisturbed habitats on either landmass is small and much more symmetrically distributed with respect to continent of origin (e.g. Old World brown trout (*Salmo trutta*), horse (*Equus caballus*), donkey, and chukar partridge (*Alectoris chukar*) in North America, and North American largemouth bass, gray squirrel (*Sciurus carolinensis*), muskrat (*Ondatra zibethica*), and mink (*Mustela vison*) in Eurasia). Even on remote oceanic islands, many of the exotic vertebrates are associated with humans in their native regions and their successful colonization must be attributed in part to their ability to exploit disturbed habitats.

To some extent this may reflect the bias that commensal species are more likely to be accidentally or intentionally introduced. However, it is still the case that these exotics tend to thrive in disturbed environments and to be much less successful in invading relatively pristine habitats. For example, in both the Hawaiian Islands (Moulton and Pimm, 1986) and New Zealand (personal observation), exotic bird species have been very successful in invading urban, suburban, and agricultural habitats and much less successful in establishing themselves in native forest. In contrast, many native species persist in relatively undisturbed habitats, but most of these natives have not been able to survive in habitats that have been substantially modified by human activities.

5.2.2 Modes and extents of invasion

The modes of colonization of most of these exotic vertebrates are known. The vast majority have been introduced, either deliberately or accidentally, by humans. The first introductions occurred as a result of the colonization and subsequent movements of primitive man. Fossil records from isolated oceanic islands are beginning to provide information on the surprisingly large impact of early humans on the biota (e.g., Olson and James, 1982; Morgan and Woods, 1986). Early human colonists not only caused the extinctions of many native species, they also brought with them the first exotics. Among the well-documented introductions of vertebrates by primitive humans are the dingo (*Canis familiaris dingo*) in Australia, the Polynesian rat (*Rattus exulans*) on many Pacific islands, and at least seven species of lizards on Hawaii (Oliver and Shaw, 1953). Of course the 'discovery' of the rest of the world by Europeans and the subsequent development of worldwide travel and commerce led to the great wave of introductions that has occurred within the last few centuries. Vertebrates were transported to foreign islands and continents accidentally in ships, planes, and their cargos. For example, the commensal rats (*Rattus rattus* and *R. norvegicus*) and the well-named house mouse (*Mus musculus*) and house sparrow (*Passer domesticus*) accompanied Europeans as they colonized the world. Many vertebrates were deliberately introduced for a wide variety of reasons: food (e.g. rabbit (*Oryctolagus cunniculus*) in Australia, goat (*Capra hircus*) on Aldabra, and pig (*Sus scrofa*) on Hawaii), sport (e.g. red deer (*Cervus elaphus*) in New Zealand, chukar partridge in North America, and rainbow trout (*Salmo gairdneri*) in Chile and New Zealand), biological control of pests (e.g. fox (*Vulpes vulpes*) to control rabbits, and giant toad (*Bufo marinus*) to control cane beetles in Australia, mongoose (*Herpestes auropunctatus*) to control rats in the Hawaiian and Antillean Archipelagos), and aesthetics (e.g. songbirds in Hawaii, house sparrow and starling in North America, and grey squirrel in Europe). Still other species were imported originally as domestic animals, furbearers, or pets, but subsequently escaped from captivity (e.g. camel, donkey, horse, and goat in Australia, several species of tropical fish and parrots in southern Florida, mink and muskrat in Europe, and horse and donkey in western North America).

It is surprising, given the disastrous impacts of some introduced species, that exotic vertebrates continue to be imported at amazingly high rates. Government agencies are gradually becoming more cautious, but they continue to introduce deliberately species for fishing, hunting, and biological control. The pet trade, perhaps the single largest source of current introductions, continues to import large numbers of exotic species with only minimal regulation in most countries.

There is such a wide range of dispersal modes, that it becomes difficult and arbitrary to define an invading or exotic species. For example, in addition to those species that have been transported across major biogeographic barriers and introduced into new areas, there are many species whose dispersal has clearly been facilitated by human activities, but which have travelled under their own power to invade new regions. Some of these have taken advantage of man-made corridors to cross previously impenetrable barriers. For example, the predaceous sea lamprey (*Petromyzon marinus*) was able to invade the upper Great Lakes, where it decimated native populations of lake trout (*Salvelinus namaycush*) and whitefish (*Coregonus* spp.), after construction of the Welland Canal enabled it to bypass Niagara Falls (Smith, 1972; Moyle, 1986). Similarly, the Suez Canal has enabled at least 30 species of fish to colonize the eastern Mediterranean Sea from the Red Sea (Por, 1971; 1977). Changes in habitat and the availability of resources have permitted many native species to expand their ranges and invade new regions. For example, when the forests of northeastern North America were cleared for agriculture in the last century, several birds and mammals from prairie habitats shifted their ranges hundreds of kilometers eastward. These include the meadowlark (*Sturnella magna*), bobolink (*Dolichonyx oryzophorus*), coyote (*Canis latrans*), and prairie deermouse (*Peromyscus maniculatus bairdii*). Similarly, within the last century in both Europe and eastern North America several songbird species have extended their ranges far to the north in apparent response to man-modified habitats, increased food availability, and perhaps climatic change. Clearing of forest and introduction of domestic livestock appears to have enabled the cattle egret (*Bubulcus ibis*) to colonize South America from Africa (and then subsequently to invade North America) and Australia from southern Asia.

The extents of invasions are enormously variable. Some species, such as the commensal black and brown rat, and house mouse, have become almost cosmopolitan as they have followed Europeans to all corners of the globe. But by no means all widespread invaders are commensal or even associated with human disturbance. The rabbit, fox, dingo, and feral cat inhabit virtually the entire Australian continent, including many virtually pristine habitats. In the 80 years since its accidental release in Czechoslovakia, the muskrat has expanded to occupy as large a geographic range in Eurasia as in its native North America. In contrast, other species, such as chital and hog deer in Australia and the crested myna and skylark in North America, have become firmly established, but only within a very local region, and they do not appear to be spreading. Even among closely related, ecologically similar species the contrasts are striking. For example, in a little over a century since its introduction into the northeastern United States, the house sparrow has spread to occupy virtually the entire North American continent and to become one of the most abundant bird species (Robbins, 1973). On the other hand, its congener, the European tree sparrow (*Passer montanus*), remains confined to a few thousand square kilometers in the central United States, where it is well established but not abundant.

How do we account for this enormous variability? To what extent are founder events, genetic adaptations, life history traits, population dynamics, interactions with other species, and relationships with the physical environment responsible for the extents of invasion by different exotic species? At present these questions have been addressed for a few vertebrate species. Studies of the giant toad and rabbit (e.g. Myers, 1971; Cooke, 1977; van Beurden, 1981) in Australia were motivated by the practical importance of assessing the probable geographic spread, ecological and economic impacts, and means of biological control for these pests. These efforts were successful in understanding the factors that contributed to the success of these species, in predicting their eventual limits of distribution, and, at least in the case of the rabbit, in attaining a certain measure of control. Studies of the house sparrow in North America (e.g. Johnston and Sealander, 1973; Johnston and Klitz, 1977) and the house mouse in Europe and North America (e.g. Berry, 1978; 1986) have elucidated some of the genetic and evolutionary processes that have accompanied colonization. It should be noted that all of these studies were *ad hoc*, made after the exotics were already enormously successful. They are valuable, but they provide few gratifying generalizations about the factors that are responsible for initial establishment, limit the extent of invasions, and account for the great variation in success among species.

5.3 CAN INVASION BIOLOGY BE MADE A PREDICTIVE SCIENCE?

5.3.1 A thought experiment with Andean sparrows

The general rules developed above suggest that at one level, it is possible to characterize attributes of successful and unsuccessful invasions and thus to make certain kinds of qualitative predictions. For example, the Andean sparrow (*Zonotrichia capensis*), a native finch widespread in both relatively pristine and highly disturbed habitats in Central and South America, is more likely to be successfully introduced into New Zealand, where land bird diversity is low, than into a climatically similar region of eastern North America or Europe, where avian diversity is much higher. Further, this temperate and high-elevation tropical species would be more likely to become established on temperate Easter Island than on low, tropical Christmas Island. Also, it would be more likely to colonize eastern Europe, where there are few ecological equivalents of sparrows, than western North America, where there are several closely related, ecologically similar species. If the Andean sparrow became established in New Zealand, it should be more successful around human habitations than in undisturbed native forest. Finally, in all of these situations the Andean sparrow would be more likely to be a successful invader than the closely related golden-crowned sparrow (*Zonotrichia atricapilla*), which appears to have narrower requirements, occurs at lower densities in restricted habitats in western North America, and is seldom found in highly human-modified environments.

But all of the above are very general, qualitative predictions. It is another matter entirely to replace the 'more likely' in any of the above statements with an accurate quantitative estimate of the probability of success. The patterns and processes discussed in the previous section may provide the basis for predicting the relative susceptibility of different regions to invasion and the relative chances of success of different kinds of exotic species, but they provide little basis for predicting the probability of success of any particular species at any given site.

What would be required to make such a specific prediction—for example, to predict whether the Andean sparrow would become established on the South Island of New Zealand? The success of this introduction should depend on: the number of birds introduced; their sex ratio, age structure, health, parasite load, and genetic constitution; the time of year of their introduction; the geology, climate, vegetation, availability of appropriate foods, and the kinds of competitors, predators, and pathogens both at their place of origin and at the release site; and many other variables (see also Ehrlich, 1986). In short, what is required is a thorough, species-specific study of the population biology and niche of the Andean sparrow, and an equally careful assessment of the biotic and physical environment at the release site. Even more information would be desirable. If establishmant were likely, it would be important to know over what area and into what habitats the new exotic would spread, what population densities it would attain, and what impact it would have on native species and on humans. This is a tall order. It requires more information than is presently available for any vertebrate species.

5.3.2 History, uniqueness, and predictability

This thought experiment demonstrates the difficulty of making invasion biology a precise, predictive science. It exposes the limitations of our ecological and evolutionary knowledge, and the difficulties of applying our limited knowledge to practical problems. Most of the problems caused by exotics are species-specific, but by definition each species is unique. It has a unique history and a unique ecological niche—a special combination of tolerances, requirements, and relationships with other organisms—that reflects the constraints of its ancestry and the influences of its past environments. This historically based uniqueness characterizes all levels of biological organization, from cells and individuals, through communities and species, to the entire biosphere. Furthermore, the environments, as well as the organisms, are unique. Each continent, ocean, island, lake, stream, and local patch of habitat has a special combination of physical conditions and biotic composition that also reflects its unique history. This kind of uniqueness renders certain kinds of prediction impossible, or at least impractical. When each unit has its own special attributes, it is impossible to predict the behavior of an individual unit without knowing all of these attributes and their dynamical consequences.

Because of this historically based uniqueness of both exotic species and their environments, the study of biological invasions has been and will continue to be the study of both general trends and special cases. The trends hold because there are systematic differences among the units. For example, islands are smaller in size, isolated by barriers to dispersal, and have a narrower range of physical conditions and fewer species than continents; geographically widespread species interact with a wider variety of physical conditions and more other species than narrowly endemic forms. But the behavior of the individual units is always somewhat unpredictable, because there are important differences even between islands in the same archipelago or species in the same genus.

5.3.3 Applications to specific problems

Basic ecologists, evolutionary biologists, and biogeographers, in search of general trends and mechanisms, will often be content to sacrifice precision for generality. They will continue to use invading species as invaluable experiments to help them understand the structure and dynamics of communities and biotas. Applied biologists, with specific problems to solve, cannot be satisfied with imprecise generalities. They will continue to deal with invading species largely on a case-by-case basis. Just as when dealing with similar problems, such as biological control of pests and preservation of rare and endangered species, they will have to take account of the unique, historically based attributes of each species and environment.

There are two lessons in this. First, just as the goals of basic and applied scientists differ, so does the kind of science that is necessary to pursue these goals. This is no grounds for lack of dialogue and cooperation between basic and applied ecologists. However, in order for such interaction to be productive, each must recognize the reality of this spectrum between general trends and unique attributes, and try to bridge the gap. This spectrum is a phenomenon of natural systems, not a creation of obstinate scientists. Second, there are limits to what both basic and applied biologists can contribute to solving practical problems. That particular problems caused by an exotic species in a certain area must always be dealt with on a case-by-case basis does not necessarily reflect on the inadequacy of basic ecological knowledge or the failure to apply general concepts to specific situations. Instead it is a necessity imposed by the historically based uniqueness of both organisms and their environments. Because of the complexity and uniqueness of these systems, the knowledge necessary to manage them precisely is vast. The public and the politicians must be understanding, patient, and willing to pay the costs.

It is in part because of the uniqueness of species and places that there is little pattern in the management of exotic vertebrates. Perceived costs and benefits of exotics and strategies for their elimination or management vary on a case-by-case basis. Some countries, such as Australia and New Zealand, devote massive efforts to strict quarantines to keep out new invaders and to control or eliminate many species that have previously invaded. Other countries, including the United States, continue to import game fishes, birds, and mammals, transplant them to new localities, and manage them for increased populations. The success of management schemes has also been enormously variable; some alien vertebrates have been eliminated, whereas others remain abundant pests despite longstanding, costly efforts to control them. Although there are many well-studied case histories, I doubt that it is useful to make generalizations.

5.4 THE PLACE OF EXOTICS IN A DISTURBED BIOSPHERE

Unless one is a fisherman, hunter, or member of an acclimatization society, there is a tendency to view all exotic vertebrates as 'bad' and all native species as 'good.' For example, most birdwatchers, conservationists, and biologists in North America view house sparrows and starlings with disfavor, if not with outright loathing; they would like to see these alien birds eliminated from the continent if only this were practical. There is a kind of irrational xenophobia about invading animals and plants that resembles the inherent fear and intolerance of foreign races, cultures, and religions. I detect some of this attitude at this conference. Perhaps it is understandable, given the damage caused by some alien species and the often frustrating efforts to eliminate or control them.

This xenophobia needs to be replaced by a rational, scientifically justifiable view of the ecological roles of exotic species. In a world increasingly beset with destruction of its natural habitats and extinction of its native species, there is a place for the exotic. Two points are particularly relevant. First, increasing homogenization of the earth's biota is inevitable, given current trends in the human population and land use. Despite efforts to prevent them, biological invasions will continue, and some proportion of them will be successful. Deliberate and accidental introductions will continue to supply many of the immigrants, but species will also invade under their own power, moving into favorable habitats that have been altered by human activities. Geographically restricted native species with sensitive requirements will continue to have high extinction rates, and those widespread, broadly tolerant forms that can live with humans and benefit from their activities will spread and become increasingly dominant.

The second point is that exotic species will sometimes be among the few organisms capable of inhabiting the drastically disturbed landscapes that are increasingly covering the earth's surface. As humans devote an increasing share of the earth's physical resources and biological productivity to supporting their own population, and do so with low efficiency, the share of resources and productivity that remains to support wildlife must diminish (Vitousek *et al.*, 1986). Urban, suburban, and agricultural areas are ecosystems. They and the less modified grazing and timber lands are managed to support human populations and certain animal and plant species that directly benefit humans. If these ecosystems are managed appropriately, they can also support wildlife, including a surprising variety of vertebrates. Sometimes these can be native species, but sometimes the choice may be between exotics or virtually no wildlife at all. Because of their large body sizes, high energy requirements, and low population densities, vertebrates in general are susceptible to local and total extinction; endemic species with narrow requirements and restricted distributions are especially sensitive. Fortunately, other vertebrates can withstand and even benefit from human activities; unfortunately many of these are alien species and many others are natives that are widely distributed, common, and often viewed as 'pests.' The term pest seems to include any species that manages to obtain a significant share of the productivity of ecosystems that are managed for anthropocentric goals; pests consume agricultural products destined for human consumption, suppress domestic plant and animal populations through competition or predation, and compete with or prey upon aesthetically desirable native species.

The role that aliens may play is illustrated by the introduced vertebrates in the grounds of the East-West Center where this conference was held on the outskirts of Honolulu in the Hawaiian Islands. The park-like surroundings were beautifully landscaped with trees, flowers, and a cascading stream. I identified more than ten birds, a lizard, a frog, and several species of fishes, all introduced. On the one hand this might be considered a depressing example of just how 'disturbed' the lowlands of Oahu have become. On the other hand, it could be considered an encouraging example of the kind of diverse and aesthetically appealing vertebrate biota such a human-modified environment can support. One thing is certain: if all of the alien vertebrates were eliminated, few if any native species would take their places. If this provides a glimpse of what the future holds for many areas throughout the world, it is not an altogether pessimistic view.

This is not to say that we should halt our efforts to conserve undisturbed habitats and native species, including safeguards to prevent the invasion of harmful exotics. In most cases native species should be preferred over aliens, because natives will be less likely to have unanticipated disastrous effects, either within highly managed ecosystems or in less disturbed adjacent habitats. But I also suspect that increasingly even the 'natives' will be invaders, in the sense that they will be broadly tolerant, common species that have expanded their habitat and geographic ranges to exploit highly disturbed environments.

It has become imperative that ecologists, evolutionary biologists, and biogeographers recognize the inevitable consequences of human population growth and its environmental impact, and that we use our expertise as scientists not for a futile effort to hold back the clock and preserve some romantic idealized version of a pristine natural world, but for a rational attempt to understand the disturbed ecosystems that we have created and to manage them to support both humans and wildlife. I have no simple recipes for how to accomplish this. For the reasons given above, it will require the collaboration of basic scientists working to understand general patterns and processes and applied scientists willing to deal with the unique attributes of particular species and places. It will be a difficult, imprecise, error-plagued effort, and if we scientists, the public, or the decision-makers expect too much too soon, all are bound to be disappointed. It will require that we remain open-minded and flexible about, among other things, the ecological roles of invading species.

ACKNOWLEDGEMENTS

The manuscript benefited from the helpful comments of P. R. Ehrlich, B. J. Fox, M. Fox, F. J. Kruger, and I. A. W. MacDonald.

REFERENCES

Atkinson, I. A. E. (1977). A reassessment of the factors, particularly *Rattus rattus* L., that influenced the decline of endemic forest birds in the Hawaiian Islands. *Pacific Sci.*, **31**, 109–33.

Atkinson, I. A. E. (1985). The spread of *Rattus* to oceanic islands and their effects on island avifaunas. In: Moors, P. J. (Ed.), *Conservation of Island Birds.* International Council for Bird Preservation, Technical Publication **3**, pp. 35–81.

Berry, R. J. (1978). Genetic variation in wild house mice: where natural selection and history meet. *Amer. Sci.*, **66**, 52–60.

Berry, R. J. (1986). Genetics of insular populations of mammals, with particular reference to differentiation and founder effects in British small mammals. *Biol. J. Linn. Soc.*, **28**, 205–30.

Brown, J. H. (1986). Two decades of interaction between the MacArthur–Wilson model and the complexities of mammalian distributions. *Biol. J. Linn. Soc.*, **28**, 231–51.

Brown, J. H., and Gibson, A. C. (1983). *Biogeography.* Mosby, St Louis. 643 pp.

Coblentz, B. E. (1978). The effects of feral goats (*Capra hircus*) on island ecosystems. *Biol. Conserv.*, **13**, 279–86.

Cooke, B. D. (1977). Factors limiting the distribution of the wild rabbit in Australia. *Proc. Ecol. Soc. Aust.*, **10**, 113–20.

de Vos, A., Manville, R. H., and van Gelder, R. G. (1956). Introduced mammals and their influence on native biota. *Zoologica*, **41**, 163–94.

Diamond, J., and Case, T. J. (1986). Overview: introductions, extinctions, exterminations, and invasions. In: Diamond, J., and Case, T. J. (Eds), *Community Ecology*, pp. 65–79. Harper and Row, New York.

Diamond, J. M., and Veitch, C. R. (1981). Extinctions and introductions in the New Zealand avifauna: cause and effect? *Science*, **211**, 499–501.

Ehrlich, P. R. (1986). Which animal will invade? In: Mooney, G. A., and Drake, J. A. (Eds), *Ecology of Biological Invasions of North America and Hawaii*, Ecological Studies 58, pp. 79–95. Springer-Verlag, New York.

Elton, C. S. (1958). *The Ecology of Invasions by Animals and Plants.* Methuen, London. 181 pp.

Falla, R. A., Sibson, R. B., and Turbott, E. G. (1966). *A Field Guide to the Birds of New Zealand.* Collins, London. 254 pp.

Fox, B. J., and Fox, M. D. (1986). The susceptibility of natural communities to invasion. In: Groves, R. H., and Burdon, J. J. (Eds), *Ecology of Biological Invasions*, pp. 57–66, Cambridge University Press, Cambridge.

Fox, M. D., and Adamson, D. (1980). The ecology of invasions. In: Recher, H. F. *et al.* (Eds), *A Natural Legacy: Ecology in Australia*, pp. 135–51. Pergamon Press, New York.

Gibb, J. A., and Flux, J. E. C. (1973). The mammals. In: Williams, G. R. (Ed.), *The Natural History of New Zealand*, pp. 334–71. A. H. and A. W. Reed, Wellington.

Holdgate, M. W., and Nace, N. W. (1961). The influence of man on the floras and faunas of the southern islands. *Polar Record*, **10**, 475–93.

Johnston, R. F., and Klitz, W. J. (1977). Variation and evolution in a granivorous bird: the house sparrow. In: Pinowski, J., and Kendeigh, S. C. (Eds.), *Granivorous Birds in Ecosystems*, International Biol. Prog. 12, pp. 15–51. Cambridge University Press, Cambridge.

Johnston, R. F., and Selander, R. K. (1973). Variation, adaptation, and evolution in the North American house sparrows. In: Kendeigh, S. C., and Pinowski, J. (Eds), *Productivity, Population Dynamics and Systematics of Granivorous Birds*, pp. 301–26. Polish Scientific Publishers, Warsaw.

King, C. (1984). *Immigrant Killers: Introduced Predators and the Conservation of Birds in New Zealand.* Oxford University Press, Auckland.

King, W. B. (1980). Ecological basis of extinction in birds. In: *Proceedings of the 17th International Congress of Ornithology*, pp. 905–11.

Kramer, R. J. (1971). *Hawaiian Land Mammals.* Charles E. Tuttle, Rutland, Vermont. 347 pp.

Long, J. L. (1981). *Introduced Birds of the World.* David and Charles, London. 528 pp.

Loope, L. L., and Snowcroft, P. G. (1985). Vegetation response within exclosures in Hawaii: a review. In: Stone, C. P., and Scott, J. M. (Eds), *Hawaii's Terrestrial Ecosystems: Preservation and Management*, pp. 377–402. Coop. Natl. Park Resources Studies Unit, University of Hawaii, Honolulu, HA.

MacArthur, R. H., and Wilson, E. O. (1967). *The Theory of Island Biogeography.* Princeton University Press, Princeton, NJ. 269 pp.

McDowall, R. M. (1978). *New Zealand Freshwater Fishes.* Heinemann, Auckland, New Zealand. 230 pp.

McKeown, S. (1978). *Hawaiian Reptiles and Amphibians.* Oriental, Honolulu, Hawaii. 80 pp.

Maciolek, J. A. (1984). Exotic fishes in Hawaii and other islands of Cceanea. In: Courtenay, W. R., and Stauffer, J. R. (Eds), *Distribution, Biology and Management of Exotic Fishes*, pp. 131–61. Johns Hopkins University Press, Baltimore, MD.

Miller, R. R. (1972). Threatened freshwater fishes of the United States. *Trans. Amer. Fish. Soc.*, **101**, 239–52.

Minkley, W. L. (1973). *Fishes of Arizona*, Arizona Game and Fish Department, Phoenix, Arizona. 293 pp.

Morgan, G. S., and Woods, C. A. (1986). Extinction and the zoogeography of the West Indian land mammals. *Biol. J. Linn. Soc.*, **28**, 167–203.

Moulton, M. P., and Pimm, S. L. (1986). The extent of competition in shaping an introduced avifauna. In: Diamond, J., and Case, T. J. (Eds), *Community Ecology*, pp. 80–97. Harper and Row, New York.

Moulton, M. P., and Pimm, S. L. (1987). Morphological assortment in introduced Hawaiian passerines. *Evol. Ecol.*, **1**, 113–24.

Moyle, P. B. (1976). Fish introductions into California: history and impact on native fishes. *Biol. Conserv.*, **9**, 101–18.

Moyle, P. B. (1986). Fish introductions into North America: patterns and ecological impact. In: Mooney, H. A., and Drake, J. A. (Eds.), *Ecology of Biological Invasions of North America and Hawaii.* Ecological Studies 58, pp. 27–43. Springer-Verlag, New York.

Myers, K. (1971). The rabbit in Australia. In: *Proc. Adv. Study Inst. Dynamics Numbers Populations*, pp. 478–506. PUDOC, Wageningen, the Netherlands.

Myers, K. (1986). Introduced vertebrates in Australia, with emphasis on the mammals. In: Groves, R. H., and Burdon, J. J. (Eds), *Ecology of Biological Invasions*, pp. 120–36, Cambridge University Press, Cambridge.

National Geographic Society (1983). *Field Guide to the Birds of North America.* National Geographic Society, Washington. 464 pp.

Nitecki, M. H. (Ed.) (1984). *Extinctions.* University of Chicago Press, Chicago.

Oliver, J. A., and Shaw, C. E. (1953). Amphibians and reptiles of the Hawaiian Islands. *Zoologica*, **38**, 65–95.

Olson, S. L., and James, H. F. (1982). Fossil birds from the Hawaiian Islands: evidence for wholesale extinction by man before western contact. *Science*, **217**, 633–5.

Orians, G. H. (1986). Site characteristics favoring invasions. In: Mooney, H. A., and Drake, J. A. (Eds), *Ecology of Biological Invasions of North America and Hawaii.* Ecological Studies 58, pp. 133–62. Springer Verlag, New York.

Pizzey, G. (1980). *A Field Guide to the Birds of Australia*, Collins, Sydney. 460 pp.

Por, F. D. (1971). One hundred years of Suez Canal: a century of Lessepsian migration: retrospect and view points. *Syst. Zool.* **20**, 138–59.

Por, F. D. (1977). *Lessepsian Migration. The Influx of Red Sea Biota into the Mediterranean by way of the Suez Canal.* Ecology Studies 23, Springer-Verlag, Berlin.

Ricklefs, R. E., and Cox, G. W. (1972). Taxon cycles in the West Indian avifauna. *Amer. Nat.*, **106**, 195–219.

Robbins, C. S. (1973). Introduction, spread, and present abundance of the Houseparrow in North America. *Ornith. Monogr.*, **14**, 3–9.

Schoener, T. W. (1974). The species-area relationship within archipelagoes: models and evidence from island birds. *Proc. 16th Intl Ornith. Congr.*, pp. 629–42.

Smith, C. W. (1985). Impact of alien plants on Hawaii's native biota. In: Stone, C. P., and Scott, J. M. (Eds), *Hawaii's Terrestrial Ecosystems: Preservation and Management*, pp. 180–250. Coop. Natl Park Resources Studies Unit, University of Hawaii, Honolulu, HA.

Smith, S. H. (1972). Factors of ecologic succession in oligotrophic fish communities of the Laurentian Great Lakes. *J. Fish. Res. Bd. Canada*, **29**, 717–30.

Strahan, R. (Ed.) (1983). *The Australian Museum Complete Book of Australian Mammals.* Angus and Robertson, London. 530 pp.

Thornton, I. (1971). *Darwin's Islands: A Natural History of the Galapagos.* Natural History Press, Garden City, NJ.

van Beurden, E. (1981). Bioclimatic limits to the spread of *Bufo marinus* in Australia: a baseline. *Proc. Ecol. Soc. Australia*, **11**, 143–9.

van Riper, S. G., and van Riper, C. (1982). *A Field Guide to the Mammals in Hawaii.* Oriental Press, Honolulu, HA.

van Riper, C., van Riper, S. G., Goff, M. L., and Laird, M. (1986). The epizootiology and ecological significance of malaria in Hawaiian landbirds. *Ecol. Monogr.*, **56**, 327–44.

Vitousek, P. M., Ehrlich, P. R., Ehrlich, A. H., and Matson, P. A. (1986). Human appropriation of the products of photosynthesis. *BioScience*, **36**, 368–73.

Williamson, M. (1981). *Island Populations.* Oxford University Press, Oxford. 286 pp.

Wilson, E. O. (1961). The nature of the taxon cycle in the Melanesian ant fauna. *Evolution*, **13**, 122–44.

Wilson, L. D., and Porras, L. (1983). The impact of man on the South Florida herpetofauna. *Univ. Kansas Mus. Nat. Hist. Spec. Publ.*, **9**, 1–89.

Zaret, T. M., and Paine, R. T. (1973). Species introduction in a tropical lake. *Science*, **182**, 449–55.

Biological Invasions: a Global Perspective
Edited by J. A. Drake *et al.*
© 1989 SCOPE. Published by John Wiley & Sons Ltd

CHAPTER 6

Aquatic Plants: Patterns and Modes of Invasion, Attributes of Invading Species and Assessment of Control Programmes

PETER J. ASHTON and DAVID S. MITCHELL

6.1 INTRODUCTION

The natural human inclination to concentrate attention on the strange and the unusual is reflected vividly in the voluminous literature devoted to those aquatic plants that possess anomalously disjunct or cosmopolitan geographical distributions (Sculthorpe, 1967). Since rivers, lakes and other water bodies are separated by tracts of land, it could be expected that aquatic plants would tend to be locally distributed within a particular land mass and that the seas and oceans would provide insurmountable barriers to dispersion between continents. Indeed, approximately 25 to 30% of the known aquatic plants are considered to be true endemics with distributions limited to a single river or portion of a river system (Sculthorpe, 1967; Cook, 1974). In contrast, however, several families and species of aquatic plants are so widely distributed over several continents, some even reaching remote oceanic islands, that they can be described justifiably as cosmopolitan. In addition, some temperate and tropical aquatic plants exhibit curious discontinuities in their ranges, over and above that due to the configuration of seas and land masses (Sculthorpe, 1967). While these continue to attract speculation and discussion, in most cases it is difficult, if not impossible, to show that the discontinuity is genuine and that the plant has not in fact been introduced in historical times to certain parts of its present-day range (Sculthorpe, 1967; Mitchell, 1974).

As a result of the mechanisms of dispersal, invasion, colonization and competition on the one hand, and changing climatic, physical and edaphic factors of the environment on the other, migrations of species continue to occur now as they occurred during geological and historical time. Aquatic floras everywhere are thus in a continual state of flux. The situation has been confounded further by man's relentless expansion of his agriculture, communications, industry and domestic life wherein he has modified existing habitats and created new, artificial habitats. Additional complications have been introduced by man's propensity for employing aquatic systems for the disposal of his waste products. Human activities that have led to the alteration, disturbance and degradation of aquatic systems have thus served to extend the range of habitats available for colonization by aquatic plants and to accelerate the natural flux of species between habitats.

In this review, we examine the processes of invasion, establishment and dispersal in aquatic plants and address three main questions:

1. What are the properties of an aquatic plant species that enable its individuals to disperse and invade new environments and what features determine their successful establishment?
2. What environmental features render a habitat vulnerable to invasion?
3. What are the most appropriate management strategies for the control of invasive aquatic plants?

In our discussion we have confined our analysis to freshwater species, omitting mention of the large and diverse brackish water and marine flora. Similarly, only vascular plants that habitually grow in flowing or standing water have been included in this review. Phreatophytes (plants which grow in high water tables, such as along streams and river banks) will be referred to where appropriate but will not be considered in detail.

6.2 THE INVASION PROCESS

The processes of plant invasion have attracted considerable attention worldwide (e.g. Elton, 1958; Connel and Slatyer, 1977; Grime,. 1979, 1985; Grubb, 1985; Groves and Burdon, 1986; Macdonald *et al.*, 1986; Mooney and Drake, 1986) and are often used to explain features of the diversity and succession of plant assemblages. However, remarkably few theoretical considerations of the invasion process seem to be generally applicable (Harper, 1977; Johnstone, 1986). Part of this problem stems from imprecise definitions of the relationship between invasion and succession. Here, we follow the views expressed by Johnstone (1986) and Breen *et al.* (1987) that the prime cause of invasion can be seen as the removal or overcoming of a barrier that has previously excluded a plant species from an area. Whilst these features can be classified on the basis of time, time is not the *cause* of biological change (Huxley, 1932); rather biological phenomena are described by their dispersion in time and space (Johnstone, 1986). In this context, therefore, plant succession is seen to be caused by sequential, but interlinked, episodes of invasion, establishment, maintenance and decline. Progressive, retrogressive and cyclical successional patterns thus reflect different modes of system instability (Johnstone, 1986).

6.2.1 Natural invasions

Natural invasions are invasions that have taken place in the absence of anthropogenic influences. They occur when an intervening barrier is removed or through the development of biotic or abiotic transporting mechanisms able to overcome the barrier in question. This leads to the suggestion that barriers and 'transporting mechanisms' oppose one another and provide powerful forces in the direction and timing of successional episodes (Breen *et al.*, 1987).

Examination of the rather scanty fossil record indicates that aquatic plant invasions have occurred naturally for several millennia (Sculthorpe, 1967). Indeed, the Pleistocene fossil record of the heterosporous aquatic fern *Azolla filiculoides* in western Europe suggests that successive waves of invasion occurred during each inter-glacial period, each new invasion restoring communities that had been eliminated during the intervening glacial periods (Moore, 1969).

The process of invasion can be seen as a series of discrete steps; these are summarized in Figure 6.1 and will be discussed in detail later. For an invasion to succeed, each step in the progression from first arrival of the invader to successful colonization of new habitats requires that a series of biotic and abiotic barriers be overcome by the invader. At each stage of the process, both chance and timing play a vital and interactive role in determining, the degree of success attained by the invader (Crawley, 1986; Gray *et al.*, 1987).

The geographical distance between the source of the invading plant and the eventual site of invasion represents perhaps the greatest barrier to natural invasions of aquatic plants. By virtue of their requirement for perennial supplies of freshwater (Sculthorpe, 1967), the adult stages of most freshwater aquatic plants cannot survive periods of transport in the absence of water or even prolonged immersion in seawater (Haller *et al.*, 1974). Thus, the production of

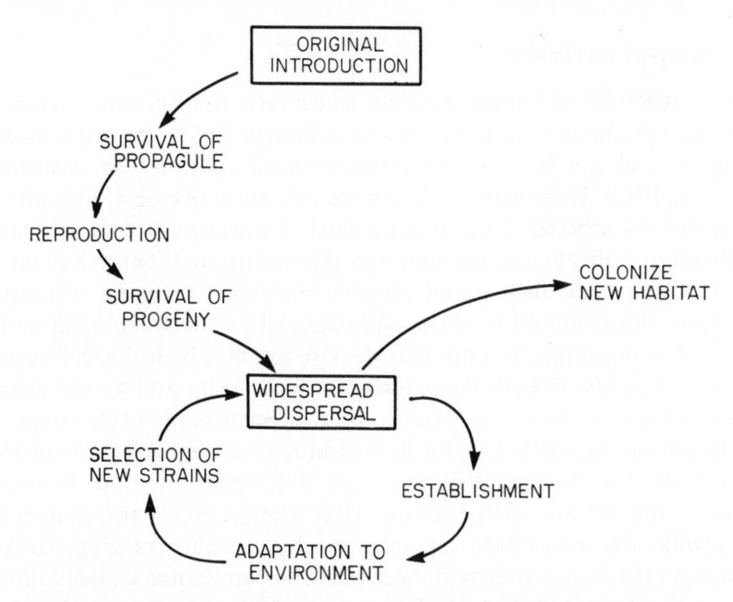

Figure 6.1. Stages in the invasion and establishment of aquatic plants

resistant seeds and propagules by many aquatic plants provides the only viable means whereby they may survive transport by biotic or abiotic mechanisms. In this regard, the presence of a small population of *Azolla filiculoides* on Inaccessible Island (37° 19′S; 12° 44′W) in the Tristan da Cunha group, located on the Mid-Atlantic Ridge, is of great interest (Ashton and Walmsley, 1984; Figure 6.2). As its name implies, the island's isolation makes it highly unlikely that this American fern was deliberately introduced by man and it has been suggested that the fern is a native species (Christensen, 1940; Wace and Dickson, 1965). However, a more likely explanation is that the fern's presence on the island might be due to the transport of spores by winds or waterfowl (Ashton and Walmsley, 1984). This possibility is supported by the several recorded instances where storm-blown American species of waterfowl have arrived at these islands and the east coast of southern Africa (Holdgate, 1965).

Natural invasions can be considered as sporadic events where the likelihood of a successful invasion is determined by interactions between chance and timing (Crawley, 1986). Against this background, man's propensity for modifying his environment and introducing plants into areas where they are not native has largely overcome the barriers to invasion that were previously imposed by geographical isolation. In turn, this has markedly increased the probability that the invasion will be successful. This is clearly evident in the dramatic increase in both the numbers and extent of aquatic plant invasions that have taken place during the 19th, and especially the 20th centuries as a direct consequence of man's activities (Sculthorpe, 1967; Mitchell, 1974; Holm *et al.*, 1977).

6.2.2 Transport mechanisms

Clearly, disruptive forces or mechanisms, which serve to inactivate, overcome or remove barriers, function in a manner that is diametrically opposed to isolating mechanisms and can therefore be conceptualized as 'transport mechanisms' (Breen *et al.*, 1987). These features determine the degree of ease or difficulty with which invasions proceed. Since hydrological characteristics are the principal determinants of all aquatic communities (Gosselink and Turner, 1978), it is logical to expect that they would also be implicated as major transporting mechanisms; this is indeed the case, especially within individual river systems. However, it is important to note that the efficacy of a hydrological transport mechanism depends on both the magnitude of the event and on the temporal scale over which the event occurs or is repeated (Breen *et al.*, 1987). Irregular or episodic high energy events (i.e. of large magnitude, such as extreme floods) bring about catastrophic changes that are quite different from those induced by seasonal or regular low energy events. High energy events therefore remove barriers while low energy events merely breach these barriers temporarily.

Wind may also act as a transport mechanism for some aquatic plants although the distance covered is likely to be short because of the relatively high seed mass

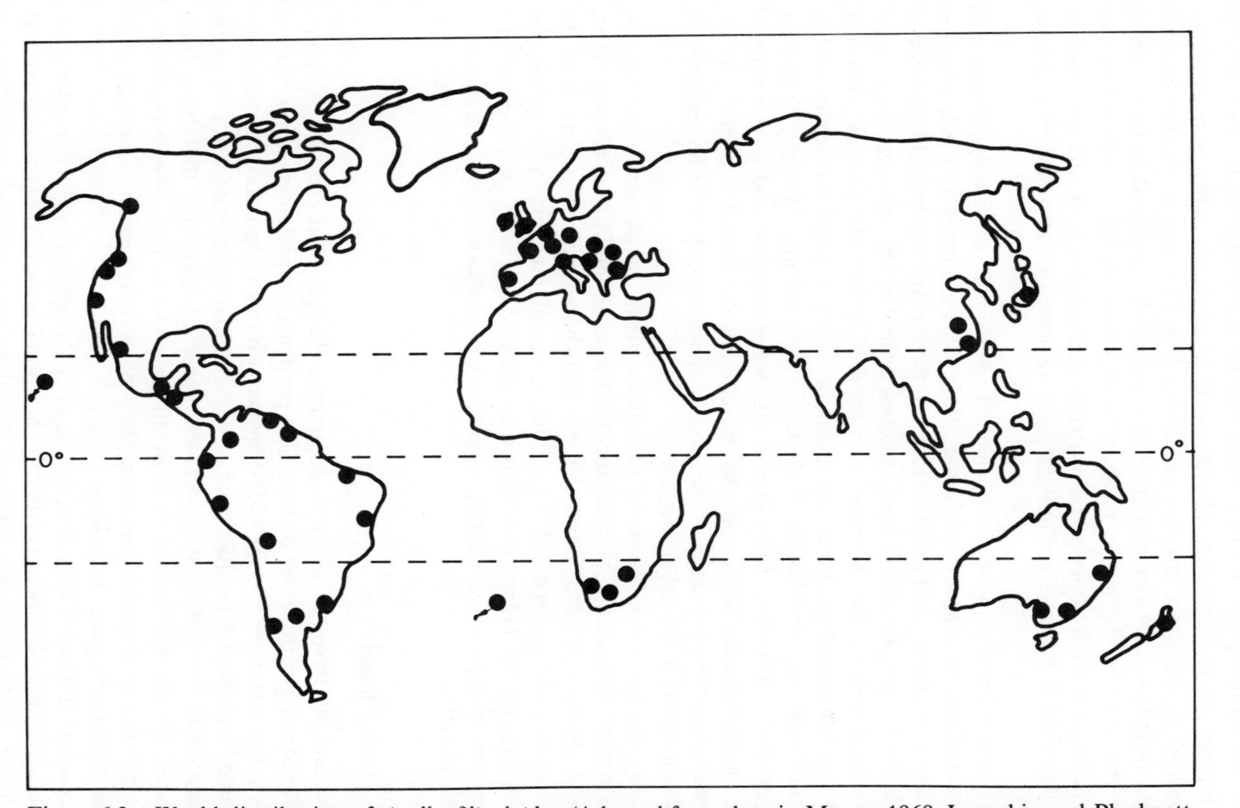

Figure 6.2. World distribution of *Azolla filiculoides*. (Adapted from data in Moore, 1969; Lumpkin and Plucknett, 1980; Ashton and Walmsley, 1984.)

and absence of structures aiding aerial transport (Sculthorpe, 1967). In addition, the likelihood that seeds will be blown onto dry terrestrial soils to desiccate and die, further limits this mode of transport. In contrast, many phreatophytes colonizing the banks of streams and rivers (e.g. *Phragmites, Typha*) produce numerous light seeds which rely on wind transport for dispersal. In these species wind transport is the major mechanism ensuring medium- and long-range dispersal to new environments and probably accounts for their wide distribution over most of the temperate and tropical regions of the world (Sculthorpe, 1967).

Animals are unquestionably the principal agents in the short range dispersal of all forms of aquatic plants (Sculthorpe, 1967). Over a century ago, in his treatise *The Origin of Species*, Darwin surmised that the remarkably wide distribution of certain freshwater plants had been achieved mainly by the carriage of seeds and vegetative fragments on the plumage and muddy feet of waterfowl (Darwin, 1872). Understandably, direct observations of such dispersal are extremely difficult to obtain (Sculthorpe, 1967) though Agami and Waisel (1986) have recently shown waterfowl to be important for the dispersal and germination of *Najas maritima* seeds. While considerable circumstantial evidence for both exo- and endozoic transport exists, these mechanisms are likely to be less effective in inter-continental dispersal (Löve, 1963).

Finally, the accidental dispersal of aquatic plants by humans must also be borne in mind. Of particular importance are the seeds and fruits of aquatic plants growing amongst irrigated crops that are distributed with the products of the harvested crop (Sculthorpe, 1967). In addition, casual transport by boats, vehicles and agricultural machinery, as well as the escape of seeds or vegetative material from ornamental ponds and botanical gardens, have also increased the spread of aquatic plants.

6.2.3 Human travel

Man's activities have brought about immense changes in global biogeography, greatly reducing continental isolation. In particular, rapidly escalating rates of inter-continental travel and the wide dispersal of human populations have largely eliminated continental isolation and have facilitated the translocation of many species. While the historical aspects of biological invasions that have been facilitated by man are dealt with in detail elsewhere in this volume (di Castri, Chapter 1), it is appropriate that we emphasize the importance of human travel in promoting aquatic plant invasions.

Since approximately 1500 AD, sporadic early European voyages of exploration and discovery were followed rapidly by the colonization of other territories and subsequent trade exchanges around the globe as road, rail and shipping transportation systems improved. Twentieth century developments in air transportation have also boosted considerably the ease, speed and scope of international travel. Coupled to this trend, man's fascination with strange and exotic organisms and the desire to develop new markets whilst surrounding himself with familiar objects in a new environment have motivated most of the deliberate introductions of aquatic plants to new territories. Typically, ports formed the major foci of human activities and associated plant introductions in these countries (Macdonald *et al.*, 1986). The process of deliberate introduction, followed by escape and subsequent spread, has been repeated in many parts of the world and species such as *Eichhornia crassipes* (Figure 6.3) and *Salvinia molesta* (Figure 6.4) are now widely distributed through the tropical and sub-tropical regions of the globe.

6.2.4 Invasion stages

Extensions to the geographic range of a plant species can only be brought about by the dispersal of breeding populations. When introduced to an area previously unoccupied by its own kind, individuals of a particular species have no chance of becoming established unless the initial immigrants form part of a 'propagule' (Figure 6.1), i.e. the minimum number of individuals able to found a reproducing population under favourable conditions (MacArthur and Wilson, 1967). Despite the fact that several aquatic plants possess some form of dormant stage in their life cycles, adapting them for dispersal, we have seen that few aquatic plants are dispersed between distant unconnected water bodies by natural mechanisms. Indeed, most initial introductions of aquatic plants to new continents have been deliberate in that the introduced species was perceived to have some special attraction and/or intended use for humans (Cook, 1985). A list of the best known invasive aquatic plants and their distribution is given in Table 6.1. In every case, man has been implicated in their deliberate or accidental introduction to continents outside their native range.

As is the case with terrestrial plants, we can recognize three main stages in a successful aquatic plant invasion, namely: the arrival of an individual or propagule, establishment of the population through reproduction, and dispersal to new localities (Figure 6.1). Subsequent adaptation of the invader to the new environment may also favour the selection of new genetic strains.

Each stage involves a series of interactions between the physical, chemical and biological features of the environment and the biological characteristics of the invader. These interactions can either promote, delay or prevent successful completion of the stage (Arthington and Mitchell, 1986). The whole invasion process is then repeated when a propagule is successfully dispersed to a new locality. Clearly, the rate at which the invasion occurs will depend on the number of individuals or propagules present in the initial inoculum and in the dispersal stage. An exponential increase in these will ensure an exponential acceleration in the apparent rate at which the invasion takes place until environmental factors become limiting (Arthington and Mitchell, 1986).

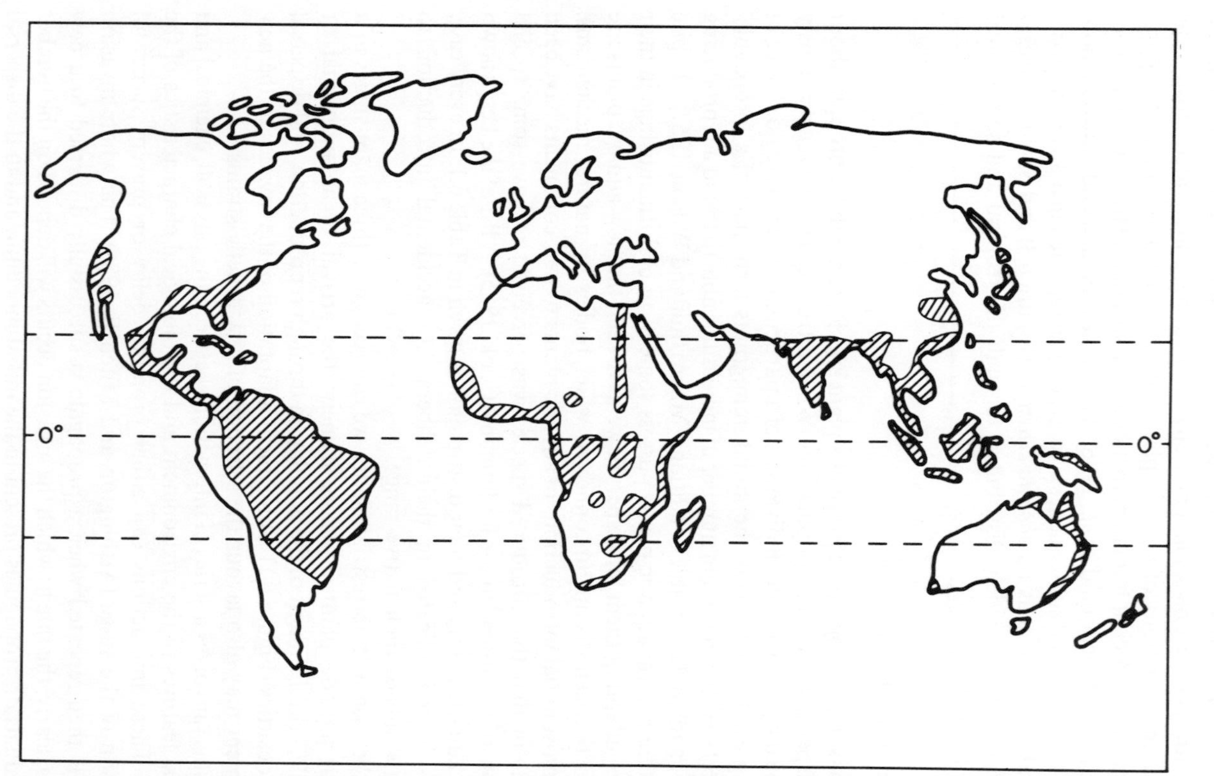

Figure 6.3. World distribution of *Eichhornia crassipes*. (Adapted from data in Sculthorpe, 1967; Robson, 1976; Holm *et al.*, 1977; Jacot Guillarmod, 1979; Gopal and Sharma, 1981; Ashton *et al.*, 1986; Gopal, 1987.)

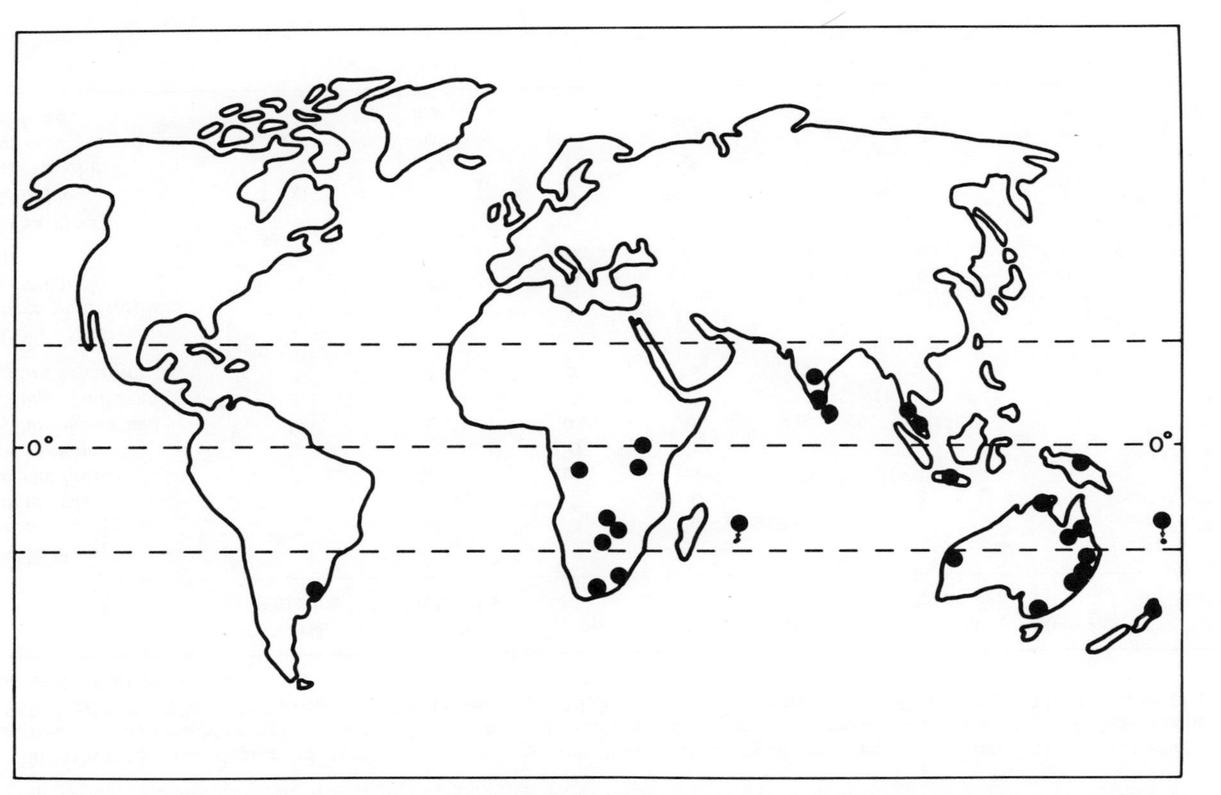

Figure 6.4. World distribution of *Salvinia molesta*. (Adapted from data in Mitchell, 1978, 1979a; Robson, 1976; Harley and Mitchell, 1981; Forno, 1983; Thomas and Room, 1986.)

Table 6.1. Distribution and degree of problem caused by the major submerged, emergent and free-floating invasive plants; x = indigenous, o = has caused minor problems, * = has caused major problems. (Data extracted from Mason, 1960; Sculthorpe, 1967; Moore, 1969; Mitchell, 1974, 1978; Robson, 1976; Holm *et al.*, 1977; Gopal and Sharma, 1981; Cook, 1985; Mitchell and Orr, 1985; Arthington and Mitchell, 1986)

Species	North America	South America	Europe	Africa	India	Southeast Asia	Australia	New Zealand
SUBMERGED								
Egeria densa	*	x	o	o			o	*
Elodea canadensis	x		*	o			*	*
Hydrilla verticillata	*	ox	ox	x	*x	ox	ox	o
Lagarosiphon major			o	x				*
Ceratophyllum demersum	*x	ox	ox	ox	ox	ox	ox	*x
Myriophyllum aquaticum	*	x	*	*		o	o	*
Myriophyllum spicatum	*		x	x	x	ox		
EMERGENT								
Alternanthera philoxeroides	*	x			o	*	*	*
Phragmites australis	*x	*x	*x	ox	ox	ox	ox	ox
Typha spp.	*x	ox	ox	ox	*x	ox	*x	ox
FREE FLOATING								
Azolla filiculoides	x	x	*	*			ox	x
Eichhornia crassipes	*	*x	*	*	*	*	*	o
Pistia stratiotes	ox	ox		*x	*x	*x	ox	o
Salvinia molesta		x		*	*	*	*	*

6.2.5 Why are some invasions successful while others are not?

Several aquatic plant species have provided spectacular examples of successful invasions when introduced into habitats with which they are not in ecological equilibrium (Sculthorpe, 1967; Mitchell, 1974). In many cases these population growths have been so rapid that the phenomena have been described as biological explosions, with the invading species completely dominating the available habitat at the expence of indigenous species (Arthington and Mitchell, 1986; Ashton *et al.*, 1986). Those aquatic plants that have spread adventively following their introduction to different parts of the world have almost invariably caused problems as weeds (Table 6.1). As a result, these species have attracted considerable attention and therefore provide the best documented cases of aquatic plant invasions. However, while the literature contains several descriptive accounts of the invasive spread of aquatic plants and short-term investigations of suitable control methods, there are few data quantifying the factors responsible for successful invasions (Mitchell, 1974). Still fewer data are available on the causes of unsuccessful invasions.

Several common features can be identified amongst those invasive aquatic plants which have demonstrated a marked capacity for adventive spread when introduced into new environments. Significantly, all of these factors relate to dispersal and reproductive mechanisms:

1. Vegetative reproduction is usually the commonest, and often the only method of reproduction.
2. Human activities are the main mechanism whereby the plants are spread both between and within continents.
3. Plants capable of very rapid rates of reproduction often become serious weeds.
4. Sexually sterile plants become locally naturalized unless they are purposely spread by humans or possess small vegetative propagules that can be widely spread by other means.

Success in the initial stages of aquatic plant invasions is dependent on reproductive capacity (Figure 6.1), provided that there is a fundamental bioclimatic match between the non-native and native environments. Opportunist R- or r- strategists (MacArthur and Wilson, 1967; Grime, 1977, 1979, 1985) with high reproductive rates are well represented among both free-floating (e.g. *Eichhornia crassipes* and *Salvinia molesta*) and submerged (e.g. *Elodea canadensis* and *Hydrilla verticillata*) invasive aquatic plants. Those plants that are capable of prolific vegetative reproduction possess a special advantage since one viable propagule (a single plant) is sufficient to start a new colony. Other attributes of the invading plant, such as the potential to reach a large population size, flexible habitat requirements and the ability to tolerate the stresses of environmental fluctuations and extremes, become more important during subsequent stages in the invasion (Arthington and Mitchell, 1986; Ashton *et al.*, 1986). Clearly, these attributes, in combination, constitute a biological strategy which predisposes a species for success as an invader However, similar combinations of attributes are not likely to be equally important in all circumstances.

Pre-adaptation to the new environment, perhaps through similarity with its native habitat, is clearly an advantage to an invader. However, such pre-adaptation only *increases* the likelihood of successful invasion; it does not *guarantee* success or decrease the possibility that the invading plant population may subsequently decline. Indeed, the *Elodea canadensis* invasions of Britain, western Europe and Australia, and the *Myriophyllum spicatum* invasions in the eastern United States, have all followed a wave pattern, with a period of rapid increase and stabilization followed by a catastrophic decline (Carpenter, 1980). In each of these cases, the mechanisms causing the population decline would seem to be a complex of interacting factors.

Aquatic ecosystems that have been disturbed by human activities appear to be particularly vulnerable to plant invasions (Cook, 1985). Nevertheless, introduced aquatic plants have also displayed considerable success in invading new ecosystems where there was no apparent disturbance but where suitable habitats and other necessary resources already existed. The *Salvinia molesta* invasion of the flood plain lakes along the Sepik River in Papua New Guinea provides an excellent example of this phenomenon (Mitchell, 1979a; Mitchell *et al.*, 1980; Thomas and Room, 1986; Arthington and Mitchell, 1986). Under these circumstances, successful invasive plants are usually R- or C-strategists (Grime, 1977, 1979) with the additional competitive advantage of freedom from the pests and diseases characteristic of their native range. They are therefore likely to experience considerably lower grazing and predator pressure and thus possess a significant competitive advantage over native species. The possibility that successful invaders may also produce allelopathic substances (Ostrofsky and Zettler, 1986), is an intriguing possibility that is, as yet, untested.

The third essential stage of every successful aquatic plant invasion is the dispersal of the organisms into suitable habitats (Figure 6.1). In the absence of this stage, the introduced organism is properly considered an 'escape' since it has not become truly naturalized (Arthington and Mitchell, 1986). This is clearly illustrated in the case of *Cyperus papyrus*, an emergent species native to Africa and western Asia, that has become established in a number of countries, including Australia. Despite its capacity for vigorous growth and relatively wide latitudinal spread in its native environment, it has shown no evidence of invasive behaviour where it has been introduced. Superficial examination indicates that the naturalized populations do not set seeds; thus the absence of effective dispersal of viable propagules is the most likely explanation for its failure to spread widely and rapidly (Arthington and Mitchell, 1986). Similarly, two other genera of widespread emergent aquatic plants, *Phragmites* and *Typha*, cause serious problems in their native distribution range as a result of rapid spread due to seed dispersal. When established in entirely new areas, however, these plants produce little seed and seldom pose serious problems unless the habitat becomes eutrophied (Aston 1973).

6.2.6 Why are some environments vulnerable while others are not?

The concept that certain environments are more vulnerable to invasion than others is a recurring theme throughout much of the ecological literature. Though not always stated explicitly, the vulnerability of a particular habitat to invasion does not imply that *any* invasive plant reaching that habitat will succeed automatically. Indeed, it is well known that while many species reach potential new habitats, few are able to invade successfully.

The ease and speed with which certain types of aquatic habitats have been invaded has provided strong support for a variety of hypotheses relating habitat susceptibility to invasion. These hypotheses have variously described the *causes* of successful invasions in terms of: simplified, or species-poor communities (Elton, 1958), poorly adapted native species (Sculthorpe, 1967), absence of predators (Mitchell, 1974; Harper, 1977), gaps generated by disturbance (Mitchell, 1974; Sousa, 1984), chemical changes (Hutchinson, 1975), competitive superiority due to tolerance of lower resource levels (Mitchell, 1974; Connell and Slatyer, 1977; Noble and Slatyer, 1980) and empty niches (Elton, 1958; Harper, 1977; Cook, 1985; Johnstone, 1986). As Johnstone (1986) notes, these concepts represent mechanisms that have facilitated invasions in certain specific circumstances and none of them appear to be applicable to *all* examples of plant invasions. Moreover, as mechanisms, they cannot be considered to be the *causes* of invasion.

Harper (1977) has suggested that all invasions occur as a function of the availability of 'safe sites' that are free of environmental hazards. These may be loosely equated with the 'empty niche' concept proposed by Elton (1958). Johnstone (1986) subsequently expanded this view, stating that 'a plant can successfully invade a site (to grow and reproduce) *only in the absence* of environmental resistance, where environmental resistance is any factor operating to decrease the intrinsic growth rate ... of the invader' (our emphasis). This categorical statement suggests that habitats are either vulnerable or resistant to invasion and takes no account of any special features possessed by the invading plant. In light of our earlier discussions (Sections 6.2.1 and 6.2.2), we can consider an 'environmental resistance' to be synonymous with a 'barrier to invasion'. Further, since we have already noted that barriers may be inactivated, overcome or removed by both biotic and abiotic transporting mechanisms, Johnstone's statement requires modification. In our view, therefore, the vulnerability of a habitat to invasion depends not only on the presence or absence of a barrier or barriers, but also on the *efficacy* of those barriers in resisting any disruptive mechanisms possessed by, or exploited by, the invader in question. Thus, a hypothetical barrier might easily prevent the ingress of species A into a specific habitat but be inadequate against species B.

Three mechanisms in particular, namely disturbance or alteration of the habitat, the absence of predators and the absence of effective competing species, have been implicated in facilitating the invasions of aquatic habitats (Sculthorpe, 1967; Mitchell, 1973, 1974; Ashton *et al.*, 1986). It is significant that habitat disturbance, for example through the impoundment of a river to form an artificial lake (Mitchell 1973) or the creation of irrigation and transport canals, has been implicated in almost all of the most spectacular examples of aquatic plant invasions. While chance and timing determine whether these mechanisms act simultaneously, sequentially, or not at all, each mechanism increases the vulnerability of a habitat to invasion, particularly by an alien species. Post-invasion successional (autogenic) changes will also modify a habitat's vulnerability to subsequent invasions.

6.3 ATTRIBUTES OF INVADING SPECIES

The r/K concept of plant life-history styles proposed by MacArthur and Wilson (1967) is somewhat over-simplified (Whittaker and Goodman, 1979) and incapable of providing a complete explanation of plant life-history phenomena. Calow and Townsend (1981) proposed that this deficiency could be resolved by the *a priori* testing of life-history strategies on the basis of optimal use of limited resources. However, as Solbrig (1981) points out, our understanding of how a plant functions in a multi-factorial environment is so limited that the optimal adaptations of a plant in any one habitat cannot easily be predicted. Thus, the classification of plant life-history phenomena by Grime (1977, 1979, 1985) into three basic categories, namely: C or competitive plants, S or stress-tolerant plants and R or ruderal plants, appears to be more practical.

6.3.1 Life forms and growth characteristics

In contrast to the terrestrial environment, aquatic environments are often held to be relatively constant thus encouraging species with perennial life cycles and a predominance of asexual reproduction (Sculthorpe, 1967; Hutchinson, 1975). While this assertion may hold true for the wet temperate and tropical latitudes, many aquatic habitats in semi-arid sub-tropical and temperate latitudes experience alternate periods of wetting and drying over both short-term (seasonal) and long-term (aseasonal) cycles (Howard-Williams and Gaudet, 1985; Mitchell and Rogers, 1985). In addition, seasonal cycles of temperature and day length, which are accentuated with increasing distance from the equator, confer further environmental variability. In these more variable habitats, therefore, annual growth patterns, desiccation resistant propagules and multiple regeneration adaptations (Grime, 1979, 1985) are likely to be the rule rather than the exception, even in those plants whose life cycles extend beyond a single annual cycle. The variability of the environment plays a major role in regulating the growth of aquatic plants and is thus critical to the success or failure of an invasion.

Free-floating invasive aquatic plants have been responsible for some of the most widespread and serious problems and are considered to be noxious weeds in most countries of the world (Table 6.1). The best-known examples, *Eichhornia crassipes, Salvinia molesta* and *Azolla filiculoides*, all have a free-floating life form, while *Ceratophyllum demersum* is usually unattached and forms large masses that drift about just below the surface of the water (Sculthorpe, 1967; Mitchell, 1974). This group of plants possesses a number of attributes which undoubtedly contribute to their success as invaders. The most important are: a capacity for extremely rapid vegetative multiplication; the ability to regenerate from relatively small portions of vegetative thallus; the complete or partial independence of sexual reproduction, which if possessed, is seldom important in the development of large populations; a growth morphology that results in the development of the largest possible area of photosynthetic tissue in relation to the whole plant, and which rapidly occupies the entire water surface or photic zone; and finally, an independence of substrate conditions and water levels (Mitchell, 1974). Where space is restricted, these plants can also initiate the accumulation of organic matter and promote the development of secondary swamp (emergent) vegetation. Thus they can be regarded as primary colonizers in aquatic ecosystems (Mitchell, 1974; Ashton *et al.*, 1986).

In contrast, submerged and emergent aquatic plants that are rooted in the sediments are dependent on a more stable hydrological regime for survival (Sculthrope, 1967). These plants can tolerate short-term changes in water level though they are unlikely to present problems in situations where rapid or extensive fluctuations occur. Indeed, water level manipulation is often used to control populations of these plants. In this group, *Phragmites, Typha, Myriophyllum aquaticum* and *Alternanthera* possess an emergent growth form while *Myriophyllum spicatum* and the hydrocharitacean genera *Egeria, Elodea, Hydrilla* and *Lagarosiphon* are all submerged plants (Sculthorpe, 1967).

Submerged species can only grow where a suitable substratum exists and where sufficient light penetrates the water column. They are therefore adversely affected by increased turbidity levels due to suspended silt and populations of planktonic algae. Clearly, these species are restricted to certain environmental conditions but, because of their close adaptation to these conditions, they can occupy all the available habitat and rapidly colonize suitable new habitats. Particular attributes possessed by submerged species include: intercellular lacunal systems to provide buoyancy to the photosynthetic organs; reduced structural rigidity; the ability to remain dormant or over-winter in unfavourable environmental conditions (Sastroutomo, 1981); the ability to regenerate from small stem fragments; the virtually complete independence of sexual reproduction; and the production of large numbers of vegetative propagules (such as tubers, turions and dormant apices) which are important in recolonization of cleared areas and invasions of new areas (Sculthorpe, 1967, Nichols and Shaw, 1986).

Emergent species possess several important attributes which include: long-lived, resistant rhizomes that can act as nutrient stores during unfavourable periods; production of tall, structurally rigid stems and leaves that shade out competing species; variable vegetative growth patterns that can prevent the ingress of competing species or rapidly exploit new colonization sites (Lovett-Doust, 1981; Hutchings and Bradbury, 1986); the production of very large numbers of minute seeds, suitable for long-range dispersal; and remarkable powers of regeneration after cutting (Westlake, 1963).

An additional feature of many invasive aquatic plants is that they are taxonomically close to species which are not invasive or resemble other plants that occupy the same habitat but are not always invasive. The reasons for this are not always clear. For example, *Salvinia molesta* belongs to a complex of four closely related South American species (Table 6.2). Two of these species, *S. molesta* and *S. herzogii*, are capable of very high growth rates but are sexually sterile and depend entirely on vegetative reproduction, thus differing from other species in the complex (Arthington and Mitchell, 1986). Indeed, laboratory studies (Mitchell and Tur, 1975; Cary and Weerts, 1983) suggest that *S. herzogii* would be as invasive as *S. molesta* if it were introduced to a new environment. Thus it would appear that the restricted distribution of *S. herzogii* and *S. molesta* in their native environment (South America) might be caused by the lack of sexual reproduction and by the complex of phytophagous insects that feed on the plants (Arthington and Mitchell, 1986).

Eichhornia crassipes and *E. azurea* provide a second example of two closely related species with similar native distribution patterns that differ in their invasiveness. The attractive flowers of both species have a similar appeal to plant

Table 6.2. Distribution and reproductive strategies of the four *Salvinia* species in the *Salvinia auriculata* complex (V = vegetative reproduction, S = sexual reproduction; adapted from Forno, 1983)

Species	Chromosome no.	Reproductive strategy	Native distribution	Alien distribution
S. auriculata	54	V and S	South and Central America	United States
S. herzogii	63	V	Southern Brazil and Northern Argentina	
S. biloba	36	V and S	Vicinity of Rio de Janeiro	—
S. molesta	45	V	Southern Brazil	Pantropical

collectors that has promoted their introduction into a number of countries as ornamentals; however, only *E. crassipes* has caused serious problems. Both species are capable of sexual and vegetative reproduction though in *E. azurea*, vegetative reproduction occurs slowly by fragmentation rather than by the rapid production of offsets as in *E. crassipes* (Barrett, 1978). Furthermore, in contrast to the free-floating habit of *E. crassipes, E. azurea* and the other six species in the genus are attached to the substrate. Clearly, the ability to undergo rapid vegetative reproduction, together with a greater degree of mobility, has favoured the spread of *E. crassipes* over that of *E. azurea* (Arthington and Mitchell, 1986; Gopal, 1987).

The members of the family Hydrocharitaceae with a similar morphology of whorls of short sessile leaves and stems with short internodes are another group of plants which may be compared. *Egeria densa, Elodea canadensis, Hydrilla verticillata* and *Lagarosiphon major* are representative of a number of species superficially so similar in morphology that they have often been confused. All are submerged species and occupy similar habitats, but have markedly different native distributions (Sculthorpe, 1967). Each of these four species has been responsible for major troublesome invasions when introduced to habitats outside of their native distribution, though other morphologically similar members of the family have not shown invasive features. Apart from the fact that each invasion has been due to vegetative growth and reproduction alone, no clear explanation exists for the mechanisms of these invasions or the lack of invasions by other members of the family (Arthington and Mitchell, 1986).

6.3.2 Morphological plasticity and the importance of rapid vegetative growth

Morphological plasticity describes the ability of a species to modify the shape and size of its vegetative structures when grown under different environmental conditions. This attribute must be distinguished from heterophylly, the presence on a single individual of two or more distinct types of leaf (Sculthorpe, (1967). Morphological plasticity is well known in aquatic and terrestrial plants, for example the morphological differences between sun-grown and shade-grown individuals of a particular species. Indeed, this feature is so well developed in some free-floating invasive aquatic plants that the different 'growth forms' of a particular species are often scarcely recognizable as belonging to the same species (Sculthorpe 1967; Hutchinson, 1975). Two examples, *Eichhornia crassipes* and *Salvinia molesta*, will suffice to illustrate this point.

Eichhornia crassipes produces two distinctive phenotypes or 'growth forms' which are characteristic of the different habitats occupied (Ashton *et al.*, 1979). The first is a small (5–8 cm high) so-called 'colonizing form', which possesses spherical float-like petioles and flattened, circular laminae. This form is found in open water situations where space is not limiting and multiplies rapidly by the production of stolons. The second is a much larger (up to 1.5 m high), so-called 'mat-form', possessing long elongated petioles and vertically arranged, orbicular laminae. This form is characteristic of dense, well-established mats and, once developed, appears to produce relatively fewer lateral stolons though the mat itself is held together by closely interlinked stolons. When either of these forms is moved from its original site in a dense mat to an open water situation, or vice versa, the plant's morphology changes within 10–20 days to suit its new habitat (Gopal, 1987).

The second example, *Salvinia molesta*, is somewhat more complicated in that three distinct phenotypes are recognizable. These are: the 'survival form', some 1–2 cm is length and bearing four to five pairs of flattened leaves that are each 5–8 mm in length, spaced a few millimeters apart on a very short rhizome; the colonizing form', which resembles the 'survival form' in that it consists of flattened pairs of leaves that are widely spaced on a short rhizome, differs in that the leaves and plants are approximately four to five times larger, the plants reaching a length of 5–10 cm; the 'mat form', with pairs of much larger (up to 5–6 cm wide) leaves that are erect, closely adpressed to each other and closely spaced on a rhizome that can reach 15–20 cm in length. Typically, the 'survival form', is found in adverse conditions, for example where nutrient supplies are low, and has very low growth rates under these conditions. The 'colonizing form' is characteristic of open water situations, where space is not limiting, and is capable of remarkably high growth rates.

The 'mat form', as its name implies, is found in established mats where all the plants are in very close contact with one another and their *in situ* growth rates are relatively low. As is the case with *Eichhornia*, when individual *Salvinia* plants are transferred from one situation to another they adapt their growth form to suit the environmental conditions.

The ability to adapt its phenotype to suit the habitat allows an invasive species to compete for a wide range of habitats and promotes dominance of the invader once a habitat has been occupied. Thus, invasive species such as *Eichhornia* or *Salvinia* possess a distinct competitive advantage over other species that lack this attribute.

In virtually every case of a biological invasion by plants, success in the initial stages depends largely on the invader's reproductive capacity. This is borne out by the spectacular rates of vegetative growth reported for a variety of invasive aquatic plants, particularly free-floating forms that are primary colonizers (Blackman, 1960; Westlake, 1963; Mitchell, 1974; Mitchell and Tur, 1975; Ashton and Walmsley, 1976; Cary and Weerts, 1983; Gopal, 1987). However, reproductive effort per unit time clearly cannot explain the *cause* of all invasions, since many non-invasive plants are equally prolific (Harper, 1977; Johnstone, 1986). Nevertheless, since the success of an invasion depends largely on the invader's ability to compete for available resources, rapid vegetative growth must be advantageous to an invader.

6.3.3 Responses to environmental cues

External environmental factors also act in a third capacity, omitted by Grime (1979), namely as 'cues' or 'signals'. This is clearly seen in cases where changes in the external environment elicit responses in plants which cannot be explained simply in terms of biomass limitation by stress or disturbance; for example, the alternate cycles of drying and wetting that the seeds of several plant species require before germination can occur (Mitchell and Rogers, 1985). Whilst other types of signals result when a stress or disturbance is removed, or is applied, the significance of adaptations to signals lies in the fact that the regulation of growth and dormancy is restricted to favourable and unfavourable periods, respectively (Mitchell and Rogers, 1985). Such adaptations are thus extremely important to invasive plants, promoting their ability to exploit favourable habitats.

The translocation of nutrients to or from underground storage organs and the reproductive events of flowering, setting seeds and the formation of dormant propagules seem to occur largely in response to environmental signals, particularly seasonal changes in photoperiod, light intensity and temperature (Mitchell and Rogers, 1985). In contrast to rooted plants, free-floating aquatic plants respond to both seasonal changes in the environment and to changes in population density (Ashton, 1977; Mitchell and Rogers, 1985). For example, increasing mat density stimulates morphological changes in *Eichhornia crassipes* and *Salvinia molesta* (Section 6.3.2) and the initiation of sporocarp production in *Azolla filiculoides* (Ashton, 1977). Similarly, the initiation of propagule germination is also regulated by environmental signals such as daylength, light intensity, temperature and oxygen saturation (Hutchinson, 1975; Haller *et al.*, 1976; Sastroutomo, 1981; Bowmer *et al.*, 1984; Nichols and Shaw, 1986). While the physiological basis of such regulation is not well understood, hormonal reactions similar to those of terrestrial seeds would appear to be operative (Weber and Nooden, 1976).

6.3.4 Competitive ability in the 'struggle for dominance'

The success of an invasive aquatic plant in a habitat reflects, initially, its genotypic capacity to adapt its morphological, physiological and reproductive strategies to avoid or overcome competition for the acquisition of resources (Breen *et al.*, 1987).

Vegetative and sexual reproductive processes are largely responsible for the success or failure of an invasive plant to survive in a particular habitat. Indeed, the true adaptation of a plant to a particular environment is defined as the capability to undergo successful sexual reproduction in that environment (Solbrig, 1981); in these terms, growth and multiplication by purely vegetative means therefore indicate tolerance of the environmental constraints, rather than adaptation. A given reproductive strategy reflects the responses elicited by several different cues, stresses and disturbances (Grime, 1979; Menges and Waller, 1983); a single species can thus display contrasting reproductive strategies when grown in different environments. Indeed, the extreme variability of most aquatic environments selects for plants that possess a high degree of reproductive plasticity, favouring species with multiple regenerative adaptations (Grime, 1979).

Each reproductive stratagem involves a trade-off between the investment of energy and materials in vegetative growth and sexual reproduction (Solbrig, 1981; Menges and Waller, 1983); seed production is thus favoured at the expense of vegetative growth, and vice versa. Whenever resources are in limited supply, sexual reproduction will tend to reduce the competitiveness and survival ability of the adult plant (Solbrig, 1981). On the other hand, the production of resistant seeds by sexual reproduction enables an invading plant such as *Eichhornia crassipes* (Matthews, 1967) to recover and recolonize the habitat after adult plants have been eliminated by catastrophic disturbance (Breen *et al.*, 1987).

In many emergent species such as *Phragmites* and *Typha*, intensive inter- and intra-specific competition for space and light promotes maximum occupancy of above- and below-ground space. The possession of a clonal or phalanx growth form (Lovett-Doust, 1981) provides a powerful barrier against encroachment by competing species, or neighbouring clones of the same species. Survival during unfavourable periods is accomplished in well-developed rhizome systems (Howard-Williams and Gaudet, 1985) which retain control of the substratum. Rapid response to environmental cues triggers mobilization of stored reserves at the onset of favourable conditions, allowing reoccupation of aerial space and expansion of the phalanx, and thus promotes local site domination, often for many years (Hutchings and Bradbury, 1986). The stoloniferous or 'guerrilla' growth form (Lovett-Doust, 1981) possessed by many invasive aquatic plants provides a high degree of parental care for the offshoot while still allowing the stolon to be lost with little or no damage to the adult plant if environmental stressess are too severe (Breen *et al.*, 1987).

Morphological plasticity and the ability to undergo rapid vegetative growth are the key attributes for success in interspecific competition between invasive aquatic plants that have the same growth form. This was shown very clearly by Bond and Roberts (1978) in the Cahora Bassa reservoir, Mozambique, where *Eichhornia crassipes, Salvinia molesta, Pistia stratiotes* and *Azolla nilotica* were in competition soon after the reservoir commenced filling. The smaller species *A. nilotica* and *S. molesta* were eliminated first and the largest and most vigorous species, *E. crassipes*, eventually dominated the flora (Figure 6.5).

6.3.5 Survival during unfavourable periods

Aquatic plant genotypes are exposed to, and thus moulded by, a variety of selective pressures with a seasonal periodicity. These pressures increase in

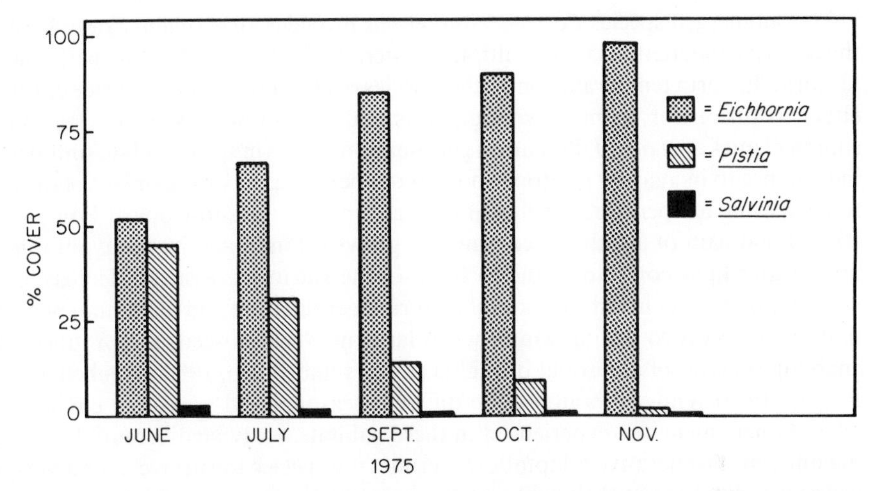

Figure 6.5. Changes in the percentage cover of *Eichhornia crassipes*, *Salvinia molesta* and *Pistia stratiotes* in macrophyte communities of the Carinde and Mucangadze Basins of Lake Cahora Bassa, Mozambique. (Redrawn from Bond and Roberts, 1978.)

magnitude with increasing distance from the equator whilst their predictability varies from a high (e.g. day length) to a low (e.g. atmospheric temperature) probability of occurrence (Mitchell and Rogers, 1985). Different species of aquatic plants can therefore be expected to respond differently to each factor or combination of factors. In addition, several other environmental phenomena, such as droughts, floods and hot or cold periods, also affect aquatic plant growth but do not have a seasonal periodicity. These events are often severe and may eliminate entire populations except for a few sheltered or well-adapted individuals (Mitchell and Rogers, 1985). This periodic screening of genotypes is the basis of 'catastrophic' selection which is an important mechanism leading to adaptive traits and speciation in aquatic plants (Lewis, 1962; Stebbins, 1974).

Several invasive aquatic plants have developed clonal populations or ecotypes as a result of their marked tendency for asexual reproduction (McNaughton, 1986; Hutchinson, 1975). Other invasive species are incapable of setting seed and depend entirely on vegetative reproduction for their survival and spread. These clonal forms allow individual species to occupy widely differing habitats and to take advantage of locally favourable growing seasons. Gopal (1987) has suggested that this may be an important feature of the success achieved by *Eichhornia crassipes*. However, it is important to note that the lack of sexual reproduction in several invasive aquatic plants indicates that these species may have spread from a single introduced clone. Their reduced gene flow and the resultant lack of genetic variability suggest that they might be particularly susceptible to an introduced biocontrol agent (Mitchell, 1974; Arthington and Mitchell, 1986; Ashton *et al.*, 1986).

The submerged species *Potamogeton crispus* provides an excellent example of the ecotypic differentiation of multiple regenerative strategies in plants with this life form. In north temperate ponds, this species over-winters beneath an ice cover either as turions or dormant young plants and seldom flowers or sets seed in summer (stuckey *et al.*, 1978). Turion germination occurs in spring or late summer in response to increasing photoperiod and temperatures in spring or in response to decreasing temperatures at the onset of autumn. In the sub-tropical Pongola River floodplain of South Africa, summer growth is inhibited by unfavourable underwater light conditions due to flood-borne silt loads. Instead, decreasing water temperatures in autumn stimulate turion germination and maximum plant biomass is recorded during winter when large numbers of seeds are produced, enabling survival of aseasonal droughts (Rogers and Breen, 1980; Mitchell and Rogers, 1985). Thus, although quite different seasonal and aseasonal environmental phenomena are experienced in these habitats, ecotypic differentiation of the different regenerative adaptations permits the species to survive over a very wide latitudinal range (Mitchell and Rogers, 1985).

The perennial rhizomes and stolons of many submerged and emergent invasive plants also provide an important survival mechanism during unfavourable conditions when much of the above-ground plant biomass may be lost or senesce (Howard-Williams and Gaudet, 1985). In addition, the possession of underground storage organs permits retention of nutrients within the plant and facilitates rapid growth at the onset of favourable environmental conditions (Sculthorpe, 1967).

6.3.6 Mechanisms facilitating intra- and inter-system transfer

Earlier in our discussion (Section 6.2.2) we emphasized that a variety of transporting mechanisms are important in the arrival and dispersal stages of aquatic plant invasions. Clearly, the efficacy of biotic and abiotic transporting mechanisms will be increased greatly, if the plant in question possesses one or more specialized adaptations that facilitate its dispersal. Indeed, most successful invasive plants possess very efficient dispersal mechanisms, such as sexual or asexual propagules, or are able to disperse widely on flood waters and regrow rapidly following fragmentation.

It is axiomatic that those aquatic plants that are capable of long-distance dispersal will already have done so: they are therefore unlikely to be recognized as invasive alien species in a particular environment. For example, the immense geographical distributions of the anemophilous genera *Phragmites* and *Typha* (Sculthorpe, 1967; Cook, 1974, 1985) suggest that these plants may have spread naturally through wind transportation of their numerous minute seeds (Cook, 1974, 1985; Breen *et al.*, 1987). While anemochory may be a characteristic of most graminaceous phreatophytes, very few truly aquatic plants possess this attribute (Sculthorpe, 1967). Nevertheless, aquatic plants that are capable of a high reproductive output, in terms of numbers of propagules, increase their chances of dispersal.

In general, therefore, plant attributes that facilitate short-range dispersal increase both the rate and chance of successful invasion. Since large seeds or propagules have a greater germination success but are less easily dispersed, trade-offs between propagule numbers, sizes and ease of dispersal are very important (Sculthorpe, 1967; Hutchinson, 1975). Additional features, such as temporary buoyancy in newly dehisced propagules or the production of fleshy fruits to attract waterfowl (Rogers and Breen, 1980), add to the efficiency of dispersal. Obviously, when dispersal in an invasive plant's native habitat is closely geared to the activities of a specific vector, then dispersal in the new habitat is likely to be very limited. Where dispersal is passive, for example by water currents, flowering and propagule production times can be extended considerably to reduce the chances of mortality. Thus, the development of hardened seed or spore coats protects these propagules from physical damage during transport by water currents and increases their chances of successful intra-system transfer (Ashton, 1977).

6.3.7 The 'perfect invader'—does it exist?

Several authors (e.g. Baker, 1965; King, 1966; Holm *et al.*, 1977; Elmore and Paul, 1983) have described and listed the attributes and characteristics of a hypothetical ideal invader. However, most of these descriptions refer to terrestrial situations and reflect particular concern with the invasion of agricultural crops by weeds. We should, therefore, be cautious when extrapolating them directly to invasions of natural or man-modified aquatic systems. Nonetheless, agricultural weeds and invasive aquatic plants do share several attributes that contribute to their success.

A logical starting point in the search for the 'perfect invader' would be to examine those aquatic species that have provided such dramatic examples of successful invasions that they are now classified as noxious aquatic weeds (Sculthorpe, 1967; Mitchell, 1974; Holm *et al.*, 1977). The most important species that warrant such consideration have been listed in Table 6.1.

Close examination of these invasive aquatic plants and the variety of habitats that each species has invaded successfully, allows assessment of each species against the criteria of the hypothetical 'perfect invader'. On balance, *Eichhornia crassipes* appears to be the closest contender for the title, further substantiating its status as the world's most troublesome water weed (Sculthorpe, 1967; Mitchell, 1974; Holm *et al.*, 1977; Gopal, 1987).

6.4 THE MANAGEMENT OF INVASIVE AQUATIC PLANTS

Excessive populations of invasive aquatic plants not only cause marked changes in the natural vegetation of aquatic ecosystems, they also inhibit or prevent the management and utilization of water resources (Mitchell, 1974; Ashton *et al.*, 1986). Despite the fact that the biological and economic consequences of these impacts are often difficult to evaluate objectively they are invariably regarded as being undesirable and therefore requiring some form of remedial action (Ashton *et al.*, 1979; Arthington and Mitchell, 1986).

Manipulation of the environment to achieve certain objectives is a well-known characteristic of modern man and forms, for example, the basis of his (largely successful) agricultural practice. However, agricultural systems are highly simplified with a greatly reduced diversity of species. The management of natural or semi-natural aquatic ecosystems that contain large numbers of species is considerably more difficult because of the complex interrelationships that exist between the various components. It is essential that these relationships be understood before management options are formulated and implemented. This situation therefore requires that research aimed at understanding the complexities and interrelationships of the aquatic environment should enjoy a high priority (Arthington and Mitchell, 1986). However, progress and the betterment of the human race cannot always wait for the result of these investigations. Management procedures therefore often have to be designed and instituted on the basis of available knowledge, however inadequate it might be. Thus, it is essential that continual environmental monitoring form part of a management programme which, in turn, must be sufficiently flexible to enable incorporation of viable alternatives if unforeseen developments occur (Mitchell, 1974). However, it must be emphasized from the outset that it is absolutely necessary to clearly define the aims and objectives of any management policy before procedures for its implementation are designed or carried out.

6.4.1 Available options

Management strategies employed against invasive aquatic plants can conveniently be divided into two basic groups, namely: protectionist and interventionist. The former group attempts to retain particular ecosystems in a hypothetical 'natural' or pristine state by preventing invasions and usually contains a strong legislative element (discussed in Section 6.4.6). Typically, these strategies involve some form of environmental or ecological management that is designed to reduce habitat disturbance. The most important invasive aquatic plants, the primary colonizers of disturbed systems, are thereby excluded.

In contrast, interventionist strategies seek to suppress or remove existing invaders from a particular habitat, thereby reducing their population size to a more 'acceptable' level and minimizing their impact on ecosystem functioning. Typical strategies include various forms and combinations of manual, mechanical, chemical and biological control techniques that confine the invader to a low level. These actions almost invariably return the habitat to an earlier successional stage, and increase the probability that the first invader may be replaced by a second, more problematic, invasive species (Cook, 1985). Because of financial considerations, complete eradication of the invader is seldom attempted.

At present, six groups of control techniques can be used to manage invasive aquatic plants:

1. Manual removal, as its name suggests, is a labour intensive and low efficiency technique that is hampered by water depth and the physical quantities of material that can be handled in a given time interval.
2. Mechanical control usually involves the use of mechanized or power-driven equipment to harvest plant material a remove it from a water body. High rates of removal are possible but the process is often inefficient in dealing with very extensive invasions where the rates of regrowth often exceed the removal capacity of the machines.
3. Chemical control involves the use of inorganic or organic herbicides, usually as a last resort, to treat extensive infestations. The herbicide is either applied directly into the plants in the case of free-floating and emergent species or into the water in the case of submerged species. A serious disadvantage is the prohibitively high cost of most chemical control programmes. Modern developments in herbicide formulation have considerably improved the performance of appropriate herbicides and at the same time have reduced many of the deleterious side effects on other aquatic organisms. Safety to other water users is of paramount importance when using herbicides to control aquatic plants.
4. Biological control techniques use one or more host-specific natural enemies from a target plant's native range to attack the plant and maintain its population at a low level. Unfortunately, the initial stages of identification, collection and screening of the organism for host specificity, followed by its breeding and dissemination on the target plant can be expensive and take several years. However, when successful, these control measures are self-sustaining. Ultimately, this technique probably holds the greatest promise in terms of minimizing the ecological impacts of control programmes.
5. Environmental manipulation involves modification of the aquatic environment, for example by lowering water levels, to disadvantage the invasive plant of interest. Significant successes have been achieved in this manner though the techniques are not entirely suitable in areas where water supplies are scarce. A further disadvantage is that unless the dead plant material is removed, a subsequent rise in water level causes rapid nutrient release and enrichment of the water.
6. Direct use of the invasive plant for economic benefit, for example as a stock feed, requires that the maintenance of a particular plant population is compatible with other uses to which the water may be put. A serious disadvantage is that the plant population serves as a source for further invasions.

So wide is the range of aquatic plant life forms and growth characteristics that a method of control appropriate to one species is often unsuitable for another species in the same habitat (Sculthorpe, 1967). In addition, the control of invasive aquatic plants often presents many specialized problems. For example, if all the aerial, rooted and submerged growing regions of a plant are not destroyed, new growth is soon produced and the problem recurs. Additional problems can arise due to the inefficient disposal of plants killed *in situ*, which can cause oxygen depletion and elevated levels of dissolved nutrients that, in turn, can promote the regrowth of surviving plant fragments.

Ideally, each problem should be treated as unique and all the local factors investigated thoroughly before a particular technique or combination of techniques is chosen. In almost every case, it is clearly advantageous to commence operations as soon as possible before the problem becomes critical and necessitates the use of drastic measures (Sculthorpe, 1967; Mitchell, 1974; Ashton *et al.*, 1979).

6.4.2 Eradication versus control and the costs of inaction

Once the successful invasion and establishment of an aquatic plant has been found to have adverse ecological and economic effects, every attempt must be made to reduce its population size and prevent further spread. The severity of the actual and potential problems posed by the invading plant influence the choice of whether to opt for total eradication of the invader or undertake a control programme to maintain the invader's population at a lower, more acceptable level. Both options employ the same techniques but differ in the degree of vigour and persistence with which they are carried out. In the short term, control is cheaper than eradication. However, maintenance of control programmes over long periods of time invariably raises their long-term cost above that of eradication.

Vigorously growing plants such as *Eichhornia* and *Salvinia* can be eradicated if the infestation is identified and dealt with at an early stage or is confined to a small body of water. However, once an infestation is well established in a large water body with inaccessible areas, eradication attempts are less likely to succeed. In either case, it is critically important that control or eradication programmes are well planned and thorough and that they are persistently carried out. Regular surveys and follow-up control of outbreaks from surviving plants or newly germinated seedlings are required for a number of years (Mitchell, 1974; Ashton *et al.*, 1979). In particular, this follow-up procedure is crucial to the success of programmes aimed at eradicating *Eichhornia crassipes* (Ashton *et al.*, 1979), whose seeds can remain viable for 10–15 years (Matthews, 1967; Gopal, 1987). The necessity to sustain expensive follow-up programmes to ensure successful control or eradication, emphasizes further the benefits of self-sustaining biological control measures (Arthington and Mitchell, 1986).

Delays in the identification of an invasive aquatic plant problem and failure to limit the plant's initial spread will allow rapid expansion of the population. This will increase the size of the problem and expand the scope, impact and costs of the remedial action required (Ashton *et al.*, 1986). Inaction, or action that is too little too late, is thus both ecologically and economically expensive.

6.4.3 Successful control programmes—why did they succeed?

Numerous control and eradication programmes directed against submerged, emergent and free-floating invasive aquatic plants in many parts of the world have been reported in the literature. Whilst considerable attention has been paid to submerged and emergent plants (e.g. Jacot Guillarmod, 1979; Bowmer *et al.*, 1984), the majority of these programmes have been aimed at the free-floating species *Salvinia molesta* and, more especially, *Eichhornia crassipes* (e.g. Sculthorpe, 1967; Gopal, 1987), a direct reflection of the latter's world-wide weed status (Gopal and Sharma, 1981). A wide variety of local ecological and economic constraints has dictated the use of an enormous array of control techniques and strategies in these programmes. As could be expected, different degrees of success have been achieved (e.g. Blackburn, 1974; Mitchell, 1974; Crafts, 1975; Soerjani, 1977; Ashton *et al.*, 1979, 1986; Mayer, 1981; Julien, 1982; Forno and Bourne, 1985; Thomas and Room, 1986; Gopal, 1987). Given the extreme variability of the data set and the fact that the 'success' of a particular programme will depend largely on its original objectives, can we identify any common features that distinguish *effective* control and eradication programmes?

Two important features have virtually guaranteed the success of programmes directed at small infestations of invasive aquatic plants, i.e. those measuring less than 20–30 hectares in extent. These were a high degree of isolation or confinement of the infestation to a single, accessible water body and the early identification of the aquatic plant problem followed by prompt and sustained remedial action (Mitchell, 1974; Mayer, 1981). The reduced cost involved in treating small infestations has favoured the choice of eradication rather than control and has promoted the widespread use of a range of herbicides in those situations involving free-floating plants (Blackburn, 1974; Newbold, 1975). Where the infestations involve submerged plants, fish such as grass carp (*Ctenopharyngodon idella*) are often used as biocontrol agents with great success (Van Zon, 1981; Julien, 1982; Wiley and Gorden, 1984). An invasive aquatic plant infestation in a river is far more difficult to control and requires extensive checks of the river catchment to reduce the frequency of reinfestation.

In contrast, the scale of operations required to deal with most of the more extensive invasions of aquatic plants (> 1000 hectares) has greatly limited the variety of control procedures that could be implemented. Typically, high rates of plant growth render the use of manual or mechanical techniques alone inappropriate; usually only chemical techniques, specific biocontrol agents or integrated programmes that cor ' 'ne several techniques stand any chance of success. Where an ecologically undesirable technique is used, for example the large-scale application of herbicides, special precautions are needed to minimize the risks of adverse ecological impacts and any deterioration in water quality (Blackburn, 1974; Mitchell, 1974).

The *Eichhornia crassipes* eradication programme at Lake Hartbeespoort, a heavily enriched, multi-purpose reservoir in South Africa, provides one of the best example of successful large-scale use of herbicides (Figure 6.6; Ashton *et al*., 1979, 1980). Extended planning, herbicide tests and inclement weather delayed initiation of the spraying programme such that the rapidly growing *E. crassipes* infestation covered 1200 hectares (60% of the lake's surface area) by the time herbicide application started. The spraying programme consisted of four separate aerial applications of herbicide supplemented with intermittent spot-treatments of marginal plants using hand-held and boat-mounted sprayers. Impacts on the environment and other water users were minimal and all adult *E. crassipes* plants in the reservoir were successfully eliminated (Ashton *et al*., 1979, 1980). A follow-up surveillance programme initiated in 1978 has monitored the lake shores for *E. crassipes* seedlings and is scheduled to continue until 1995 to

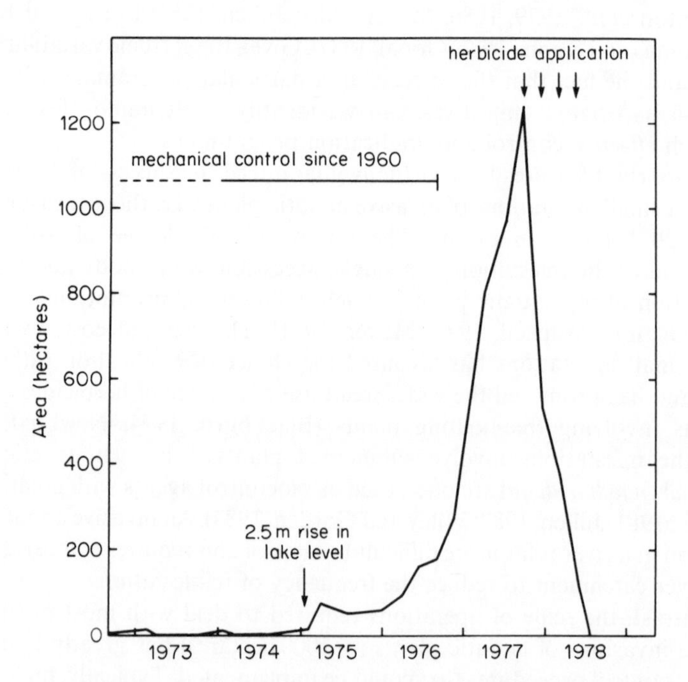

Figure 6.6. Changes in the area covered by *Eichhornia crassipes* on Lake Hartbeespoort, South Africa, before and after herbicide application. (Adapted from data in Ashton *et al*., 1979.)

minimize the risk of reinfection. Concomitant checks are made of streams and rivers in the catchment above the lake to avoid the development of new *E. crassipes* infestations (Ashton *et al.*, 1979, 1980). In the case of Lake Hartbeespoort, success was entirely dependent on careful planning, diligent and sustained implementation of the chosen techniques and the maintenance of an efficient, vigilant follow-up operation over many years (Ashton *et al.*, 1986).

The two biological control programmes directed against *Salvinia molesta* in Australia (Room *et al.*, 1981, 1984) and Papua New Guinea (Mitchell, 1981; Thomas and Room 1985, 1986) are outstanding examples of the success that can be achieved against both small and large infestations with a host-specific control agent.

The control agent in question, a Brazilian curculionid beetle found feeding only on *Salvinia molesta*, was initially identified as *Cyrtobagous singularis* and introduced to Australia as a potential biocontrol agent against *S. molesta* (Forno and Bourne, 1985). When released on Lake Moondarra in Australia, the beetle destroyed a 200 hectare *S. molesta* infestation in 14 months (Figure 6.7; Room *et al.*, 1981). Subsequent studies have shown the beetle to be a separate species, *Cyrtobagous salviniae* Calder and Sands, and it is being used with great success to control *S. molesta* elsewhere in Australia (Room *et al.*, 1984) as well as in Namibia

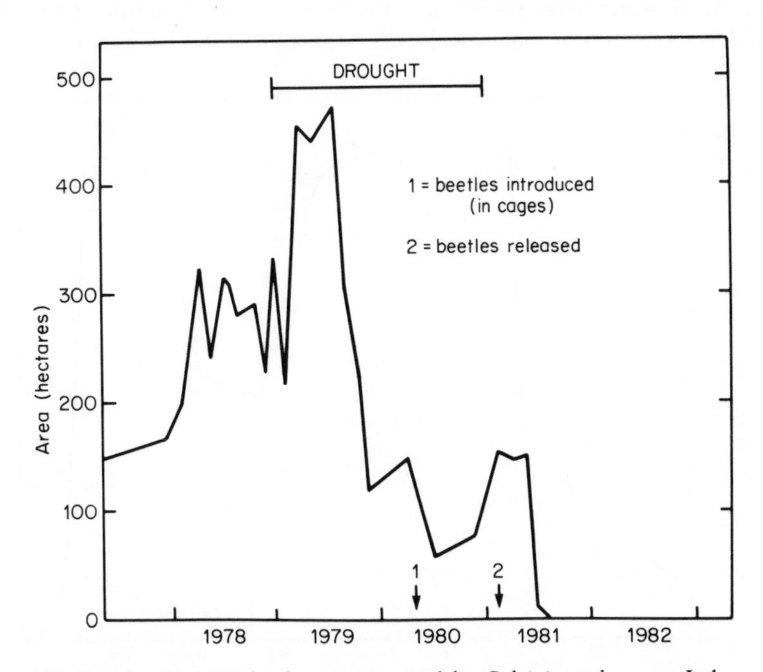

Figure 6.7. Changes in the area covered by *Salvinia molesta* on Lake Moondarra, Australia, before and after introduction of *Cyrtobagous salviniae*. (Redrawn from Thomas and Room, 1986.)

(Schlettwein, 1985), India (Room and Thomas, 1986) and Papua New Guinea (Thomas and Room, 1986).

The rapid control exerted by *C. salviniae* on the *S. molesta* infestation in the Sepik River floodplain lakes in Papua New Guinea has provided the most spectacular example so far of successful biological control of an aquatic weed (Figure 6.8; Thomas and Room, 1985, 1986). Within a two year period, *C. salviniae* destroyed an estimated two million tonnes of *S. molesta*, reducing the plant's cover from approximately 200 km^2 to 2 km^2 (Thomas and Room, 1986).

The success of *C. salviniae* in controlling *S. molesta* contrasts markedly with the failures of other insects, particularly the related *C. singularis*, the pyralid moth *Samea multiplicalis* and the lesser success of the aquatic acridid grasshopper *Paulinia acuminata.* Intensive investigations have shown that the success of *C. salviniae* can be attributed to the deliberate dispersal of the beetles over the whole *S. molesta* infestation, the lack of predators, parasites and diseases in its new environment, and the selective feeding behaviour of both the larvae within the rhizome and the adults that destroy *Salvinia* growing points (Thomas and Room, 1986). This feeding pattern has considerable significance for *S. molesta* which relies entirely on vegetative propagation (Mitchell, 1974). High nitrogen levels within *S. molesta* plants increase their 'palatability' to *C. salviniae* and, together with elevated water temperatures, amplify the beetle's feeding rate and promote rapid development of *C. salviniae* populations (Forno and Bourne, 1985; Room

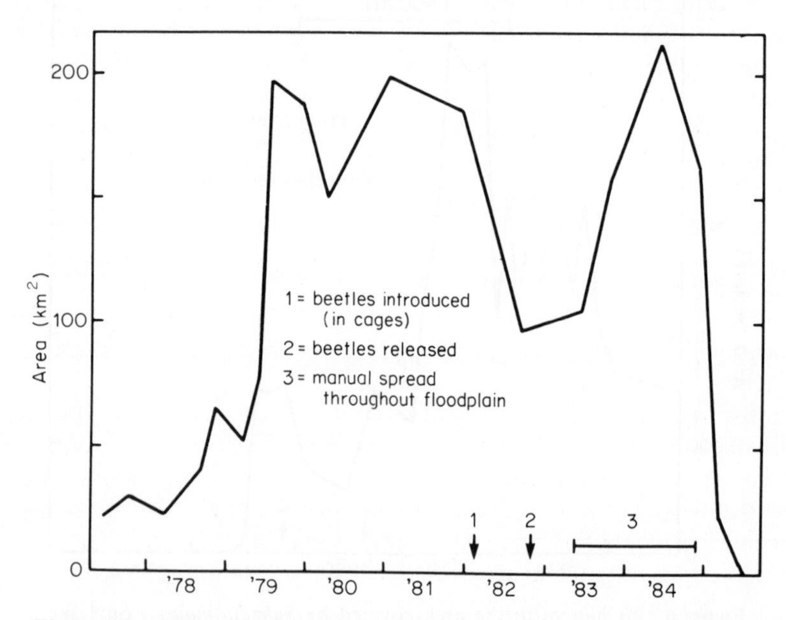

Figure 6.8. Changes in the area covered by *Salvinia molesta* on 100 lakes in the Sepik River floodplain, Papua New Guinea, before and after introduction of *Cyrtobagous salviniae.* (Redrawn from Thomas and Room, 1986.)

and Thomas, 1985; Thomas and Room, 1986). *C. salviniae* is thus a highly cost-effective and environmentally sound means of biological control against *S. molesta* invasions in the tropics. However, the insect's sensitivity to lower temperatures precludes its use in cool temperate areas where *S. molesta* can still grow (Cary and Weerts, 1983; Room and Thomas, 1986).

Another example of successful biological control of an invasive aquatic plant has been the use of the flea beetle (*Agasicles hygrophila*) against alligator weed (*Alternanthera philoxeroides*) in Australia and several of the southeastern states of the United States. As in the case of *Cyrtobagous* and *Salvinia*, the *Agasicles* success can be largely attributed to its host-specific selective feeding behaviour and possession of similar environmental tolerances to those of *Alternanthera* (Andres, 1977; Julien, 1982).

6.4.4 Unsuccessful control programmes—why did they fail?

In our preceding discussion we emphasized the importance of careful planning, use of appropriate techniques and the need to maintain a vigilant follow-up operation. Indeed, failure to comply with one or more of these three requirements has led to the failure of several attempts to chemically control or eradicate infestations of invasive plants.

For example, at Lake Bon Accord, South Africa, two aerial herbicide applications reduced an *Eichhornia crassipes* infestation that covered the entire surface of the lake (200 hectares) to less than one hectare. However, failure to maintain a follow-up programme resulted in complete reinfestation of the impoundment within a period of 8 months (Ashton *et al.*, 1980). A second series of herbicide applications, this time combined with an efficient follow-up operation, has since eradicated *E. crassipes* from this lake (Ashton *et al.*, 1986). In another, more tragic example, a 30-year period (1952–1982) of successful *E. crassipes* control with the herbicide 2, 4-D on Lake McIlwaine, Zimbabwe (Jarvis *et al.*, 1982), has since been negated by a failure to maintain the required follow-up programme (P. J. Ashton, personal observation). In those situations where funds are limited, the increasing costs of both herbicides and the equipment for their application can lead to a loss of continuity in a chemical control programme and thus allow the invasion to return.

Another very important feature that can thwart attempts to chemically control invasive aquatic plants, is the problem of inaccessible terrain. For example, early attempts at aerial applications of herbicide to control the *Eichhornia crassipes* infestation in the newly flooded Lake Brokopondo, Surinam, were frustrated by the highly indented shoreline, dense riparian forests and the shelter afforded to *E. crassipes* by partly inundated trees (Van Donselaar, 1968; Leentvaar, 1973). Similarly, attempts to chemically control the *Salvinia molesta* infestation in the rivers, lakes and swamps of northern Botswana and the eastern Caprivi area of Namibia were also largely unsuccessful. Here, control of *S. molesta* plants

growing in the open water areas was effective, while plants sheltered by the dense *Phragmites* reedswamps were unaffected by the herbicide and rapidly reinvaded the cleared areas (Edwards and Thomas, 1977).

An important point to remember is that the large-scale use of herbicides to control extensive aquatic plant infestations is ecologically undesirable; such a course of action should be reserved for crisis conditions only (Ashton *et al.*, 1979, 1980). Furthermore, an inadequate knowledge of the invasive plant's ecology can lead to the choice of an inappropriate herbicide, for example one which cannot control all of the invader's life stages. Chemical control techniques are thus seldom effective in the long term unless they are integrated with other control techniques (e.g. manual or mechanical removal, ecosystem manipulation, biological control; Soerjani, 1977).

Biological control is widely considered to be the ideal solution to all invasive aquatic plant problems. Indeed, this view is strongly reinforced by the spectacular successes achieved, for example, against *Salvinia molesta.* However, a recent survey (Julien, 1982) has shown that less than half of the programmes undertaken have been effective and the likelihood of success in a particular programme can rarely be predicted. The use of an inappropriate organism or deployment of the organism in an unfavourable environment appear to be the most frequent reasons for the failure of biological control programmes (Julien, 1982; Gopal, 1987).

Biological control programmes directed against *Eichhornia crassipes* in India and North America using the Chinese grass carp (*Ctenopharyngodon idella*) provide good examples of the use of an inappropriate organism. These (and other) herbivorous fish fed preferentially on submerged aquatic plants (Van Zon, 1981; Wiley and Gorden, 1984), and consumed *E. crassipes* only as a last resort (Mehta and Sharma, 1972). Thus, their preferential feeding pattern, combined with a poor reproductive performance outside of their native range, necessitated very high stocking rates (Baker *et al.*, 1974), which rendered this species totally unsuited for *E. crassipes* control in most instances.

In contrast, an organism that is a successful biocontrol agent in one environment may prove to be far from effective when it is introduced elsewhere. A case in point is provided by the curculionid beetle *Neochetina eichhorniae* which has achieved a reasonable degree of success as a biological control agent against *E. crassipes* in Australia (Harley and Wright, 1984; Wright, 1984) and the United States (Manning, 1979). However, this organism appears to have had little impact on *E. crassipes* in South Africa despite the fact that it has maintained large populations at several sites for many years (P. J. Ashton, personal observation). In this case, it is possible that climatic and local environmental differences may have altered the host plant's chemistry and cuticle structure, thereby reducing its palatability and decreasing its susceptibility to the introduced beetles (Wright, 1984). However, the effects of *N. eichhorniae* on *E. crassipes* can be augmented when a plant pathogen such as the fungus *Cercospora rodmanii* is inoculated

onto the insect-damaged plants. Trials in the southeastern United States have shown that this and other combinations of insects and plant pathogens have considerable potential for the biological control of *E. crassipes* and deserve far greater attention in future (Freeman, 1977).

The importance of local environmental conditions and natural successional changes in the aquatic vegetation should not be underestimated. Indeed, successional changes can mask or amplify the impact of an introduced biological control agent and thereby complicate assessments of the control programme's efficacy. The biological control programme directed against the *Salvinia molesta* infestation on Lake Kariba, Zimbabwe, highlighted this dilemma. After closure of the dam wall in December 1958, the rising waters of the new lake inundated huge areas of vegetation. Scattered *S. molesta* colonies began to proliferate rapidly in the calm, nutrient-enriched waters and the mats became securely anchored amongst partly inundated trees. The area covered by the plant rose to a peak in May 1962 when it occupied 1003 km^2 (21.5%) of the lake's surface (Figure 6.9; Mitchell and Rose, 1979; Marshall and Junor, 1981). Subsequently, the area decreased and then fluctuated between about 400 and 800 km^2. Approximately 2 years after the introduction of the Trinidad strain of an aquatic acridid grasshopper (*Paulinia acuminata*), there was a dramatic drop in the area covered by *S. molesta*, to approximately 100 km^2 (Mitchell and Ross, 1979). However,

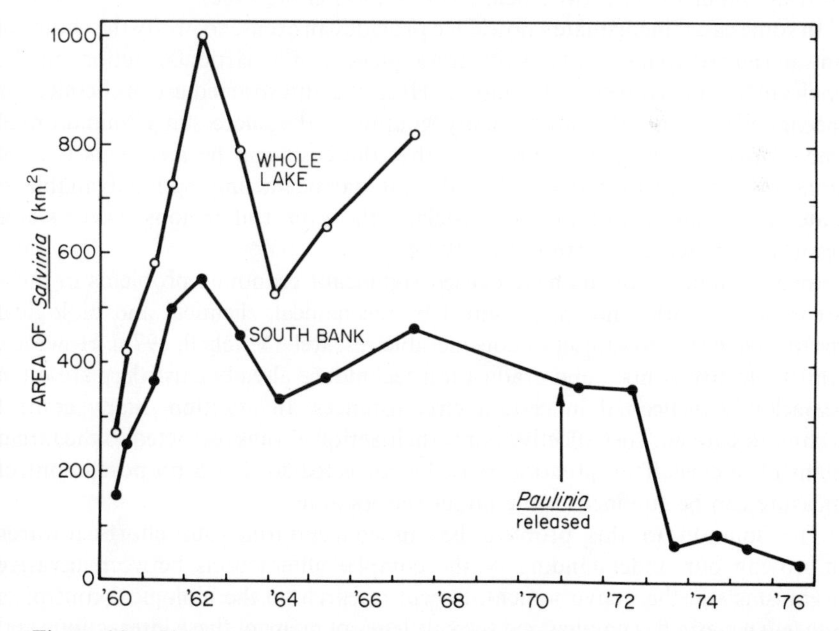

Figure 6.9. Changes in the area covered by *Salvinia molesta* on Lake Kariba, Zambia/Zimbabwe, before and after release of *Paulinia acuminata*. (Redrawn from Mitchell and Rose, 1979.)

while *P. acuminata* did feed actively on the *S. molesta* plants, a combination of environmental factors was also implicated in the plant's decline. As the lake matured, decomposition of the trees that had previously anchored *Salvinia* mats plus increasing competition from native aquatic plants and the declining nutrient content of the lake water were considered to be the main factors responsible (Mitchell and Ross, 1979; Marshall and Junor, 1981). Thus, while the introduction of *P. acuminata* as a biological control agent may have contributed to reducing the *S. molesta* population, biological control alone was not effective.

6.4.5 Can programme success be predicted?

The ability to predict the outcome of particular management strategies is one of the prime objectives of water managers concerned with counteracting the adverse effects of invasive aquatic plants. Ideally, accurate predictions regarding the consequences of a particular action require a sound understanding of the system under consideration and the ecology of the invasive plants as well as the likely impacts on the environment of the proposed course of action. Sadly, however, our present knowledge of aquatic ecosystems is often inadequate to meet these criteria and formulate ecologically sound management programmes (Mitchell, 1979b); many control programmes continue to be implemented on a 'best guess' or trial and error basis (Mitchell, 1974; Ashton *et al*., 1986).

In some cases, inadequate knowledge provides an excuse to justify the choice of an alternative scenario where the consequences of a particular action on the environment are conveniently ignored. Here, the unrestrained use of chemical or mechancial techniques can virtually guarantee the successful elimination of undesirable aquatic plants, provided that the costs can be met. This type of situation is obviously undesirable and yet it may be the only option available to water management authorities, especially those in arid regions where water resources are scarce (Ashton *et al*., 1986).

Invasive aquatic plants have caused significant economic problems in most areas of the world and their control by mechanical, chemical and biological means has been the subject of considerable research (Mitchell, 1978). However, whilst effective control and eradication techniques already exist they are often remarkably ineffectual in certain circumstances. In addition, many control techniques are not cost-effective nor is their action always restricted to the target plant of interest. This situation must be corrected so that a proposed control measure can be confined to the target species alone.

The solution to this problem lies in concentrating our efforts towards improving our understanding of the complex interactions between invasive organisms and their environment. Recent research on the biological control of *Salvinia molesta* has emphasized the subtleties of many of these interactions and has highlighted their importance as determinants of successful control (Thomas and Room, 1986).

Therefore, in the case of *Salvinia molesta* invasions, our improved understanding now enables us to make reasonable predictions as to whether or not *Cyrtobagous salviniae* will be an effective biological control agent in a particular environment. While this is a tremendous achievement in itself, further studies are still required to refine this predictive ability for *S. molesta*. A wealth of biological information is available on another notorious invader *Eichhornia crassipes* (Gopal, 1987), a species which has been controlled or eradicated successfully in a number of different countries. However, despite these successes, we still cannot make accurate predictions as to whether a particular management option will succeed or not.

6.4.6 The role of legislation

Legislation plays a vital role in two major arenas of invasive aquatic plant control: namely, avoiding the establishment of invasive plants, and determining the means and manner by which such plants may safely be controlled or eradicated.

Legislation preventing the establishment of invasive plants aims to prevent entry into a country and restrict spread within the country. For both components, specific plants are identified as being undesirable on the basis of their known invasive behaviour. However, legislation seldom prevents the entry of a 'new' species unless it is recognized as a problem in another country. Legislation cannot prevent the unauthorized or accidental introduction of invasive plants.

Where problems have been experienced due to invasive aquatic plants, legislation restricting their spread within the country has received considerable attention. In general terms, such legislation is only feasible when humans are the main, if not the only, dispersal agent (Arthington and Mitchell, 1986). A far more common course of action is to promulgate dangerously invasive plants as 'noxious weeds' and promote public participation in their control. For example, public participation in the dispersal of *Cyrtobagous salviniae* over the Sepik River floodplain was vital to the successful control of *Salvinia molesta* (Thomas and Room, 1986). Noxious weed legislation is important in that it not only empowers water management authorities to control the plant wherever it may be found, it makes such control obligatory (Ashton *et al*., 1986).

The second important role of invasive aquatic plant legislation lies in its regulation of the control techniques that may be used in particular situations and in the stipulations as to the safety precautions that are required For example, the use of herbicides in or near water supplies used for human or stock consumption is a highly sensitive issue and all possible precautions must be taken to ensure the safety and well-being of every water user (Bates, 1976; Ashton *et al*., 1980).

Regional and international legislation relating to invasive plants is gradually changing as we improve our knowledge of the invasive characteristics of particular plant groups (Navaratnam and Catley, 1986). The apparent trend towards a greater stress on preventive measures, away from a more retrospective stance where legislation is only enacted against a particular plant after it has caused a problem, is worthy of special emphasis. At present, the most urgent requirement for improved legislation relates to the nursery, aquarium and aquaculture trades who are responsible for the importation and sale of a considerable number of aquatic plants. Far stricter controls are required to limit their trade in invasive aquatic plants (Ashton *et al.*, 1986).

6.4.7 Integrated control strategies

Our preceding discussions have emphasized the fact that many of the methods currently used to control or eradicate invasive aquatic plants are not always successful when used individually. Furthermore, many of the more effective techniques have undesirable effects on the environment and other aquatic biota and therefore should be used with extreme caution. Ideally, control or eradication strategies must strive to minimize adverse ecological impacts. Such sensitivity to ecological constraints requires the use of an array of techniques, each carefully integrated to coincide with different phases of the target plant's life cycle and, most importantly, geared to the dynamics of the affected ecosystem. This type of approach is increasingly being referred to as 'integrated control'.

Integrated control programmes have a definite advantage over more conventional single-option programmes in that they are highly flexible and are thus better able to deal with unexpected situations. Furthermore, integrated control tends to retain biotic diversity, and thus has the potential to stabilize the system and render it less prone to further invasions. In contrast, the use of a single method, particularly chemical or mechanical techniques, tends to interrupt the existing successional trend by an abrupt simplification of the system. The system reverts to an earlier, less stable successional stage that is once again susceptible to invasion.

Some of the most pressing aquatic plant problems today concern the need to control invasions of nature reserves and natural ecosystems as well as artificial water supply reservoirs where external perturbations must be kept to an absolute minimum. Both types of problems would benefit by the use of integrated control strategies.

To date, several control and eradication programmes directed at specific target plants have employed a range of techniques to accomplish their objectives (Mitchell, 1978; Ashton *et al.*, 1980; Gopal, 1987). However, very few programmes have deliberately integrated a variety of control techniques with the dynamics of the plant and ecosystem. Once again, the greatest barrier to the realization of this ideal is our inadequate knowledge of plant–environment interactions and ecosystem dynamics.

6.5 CONCLUSIONS

It has been stressed that the art and science of controlling water flows is probably the one branch of engineering science that has contributed most to the development of civilization (Baxter and Glaude, 1980). In the process, a wide range of new aquatic ecosystems has been created and numerous existing systems have been modified. Unfortunately, man's skill at, and knowledge of engineering systems have not been matched by his understanding of the intricacies of biological systems and his ability to maintain them in a stable condition (Mitchell, 1974). Coupled to this, the tremendous increase in international travel dramatically increased the ease with which invasive plants are transported around the globe. Consequently, man is not only responsible for the introduction of invasive plants to an area, he also creates conditions which promote their excessive growth. Indeed, most successful aquatic plant invaders are primary colonizers of disturbed habitats (Cook, 1985) and are well represented in the early successional stages of plant communities.

In the light of man's frequent involvement in aquatic plant introductions, it is not surprising that the absence of a long-distance dispersal mechanism does not hinder a plant's invasive potential. Instead, the possession of an efficient short-range dispersal mechanism is sufficient to increase both the likelihood and rate of successful invasion. Another essential attribute of successful invaders is their high reproductive output. Here, the ability to undergo rapid vegetative reproduction is far more important than sexual reproduction. However, the ability to multiply rapidly also places constraints on an invasive plant since the available resources of light, nutrients and space become limited with time unless the plant can regulate its own population size (Gopal, 1987). Invaders such as *Eichhornia crassipes* and *Salvinia molesta* have largely achieved this with their high degree of morphological plasticity, competitive ability and tolerance of a wide range of environmental conditions. Indeed, their success as invaders can be gauged both by their wide distribution and by the attention they have attracted. However, whilest we know most of the attributes that increase the probability of notorious species such as *E. crassipes* and *S. molesta* becoming successfully established in a new environment, we are still unable to make accurate predictions about individual situations.

Sadly, our knowledge of the biology and ecology of other invasive aquatic plants is far less extensive, particularly with regard to their responses to environmental cues and their interactions with the biotic and abiotic features of new environments. This situation is further confounded by the interacting effects of chance and timing (Crawley, 1986) which can either amplify, modify or remove invasion barriers (Breen *et al.*, 1987). Consequently, our predictive ability is limited to the rather banal generalization that the probability of an invasion succeeding will increase as the bioclimatic match between the plant's native and invaded environments improves.

Excessive populations of invasive aquatic plants are a direct consequence of, and an expanding obstruction to, man's increased management and utilization of water resources (Mitchell, 1974). Inevitably, such populations are considered to be undesirable and therefore requiring remedial action. While a wide variety of management options are available, the choice of technique requires a sound understanding of the ecosystem under consideration and the ecology of the invasive plant as well as the likely impacts on the environment of the proposed course of action.

Preference should therefore be given to a stratagem that is both ecologically sound and cost-effective (Mitchell, 1974; Ashton *et al.*, 1986). In this regard, the use of specific biological control agents against target plants is now receiving increased attention as a result of the successes that have been achieved against *Salvinia molesta* (Thomas and Room, 1986). Yet another approach, the use of plant pathogens in combination with phytophagous insects, is also receiving greater attention (Freedom, 1977). Whatever option is chosen, the success or failure of a control programme depends on careful planning, diligent and sustained implementation of the chosen techniques and maintenance of efficient and vigilant follow-up operations. Failure to comply with one more of these requirements will lead to the failure of the programme.

Ultimately, the use of biological control techniques and the development of carefully integrated control programmes hold the greatest promise for the future. Ideally, integrated control programmes against invasive aquatic plants should form part of an overall catchment management programme (Van Schayck, 1986). Unfortunately, our presently inadequate knowledge prevents realization of this ideal. However, since the scale and scope of aquatic plant invasions are increasing world-wide, urgent attention must be given to remedying this deficiency. In addition, it is particularly important that everyone who is involved with the ecology and management of invasive aquatic plants should adopt a far broader outlook and appreciation of the problems involved. We hope that this review will stimulate such an approach.

ACKNOWLEDGEMENTS

Financial support for this work was provided by: the SCOPE International Synthesis Symposium on Biological Invasions, the CSIR's Foundation for Research Development through the Invasive Biota Working Group and the CSIR's National Institute for Water Research. Grateful thanks are due to each of these organizations.

REFERENCES

Agami, M., and Waisel, Y. (1986). The role of ducks in distribution and germination of seeds of the submerged hydrophyte *Najas maritima* L. *Oecologia*, **68**, 473–5.

Andres, L. A. (1977). The economics of biological control of weeds. *Aquat. Bot.*, **3**, 111–23.
Arthington, A. H., and Mitchell, D. S. (1986). Aquatic invading species. In: Groves, R. H., and Burdon, J. J. (Eds.), *Ecology of Biological Invasions, An Australian Perspective*, pp. 34–53. Australian Academy of Science, Canberra.
Ashton, P. J. (1977). Factors affecting the growth and development of *Azolla filiculoides* Lam. In: *Proceedings of the Second National Weeds Conference of South Africa, Stellenbosch University, South Africa, 2–4 February 1977*, pp. 249–68. A. A. Balkema, Cape Town.
Ashton, P. J., Appleton, C. C., and Jackson, P. B. N. (1986). Ecological impacts and economic consequences of alien invasive organisms in southern African aquatic ecosystems. In: Macdonald, I. A. W., Kruger, F. J., and Ferrar, A. A. (Eds), *The Ecology and Management of Biological Invasions in Southern Africa*, pp. 247–57. Oxford University Press, Cape Town.
Ashton, P. J., Scott, W. E., and Steyn, D. J. (1980). The chemical control of the water hyacinth (*Eichhornia crassipes* (Mart.) Solms. *Progr. Water Technol.*, **12** (Toronto), 865–82.
Ashton, P. J., Scott, W. E., Steyn, D. J., and Wells, R. J. (1979). The chemical control programme against the water hyacinth *Eichhornia crassipes* (Mart.) Solms on Hartbeespoort Dam: historical and practical aspects. *South Afr. J. Sci.*, **75**, 303–6.
Ashton, P. J., and Walmsley, R. D. (1976). The aquatic fern *Azolla* and its *Anabaena* symbiont. *Endeavor*, **35**, 39–43.
Ashton, P. J., and Walmsley, R. D. (1984). The taxonomy and distribution of *Azolla* species in southern Africa. *Bot. J. Linn. Soc.*, **89**, 239–47.
Aston, H. I. (1973). *Aquatic Plants of Australia.* Melbourne University Press, Melbourne. 368 pp.
Baker, G. E., Sutton, D. L., and Blackburn, R. D. (1974). Feeding habits of the White Amur on water hyacinth. *Hyacinth Control Journal*, **12**, 58–62.
Baker, H. G. (1965). Characteristics and modes of origin of weeds. In: Baker, H. G., and Stebbins, G. L. (Eds), *The Genetics of Colonizing Species*, pp. 147–68. Academic Press, London.
Barrett, S. C. H. (1978). Floral biology of *Eichhornia azurea* (Swartz) Kunth (Pontederiaceae). *Aquat. Bot.*, **5**, 217–28.
Bates, J. A. R. (1976). Pesticides Safety Precautions Scheme—the registration of aquatic herbicides. In: Robson, T. O., and Fearon, J. H. (Eds), *Aquatic Herbicides.* Monograph No. 16, pp. 25–38. British Crop Protection Council, Cambridge.
Baxter, R. M., and Glaude, P. (1980). Environmental effects of dams and impoundments in Canada: experience and prospects. *Can. Bull. Fish. Aquat. Sci.* **205**, 1–34.
Blackburn, R. D. (1974). Chemical control. In: Mitchell, D. S. (Ed.), *Aquatic Vegetation and its Use and Control*, pp. 85–98. UNESCO, Paris.
Blackman, G. E. (1960). Responses to environmental factors by plants in the vegetative phase. In: Zarrow, M. X. (Ed.), *Growth of Living Systems*, pp. 525–56. Basic Books, New York.
Bond, W. J., and Roberts, M. G. (1978). The colonization of Cabora Bassa, Mozambique, a new man-made lake, by floating aquatic macrophytes. *Hydrobiologia*, **60**, 243–59.
Bowmer, K. H., Mitchell, D. S., and Short, D. L. (1984). Biology of *Elodea canadensis* Mich. and its management in Australian irrigation systems. *Aquat. Bot.*, **18**, 231–8.
Breen, C. M., Rogers, K. H., and Ashton, P. J. (1987). The role of vegetation processes in swamps and flooded plains. In: Symoens, J. J. (Ed.), *Vegetation of Inland Waters*, pp. 223–47, Dr W. Junk, Dordrecht.
Calow, P., and Townsend, C. R. (1981). Energy, ecology and evolution. In: Townsend, C. R., and Calow, P. (Eds), *Physiological Ecology: An Evolutionary Approach to*

Resource Use, pp. 3–19. Blackwell Scientific Publications, Oxford.
Carpenter, S. R. (1980). The decline of *Myriophyllum spicatum* in a eutrophic Wisconsin lake. *Can. J. Bot.*, **58**, 527–35.
Cary, P. R., and Weerts, P. G. J. (1983). Growth of *Salvinia molesta* as affected by water temperature and nutrition. I. Effects of nitrogen level and nitrogen compounds. *Aquat. Bot.*, **16**, 163–72.
Christensen, C. (1940). The pteridophytes of Tristan da Cunha. *Results of the Norwegian Scientific Expedition to Tristan da Cunha*, **6**, pp. 114–16.
Connell, J. H., and Slatyer, R. O. (1977). Mechanisms of succession in natural communities and their role in community stability and organization. *Amer. Nat.*, **111**, 1119–44.
Cook, C. D. K. (1974). *Water Plants of the World: A Manual for the Identification of the Genera of Freshwater Macrophytes.* Dr W. Junk, The Hague. 561 pp.
Cook, C. D. K. (1985). Range extensions of aquatic vascular plant species. *J. Aquat. Plant Manage.*, **23**, 1–6.
Crafts, A. S. (1975). *Modern Weed Control.* University of California Press, Berkeley. 440 pp.
Crawley, M. J. (Ed.) (1986). *Plant Ecology.* Blackwell Scientific Publications, Oxford. 508 pp.
Darwin, C. (1982). *The Origin of Species by Means of Natural Selection*, 6th Edn. John. Murray, London. 458 pp.
Edwards, D., and Thomas, P. A. (1977). The *Salvinia molesta* problem in the northern Botswana and eastern Caprivi area. In: *Proceedings of the Second National Weeds Conference of South Africa, Stellenbosch University, South Africa, 2–4 February 1977*, pp. 221–37. A. A. Balkema, Cape Town.
Elmore, C. D., and Paul, R. N. (1983). Composite list of C4 weeds. *Weed Sci.*, **31**, 686–92.
Elton, C. S. (1958). *The Ecology of Invasions by Animals and Plants.* Methuen, London. 181 pp.
Forno, I. W. (1983). Native distribution of the *Salvinia auriculata* complex and keys to species identification. *Aquat. Bot.*, **17**, 71–83..
Forno, I. W., and Bourne, A. S. (1985). Feeding by adult *Cytobagous salviniae* on *Salvinia molesta* under different regimes of temperature and nitrogen content, and effects on plant growth. *Entomophaga*, **30**, 279–86.
Freeman, T. E. (1977). Biological control of aquatic weeds with plant pathogens. *Aquat. Bot.*, **3**, 175–84.
Gopal, B. (1987). Water Hyacinth. *Aquatic Plant Studies—1.* Elsevier Science Publishers, Amsterdam. 471 pp.
Gopal, B., and Sharma, K. P. (1981). *Water-Hyacinth: The Most Troublesome Weed in the World.* Hindasia Publishers, Delhi. 229 pp.
Gosselink, J. G., and Turner, R. E. (1978). The role of hydrology in freshwater wetland ecosystems. In: Good, R. E., Whigham, D. F., and Simpson, R. L. (Eds), *Freshwater Wetlands: Ecological Processes and Management Potential*, pp. 21–47 Academic Press, New York.
Gray, M. J., Crawley, M. J., and Edwards, P. J. (Eds) (1987). *Colonization, Succession and Stability.* Blackwell Scientific Publications, Oxford. 330 pp.
Grime. J. P. (1977). Evidence for the existence of three primary strategies in plants and its relevance to ecological and evolutionary theory. *Amer. Nat.*, **111**, 1169–94.
Grime, J. P. (1979). *Plant Strategies and Vegetation Processes.* John Wiley & Sons, Chichester. 222 pp.
Grime, J. P. (1985). Towards a functional description of vegetation. In: White, J. (Ed.), *The Population Structure of Vegetation*, pp. 501–14. Dr W. Junk, Dordrecht.
Groves, R. H., and Burdon, J. J. (Eds.) (1986). *Ecology of Biological Invasions: An Australian Perspective.* Australian Academy of Science, Canberra. 166 pp.

Grubb, P. J. (1985). Plant populations and vegetation in relation to habitat, disturbance and competition: problems of generalization. In: White, J. (Ed.), *The Population Structure of Vegetation*, pp. 595–620. Dr W. Junk, Dordrecht.
Haller, W. T., Miller, J. L., and Garrod, L. A. (1976). Seasonal production and germination of *Hydrilla* vegetative propagules. *J. Aquat. Plant Manage.*, **14**, 26–9.
Haller, W. T., Sutton, D. L., and Barlowe, W. C. (1974). Effects of salinity on growth of several aquatic macrophytes. *Ecology*, **55**, 891–4.
Harley, K. L. S., and Mitchell, D. S. (1981). The biology of Australian weeds. 6. *Salvinia molesta* D. S. Mitchell, *J. Aust. Inst. Agr. Sci.*, **47**, 67–76.
Harley, K. L. S., and Wright, A. D. (1984). Implementing a program for biological control of water hyacinth, *Eichhornia crassipes.* In: Thyagarajan, G. (Ed.), *Proceedings of the International Conference on Water Hyacinth*, pp. 58–69. UNEP, Nairobi.
Harper, J. L. (1977). *Population Biology of Plants.* Academic Press, London. 892 pp.
Holdgate, M. W. (1965). The biological report of the Royal Society expedition to Tristan da Cunha, 1962. Part III. The fauna of the Tristan da Cunha Islands. *Phil. Trans. R. Soc., Lond., S B*, 249, 361–402.
Holm, L. G., Plucknett, D. L., Pancho, J. V., and Herberger, J. P. (1977). *The Worst Weeds: Distribution and Biology.* University Press of Hawaii, Honolulu. 609 pp.
Howard-Williams, C., and Gaudet, J. J. (1985). The structure and functioning of African swamps. In: Denny, P. (Ed.), *The Ecology and Management of African Wetland Vegetation*, pp. 153–76. Dr W. Junk, Dordrecht.
Hutchings, M. J., and Bradbury, K. (1986). Ecological perspectives on clonal herbs. *Bioscience*, **36**, 178–82.
Hutchinson, G. E. (1975). *A Treatise on Limnology. Vol. III—Limnological Botany.* John Wiley and Sons, New York. 660 pp.
Huxley, J. S. (1932). *Problems of Relative Growth.* Methuen, London. 276 pp.
Jacot Guillarmod, A. F. M. G. (1979). Water weeds in southern Africa. *Aquat. Bot.*, **6**, 377–91.
Jarvis, M. J. F., Mitchell, D. S., and Thornton, J. A. (1982). Aquatic macrophytes and *Eichhornia crassipes.* In: Thornton, J. A., and Nduku, W. K. (Eds.), *Lake McIlwaine. The Eutrophication and Recovery of a Tropical African Man-Made Lake*, pp. 137–44. Dr W. Junk, The Hague.
Johnstone, I. M. (1986). Plant invasion windows: a time-based classification of invasion potential. *Biol. Rev.*, **61**, 369–94.
Julien, M. H. (1982). *Biological Control of Weeds—A World Catalogue of Agents and Their Target Weeds.* Commonwealth Agricultural Bureau, Slough, United Kingdom. 108 pp.
King, L. J. (1966). *Weeds of the World: Biology and Control.* Interscience Publishers, New York. 526 pp.
Leentvaar, P. (1973). Lake Brokopondo. In Ackermann, W. C., White, G. F., and Worthington, E. B. (Eds.) *Manmade Lakes: Their Problems and Environmental Effects*, pp. 186–196. American Geophysical Union, Washington DC.
Lewis, H. (1962). Catastrophic selection as a factor in speciation. *Evolution*, **16**, 257–71.
Löve, D. (1963). Dispersal and survival of plants. In: Löve, A., and Löve, D. (Eds), *North Atlantic Biota and their History*, pp. 189–205. Pergamon Press, Oxford.
Lovett-Doust, L. (1981). Population dynamics and local specialization in a clonal perennial (*Ranunculus repens*). 1. The dynamics of ramets in contrasting habitats. *J. Ecol.*, **69**, 743–55.
Lumpkin, T. A., and Plucknett, D. L. (1980). *Azolla*: Botany, physiology and use as a green manure. *Econ. Bot.*, **34**, 111–53.
MacArthur, R. H., and Wilson, E. O. (1967). *The Theory of Island Biogeography. Monographs in Population Biology.* Princeton University Press, Princeton, New Jersey. 203 pp.

Macdonald, I. A. W., Kruger, F. J., and Ferrar, A. A. (Eds) (1986). *The Ecology and Management of Biological Invasions in Southern Africa*. Oxford University Press, Cape Town. 324 pp.
McNaughton, S. J. (1966). Ecotype functions in the *Typha* community-type. *Ecological Monographs*, **66**, 297–325.
Manning, J. H. (1979). Establishment of water hyacinth weevil populations in Louisiana. *J. Aquat. Plant Manage.*, **17**, 39–41.
Marshall, B. E., and Junor, F. J. R. (1981). The decline of *Salvinia molesta* on Lake Kariba. *Hydrobiologia*, **83**, 477–84.
Mason, R. (1960). Three waterweeds of the family Hydrocharitaceae in New Zealand. *N. Z. J. Sci.*, **3**, 382–95.
Matthews, L. J. (1967). Seedling establishment of water hyacinth. *Pest Articles and New Summaries* (*PANS*), **13**, 7–8.
Mayer, H. G. (1981). Chemical control of water hyacinth (*Eichhornia crassipes* Solms) considering least environmental disturbance. *Acta Hydrochim. Hydrobiol.*, **9**, 57–68.
Mehta, I., and Sharma, R. K. (1972). Control of aquatic weeds by the Amur in Rajasthan, India. *Hyacinth Control Journal*, **10**, 16–19.
Manges, E. S., and Waller, D. M. (1983). Plant strategies in relation to elevation and light in floodplain herbs. *Amer. Nat.*, **122**, 454–73.
Mitchell, D. S. (1973). Aquatic weeds in man-made lakes. In: Ackermann, W. C., White, G. F., and Worthington, E. B. (Eds.), *Manmade Lakes: Their Problems and Environmental Effects*, pp. 606–11. American Geophysical Union, Washington DC.
Mitchell, D. S. (Ed.) (1974). *Aquatic Vegetation and its Use and Control*. UNESCO, Paris. 135 pp.
Mitchell, D. S. (1978). *Aquatic Weeds in Australian Inland Waters*. Australian Government Publishing Service, Canberra: 189 pp.
Mitchell, D. S. (1979a). *The Incidence and Management of Salvinia molesta in Papua New Guinea*. Office of Environment and Conservation and Department of Primary Industry, Port Moresby. 51 pp.
Mitchell, D. S. (1979b). Formulating aquatic weed management programs. *J. Aquat. Plant Manage.*, **17**, 22–4.
Mitchell, D. S. (1981). The management of *Salvinia molesta* in Papua New Guinea. In: Delfosse, E. S. (Ed.), *Proceedings of the Fifth International Symposium on Biological Control of Weeds*, Brisbane, Australia, 22–29 July 1980, pp. 31–4. CSIRO, Melbourne.
Mitchell, D. S., and Orr, P. T. (1985). *Myriophyllum* in Australia. In: *Proceedings of the First International Symposium of Watermilfoil* (*Myriophyllum spicatum*) *and Related Haloragaceae Species*, Vancouver, British Columbia, Canada, 23–24 July 1985, pp. 27–33. Aquatic Plant Management Society, Vicksburg, USA.
Mitchell, D. S., Petr, T., and Viner, A. B. (1980). The water fern *Salvinia molesta* in Sepik River, Papua New Guinea. *Environm. Conserv.*, **7**, 115–22.
Mitchell, D. S., and Rogers, K. H. (1985). Seasonality–aseasonality of aquatic macrophytes in southern hemisphere inland waters. *Hydrobiologia*, **125**, 137–50.
Mitchell, D. S., and Rose, D. J. W. (1979). Factors affecting fluctuations in extent of *Salvinia molesta* on Lake Kariba. *Pest Articles and News Summaries* (*PANS*), **25**, 171–7.
Mitchell, D. S., and Tur, N. M. (1975). The rate of growth of *Salvinia molesta* (*S. auriculata* Auct.) in laboratory and natural conditions. *J. Appl. Ecol.*, **12**, 213–25.
Moony, H. A., and Drake, J. A. (Eds) (1986). *Ecology of Biological Invasions of North America and Hawaii*. Springer-Verlag, New York. 330 pp.
Moore, A. W. (1969). *Azolla*: Biology and agronomic significance. *Bot. Rev.*, **35**, 17–34.
Navaratnam, S., and Catley, A. (1986). Quarantine measures to exclude plant pests. In: Groves, R. H., and Burdon, J. J. (Eds), *The Ecology of Biological Invasions: An*

Australian Perspective, pp. 106–12. Australian Academy of Science, Canberra.

Newbold, C. (1975). Herbicides in aquatic systems. *Biol. Conserv.*, **7**, 97–118.

Nichols, S. A., and Shaw, B. H. (1986). Ecological life histories of the three aquatic nuisance plants, *Myriophyllum spicatum, Potamogeton crispus* and *Elodea canadensis. Hydrobiologia*, **131**, 3–21.

Noble, I. R., and Slatyer, R. O. (1980). The use of vital attributes to predict successional changes in plant communities subject to recurrent disturbances. *Vegetatio*, **43**, 5–21.

Ostrofsky, M. L., and Zettler, E. R. (1986). Chemical defences in plants. *J. Ecol.*, **74**, 279–87.

Robson, T. O. (1976). A review of the distribution of aquatic weeds in the tropics and subtropics. In: Varshney, C. K. and Rzoska, J. (Eds.), *Aquatic Weeds in South East Asia*, pp. 25–30. Dr W. Junk, The Hague.

Rogers, K. H., and Breen, C. M. (1980). Growth and reproduction of *Potamogeton crispus* in a South African lake *J. Ecol.*, **68**, 561–71.

Room, P. M., Forno, I. W. and Taylor, M. F. J. (1984). Establishment in Australia of two insects for biological control of the floating weed *Salvinia molesta. Bull. Entomol. Res.*, **74**, 505–16.

Room, P. M., Harley, K. L. S., Forno, I. W., and Sands, D. P. A. (1981). Successful biological control of the floating weed *Salvinia. Nature* (*London*). **294**, 78–80.

Room, P. M., and Thomas, P. A. (1985). Nitrogen and establishment of a beetle for biological control of the floating weed *Salvinia* in Papua New Guinea. *J. Appl. Ecol.*, **22**, 139–56.

Room, P. M., and Thomas, P. A. (1986). Population growth of the floating weed *Salvinia molesta*: field observations and a global model based on temperature and nitrogen. *J. Appl. Ecol.*, **23**, 1013–28.

Sastroutomo, S. S. (1981). Turion formation, dormancy and germination of curly pondweed, *Potamogeton crispus* L. *Aquat. Bot.*, **10**, 161–73.

Schlettwein, C. H. G. (1985). Distribution and densities of *Cyrtobagous singularis* Hustache (Coleoptera Curculionide) on *Salvinia molesta* Mitchell in the Eastern Caprivi Zipfel. *Madoqua*, **14**, 291–3.

Sculthorpe, C. D. (1967). *The Biology of Aquatic Vascular Plants.* Edward Arnold, London. 610 pp.

Soerjani, M. (1977). Integrated control of weeds in aquatic areas. In: Fryer, J. D. and Matsunaka, S. (Eds.), *Integrated Control of Weeds*, pp. 121–51. University of Tokyo Press, Tokyo.

Solbrig, O. T. (1981). Energy, information and plant evolution. In: Townsend, C. R. and Calow, P. (Eds), *Physiological Ecology: An Evolutionary Approach to Resource Use*, pp. 274–99. Blackwell Scientific Publications, Oxford.

Sousa, W. P. (1984). The role of disturbance in natural communities. *Ann. Rev. Ecol. Syst.*, **15**, 353–91.

Stebbins, G. L. (1974). *Flowering Plants, Evolution Above the Species Level.* Edward Arnold, London, 399 pp.

Stuckey, R. L., J. R. Wehrmeister, and R. J. Bartolotta (1978). Submersed aquatic vascular plants in ice-covered ponds of central Ohio. *Rhodora*, **80**, 575–80.

Thomas, P. A., and Room, P. M. (1985). Towards biological control of *Salvinia* in Papua New Guinea. In: *Proceedings of the Sixth International Symposium on Biological Control of Weeds, Vancouver, Canada, 19–25 August 1984*, pp. 567–74. Canada Department of Agriculture, Ottawa.

Thomas, P. A., and Room, P. M. (1986). Taxonomy and control of *Salvinia molesta. Nature* (*London*), **320**, 581–4.

Van Donselaar, J. (1968). Water and marsh plants in the artificial Brokopondo Lake

(Surinam, S. America) during the first three years of its existence. *Acta Botan. Neerland.*, **17**, 183–96.

Van Schayck, C. P. (1986). The effect of several methods of aquatic plant control on two bilharzia-bearing snail species. *Aquat. Bot.*, **24**, 303–9.

Van Zon, J. C. J. (1981). Status of the use of grass carp (*Ctenopharyngodon idella* Val.). In: Delfosse, E. S. (Ed.), *Proceedings of the Fifth International Symposium on Biological Control of Weeds, Brisbane, Australia, 22–29 July 1980*, pp. 249–60. CSIRO, Melbourne.

Wace, N. M., and Dickson, J. H. (1965). The biological report of the Royal Society expedition to Tristan da Cunha, 1962. Part II. The terrestrial botany of the Tristan da Cunha Islands. *Phil. Trans. R. Soc. Lond. B*, **249**, 273–360.

Weber, J. A., and Nooden, L. D. (1976). Environmental and hormonal control of turion formation in *Myriophyllum verticillatum*. *Amer. J. Bot.*, **63**, 936–44.

Westlake, D. F. (1963). Comparisons of plant productivity. *Biol. Rev.*, **38**, 385–425.

Whittaker, R. H., and Goodman, D. (1979). Classifying species according to their demographic strategy. I. Population fluctuations and environmental heterogeneity. *Amer. Nat.*, **113**, 185–200.

Wiley, M. J., and Gorden, R. W. (Eds.) (1984). *Biological Control of Aquatic Macrophytes by Herbivorous Carp.* Aquatic Biology Technical Report 1984 (11). Illinois Natural History Survey, Champain, Illinois. 129 pp.

Wright. A. D. (1984). Effect of biological control agents upon water hyacinth in Australia. In: Thyagarajan, G. (Ed.), *Proceedings of the International Conference on Water Hyacinth*, pp. 823–33. UNEP, Nairobi.

Biological Invasions: a Global Perspective
Edited by J. A. Drake *et al.*
© 1989 SCOPE. Published by John Wiley & Sons Ltd

CHAPTER 7
Temperate Grasslands Vulnerable to Plant Invasions: Characteristics and Consequences

RICHARD N. MACK

7.1 INTRODUCTION

From at least the time of de Candolle's publication of *Géographie Botanique Raisonnée* in 1855 the great disparities in the extent of plant invasions worldwide have been the subject of continuing observation, conjecture, and analysis (Hooker, 1860; Darwin, 1872; Wallace, 1892; Salisbury, 1933; Crawley, 1986). Even though much of this inquiry is still based on anecdotes (Thellung, 1912; Ridley, 1930; Elton, 1958), several inextricably linked central questions have emerged. How has evolution seemingly predisposed the native floras in some regions such that members of these floras have repeatedly become naturalized elsewhere? Do these successful invaders come largely from communities less likely to be invaded themselves?

Clear answers are obviously elusive because any taxon reflects the often orthogonal forces in its phylogeny (Gould and Lewontin, 1979; Harper, 1981), a problem only exacerbated for comparisons among communities. But if we are to answer these questions satisfactorily, especially if the answers are to predict the course of future invasions (Forcella, 1985a; Forcella and Wood, 1984; Regal, 1986), sound experimentation and manipulation will stem in part from assembling observations of invasions with as few conflicting (and untestable) explanations as possible. One synthetic approach for organizing these observations relies on using a single biome that displays a wide range of responses to plant invasions (Naveh, 1967; Moore, 1983; Kruger *et al.*, this volume). Despite all the environmental variation within such a large biogeographic unit, the constituent communities still share similar climate, life forms, and resultant physiognomy, thereby eliminating at least some explanations for communities' varying vulnerabilities to invasion.

Temperate grasslands* provide examples to simultaneously explore these

*I define temperate grasslands (steppes) as naturally occurring treeless communities in middle latitudes (between 30° and about 50°); perennial grasses usually dominate these communities because soil moisture or effective precipitation is unfavorable to support the natural establishment of trees on level ground. With further aridity or saline soils shrubs may predominate. Fire and the role of animals do not in themselves explain the distribution of these grasslands.

central questions and their specific components. The temperate grasslands in Australia, South America and western North America (the Intermountain West and California's Central Valley) are among the extreme examples of what Elton (1958) has called, 'the great historical convulsions' of the earth's biota: massive changes in the species composition of once vast communities through the transoceanic transport of alien organisms and their subsequent incursion into new ranges.

In less than 300 years (and in most cases, little more than 100 years) much of the temperate grassland outside Eurasia (a collective area of 2.0×10^6 km^2 or more than the combined areas of Spain, France, Portugal, Belgium, the Netherlands, and West Germany) has been irreparably transformed by human settlement and the concomitant introduction of alien plants. Few other changes in the distribution of the earth's biota since the end of the Pleistocene have been as radical (cf. the spread and proliferation of humans) or as swift (cf. plant and animal invasions on remote islands).

Not all temperate grasslands have proven so vulnerable to plant invasions. The grasslands in Eurasia and especially in southern Africa and central North America have also received many alien plants, but these communities have changed comparatively little, except where plowed (Costello, 1944; Henderson and Wells, 1986). Consequently, within one major biogeographic unit, with representatives on six continents we can compare communities extensively transformed with those little altered by plant invasions.

This chapter is divided into three sections. The first section discusses those readily identifiable traits that made four temperate grasslands vulnerable to massive plant invasions and naturalizations. The second, longer section illustrates consequences of these characteristics in the transformation of the vegetation, while the third section deals with some potential hazards for these vulnerable grasslands.

7.2 TEMPERATE GRASSLANDS VULNERABLE TO PLANT INVASION: THE CHARACTERISTICS

The two quintessential characteristics of temperate grasslands vulnerable to plant invasion are the lack of large, hooved, congregating mammals in the Holocene or longer and dominance by caespitose grasses. These characteristics are so intrinsically related that a clear chronology of cause and effect cannot always be separated. For example, in each of these grasslands caespitose grasses persisted because large mammalian grazers were either rare or missing altogether. Yet in western North America, the phenology of these tussock grasses ensured that bovids remained so sparse that they had been almost extirpated before Europeans arrived.

7.2.1 The lack of large, hooved, congregating mammals

Each vulnerable grassland lacked sufficient large mammalian grazers to affect selection in perennial grasses. The lack of placental mammals in Australia (other than rodents, bats, and the dingo) before the introduction of many mammals with European colonization needs no elaboration (Archer, 1981). The largest native grazers were the red and the Eastern gray kangaroos; both species' adults weigh < 100 kg (Frith and Calaby, 1969). South America's native fauna includes placental mammals, but lacks bovids and sheep (Simpson, 1980), major phylogenetic lines for large, congregating grazers. The Holocene distribution of large grazers in North America represents a curious alternative situation: enormous herds of bison were supported on the Great Plains, yet these large (500 kg), congregating animals occurred only in small, isolated herds in the Intermountain West (Mack and Thompson, 1982) and probably none occurred in California's Central Valley by historic time (Roe, 1951). Elk, deer, and antelope are still prevalent but were apparently never abundant.

The low numbers of these animals in the Intermountain West is reflected in the Holocene fossil record (Schroedl, 1973), in the independent accounts of 19th century explorers (Mack and Thompson, 1982 and references therein), and even in the insect fauna. No members of *Onthophagus*, a widespread genus of dung beetles, are native to western North America, even though 34 species of this genus occur east of the Rockies (Howden, 1966). The phenology of the dominant C_3 caespitose grasses may account for this paucity of bison; in both vulnerable grasslands in western North America the native grasses on zonal soils are all vegetatively dormant by early summer when lactating bison need maximum green forage. The only green forage then available would have been limited to river valleys, thereby placing a severe constraint on the size and distribution of herds (Mack and Thompson, 1982).

7.2.2 The dominance of caespitose grasses

Why are communities dominated by caespitose grasses often so vulnerable to plant invasion? Much of the explanation stems from the morphology and phenology of caespitose or tussock grasses (Mack and Thompson, 1982). Tussock grasses develop by intervaginal tillering in which the emerging tillers remain erect inside the leaf sheath. A tightly-packed cluster of such erect tillers is much more exposed to grazing by ungulates than the low, sprawling form of rhizomatous grasses. In western North America the apical meristem in caespitose grasses becomes elevated as it resumes growth in late winter. Thus, the meristem is placed in jeopardy throughout its growing season to removal by grazers (Branson, 1953; Heady, 1975). Without the ability to readily produce axillary buds, the population of a caespitose grass persists on a site exclusively through sexual reproduction. But in a grazing environment the flowering tillers are often removed. The seedlings of these grasses are also sensitive to grazing. In trials comparing the survival of the seedlings of *Agropyron spicatum*, a dominant caespitose grass in the Intermountain West, with seedlings of the aggressive alien *Bromus tectorum*, *Agropyron spicatum* seedlings routinely died if grazed once by voles. Seedlings of *Bromus tectorum* persisted even if similarly grazed every 2 weeks (Pyke, 1983).

Rhizomatous or turf grasses display alternative development: extravaginal tillering in which the emerging stem splits the leaf sheath. Consequently, the grass develops a sprawling habit because tillers remain prostrate. This crucial difference among grasses in their morphology has probably been under strong selection by grazers; grazed or clipped populations repeatedly display a prostrate habit under genetic control (Warwick and Briggs, 1978; Gray and Scott, 1980). The extent to which large mammalian grazing may have influenced natural selection in grasses is illustrated by the morphology of *Poa pratensis*, a rhizomatous grass now widely distributed as forage in temperate grasslands. *Poa pratensis* displays (1) bud height at ground level throughout the growing season, (2) underground rhizomes, (3) high shoot density, (4) short leaves and usually short stature, (5) a conduplicate stem and leaf, (6) tolerance of puddling (i.e., being covered by mud), and (7) a low flowering to vegetative culm ratio (Mack and Thompson, 1982 and references therein).

The hazards created for a caespitose grass by some grazers are not limited to herbivory; ungulates the size of modern cattle can cause much damage to herbaceous plants with their hooves. The damaged turf of rhizomatous grasses is readily replaced by vegetative propagation, an ability caespitose grasses usually lack (Jewiss, 1972). Such damage is intensified locally by large, gregarious animals such as bison through their wallowing and milling about. The native Holocene grazers in the vulnerable grasslands were either too light (e.g. the guanaco in South America, kangaroos in Australia, deer and antelope in North America) or too sparse (such as elk and bison in western North America) to affect selection in grasses through trampling.

In caespitose grasslands trampling can also alter community composition by destroying the matrix of small plants between the tussocks. For example, in undisturbed communities in the Intermountain West this matrix is occupied by herbaceous perennials, annuals, shrubs, and a cryptogamic crust composed of mosses, lichens, and liverworts. If intact, the crust restricts seedling establishment by preventing the radicle's contact with the mineral soil. But once the crust is broken by hooves, the mineral soil is exposed and seedlings may readily establish (Daubenmire, 1970). Once European settlers arrived, alien plants began to colonize these new and renewable sites of disturbance. Whether through grazing or trampling, or both, the common consequences of the introduction of livestock in the four vulnerable grasslands were destruction of the native caespitose grasses, dispersal of alien plants in fur or feces, and continual preparation of a seed bed for aliens that evolved with large mammals in Eurasia and Africa (Roseveare, 1948; Burcham, 1957; Mack, 1981).

The caespitose habit does not always spell vulnerability to grazing and trampling, although that outcome commonly occurs. The few successful forage grasses naturalized on arid range lands in the Intermountain West are caespitose, most notably the *Agropyron cristatum–Agropyron desertorum* complex. They are native to arid central Asia where strong selection by native grazers has long operated. Among other features, these grasses have flexible resource allocation in which tillers are replaced while new root growth is curtailed (Caldwell *et al.*, 1981; Richards, 1984). These species present intriguing cases in which features have arisen that largely compensate for the possession of the caespitose habit in a grazing, trampling environment.

The comparative invulnerability of the other two temperate grasslands settled by Europeans, the Great Plains of North America and the South African veld, seems tied to their dominance by both rhizomatous and caespitose grasses tolerant of grazing. For the Great Plains this resistance to grazing correlates with the conspicuous replacement of stipoid grasses by genera from both Asia and southern North America (e.g. the rhizomatous genera *Buchloe* and *Bouteloua*), a replacement not duplicated in other temperate grasslands in the western hemisphere. Descendants of these Tertiary immigrants along with genera already present (e.g *Andropogon*, *Sorghastrum*, *Panicum*) (Stebbins, 1975) likely underwent the strong selection by bison beginning in middle Pleistocene that led to their later resistance to livestock (Guthrie, 1970; Mack and Thompson, 1982). The other grasslands in the western hemisphere received different genera, such as *Agropyron* and *Festuca* in western North America and *Piptochaetium* in South America. Why did these other grasslands either not receive or not retain the same late Tertiary immigration as the Great Plains? The answer for western North America may lie in the apparent inability of the descendants of this Tertiary immigration, particularly the C_4 species, to tolerate summer drought (Teeri and Stowe, 1976; Ehleringer, 1978), perhaps the overriding climatic feature of both the Intermountain West and California's Central Valley.

7.3 TEMPERATE GRASSLANDS VULNERABLE TO INVASION: THE CONSEQUENCES

Despite the wide geographic separation of most of the vulnerable grasslands from each other, the chronologies of their transformations by alien plants are remarkably consistent. Each was colonized by western Europeans (mainly Britons and Spaniards) and later intensively settled by their descendants (Australians, Argentineans, and Americans). And each region was eventually developed to raise the cereals, legumes, and forage crops of western Europe. This similarity in agriculture coupled with these regions' similarity in evolutionary history (and the resultant habit of the dominant grasses), overrode many regional differences in climate, soils, topography, and native flora. As a result, each now share much the same alien flora.

7.3.1 The temperate grasslands of South America

Temperate grasslands in South America are restricted to the southern end of the continent but include some of the highest latitude (about 54°) grasslands altered by alien plants. These treeless communities are readily divided into a humid grassland in Uruguay and northern Argentina (centering on the pampas south of Buenos Aires) and an arid steppe in Patagonia to the south. Together these two areas occupy over 1 million km^2 (Soriano, 1979).

The climate of the pampas is characterized by warm often humid summers and mild winters; temperatures usually remain well above freezing throughout the year (Trewartha and Horn, 1980). Annual precipitation can be as much as 1000 mm in the humid pampas near Buenos Aires, but declines markedly in the interior (Trenque Lauquen, 650 mm) and southward (Bahia Blanca, 635 mm). Although there is usually no droughty season, summer receives more precipitation than winter (Rumney, 1968).

The conspicuous dominants in the original pampean vegetation are all caespitose grasses (Cabrera, 1971). Prominent xeric grasses in the pampas south of Buenos Aires form the tallest layer in a multi-layered community and include *Agropyron laguroides*, *Aristida murina*, *Briza subaristata*, *Panicum bergii*, *Piptochaetium bicolor*, *Stipa hyalina*, and *Stipa papposa* (Parodi, 1930). Herbs, such as *Asclepias mellodora*, *Baccharis cordifolia*, *Heimia salicifolia*, *Melica macra*, *Sphaeralcea miniata*, and *Verbena chamaedryfolia*, occupy the shorter layers (Parodi, 1930; Soriano, 1979). With increasing aridity to the west, *Poa ligularis*, *Stipa ichu*, *Stipa tenuissima*, and *Stipa trichotoma* persist (Soriano, 1979).

Patagonia lies south of the pampas and is characterized by lower rainfall (< 500 mm per year) and lower plant coverage than in the pampas, but caespitose grasses are still conspicuous. *Stipa chrysophylla*, *Stipa humilis*, and *Stipa speciosa* along with *Bromus macranthus*, *Hordeum comosum*, and *Poa ligularis* and a shrub, *Mulinum spinosum*, tolerate these xeric conditions (Soriano, 1979). Further south a steppe dominated by *Festuca pallescens* borders the subantarctic forest.

7.3.1.1 The invasion of 'European cardoon and a tall thistle'

The pampas was the site of the earliest documented transformation of a landscape by alien plants. In 1833 HMS Beagle reached Bahia Blanca on the coast of central Argentina, and Darwin elected to explore the country north to Buenos Aires on horseback (Darwin, 1898). In the *Origin of Species* he remarked that the European cardoon (*Cynara cardunculus*) and a tall thistle (*Silybum marianum*) '... are now the commonest [plants] over the whole plains of La Plata, clothing square leagues of surface almost to the exclusion of every other plant...' (Darwin, 1872).

Darwin reckoned the new range of cardoon was already vast because even in southern Uruguay he found, 'very many (probably several hundred) square miles are covered by one mass of these prickly plants, and are impenetrable by man or beast. Over the undulating plains, where these great beds occur, nothing else can now live.' In the Banda Oriental, for example, the few estancias were restricted to valleys because these aliens dominated the zonal soil (Darwin, 1898).

The scenes Darwin reported had probably arisen in less than 75 years von Tschudi (1868) claims the cardoon was not in Argentina before about 1769. Hudson, who lived south of Buenos Aires in the 1840s, gives a graphic description of the environmental transformation caused by these alien plants. 'In places the land as far as one could see was covered with a dense growth of cardoon thistles, or wild artichoke, of a bluish or grey–green colour, while in other places the giant thistle [*S. marianum*] flourished... standing when in flower six to ten feet high' (Hudson, 1923). Prominence of *S. marianum* varied from year to year, but in a 'thistle year' the plants became so prominent they confined travel on horseback to narrow cattle tracks. Darwin (1898) learned that during such years robbers hid among the thistles. When he asked whether robbers were numerous in the area, the oblique answer was, 'The thistles are not up yet.'

By December the thistles' vegetative growth was dead, but the combustibility of the stalks greatly enhanced the danger of fire. The dried stalks of the cardoon and the giant thistle were nevertheless the main, if not the only, source of fuel in the treeless pampas, and large piles of the highly flammable stalks were stacked outside each hut (Hudson, 1923). In addition to altering the frequency and intensity of fire, their thick roots also influenced the structure of the stoneless soil (Haumann, 1927) and may have eliminated native plants locally by shading.

Referring to the cardoon, Darwin (1898) doubted, '... whether any case is on record, of an invasion on so grand a scale of one plant over the aborigines.' Even after myriad invasions over the last century and a half, the impact of these two composites in Argentina rivals the role of any other alien plants anywhere, including *Eichhornia crassipes* and *Salsola kali*. They were eventually controlled only with the extensive plowing of the pampas at the end of the 19th century, although both are still regarded as serious problems in Argentina (Marzocca, 1984).

7.3.1.2 Farming brings new invasions

von Tschudi (1868) claimed *C. cardunculus* arrived in Argentina in the hide of a donkey. Even if apocryphal, this example illustrates that many early plant immigrants probably arrived with livestock, and for 250 years these flat plains were grazed but not extensively plowed. But beginning in the mid-19th century immigrant farmers were encouraged to raise alfalfa as a means of raising even more livestock. Transforming the pampas from pasturage to farm land greatly expanded the opportunity for alien plant entry and establishment. Berg (1877) lists over a 100 vascular plants as adventive near Buenos Aires and in Patagonia; many species in his list are common contaminants of seed lots (e.g. *Capsella bursa-pastoris, Chenopodium album, Marrubium vulgare, Stellaria media*) and were likely spread with farming.

Argentina has always had strong commercial ties to western Europe, and it is not surprising that most alien plants in Argentinean grasslands originated in Eurasia. Parodi (1930) considered 60 species as naturalized (as opposed to merely adventive) in the Pergamino District near Buenos Aires. His extensive list includes such well-known plant invaders as:

Avena fatua	*Avena ludoviciana*	*Brassica nigra*
Bromus hordaceus	*Centaurea melitensis*	*Poa annua*
Cynodon dactylon	*Erodium malacoides*	*Festuca myuros*
Lactuca serriola	*Lolium multiflorum*	*Marrubium vulgare*
Medicago lupulina	*Nasturtium officinale*	*Rumex crispus*
Portulaca oleracea	*Convolvulus arvensis*	*Silybum marianum*
Sisymbrium officinale	*Stellaria media*	*Urtica urens*

Parodi considered 16 species, including *Avena fatua, Avena ludoviciana, Briza minor, Bromus hordaceus, Festuca myuros, Genraium dissectum*, and *Lolium multiflorum* so common in fields as to be indistinguishable in their roles from natives. *Lolium multiflorum* replaces alien dicots such as *Conyza bonariensis* and *Carduus acanthoides* in succession in abandoned agricultural fields, although it may be replaced later by native perennial grasses (D'angela *et al.*, 1986).

The introduction, spread and persistence of many aliens can certainly be attributed to their close association with crops. The seeds of *Avena fatua* are common contaminants of wheat and barley, and seeds of *Lactuca serriola* and *Stellaria media* often occur in seed lots of alfalfa (Mack, 1986). Many annual bromes undoubtedly arrived in the pampas in contaminated wheat and are now naturalized (e.g. *Bromus japonicus, Bromus mollis, Bromus racemosus*) (Parodi, 1947).

Alien annual bromes are also prominent in the semi-arid steppe of Patagonia. Perhaps the most widespread alien in this group is *Bromus tectorum*, which has occupied many disturbed sites and occurs along roadways (Soriano, 1956a). Other alien bromes now naturalized in this most southerly temperate grassland include *Bromus rigidus, Bromus rubens, Bromus secalinus* and *Bromus sterilis* (Parodi, 1947). Although alien grasses are prominent, other Eurasian aliens such as *Capsella bursa-pastoris, Erodium cicutarium, Erysimum repandum, Microsteris gracilis, Sisymbrium altissimum*, and *Taxaracum officinale* are also common (Soriano, 1956a, 1956b).

More recent immigrants pose further threats in the pampas and Patagonia. Marzocca (1984) lists several dozen aliens deemed so serious as to be officially considered 'plagues of agriculture' in Argentina. His list includes *Carduus nutans, Centaurea calcitrapa, Cyperus rotundus, Kochia scoparia, Salsola kali*, and *Sorghum halepense*. He also lists *Diplotaxis tenuifolia* which was deliberately imported as a source of nectar for honeybees but is now spreading along roadsides in Patagonia (G. de Fosse, personal communication). *Senecio madagascariensis* typifies many new plant immigrants that may become serious invaders. Although found in the 1940s around Bahia Blanca, the alien has since spread to the southern pampas where it inhabits both agricultural fields and disturbed sites (Verona *et al.*, 1982). The massive transformations of Argentinean vegetation that began with the introduction of the 'European cardoon and a tall thistle' clearly continue.

7.3.2 Grasslands in the Intermountain West of North America

I have recently reviewed the consequences of alien plant invasions in the intermountain grasslands of western North America (Mack, 1981, 1984, 1986). Consequently, the purpose of my brief comments here is to provide a basis for comparison with these other vulnerable grasslands.

The grasslands of the Intermountain West occur in a geomorphologically diverse region bounded by the Rocky Mountains to the east and the Cascade–Sierra Nevada Ranges to the west. To the north these grasslands border coniferous forest at the convergence of the Cascade and Rocky Mountains in central British Columbia; to the south they form a regional ecotone with desert in southern Nevada. The region's geomorphological diversity includes both broad plateaus drained by large rivers (principally the Columbia and the Snake) and many small forested mountain ranges in Oregon, Utah, and especially Nevada. Consequently, the area originally supporting temperate grassland appears in map view as a matrix surrounding these isolated uplands (Daubenmire, 1969).

The regional climate is influenced principally by the annual movement of the prevailing Westerlies and the reduction of the moisture in these air masses as they traverse the Cascade Range and the Sierra Nevada. As a result precipitation, including snow, is received primarily in autumn and winter. Summers are hot and dry (Daubenmire, 1970).

Where undisturbed, the temperate grasslands are either dominated solely by caespitose grasses (*Agropyron spicatum, Festuca idahoensis, Stipa comata*, and *Poa secunda*) or these grasses share dominance with drought-tolerant shrubs (principally *Artemisia tridentata* but also *Sarcobatus vermiculatus, Chrysothamnus nauseosus* and *Atriplex confertifolia*). Prominence of shrubs is usually in inverse relation to effective precipitation. The matrix between caespitose grasses and shrubs in these communities is occupied by annuals and perennial herbs, particularly composites (e.g. *Balsamorhiza sagittata, Helianthella uniflora*) and legumes (*Astragalus* spp. and *Lupinus* spp.). In addition the soil surface is covered by a thin cryptogamic crust (Daubenmire, 1969, 1970).

There were few permanent European settlements in this vast region until 1870. But the repeated discovery of gold soon after the American Civil War, followed by recognition of the potential of the Columbia and Snake River drainages for growing wheat, sparked rapid and intensive settlement from 1870 to 1890

(Meinig, 1968). Tracts unsuitable for crops were rapidly converted to pasturage. Even in areas never plowed, livestock soon destroyed the native communities. Here as in other vulnerable grasslands, this continual disturbance greatly facilitated the establishment of aliens (Mack, 1981).

Alien plants appeared with the first European settlements and became more diverse and conspicuous as settlement increased. *Agropyron repens, Hordeum pusillum*, and *Erodium cicutarium* appear in the few lists of plants collected before 1850 (Mack, 1986). Some, such as *Erodium cicutarium*, may have arrived as seeds attached to livestock. But the entry for most was through contaminated lots of crop seeds. Repeatedly agriculturists of the era complained about the accidental (or even deliberate) contamination of seed lots. By 1900 Seed Purity Laws had been enacted, but many noxious aliens had already entered the region; many more arrived later. For example, by 1929 the alien flora within five counties in eastern Washington and adjacent Idaho had grown to 200 species (Mack, 1986).

The speed and extent of these regional invasions was facilitated by a railroad system established simultaneously with the wave of human immigrants in the late 19th century. Many aliens were first recorded along railroad tracks (e.g. *Bromus tectorum* and *Salsola kali*) (Mack, 1981, 1986). As a result of this effective means of dispersal, some aliens filled their new ranges in as little as 40 years, such as *Bromus tectorum* which covered $> 200\,000\ km^2$ between *ca.* 1890 and 1930 (Mack, 1981).

Although many aliens arrived and became naturalized, probably less than a dozen have become community dominants (e.g. *Avena fatua, Bromus tectorum, Cirsium vulgare, Elymus caput-medusae, Hologeton glomeratus, Poa pratensis, Salsola kali*, and *Sisymbrium altissimum*) (Young *et al.*, 1972; Yensen, 1981). Yet these species along with the natives, *Chrysothamnus nauseosus, Matricaria matricaroides, Plantago patagonica*, and *Achillea lanulosum*, have irreparably altered much of the regional vegetation in less than 50 years (Daubenmire, 1970). Neither the areal extent of this grassland, nor its floristic diversity, nor its interruption by mountains has had any measurable effect in retarding the spread of aliens. The outlook is only for more change in the vegetation as new immigrants (e.g. *Aegilops cylindrica, Euphorbia esula, Centaurea* spp., *Isatis tinctoria, Ventanata dubia*) displace the old through the same agencies that fostered earlier invasions: disturbance and transport.

7.3.3 Grasslands in the Central Valley of California

Between the Coast Ranges of California and the Sierra Nevada lies the Central Valley. This immense elongate trough is the drainage for two major rivers, the Sacramento and the San Joaquin, and their tributaries. Although part of the Valley originally contained marshes along these rivers, the zonal soils supported temperate grasslands. The Valley's climate is influenced by much the same forces that determine climate in the Intermountain West: the annual movements of the prevailing Westerlies and the orographic effect caused by the mountain ranges to the west. In summer the Westerlies move inland far north of California, and little precipitation is received; summers are hot and mostly dry. Most precipitation falls in winter when the Valley lies within the storm track of the Westerlies. In any season however precipitation is reduced as it passes over the Coast Ranges (Major, 1977).

Other than remote islands, the vegetation in few other areas has been altered so completely by plant invasions. Consequently conjecture surrounds most of our knowledge of these grasslands, although the few fragments of putative pristine vegetation suggest *Stipa pulchra* was the chief dominant. It was associated with other perennial grasses including, *Aristida divaricata, Elymus glaucus, Festuca idahoensis, Koeleria cristata, Muhlenbergia rigens, Stipa cernua*, and *Stipa coronata*; all are caespitose (Heady, 1977; Daubenmire, 1978; Bartolome *et al.*, 1986). Among the native perennial grasses only *Elymus triticoides* is a sod-former (Burcham, 1957). Shrubs were not conspicuous except on alkali or saline soils, such as in the southern San Joaquin Valley (Heady, 1977).

The invasions of alien plants began with Spanish settlement in 1769, and by 1823 Spanish missions dotted the California coast as far north as Sonoma (Parish, 1920). Hendry's (1931) dissections of the adobe bricks used to build these missions reveal aliens, such as *Avena fatua, Brassica nigra, Centaurea melitensis, Chenopodium album, Hemizonia congesta, Hordeum pusillum*, and *Poa annua*, had entered California by 1824; *Erodium cicutarium, Rumex crispus*, and *Sonchus asper* had entered even earlier. The extent to which feral herds were responsible for spreading alien plants through the Central Valley cannot be accurately established. By at least 1772 the Spanish were entering the Valley, and it is likely they left livestock behind in their travels (Burcham, 1957). These animals and their descendants could have contributed to the introduction and dissemination of aliens. For example, *E. cicutarium* could have been easily spread by animals because its corkscrew awn tenaciously adheres to fur.

By whatever mechanisms, *Avena fatua* gained prominence early. Leonard traversed the Central Valley in 1833 and saw, 'a large prairie covered with wild oats—which at this season of the year when nothing but the stock remains, has much the appearance of common oats' (Leonard, 1934). The prominence of wild oats was independently cited by Bryant (1848) who travelled along the Mokelumne River, a tributary of the San Joaquin River, in 1846 and 1847.

The role Spanish settlements played in speading alien plants was dwarfed by the human immigration sparked by the discovery of gold in the Central Valley in 1848. Not only did this event enormously increase the opportunity for alien plant entry in contaminated seed lots, imported forage, and packing materials, but American settlers also created a huge new market for the Central Valley's livestock. The huge numbers of cattle and sheep recorded for the Central Valley from the mid-19th century onward testify to the tremendous grazing and trampling these animals must have exerted on the communities dominated by caespitose grasses (Burcham, 1957). In a pattern repeated 25 years later in the Intermountain West, the rapid destruction of native grasslands by livestock sparked appeals for the importation of forage species (Bolander, 1866; Cronise,

1868). Such appeals were soon answered with the deliberate introduction of *Cynodon dactylon, Lolium perenne, Medicago lupulina*, and *Sorghum halepense* (Robbins, 1940).

Concomitant with the massive increase of livestock the native grasses entered a devastating cycle of human-induced changes and natural calamities. Both seedlings and adult plants of these native grasses were destroyed by heavy trampling and grazing; losses not readily replaced because the native grasses lack a large seed bank. Episodes of both severe drought and flooding between 1828 and the 1860s and the continual activity of livestock further combined to disrupt the age structure within these slow-growing plants. In contrast the alien annuals, such as *Avena fatua, Bromus rubens*, and *Festuca myuros* were much more tolerant of grazing. Even if devastated by the vagaries of weather, recruits of the alien species could emerge from a persistent seed bank (Baker, 1978).

Replacement of the native plants with aliens was not caused totally by livestock. By 1865 much of the Central Valley was devoted to the production of wheat, barley, and alfalfa (Brewer, 1883a), and this agriculture allowed many new aliens to be accidentally introduced. Many species in Robbins' (1940) list of early plant immigrants are common contaminants in the seed lots of legumes (e.g. *Anthemis cotula, Capsella bursa-pastoris, Plantago lanceolata, Plantago major, Portulaca oleracea, Rumex acetosella, Rumex crispus, Stellaria media*) or cereals (e.g. *Bromus secalinus, Convolvulus arvensis, Festuca myuros, Poa annua*), or both (e.g. *Sonchus asper*).

By 1880 transformation of the vegetation in the Central Valley to an annual grassland was virtually complete. Robbins (1940) lists 91 alien species established in California by 1860. His list includes composites such as *Cirsium arvense*, grasses such as *Avena fatua, Bromus rubens, Bromus rigidus*, and *Cynodon dactylon*, and other commonly dispersed species (*Marrubium vulgare, Verbascum thapsus*, and *Portulaca oleracea*).

The early plant immigrants have not always retained their role as dominants as newer immigrants reached the Central Valley. According to Burcham (1957), the dominance of alien plants in California, including the Central Valley, has gone through four phases.

Phase I dominants: (pre-1860)	*Avena fatua* and *Brassica* sp. (*nigra*?).
Phase II dominants and associates: (*ca.* 1860–1880)	*Erodium cicutarium, Hordeum leporinum, Bromus* spp., *Gastridium ventricosum*
Phase III dominants: (early 1880s)	*Hordeum leporinum, Hordeum hystrix, Bromus rubens, Madia sativa, Centaurea melitensis*
Phase IV dominants: (modern)	*Elymus caput-medusae, Aegilops triucialis, Brachypodium distachyon*

Heady (1977) concurs with this basic scheme or replacement, while Baker (1978) views a three-phase transformation.

The evidence for some of Burcham's phases is equivocal. Contemporary accounts agree *Avena fatua* was the most conspicuous and probably the most prominent alien from 1840 to 1860 (Bolander, 1866; Brewer, 1883b). But by 1868, Cronise reported wild oats were, 'fast disappearing' in California due to overgrazing. In documenting the slow rise to prominence of *Erodium cicutarium* (filaree) Brewer (1883b) mentions Nutall collected this species far in the interior of the state by 1836. Although Brewer claims filaree increased until perhaps 1865 to 1870 (simultaneous with the decline in *Avena fatua*), he clearly states that in comparison to wild oats, filaree, 'never had such possession of the soil anywhere.' *Hordeum leporinum* had arrived by the 1860s, but Brewer describes it only as occurring where overgrazing is too severe for either filaree or wild oats.

The case for Burcham's Phase III occurring in the Central Valley is not substantiated by his references: neither Brewer (1883b) nor Hilgard *et al.* (1882) mentions these aliens as dominant at that time. Instead Hilgard's (1891) discussion of important aliens in California is a more reliable account from which the relative status of aliens in the Central Valley can be gleaned. Apparently *Erodium cicutarium* was still prominent along with *Brassica nigra, Bromus mollis, Centaurea melitensis, Centaurea solstitialis, Eremocarpus setigerus, Erigeron canadense, Lolium temulentum*, and *Silene gallica*.

As Robbins (1940) points out, the absence of some aliens in Hilgard's list that became important in the 20th century, such as *Salsola kali*, suggests these aliens arrived or gained prominence more recently, or both. Late entry is perhaps best documented for *Elymus caput-medusae*, a dominant in Burcham's Phase IV. It was first recorded on the west coast in 1887 (McKell *et al.*, 1962) but is now an important alien in the Valley (Heady, 1977; Baker, 1978). Whatever the chronology of dominance, the vegetation in the Central Valley has long borne virtually no resemblance to the pre-settlement plant communities. The region illustrates the current extreme to which a temperate grassland can be altered by disturbance and plant introductions.

7.3.4 Australian temperate grasslands

The temperate grasslands of Australia defy clear delineation because so much of the continent, although arid is not temperate, and much temperate forest has been converted to grassland. Perhaps for these reasons Moore (1970a) and his collaborators view the extent of temperate grasslands as confined to a few small areas in eastern South Australia, and the tablelands in Victoria and New South Wales. Groves and Williams (1981) largely share this view. Much larger areas in southern Australia are, however, routinely considered shrub steppe and are environmentally similar to the arid shrub steppe; i.e. temperate, grassland, in the Intermountain West and Patagonia. From the standpoint of the consequences of alien plant invasions for these treeless areas (as well as much grassland created from sclerophyllous woodland and eucalypt forest), they will be considered collectively in my discussion.

The interaction between precipitation and temperature dictates the extent of grasslands in Australia. Precipitation for most of these grasslands results from systematic cyclonic storms derived from maritime polar air; the bulk of this precipitation falls as rain in winter. Annual rainfall ranges from > 600 mm in New South Wales to about 250 mm along the diffuse border between shrub-steppe and desert in the center of the continent. Although frost may be frequent, it is rarely persistent; monthly average temperatures in summer can exceed 25 °C (Rumney, 1968; Beadle, 1981).

Here as in the other three invaded temperate grasslands I have considered, the dominant native grasses are all caespitose (Beadle, 1981), and especially in southeastern Australia they are now largely confined to railroad right of ways, cemeteries, and other sites long protected from grazing (Groves, 1965). Perhaps the most wide ranging caespitose grass, even if not dominant throughout its range within the mesic to humid grasslands, is *Themeda australis. Dichelachne crinita, Eryngium rostratum, Poa caespitosa* sensu lato, *Stipa aristiglumis*, and *Stipa bigeniculata* were common associates or dominant locally (Patton, 1936; Groves, 1965; Connor, 1963, Moore, 1970b). Large areas in southern Australia are dominated by chenopodiaceous shrubs with a matrix of perennial grasses and smaller forbs. The most common shrubs are *Atriplex vesicaria, Kochia sedifolia, Kochia pyramidata*, and *Kochia astrotricha* (Perry, 1970; Beadle, 1981).

European settlement of temperate grasslands in southeastern Australia began by at least 1830 (Willis, 1964) and was largely completed in New South Wales by 1880 (Moore and Biddiscombe, 1964). The accompanying introduction of livestock led to a pronounced and usually rapid shift in these grasslands from dominance by perennial caespitose grasses, which produce most vegetative growth in summer, to communities dominated by alien annual grasses active in winter. The phenology of the wide ranging dominant, *Themeda australis*, has been often cited as one reason for this rapid species replacement (Moore, 1970b). In southeastern Australia maximum vegetative growth in *Themeda australis* occurs in summer when soil moisture is lowest (Groves, 1965). Consequently, this perennial may have been at a competitive disadvantage with aliens for which vegetative growth coincided with more abundant soil moisture in the winter.

As with the reconstruction of any 19th century events in sparsely settled grasslands, the epidemiologies of the early plant invasions in southern Australia are sketchy. Some of the earliest alien plant populations were established at Sydney, and many of the species Brown found at this settlement in 1802–1804 were candidates for introduction in the temperate grasslands to the west: *Cotula coronopifolia, Poa annua, Plantago major, Silene gallica*, and *Urtica urens* (Britten, 1906). Kloot (1983) estimates there were about 100 species naturalized in South Australia by 1855. Although not all reached the interior upon arrival, his list includes many weeds now common in temperate grasslands worldwide (e.g.

Avena fatua, Centaurea melitensis, Cynodon dactylon, Cynara cardunculus, Galium aparine, Rumex acetosella). *Silybum marianum* must have arrived some time earlier because its spread along with the contemporaneous invasion of *Cirsium vulgare* prompted the Thistle Act of 1851. These species were but a small fraction of the alien flora that was to descend upon southern Australia once human immigration soared in the remainder of the 19th century.

From the outset of settlement in southern Australia much of these temperate grasslands have been intensively managed as pastures rather than range land for cattle and sheep. This intense management has involved not only the deliberate introduction of numerous pasture grasses from Europe (e.g. *Poa pratensis, Lolium perenne*) but also the widespread application of superphosphate and the accumulation of soil N through the introduction of *Trifolium* spp. These practices in conjunction with alteration of the fire frequency and the continual disturbance caused by livestock have caused permanent species replacements. For a native caespitose grass the combination of fire followed by continued grazing may eliminate so many vegetative shoots that regeneration is thwarted (Groves and Williams, 1981).

Moore (1970a) has outlined the conversion as following a predictable scheme:

Tall warm season perennial grasses
↓—grazing
Cool season perennial grasses
↓—grazing
Short cool season perennial grasses
↓—grazing
Short cool season perennial grasses
| Species from more arid communities ↙
Short warm season perennial grasses
↓—grazing
Short cool season perennial species
|
Short warm season perennial species
| Mediterranean annuals ↘
Introduced cool season annuals
↓—grazing and superphosphate
Introduced cool season annuals
|
Native and introduced warm season annuals
↓
Introduced nitrophilous annuals and biennials

One of the early and most widely introduced aliens in both natural and derived grasslands in Australia has been *Trifolium subterraneum* and allied species. By raising the N or P level in the soil, clovers permit the entry of noxious aliens such as *Carduus pycnocephalus*, *Cirsium vulgare*, and *Silybum marianum* (Moore, 1959; Michael, 1970).

Even with differences in agricultural practice, the temperate grasslands in Australia share many of the same invaders naturalized elsewhere. In addition to those already mentioned in this section, species naturalized in Australia and shared among vulnerable temperate grasslands include *Avena ludoviciana*, *Erodium cicutarium*, *Hordeum glaucum*, *Hordeum leporinum*, *Stellaria media*, and *Xanthium* spp. (Michael, 1970; Beadle, 1981; Michael, 1981). Other aliens have become more prominent in Australia than elsewhere, such as *Echium plantagineum*, *Nasella trichotoma*, and *Oxalis pes-caprae*. Whatever the circumstances, the range of these aliens and newer immigrants is likely to expand (e.g. *Carduus nutans*, Medd and Smith, 1978). Kloot (1983) estimates five to six species per year become naturalized in South Australia alone, the same rate since *ca*. 1836. As a result of this continuing onslaught of aliens, Williams (1985) views the temperate grasslands along with other Australian vegetation as not yet in a 'steady state,' but certain to experience further consequences from plant invasions.

7.4 NEW POTENTIAL HAZARDS

At least as early as Columbus' second voyage to the western hemisphere, Europeans and their colonial descendants have made transoceanic introductions of plants (Crosby, 1952). Temperate grasslands are superbly well suited for growing most cereals, animal forage, and many other crops. As a result, these regions have probably been subjected more to the consequences of this transoceanic traffic than any other biome, with the possible exception of tropical islands. The extent and source of these plant introductions reflect in part the dietary preference of Europeans, but also the origin of many food and forage species in Eurasia. Temperate North America, for example, has contributed surprisingly few important cultivated crops (Hodge and Erlanson, 1956) or forage species (Hartley and Williams, 1956). Whenever the native flora in North America or elsewhere was found lacking in a food or forage crop, plants have been imported and established by governments or private ventures (Wickson, 1887; Ryerson, 1976; Williams, 1985).

These deliberate introductions have sometimes turned disastrous. Vulnerable temperate grasslands have repeatedly been the site of plant immigrants that escaped the bounds of crop field or pasture and are now obvious hazards to the native vegetation. *Sorghum halepense* was deliberately introduced in Argentina and the United States as a forage crop (Wickson, 1891) and is now a serious pest (Holm *et al*., 1977). *Bromus tectorum* was introduced at least once in the

Intermountain West for the same purpose (Mack, 1981), and *Salsola kali* may have been deliberately introduced for the same purpose in the Dakotas (Dewey, 1894). Other aliens with medicinal or even ornamental value have readily established outside cultivation in a new range (e.g. *Datura stramonium, Marrubium vulgare, Nepeta cataria, Silybum marianum*).

The potential for further detrimental introductions, in a sense for history to repeat itself, arises in several areas of current agricultural research and commerce. Even today insufficient regard is often given to the potential hazard created by deliberate releases, as illustrated by *Kochia prostrata.* Introduced in the United States in the early 1960s because of its reputed value as forage in the Soviet Union (Keller and Bleak, 1974), this small shrub is now being released for commercial use in the western United States (Stevens *et al.*, 1985). Evaluation of this alien for release has apparently centered on its ability to tolerate its new range and its palatability for livestock (Keller and Bleak, 1974). The only detrimental feature investigated (and resolved) dealt with the possibility the plant might produce toxic amounts of oxalate (Davis, 1979). Its potential to invade sites for which it was not intended has apparently never been critically evaluated, although the shrub can spread rapidly by seed which it produces in huge quantities (about 1700 kg/ha/year) (Stevens *et al.*, 1985).

What effect could *Kochia prostrata* have on native grasslands? Stevens *et al.* (1985) claim the shrub does not reduce the density of established perennials but will reduce the dominance of *Bromus tectorum* and *Halogeton glomeratus.* Reports of competition with such aggressive aliens are nonetheless disturbing because the competitive ability of the seedlings of perennial grasses such as *Agropyron spicatum* is much lower than is the ability of *Bromus tectorum* (Daubenmire, 1970). Rather than an advantage, the ability of *Kochia prostrata* to colonize the space between caespitose grasses (Stevens *et al.*, 1985) may hasten the demise of native communities. Without acquired pests or parasites (Moore *et al.*, 1982) little may check its spread in western North America. The lack of assessment of the potential detrimental features of *Kochia prostrata* is surprising; another alien congener, *Kochia scoparia*, once also thought to be a desirable forage species, has instead become an invasive weed (Forcella, 1985b).

Ridley's (1930) tome on plant dispersal is testimony to the myriad means by which plants have been transported to new ranges; yet bizarre means still arise. Seed mixtures of both aliens and natives are now commercially available in the United States under such names as 'native wildflower mixtures' and 'Wildflower Meadow.' One seed purveyor even offers his wares (e.g. *Digitalis purpurea* and *Papaver rhoeas*) in packets prepared as attractive postcards. Consequently, the buyer may unwittingly disseminate aliens including, *Centaurea cyanus, Cosmos sulphureus, Cichorium intybus, Festuca ovina,* and unknown seed contaminants on abandoned ground in the misguided notion that such assemblages recreate an aesthetic, yet natural, community. The spread of plants without regard to their invasive ability occasionally verges on negligence. For example, *Hedychium gardnerianum* (Kahili ginger) presents a serious eradication problem in Hawaii (Smith, 1985), yet seeds and cuttings of this aggressive alien are still sold locally for outdoor ornamental plantings. Although few serious naturalizations will probably arise in temperate grasslands from the dissemination of such commercial seeds (and their seed contaminants), this practice is nonetheless ludicrous and should be stopped.

Great care should even be exerted in selecting the parentage of hybrids planned for introduction. Further hybridization with natives may occur after release with the offspring displaying detrimental features of the parents (Baker, 1972). For example, fertile hybrids between the native *Agropyron spicatum* and the alien *Agropyron repens* (quackgrass) are being released in the Intermountain West. The two hybrid strains so far produced incorporate the nutritional value and climatic tolerance of *Agropyron spicatum* with the ability of *Agropyron repens* to tolerate livestock and to spread via rhizomes (Asay and Hansen, 1984). But *Agropyron repens* seems a poor choice for a parent; it is a competitively aggressive weed in many agricultural situations (Holm *et al.*, 1977) and is one of the few plants for which claims of allelopathic potential may be warranted (Buchholtz, 1971). Despite assurances that the vegetative spread inherited from *Agropyron repens* is '...under genetic control and can be succesfully altered through selection' (Asay, 1983), the ominous potential exists of vegetatively-spreading weedy hybrids forming in nature.

The potential for uncontrolled hybridization also exists as new genotypes of forage species are created through techniques in recombinant DNA genetics. Alfalfa is one of the few forage species from which a hardy vascular plant can be grown from callus (Bingham *et al.*, 1975): a critical step in producing a new recombinant product. Consequently, it may provide the first genetically-altered material released in grasslands (e.g. a genotype with enhanced disease or herbicide resistance). Alfalfa is however an obligate outcrosser, and there is the unevaluated possibility that such an alfalfa recombinant could hybridize with weedy relatives, thereby significantly extending the range of any wild hybrids (M. Kahn, personal communication). The benefits of creating new genotypes by these techniques are obvious, and this approach should be fostered. But any releases should be tempered with knowledge that new genotypes are functionally similar to alien plants; potential consequences of their release need not be unforeseen given the history of agricultural invaders (Regal, 1986).

Any evaluation is long overdue on the merits of deliberately introducing plants into these four vulnerable grasslands; deliberate plant releases have occurred for at least 100 years in each region, the releases occur under governmental aegis (adding to the inertia against change), their effects (both beneficial and detrimental) are largely irreversible, and more releases are imminent. Rather than attempt a ban on future plant introductions, attempts should be made to develop guidelines to identify an invasive plant before release. In addition to an evaluation of the ecological traits of any proposed plant introduction (e.g.

dispersal mode, habitat specificity, reproductive potential, response to disturbance, competitive ability with natives), such guidelines will need to reflect a move beyond the largely *post hoc* explanations of naturalizations that now dominate the biology of invasions (Harper, 1965; Harper as cited in Holdgate, 1986). Our current understanding is uncomfortably reminiscent of the epigram often told about economics: any economist can explain in detail the causes of the last business recession, but no economist can predict the timing and cause of the next one. By this standard both economics and the biology of invasions appear as disciplines, but not yet sciences. One informative approach in the biology of invasions would involve comparative studies of the population biology of closely related, congeneric pairs living together: one naturalized, the other a less prevalent native or alien. The necessity of such investigations in vulnerable temperate grasslands is apparent because these regions will continue to attract much of the effort for plant introduction.

7.5 SUMMARY AND CONCLUSIONS

Each temperate grassland has had one of two evolutionary histories: long-term inhabitation by large, hooved, congregating mammalian grazers that imparted a resistance to plant invasions or no recent association with these grazers and correspondingly low resistance to plant invasion and community alteration. Resistance is expressed chiefly in the prominence and proliferation of rhizomatous grasses, the products of strong selection by continual grazing and trampling. In contrast those grasslands destined to be destroyed by plant invasions (grasslands in Argentina and Uruguay, the Intermountain West of western North America, the Central Valley of California, and southern Australia) were all dominated by caespitose grasses before European settlement. Both as adults and juveniles, these grasses display little tolerance of either grazing or trampling and are quickly eliminated from a community when disturbance is suddenly increased.

The level of disturbance increased in all temperate grasslands outside Eurasia with colonization by Europeans. The four vulnerable grasslands were rapidly colonized from *ca.* 1780 to *ca.* 1880. Each was first settled as pasturage for livestock but was soon converted in varying degrees to cereal agriculture. The inherent vulnerability of these regions to continual disturbance coupled with the massive entry and spread of aliens in contaminated seed lots, ensured the replacement of native species by aliens was intense, pervasive, and swift.

With so many similarities in the history and character of settlement, these vulnerable grasslands share many alien species, including *Silybum marianum, Cynara cardunculus, Salsola kali, Cirsium vulgare, Avena fatua,* and annual bromes. Although differing in ecological amplitude, all prominent aliens share the tolerance or requirement of continual disturbance as found in agro-ecosystems, a feature missing among the natives. Many of these invaders originated in the grasslands of Eurasia; the same selection forces that predisposed the Eurasian grasslands to resist invasion also produced many aggressive invaders.

Perhaps a third of the naturalized flora in these vulnerable grasslands were introduced as forage, or for medicinal or ornamental purposes. There are clearly enormous risks associated with deliberately releasing alien plants in new ranges. Although plant introductions have long been encouraged by the governments in Argentina, Australia, and the United States, the overall wisdom of this practice for pasturage has never been formally discussed, much less critically evaluated. Any discussion on future conservation in these grasslands, however, may well be made moot by the continuing release of invasive cultivars.

It is the irony of temperate grasslands that some have produced many of the most aggressive plant invaders, while others are among the most susceptible communities to invasions. This extreme dichotomy illustrates that floras need not evolve on remote islands to reflect the lack of those selective forces, such as continual disturbance, suddenly imposed by humans. If the invasions of remote islands (Elton, 1958; Moore, 1983; Smith, 1985) and these invaded temperate grasslands are any guide, predicting vulnerable communities and future invaders will improve by being alert to those changes in an environment that coincide with the evolutionary history of potential invaders but are clearly unique to the natives.

ACKNOWLEDGMENTS

I thank K. Asay, R. A. Black, R. Boo, G. de Fosse, R. A. Distel, J. Engle, A. J. Gilmartin, M. Kahn, J. Kraus, M. Jennings, L. Novara, K. J. Rice, J. Richards, L. Rieseberg, R. Robberecht, J. Romo, D. Soltis, and K. Stack for their assistance at various stages in the writing of this paper. I thank J. Drake, R. Groves, R. Hobbs, and I. Noble for critical review of the manuscript.

REFERENCES

Archer, M. (1981). A review of the origins and radiations of Australian mammals. In: Keast, A. (Ed.), *Ecological Biogeography of Australia*, Vol. 3, Part 6, pp. 1435–88. W. Junk, The Hague.

Asay, K. H. (1983). Promising new grasses for range seedings. In: Monsen, S. B., and Shaw, N. (Eds.), *Managing Intermountain Rangelands–Improvement of Range and Wildlife Habitats*, pp. 110–14. USDA—For. Ser. Gen. Tech. Rep. INT–157.

Asay, K.H., and Hansen, W. T. (1984). Prospects for genetic improvement in the Quackgrass x Bluebunch Wheatgrass hybrid. *Crop Sci.*, **24**, 743–5.

Baker, H. G. (1972). Migrations of weeds. In: Valentine, D. H. (Ed.), *Taxonomy, Phytogeography, and Evolution*, pp. 327–47. Academic Press, New York.

Baker, H. G. (1978). Invasion and replacement in Californian and neotropical grasslands. In: Wilson, J. R. (Ed.), *Plant Relations in Pastures*, pp. 368–84. CSIRO East Melbourne, Australia.

Bartolome, J. W., Klukkert, S. E., and Barry, W. J. (1986). Opal phytoliths as evidence for displacement of native Californian grassland. *Madrono*, **33**, 217–22.

Beadle, N. C. W. (1981). *The Vegetation of Australia*, Cambridge University Press, Cambridge, 690 pp.

Berg, C. (1877). Enumeracion de las plantas Europeas. *An. Soc. Cien. Argent.*, **3**, 183–206.

Bingham, E. T., Hurley, L. V., Kaatz, D. M., and Saunders, J. W. (1975). Breeding alfalfa which regenerates from callus tissue in culture. *Crop Sci.*, **15**, 719–21.

Bolander, H. N. (1866). The grasses of the state. In: *Appendix to Journals of Senate and Assembly of the 16th session of the Legislature of the State of California (1865–66)*, Vol. III, pp. 131–45.

Branson, F. A. (1953). Two new factors affecting resistance of grasses to grazing. *J. Range Manage.*, **6**, 165–71.

Brewer, W. H. (1883a). Report on the cereal production of the United States. In: *Report on the Productions of Agriculture as returned at the Tenth Census.* U S Dept. Inter. Census Office. 47th Congress. 2nd session. Misc. Doc. 42. Part 3. 173 pp.

Brewer, W. H. (1883b). Pasture and forage plants. In: *Report on Cattle, Sheep, and Swine. Report on the Productions of Agriculture as returned at the Tenth Census*, pp. 5–10. US Dept. Inter. Census Office. 47th Congress. 2nd session. Misc. Doc. 42. Part 3.

Britten, J. (1906). Introduced plants at Sydney, 1802-4. *J. Bot.*, **44**, 234-5.

Bryant, E. (1848). What I Saw in California, 2nd Edn. D. Appleton, New York, 455 pp.

Buchholtz, K. P. (1971). The influence of allelopathy on mineral nutrition. In: US National Commission for I. B. P. (Eds.), *Biochemical Interactions Among Plants*, pp. 86–9. National Academy of Sciences, Washington, DC.

Burcham, L. T. (1957). *California Range Land*. Dept. Nat. Resour. (California), Div. For., Sacramento, 261 pp.

Cabrera, A. L. (1971). Fitogeografia de la Republica Argentina. *Boln. Soc. Argent. Bot.*, **14**, 1–42.

Caldwell, M. M., Richards, J. H., Johnson, D. A., Nowak, R. S., and Dzuree, R. S. (1981). Coping with herbivory: Photosynthetic capacity and resource allocation in two semiarid *Agropyron* bunchgrasses. *Oecologia (Berlin)*, **50**, 14–24.

Connor, D. J. (1966). Vegetation studies in north-west Victoria II. The Horsham area. Proc. R. Soc. Victoria, 79, 637–47.

Costello, D. F. (1944). Natural revegetation of abandoned plowed land in the mixed prairie association of northeastern Colorado. *Ecology*, **25**, 312–26.

Crawley, M. J. (1986). What makes a community invasible? In: Gray, A. J., Crawley, M. J., and Edwards, P. J. (Eds.), *Colonization, Succession and Stability*. Blackwell Scientific Publications, Oxford.

Cronise, T. F. (1868). *The Natural Wealth of California*, H. H. Bancroft, San Francisco, 696 pp.

Crosby, A. W. (1952). *The Columbian Exchange: Biological and Cultural Consequences of 1492.* Greenwood Press, Westport, Connecticut, 268 pp.

D'angela, E., Leon, R. J. C., and Facelli, J. M. (1986). Pioneer stages in a secondary succession of a pampean subhumid grassland. *Flora*, **178**, 261–70.

Darwin, C. (1872). *The Origin of Species*, Vol. 1, 6th Edn, John Murray, London, 365 pp.

Darwin, C. (1898). *Journal of Researches into the Natural History and Geology of the Countries visited during the Voyage of H. M. S. Beagle Round the World, under the Command of Capt. Fitz Roy, R. N.* D. Appleton, New York, 519 pp.

Daubenmire, R. (1969). Ecologic plant geography of the Pacific Northwest. *Madrono*, **20**, 111–28.

Daubenmire, R. (1970). *Steppe Vegetation of Washington.* Wash. Agric. Exp. St. Tech. Bull. 62. 131 pp.

Daubenmire, R. (1978). *Plant Geography*. Academic Press, New York, 338 pp.
Davis, A. M. (1979). Forage quality of prostrate Kochia compared with three browse species. *Agron. J.*, **71**, 822–4.
de Candolle, A. (1855). *Géographie Botanique Raisonnée*, Vol. 2. Librairie de Victor Mason, Paris, 1365 pp.
Dewey, L. H. (1894). *The Russian Thistle*. US. Dept. Agric., Div. Botany Bull. 15. 26 pp.
Ehleringer, J. R. (1978). Implications of quantum yield differences on the distribution of C3 and C4 grasses. *Oecologia (Berlin)*, **31**, 255–67.
Elton, C. S. (1958). *The Ecology of Invasions by Animals and Plants*. Methuen, London, 181 pp.
Forcella, F. (1985a). Final distribution is related to rate of spread in alien weeds. *Weed Res.*, **25**, 181–91.
Forcella, F. (1985b). Spread of kochia in the Northwestern United States. *Weeds Today*, **16**(4), 4–6.
Forcella, F., and Wood, J. T. (1984). Colonization potentials of alien weeds are related to their 'native' distributions: Implications for plant quarantine. *J. Aust. Inst. Agric. Sci.*, **50**/1, 35–41.
Frith, H. J., and Calaby, J. H. (1969). *Kangaroos*. Humanities Press, New York, 209 pp.
Gould, S. J., and Lewontin, R. C. (1979). Spandrels of San-Marco and the Panglossian paradigm—a critique of the adaptationist program. *Proc. R. Soc. London, B.*, **205**, 581–98.
Gray, A. J., and Scott, R. (1980). A genecological study of *Puccinellia maritima* Huds. (Pal.) I. Variation estimated from single-plant samples from British populations. *New Phytol.*, **85**, 89–107.
Groves, R. H. (1965). Growth of *Themeda australis* tussock grassland at St. Albans, Victoria. *Aust. J. Bot.*, **13**, 291–302.
Groves, R. H., and Williams, O. B. (1981). Natural grasslands. In: Groves, R. H. (Ed.), *Australian Vegetation*, pp. 293–316. Cambridge University Press, Cambridge, UK.
Guthrie, R. D. (1970). Bison evolution and zoogeography in North America during the Pleistocene. *Q. Rev. Biol.*, **45**, 1–15.
Harper, J. L. (1965). Establishment, aggression, and cohabitation in weedy species. In: Baker, H. G., and Stebbins, G. L. (Eds.), *The Genetics of Colonizing Species*, pp. 245–65. Academic Press, New York.
Harper, J. L. (1981). After description. In Newman, E. I. (Ed.), *The Plant Community as a Working Mechanism*. Special Publication of the British Ecological Society, No. 1, pp. 11–25. Blackwell Scientific Publications, Oxford.
Hartley, W., and Williams, R. J. (1956). Centres of distribution of cultivated pasture grasses and their significance for plant introduction. *Proc. Int. Grassl. Congr.*, 7th, pp. 190–9.
Hauman, L. (1927). Les modifications de la flore Argentine sous l'action de la civilisation. *Mem. Acad. R. Belg. 2nd series*, **9**, 1–100.
Heady, H. F. (1975). *Rangeland Management*. McGraw-Hill, New York, 460 pp.
Heady, H. F. (1977). Valley grassland. In: Barbour, M. G., and Major, J. (Eds.), *Terrestrial Vegetation of California*, pp. 491–514. Wiley-Interscience, John Wiley, New York.
Henderson, L., and Wells, M. J. (1986). Alien plant invasions in the grassland and savanna biomes. In: Macdonald, I. A. W., Kruger, F. J., and Ferrar, A. A. (Eds.), *The Ecology and Management of Biological Invasions in Southern Africa*, Oxford University Press, Cape Town.
Hendry, G. W. (1931). The adobe brick as a historical source. *Agric. Hist.*, **5**, 110–27.
Hilgard, E. W. (1891). The weeds of California. In: *Report of the Work*, pp. 238–52. Univ. Calif., Berkeley, Agric. Exp. St.

Hilgard, E. W., Jones, T. C., and Furnas, R. W. (1882). Report on the climatic and agricultural features and the agricultural practice and needs of the arid regions of the Pacific Slope, with notes on Arizona and New Mexico. US Dept Agric. [Department Reports No. 20]. 182 pp.
Hodge, W. H., and Erlanson, C. O. (1956). Federal plant introduction—a review. *Econ. Bot.*, **10**, 299–334.
Holdgate, M. W. (1986). Summary and conclusions: Characteristics and consequences of biological invasions *Phil. Trans. R. Soc. Lond. B.*, **314**, 733–42.
Holm, L. G., Plucknett, D. L., Pancho, J. V., and Herberger, J. P. (1977). *The World's Worst Weeds.* University Press of Hawaii, Honolulu, 609 pp.
Hooker, J. D. (1860). *The Botany of the Antarctic Voyage*, Part III, *Flora Tasmaniae*, Vol. I Dicotyledones. L. Reeve, London, 799 pp.
Howden, H. F. (1966). Some possible effects of the Pleistocene on the distributions of North American Scarabaeidae (Coleoptera). *Can. Entomol.*, **98**, 1177–90.
Hudson, W. H. (1923). *Far Away and Long Ago.* J. M. Dent, London. AMS Press reprint, New York (1968), 332 pp.
Jewiss, O. R. (1972). Tillering in grasses—its significance and control. *J. Br. Grassl. Soc.*, **27**, 65–82.
Keller, W., and Bleak, A. T. (1974). Kochia prostrata: a shrub for western ranges? *Utah Sci.*, **35**, 24–5.
Kloot, P. M. (1983). Early records of alien plants naturalised in South Australia. *J. Adelaide Bot. Gard.*, **6**, 93–131.
Leonard, Z. (1934). *Narrative of the Adventures of Zenas Leonard, Written by Himself.* (Quaife, M. M., Ed.). Lakeside Press, Chicago, 278 pp.
Mack, R. N. (1981). Invasion of *Bromus tectorum* L. into western North America: An ecological chronicle. *Agro-Ecosystems*, **7**, 145–65.
Mack, R. N. (1984). Invaders at home on the range. *Nat. Hist.* **93**, 40–7.
Mack, R. N. (1986). Alien plant invasion into the Intermountain West: A case history. In: Mooney, H. A., and Drake, J. A. (Eds.), *Ecology of Biological Invasions of North America and Hawaii*, pp. 191–213. Springer, New York.
Mack, R. N., and Thompson, J. N. (1982). Evolution in steppe with few, large, hooved mammals, *Amer. Nat.*, **119**, 757–73.
Major, J. (1977). California climate in relation to vegetation. In: Barbour, M.G., and Major, J. (Eds.), *Terrestrial Vegetation of California*, pp. 11–74. Wiley-Interscience, John Wiley, New York.
Marzocca, A. (1984). *Manual de Malezas.*, 3rd Edn. Editorial Hemisferio Sur, Buenos Aires, 580 pp.
McKell, C. M., Robison, J. P. and Major, J. (1962). Ecotypic variation in Medusahead, an introduced annual grass. *Ecology*, **43**, 686–98.
Medd, R. W., and Smith, R. C. G. (1978). Prediction of the potential distribution of *Carduus nutans* (Nodding Thistle) in Australia. *J. Appl. Ecol.*, **15**, 603–12.
Meinig, D. (1968). *The Great Columbia Plain.* University of Washington Press, Seattle, 576 pp.
Michael, P. W. (1970). Weeds of grasslands. In: Moore, R. M. (Ed.) *Australian Grasslands*, pp. 349–60. Australian National University Press, Canberra.
Michael, P. W. (1981). Alien plants. In: Groves, R. H. (Ed.) *Australian Vegetation*, pp. 44–64. Cambridge University Press, Cambridge.
Moore, D. M. (1983). Human impact on island vegetation. In: Holzner, W., Werger, M. J. A., and Ikusima, I. (Eds.), *Man's Impact on Vegetation*, pp. 237–46. Junk, The Hague.
Moore, R. M. (1959). Ecological observations on plant communities grazed by sheep in

Australia. In: Keast, A., Crocker, R. L., and Christian, C. S. (Eds.), *Biogeography and Ecology in Australia*, pp. 500–16. Junk, The Hague.
Moore, R. M. (1970a). Australian grasslands. In: Moore, R. M. (Ed.), *Australian Grass lands*, pp. 87–100. Australian National University Press, Canberra.
Moore, R. M. (1970b). *Australian Grasslands*. Australian National University Press, Canberra, 455 pp.
Moore, R. M., and Biddiscombe, E. F. (1964). The effects of grazing on grasslands. In: Barnard, C. (Ed.), *Grasses and Grasslands*, pp. 221–35. Macmillan, London.
Moore, T. B., Stevens, R., and McArthur, E. D. (1982). Preliminary study of some insects associated with rangeland shrubs with emphasis on *Kochia prostrata*. *J. Range Manage.*, **35**, 128–30.
Naveh, Z. (1967). Mediterranean ecosystems and vegetation types in California and Israel. *Ecology*, **48**, 445–59.
Parish, S. B. (1920). The immigrant plants of southern California. *South. Calif. Acad, Sci. Bull.*, **19**, 3–30.
Parodi, L. R. (1930). Ensayo fitogeografico sobre el partido de Pergamino. *Rev. Fac. Agron. Vet. Univ. Buenos Aires*, **7**, 65–271.
Parodi, L. R. (1947). Las gramineas del genero Bromus adventicias en La Argentine. *Rev. Argent. Agron.*, **14**, 1–19.
Patton, R. T. (1936). Ecological studies in Victoria. IV. Basalt plains association. *Proc. R. Soc. Victoria*, **48**, 172–90.
Perry, R. A. (1970). Arid shrublands and grasslands. In: Moore, R. M. (Ed.) *Australian Grasslands.*, pp. 246–59. Australian National University, Canberra.
Pyke, D. A. (1983). Demographic responses of *Bromus tectorum* and seedlings of *Agropyron spicatum* to grazing by cricetids. Ph.D. thesis, Washington State University, Pullman.
Regal, P. J. (1986). Models of genetically engineered organisms and their ecological impact. In: Mooney, H. A., and Drake, J. A. (Eds.), *Ecology of Biological Invasions of North America and Hawaii*, pp. 111–29. Springer, New York.
Richards, J. H. (1984). Root growth response to defoliation in two *Agropyron* bunch-grasses: Field observations with an improved root periscope. *Oecologia* (*Berlin*), **64**, 21–5.
Ridley, H. N. (1930). *The Dispersal of Plants Throughout the World*. L. Reeve, Kent., 744 pp.
Robbins, W. W. (1940). *Alien Plants Growing without Cultivation in California*. Univ. Calif., Berkeley, Agric. Exp. St. Bull. 637. 128 pp.
Roe, F. G. (1951). *The North American Buffalo: A Critical Study of the Species in its Wild State*. University of Toronto Press, Toronto, 957 pp.
Roseveare, G. M. (1948). *The Grasslands of Latin America*. Imp. Bur. Pastures Field Crops, Aberystwyth, Bull. 36. 271 pp.
Rumney, G. R. (1968). *Climatology and the World's Climates*. Macmillan, New York, 656 pp.
Ryerson, K. A. (1976). Plant introductions. *Agric. Hist.*, **50**, 248–57.
Salisbury, E. J. (1933). The East Anglian flora: A study in comparative plant geography. *Trans. Norfolk & Norwich Naturalist's Soc.*, **13**, 191–263.
Schroedl, G. F. (1973). The archaeological occurrence of bison in the southern plateau. Ph.D. thesis, Washington State University, Pullman.
Simpson, G. G. (1980). *Splendid Isolation: The Curious History of South American Mammals*. Yale University Press, New Haven, pp. 266.
Smith, C. W. (1985). Impact of alien plants on Hawai'i's native biota. In: Stone, C. P., and Scott, J. M. (Eds.), *Hawai'i's Terrestrial Ecosystems: Preservation and Management*, pp. 180–250. Cooperative National Park Resources Studies Unit, University of Hawaii, Honolulu.

Soriano, A. (1956a). Los distritos floristicos de la Provincia Patagonica. *Rev. Invest. Agric.*, **10**, 323–47.

Soriano, A. (1956b). Aspectos ecologicos y pasturiles de la vegetacion patagonica relacionados con su estado y capacidad de recuperacion. *Rev. Invest. Agric.*, **10**, 349–72.

Soriano, A. (1979). Distribution of grasses and grasslands of South America. In: Numata, M. (Ed.), *Ecology of Grasslands and Bamboolands in the World*, pp. 84–91. Junk, The Hague.

Stebbins, G. L. (1975). The role of polyploid complexes in the evolution of North American Grasslands. *Taxon*, **24**, 91–106.

Stevens, R., Jorgensen, K. R., McArthur, E. D., and Davis, J. N. (1985). 'Immigrant' forage Kochia. *Rangelands*, **7**(1), 22–3.

Teeri, J. A., and Stowe, L. G. (1976). Climatic patterns and the distributions of C4 grasses in North America. *Oecologia (Berlin)*, **23**, 1–12.

Thellung, A. (1912). La flore adventice de Montpellier. *Mém. Soc. Natn. Sci. Nat. Math. Cherbourg*, **38**, 57–728.

Trewartha, G. T., and Horn, L. H. (1980). *An Introduction to Climate*, 5th Edn. McGraw-Hill, New York, 416 pp.

Tschudi, J. J., von. (1868). *Reisen durch Sudamerika*, Vol. 4. Leipzig: Verlag F. A. Brockhaus. Stuttgart: Omnitypie—Gesellschaft Nachf. Leopold Zechnall reprint; 1971. 320pp.

Verona, C. A., Fernandez, O. N., Montes, L., and Alonso, S. I. (1982). Problematica agroecologica y biologia de Senecio madagascariensis Poiret (Compositae). I. Problematica agroecologica y biologia de la maleza. *Ecologia (Argentina)*, **7**, 17–30.

Wallace, A. R. (1892). *Island Life*, 2nd Edn. Macmillan, London, 563 pp.

Warwick, S. I., and Briggs, D. (1978). The genecology of lawn weeds. I. Population differentiation in *Poa annua* L. in a mosaic environment of bowling green lawns and flower beds. *New Phytol.*, **81**, 711–23.

Wickson, E. J. (1887). Report on grasses, forage plants, and cereals. Appendix II. In *Supplement to the Biennial Report of the Board of Regents*, pp. 81–107. Univ. Calif., Berkeley, Agric. Exp. St.

Wickson, E. J. (1891). Grasses and forage plants. In: *Report of Work*, pp. 201–20. Univ. Calif., Berkeley, Agric. Exp. St.

Williams, O. B. (1985). Population dynamics of Australian plant communities, with special reference to the invasion of neophytes. In: White, J. (Ed.), *The Population Structure of Vegetation*, pp. 623–35. Junk, The Hague.

Willis, J. H. (1964). Vegetation of the basalt plains in western Victoria. *Proc. R. Soc. Victoria*, **77**, 397–418.

Yensen, D. (1981). The 1900 invasion of alien plants into southern Idaho. *Great Basin Nat.*, **41**, 176–83.

Young, J. A., Evans, R. A., and Major, J. (1972). Alien plants in the Great Basin. *J. Range Manage.*, **25**, 194–201.

Biological Invasions: a Global Perspective
Edited by J. A. Drake *et al.*
© 1989 SCOPE. Published by John Wiley & Sons Ltd

CHAPTER 8

The Characteristics of Invaded Mediterranean-climate Regions

FRED J. KRUGER, G. J. BREYTENBACH, IAN A. W. MACDONALD, AND D. M. RICHARDSON

8.1 INTRODUCTION

Invasions by introduced organisms of the mediterranean regions of the world have held a special intrigue, because of the contrasts among the regions in the manifestation of the invasions, because the invasions seem to have been unusually severe for continental situations, and because mediterranean regions provide useful test cases.

This is a brief synthesis of the degrees and patterns of invasion, with emphasis on California, Chile, Australia, and South Africa. We explore possible explanations for any differences in the invasions between the various regions by reference to history, the physical environment, the patterns of disturbance, and factors associated with the biotic communities. Cases of invasive species are analysed to find reasons for differences in their behaviour on different continents. The Mediterranean Basin is not included in the analysis. Di Castri (this volume) summarizes historical information pertaining to that region.

For this review, we took mediterranean California to be equivalent to the Californian Floristic Province (Barbour and Major, 1977). The equivalent in Chile was taken to include the region of the matorral and espinal of the Coastal Ranges, the Central Valley, and the Andean foothills (Rundel, 1981). In Australia, the southern open-forests, southern sclerophyll communities, and mallee (Specht, 1981) were included, and in South Africa, the Fynbos Biome (Rutherford and Westfall, 1986).

8.2 PATTERNS OF INVASION IN DIFFERENT MEDITERRANEAN ECOSYSTEMS

8.2.1 Degrees and patterns of invasion in the different regions

Mediterranean ecosystems are often occupied by introduced plant species to the point that the native plant species are almost entirely displaced from their original habitat (Mooney, Hamburg, and Drake, 1986; Macdonald and Richardson, 1986). In California, Chile, and many Australian habitats, the displacement has been by many species of introduced herbs, mainly annuals. In California, this extreme invasion has occurred over an area of about nine million hectares of grassland and woodland (Heady, 1977). In the fynbos of the Cape Province of South Africa, about 800 000 hectares have been invaded by a total of around 20 species, mainly trees and shrubs (Macdonald and Richardson, 1986). A total of 674 introduced plant species have been recorded as naturalized in the state of California, with a native flora of about 5000 species (Mooney *et al.*, 1986), and 367 in the region of the Fynbos Biome of the southwestern Cape, with about 8500 native species. South Australia and Victoria have naturalized floras of 517 and 550 species, respectively (Specht, 1981).

Despite the extensive invasion of these natural landscapes, there are few data for a quantitative comparison between the mediterranean regions of the different continents, nor between regions within continents. Macdonald (1984) found that, with some exceptions, fynbos ecosystems were more extensively invaded by introduced plants than were other South African ecosystems. Fox and Fox (1986) concluded from a study of reserves in New South Wales that undisturbed heathlands were more severely invaded by introduced plants than were other ecosystems, though not where disturbed ecosystems were compared. Macdonald, Powrie and Siegfried (1986) analysed data on the sizes of invasive floras and faunas in different reserves in South Africa as well as data on the frequency of remains of the house mouse *Mus musculus* and black rat *Rattus rattus* in owl pellets. They found little evidence that the plant invasions of the fynbos were greater than those in other biomes. However the number of species of introduced birds and mammals was higher in fynbos nature reserves. The grasslands of the American Midwest are as much invaded as those of the chaparral region (Mack, this volume, cf. Heady, 1977 and Mooney *et al.*, 1986), but the mediterranean region in California tends to have a larger invasive flora than adjacent biomes (Figure 8.1). It seems that mediterranean habitats are often distinguished by a higher degree of invasion than are the habitats of other continental regions, but that this is not always so.

The relationship between the numbers of introduced plant species and natives in the floras of California is compared with that of the Cape fynbos in Figure 8.1. The Cape data set is biased, being from reserves, but includes several observations for areas which were previously disturbed and only recently reserved. Equivalent areas contain much smaller total floras in California than in the fynbos (Kruger and Taylor, 1979), so that areal extent should be considered when making this comparison, but the ratio of introduced species to natives tends to be consistently higher in California than in the Cape, with a few exceptions. The heavily disturbed Tygerberg Nature Reserve, with relatively fertile soils, has as high a proportion of introduced species (Table 8.4) as do many Californian areas, and in California the San Dimas Experimental Forest, which includes areas sown

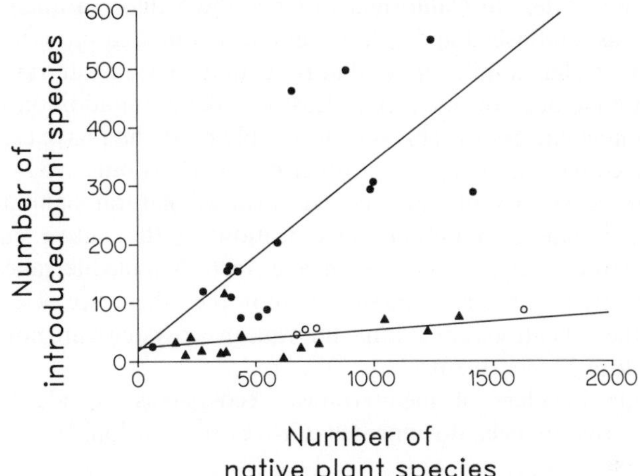

Figure 8.1. Relationships between numbers of naturalized introduced species and numbers of native species within defined land areas in California and the Cape. Dots represent data for mediterranean sites in California (r^2 for the linear regression $= 0{,}54$), circles are for non-mediterranean sites in California, and triangles are for mediterranean sites in the Cape ($r^2 = 0.14$) (See text for further explanation.)

to pasture grasses but which is mostly well protected, has a relatively low proportion of introduced to native species (77 to 442; Mooney and Parsons, 1973). Nevertheless, even small fynbos reserves that have had a history of disturbance, such as the Cape Flats Nature Reserve and Goukamma, have relatively fewer invasive plant species that do California areas. And in California, areas such as the Santa Monica Mountains, most of which are incorporated in National Forests, have high proportions of introduced species in their floras. The higher ratio of invasive plant species to native plant species in the relatively poorer floras of California is striking.

In the case of animals, even less comparative data are available. Several species of reptiles, birds, and mammals have naturalized within all the mediterranean regions but most are commensal with humans. Among birds, such cosmopolitan commensals as the feral pigeon *Columba livia* and the house sparrow *Passer domesticus* have succeeded in all the regions. The European starling *Sturnus vulgaris* is now abundant in the Cape, California, and the southern and eastern portions of Australia. It is being actively prevented from spreading into southwestern Australia and has not yet been introduced to Chile (Long, 1981). Birds which are not commensals have succeeded on some continents but not on others. No gamebird species from other continents have successfully established in the Cape (Brooke, Lloyd, and de Villiers 1986). By contrast, both the chukar partridge *Alectoris chukar* and the ring-necked pheasant *Phasianus colchicus* have become invaders in California. In Chile the California quail *Lophortyx californicus* has established and spread but the ring-necked pheasant has failed there as in Australia, despite repeated introductions. The mediterranean regions of Australia have been invaded by at least two doves (in addition to the feral pigeon) and nine passerines. This is considerably more than have established in any of the mediterranean regions of other continents (Long, 1981).

One or more species of mammal has invaded natural habitats on each continent and caused noticeable modification to the native communities (Table 8.1). More species of mammal invade Australian mediterranean regions, and each mostly in greater abundance, than in the others (see below), though, once more, the indications are that mediterranean ecosystems are not singular in this respect (e.g. Myers, 1986).

Invertebrate invaders of mediterranean ecosystems outside transformed habitats have seldom been documented (Groves and Burdon, 1986; Macdonald

Table 8.1. List of invasive mammals in the four mediterranean-type regions. I: invasive, widespread; i: invasive, local; S: introduced, widely established but not invasive in untransformed habitats; s: as for S, but local; F: introduced, but failed to establish (including domestic animals that have not become feral). Collated from information in Brooke *et al.* (1986), Mooney *et al.* (1986), Myers (1986), and Fuentes *et al.* (1984) and references therein

Common name	Scientific name	Region			
		California	Chile	Cape	Australia
Virginia opossum	*Didelphis virginiana*	I			
European rabbit	*Oryctolagus cuniculus*	F	I	F	I
European hare	*Lepus europeus*		I		I
Grey squirrel	*Sciurus carolinensis*	S		S	F
Fox squirrel	*Sciurus niger*	s			
Brown rat	*Rattus norvegicus*	S	S	s	S
Black rat	*Rattus rattus*	S	S/I	S	I
House mouse	*Mus musculus*	S	S/I	S	I
Feral dog	*Canis familiaris*	S	?	s	I
Red fox	*Vulpes vulpes*				I
Domestic cat	*Felis catus*	F	F	F	I
Donkey	*Equus asinus*	i		i	i
Feral horse	*Equus caballo*	I		i	I
Feral pig	*Sus scrofa*	I	F	s	I
Feral cattle	*Bos taurus*	i	F	i	i
Fallow deer	*Dama dama*			s	I
Red deer	*Cervus elaphus*			F	I
Sambar	*Cervus unicolor*				I
Hog deer	*Axis porcinus*				i
Feral sheep	*Ovis ammon musimon*	F	F	F	I
Feral goat	*Capra hircus*	F	F	F	I
Barbary sheep	*Ammotragus lervia*	I			
Himalayan tahr	*Hemitragus jemlahicus*	I		i	

and Richardson, 1986; Mooney *et al.*, 1986). The unaided spread of the Argentine ant *Iridomyrmex humilis* into relatively undisturbed fynbos communities from foci associated with human occupation or traffic (De Kock, unpublished; Donelly and Gilliomee, 1985; Macdonald and Richardson, 1986) has disrupted both the native ant communities and the patterns of recruitment in native species of myrmecochorous plants (Bond and Slingsby, 1984). This has apparently not occurred elsewhere (e.g. Majer, 1978). The means by which this species, which has a low capacity for dispersal, spreads from infected foci into communities rich in ant species is not known, but the potential exists for extensive and severe impacts on natural ecosystems.

Overall, therefore, there are substantial differences among the regions with regard to the invasions that have occurred, as well as some general similarities.

8.2.2 Invasive floras: correspondences and divergences in different regions

In all the regions discussed here, though to a lesser extent in Chile, the major vegetation patterns are strongly determined by edaphic factors (e.g. Raven 1977; Specht and Moll, 1983). There are marked disparities in the kind and degree of invasion of the different vegetation formations.

The forests of California are not much invaded by introduced plants, despite extensive and regular disturbance by forest operations including clearfelling (Mooney *et al.*, 1986). Coastal dune vegetation, salt marshes, valley grasslands, and oak woodlands are often heavily invaded. In the coastal dunes, up to 43% of the species in a community may be introduced, and the spread of the marram grass *Ammophila arenaria* has caused marked change to coastal dune landscapes in northern California (Barbour *et al.*, 1981). Introduced grasses and forbs, principally annuals, dominate the communities in the valley grasslands (McNaughton, 1968; Gulmon, 1977) and similarly occupy the understorey in oak woodlands (Heady, 1977). The composition of these communities varies along environmental gradients (Janes in Heady, 1977), with succession after fire and with variable grazing pressure (Heady, 1977), and from year to year (Gulmon, 1977, 1979), but the composition at a site shows no net change over time (Gulmon, 1977, 1979) and little or no recovery of the original native composition occurs if the grassland is protected against grazing and other factors of disturbance (Heady, 1977; Mack, this volume).

Mature California chaparral has few or no invaders other than those found in mechanically disturbed sites, although many introduced annuals and a few perennials may be found among the post-fire herbaceous flora. In post-fire chaparral in San Diego County, Keeley *et al.*, (1981) recorded 11 introduced species among 99 herbaceous species present after fire. Grasses such as *Bromus mollis* and *Bromus rubens* and forbs such as *Centaurea cyanus* were sometimes prominent, though most species had less than 0.1% cover and all disappeared within a few years after fire. *Lolium multiflorum* dominated in some cases, but this was where it had been sown as a soil conservation measure. Naturalized shrubs from the Mediterranean Basin, such as *Cistus* spp., have not invaded chaparral (Montgomery and Strid, 1976). The mature chaparral, which is relatively poor in plant species, is thus singularly free of introduced species.

The pattern of invasion of California grasslands is repeated in Chile, with most of the introduced species being shared between the regions (Gulmon, 1977; see later). However, unlike California chaparral, the Chilean matorral is extensively occupied in its mature state by elements of the same exotic flora, evidently because the prevalent grazing and burning of the matorral and the extensive wood-cutting maintains open canopies (Keeley and Johnson, 1977). But the richest representation of exotic herbs is found soon after fire and the abundance and diversity of introduced species declines thereafter (Keeley and Johnson, 1977).

Few data are available for the whole mediterranean region of Australia. In South Australia, pronounced differences are found in the relative degree of invasion of the different major ecosystems there. Relative to the native flora, the most invasive species are found in the coastal and the savanna land systems, and the least in the sclerophyll and the mallee land systems—the arid land system being not strictly mediterranean in climate (Table 8.2; Specht, 1972). This pattern correlates broadly with the abundance of the invasive plants. Specht (1972) argued that the plant species introduced South Australia were not pre-adapted to low soil fertility and hence were excluded from the infertile sclerophyll and mallee land systems, except where they were admitted along roadsides and similar disturbed sites. Soil fertility as a determinant of the form of plant invaders is discussed later.

Table 8.2. Numbers of naturalized introduced plant species in the different major land systems of South Australia, relative to the sizes of the native floras (after Specht, 1972)

Land system	Number of plant species	
	Native	Introduced
Mediterranean		
Coastal	56	22
Sclerophyll	344	18
Savanna	304	127
Mallee	424	40
Not Mediterranean		
Arid	231	5

Data from various sources have been used to derive the illustrative cases summarized in Table 8.3. The invasive populations in California and Chile are dominated by annual and biennial herbs, especially grasses of the genera *Avena*, *Bromus*, *Festuca*, and *Hordeum*, and forbs in the Geraniaceae and Asteraceae. In Australia and the Cape, these forms invade only on moderately fertile substrates, i.e. moderately leached soils in the terminology of Specht and Moll (1983). These patterns have been well documented for South Australia by Specht (1972). Grasslands and *Eucalyptus* woodlands on more fertile soils are widely invaded by

Table 8.3. List of prominent invasive plant species of the four mediterranean-type regions, with indications of their distributions. Information is derived from McNaughton (1968), Specht (1972), Gulmon (1977), Bridgewater and Backshall (1981), Macdonald and Jarman (1984), J. A. Vlok (unpublished), and Wells *et al.* (1986). The symbol I indicates that the species has been recorded as introduced in the region but has not been observed to spread, + that the species has naturalized but does not spread except in markedly disturbed habitats (road cuts and similar), and + + that the species has been observed to spread widely in natural vegetation. The columns headed 'low' and 'mod' for Australia and the Cape refer to substrates with low and moderate nutrient supplies, respectively, corresponding with the highly leached and moderately leached categories of Specht and Moll (1983)

Species	Region					
	California	Chile	Australia		Cape	
			low	mod	low	mod
Shrubs and trees						
Acacia cyclops and congeners	I	I	—	—	+ +	+
Pinus halepensis	I	—	?	?	+	+ +
Pinus pinaster	?	—	I	?	+ +	I
Pinus radiata	—	I	+	+ +	+ +	+
Hakea sericea and congeners	I	—	—	—	+ +	I
Cytisus scoparius	+	?	I	+	I	I
Annual and biennial herbs						
Aira caryophyllea	+ +	+ +	I	+	I	+ +
Anagallis arvensis	+	+	I	+ +	I	+ +
Arctotheca calendula	I	I	I	+ +	—	—
Avena barbata and congeners	+ +	+ +	I	+ +	I	+ +
Brachypodion distachyon	+ +	?	I	+	?	?
Briza maxima and *B minor*	+ +	+ +	I	+ +	I	+ +
Bromus mollis and congeners	+ +	+ +	I	+ +	I	+ +
Echium lycopsis	I	I	I	+ +	I	+ +
Erodium cicutarium, *E. moschatum*, and congeners	+ +	+ +	I	+ +	I	+ +
Hypochoeris glabra	+ +	+ +	I	+ +	I	+ +
Lolium multiflorum and congeners	+ +	+ +	I	+ +	I	+ +
Vicia bengalensis	?	?	I	+ +	I	+ +

herbaceous species from the Mediterranean Basin and South Africa, and some trees and shrubs such as *Olea europaea, Lavandula stoechas*, and *Lycium ferocissimum* (lately, also *Pinus radiata*—Van der Sommen, 1986). Mallee vegetation has been little invaded, as also the heaths and the *Eucalyptus* woodland with sclerophyllous understories. The sclerophyll land system is now being invaded by *Pinus radiata* but principally on more fertile sites (Van der Sommen, 1986), and most invasions are by herbaceous species from Europe and South Africa. One feature of the exotic herbaceous flora of Australia is the abundance of geophytes from South Africa. This life form has also invaded elsewhere (e.g. *Iris pseudacorus* in California—Raven and Thomas, 1970) but then few species have been involved.

Where the habitats in the four different regions are similar in climate and substrate, the same species of plant invaders tend to be found. Gulmon (1977) has analysed the composition of grasslands in Chile and California. Twenty-seven of the 48 introduced species recorded in her samples for both continents have been recorded as established in natural or nearly natural ecosystems in South Africa by Wells *et al.* (1986) and by Macdonald (personal communication) and Vlok (unpublished), and of these 24 were recorded for lowland habitats, with more fertile soils, in the fynbos biome. Thirty-four species were shared between California and Chile, and 16 are common to all three regions. Some of the dominant species in her samples have been recorded by Vlok (unpublished) in a recent survey of reserves in Coastal Renosterveld and other coastal vegetation. There included *Avena barbata, A. fatua, Erodium moschatum, Lolium multiflorum*, and *Vulpia* (*Festuca*) *myuros*. The level of invasion correlated with degree of disturbance, i.e. cultivation history in this case.

There are similar correspondences between the Cape and California in the identity of dominant invasive plants reported for California grasslands by McNaughton (1968). However, among the dominants in Gulmon's samples are some species such as *Aira caryophyllea* and *Festuca dertonensis* (annuals in the Poaceae), and *Brassica nigra* (Brassicaceae) which are recorded for South Africa but not as dominants. Many of the dominant invaders sampled by McNaughton (1968) and Gulmon (1977) are not recorded as naturalized in South Africa (Wells *et al.*, 1986). Gulmon (1977) also emphasized the differences between Chile and California that were evident in the identity and abundances of the plants sampled, despite overall convergence. For example, *Festuca dertonensis* was dominant in Chile, but rare or absent in California. *Avena fatua* was among the dominants in California, but not recorded in her sample from Chile (though present elsewhere E. R. Fuentes, personal communication). Thus, there are as many substantial differences in the patterns of plant invasions of like habitats on the different continents as there are similarities, which may be owing to the detailed differences between continents in history, climate, substrate fertility, and disturbance regime (e.g. Gulmon, 1977).

In the fynbos region, the most prominent plant invaders are trees and shrubs

(Macdonald and Richardson, 1986). By their nature these species persist through the course of succession after fire. None except for *Pinus radiata* is found to be invasive on the other three continents, even though some, such as species of *Acacia* and *Pinus*, have been introduced to California (Mathias and McClintock, 1963) and to a lesser extent, Chile (F. J. Kruger, personal observation). *Acacia cyclops, A. dealbata, A. longifolia, A. saligna, Eucalyptus lehmannii, Leptospermum laevigatum*, and *Pinus halepensis* are widely used as ornamentals in southern California, in various climatic conditions and in situations adjoining a diversity of natural habitats. The cultivated trees flower and fruit in abundance, yet nowhere do they invade the adjoining natural vegetation. Even on mechanically disturbed sites, only *A. saligna* was noted to recruit young plants (F. J. Kruger, personal observation). In the Cape these invaders are confined entirely, or virtually so, to areas with severely nutrient-deficient soils (Macdonald and Richardson, 1986), habitats which are not found in Chile or California (see also later).

Therefore, there is a tendency for similar plant invasions to occur in relatively fertile habitats on the different continents and the differences in abundance may reflect different disturbance histories (see later) or differences in the details of environmental regimes, such as the length of summer drought or the quantity of winter rainfall (Jackson, 1985). Only Australia and South Africa share extensive infertile habitats in the mediterranean-climate zones, and there are as yet few correspondences between the invasive floras in these habitats on the two continents. Major discrepancies between regions may be due to historical accident. An example is *Arctotheca* (*Cryptostemma*) *calendula*, the Cape weed, which is abundant in mediterranean Australia but absent from California where it apparently has the potential to invade (Heady, 1977).

8.3 HISTORICAL AND CULTURAL FACTORS

The human history of the Cape region differs substantially from that of the other regions (Deacon, 1986). Sparse populations of Early Stone Age people occupied the area from about 1.0–1.5 million years before the present (BP). Changes in technology at about 150 000 year BP, at the start of the Middle Stone Age, included the control of fire. These hunter-gatherers probably had relatively little influence on the biota. For example, the extinctions of plant and animal species that occurred during this time were due to climate change rather than human activity (Deacon, 1986). Khoi pastoralists arrived on the scene at about 2 000 years BP, bringing sheep, cattle, goats, and dogs to the area for the first time. Although the extinction of at least one species of antelope, the bluebuck *Hippotragus leucophaeus*, has been ascribed to the competition with the new grazers (Deacon, 1986), the paleoecological record does not reflect major changes in the flora which correlate with the arrival of pastoralism (e.g. Scholtz, 1986).

In addition to the domesticated animals of the pastoralists, these early people introduced at least some species of plants, such as *Ricinus communis*, dated to at least 3 000 years BP, and *Medicago polymorpha* (about 1 200 years BP). Cape ecosystems have had a long history of gradual evolution with humankind so that any ecological study must account for the effects of people on the habitat, but the per-European settlements evidently did not have the marked and abrupt ecological consequences as did similar settlements on other continents.

The next human colonists in the Cape were the Europeans who settled in 1652. Their population reached about 1300 after 50 years and 150 years passed before the whole of the fynbos region as settled. Agricultural development was slow until the diamond and gold mines in the interior created a demand for agricultural produce from around 1870 onward. The impact of the changes wrought to vegetation by new grazing patterns was apparent from within about a century of settlement, although this involved the replacement of native perennial grasslands by native shrubs in genera such as *Elytropappus* and *Relhania* rather than invasions of introduced herbs (Cowling, 1983; Scholtz, 1986).

In Australia the arrival of the Aborigines at about 40000 years BP resulted in marked changes to the ecosystems, principally because of the customary use of fire to regulate the condition of vegetation and animal populations (Kershaw, 1986). Aborigines introduced no plants (Groves, 1986a), but brought the dingo *Canis familiaris dingo* about 3500 years BP and this evidently contributed to the extinction of native predators such as the thylacine (Myers, 1986). European settlement was relatively late, from about 1830 onward. Population densities have remained low to the present day. Western Australia, for example, has a population density of less than 0.6 per square kilometre with 70% of the people living in the Perth region. However, this settlement had profound effects on ecosystems. The grazers they introduced differed markedly from native herbivores in their foraging patterns (Specht, 1972; Myers, 1986). Native animal communities were disrupted through the persecution of both predators and grazing animals (Myers, 1986). There was extensive deforestation to develop pasture land, and widespread propagation of introduced pasture species, mainly of Mediterranean origin but also from southern Africa and elsewhere, as well as fertilization of pastures (Specht, 1972, 1981; Moore, 1975).

Chile and California were settled only from about 12 000 years BP, perhaps earlier (Bray, 1986), by small populations of Indians who later developed forms of shifting cultivation for the production of several different crop species. Burning of the shrublands was evidently general practice. In Chile, Indians introduced domesticated camelids shortly before European settlement. There the Indian economy was evidently little affected by the Incan conquest in the late 15th century, except near irrigation settlements (Aschmann, 1977). Spanish colonists arrived in Chile in the first half of the 16th century (Mooney *et al.*, 1972; Aschmann and Bahre, 1977). Cattle ranching developed rapidly and complaints of over-grazing were noted from an early stage (Bahre, 1979). European settlement was accompanied by local deforestation stimulated by the needs of mining, construction, and fuel (Mooney *et al.*, 1972; Bahre, 1979). Extensive agriculture in the Chilean Central Valley with sheep herding and wood-gathering on the matorral-covered slopes are patterns which have persisted to the present day (Cañas *et al.*, 1982), and the structure of the natural vegetation has been severely modified almost everywhere (Aschmann and Bahre, 1977; Bahre 1979).

California was effectively settled by the Spaniards only in 1769, and European populations did not grow until toward the end of the century. Replacement of the herbaceous layers in lowland vegetation by Mediterranean herbs occurred rapidly from the middle of the 19th century, in association with the growth and intensification of stock ranching (Aschmann and Bahre, 1977; Jackson, 1985).

Thus, all these regions were occupied by aboriginal populations who managed fire but husbanded relatively few domesticated plants or animals. European settlers effectively displaced these economies and, besides bringing a diversity of domesticated and other organisms, introduced new regimes of disturbance to the natural ecosystems. The similarities of climate with that of the Mediterranean Basin and the traffic between Europe and the colonies, together with the availability of a weed flora pre-adapted to the new environments through the selective influence of millennia of human activity (Naveh, 1967, 1975; Raven, 1977; Jackson, 1985), led naturally to the spread of herbaceous plants from this source in the new regions. However, there were important differences among the regions in the patterns of introduction, owing partly to differences in the natural resources of the newly settled areas, and partly to cultural differences between the settler societies. The colonial links through Britain between Australia and South Africa, for example, may have ensured earlier exchange of many species of trees and shrubs than would have been the case for other continents (e.g. Shaughnessy, 1986). The direct traffic between Spain and California and Chile would also have biased the introduction of Mediterranean plants to these regions, relative to the Cape and Australia (Groves, 1986b).

In the Cape, little forest was available as a source of timber and fuel and there were few trees for shelter. Trees were immediately introduced from Europe. These included *Pinus pinaster*, now a major weed (Shaughnessy, 1986). Many other trees and shrubs were introduced in the first half of the 19th century and subsequently. There is at least tentative evidence of a progressive degree of naturalization among the introduced flora, so that the time-course of introduction in itself must be factored into any analysis of the causes of invasion (Baker, 1986; Kruger, Richardson, and Van Wilgen, 1986), together with patterns of cultivation (Shaughnessy 1986). On the other continents the settlers had less need to plant trees and shrubs, so that fewer species were introduced, and these later, than was the case in the Cape. This must explain part of the differences between regions in their invasive floras.

Australian environments are marked by the diversity and abundance of introduced vertebrate animals as well as by their extensive alien flora (Myers, 1986). These animals were actively encouraged by acclimatization societies in the second half of the 19th century. Similar societies never occurred in South Africa, where introduced vertebrates are rare outside sites of human occupancy and the aquatic habitats (Brooke *et al.*, 1986), so that differences between these two regions in the extent of invasion by exotic vertebrates may in part be owing to cultural and historical factors. Among the mammals listed by Myers (1986) as being invasive in Australia, only feral cattle, horses and donkeys have succeeded in invading natural habitats in the other regions and then on a small scale (Table 8.1), though the rabbit and the European hare are widespread in Chile. Species such as the pig, rabbit, and fallow deer, which are widely established in Australia and elsewhere in natural habitats, have failed to establish feral populations in the Cape in spite of repeated introduction (Brooke *et al.*, 1986). There is thus a greater rate of success in establishment of feral populations of introduced mammals in Australia than in South Africa, in mediterranean environments as well as others.

The profound transformation of ecosystems in these regions from the conditions of 500 years ago has been strongly patterned by topography in all cases except Australia. In the Cape, up to 85% of the lowland vegetation on fertile soils, the Coastal Renosterveld, has been converted to agricultural fields (Moll and Bossi, 1984). By contrast, as little as 10–20% of the Mountain Fynbos, on rugged terrain with highly leached soils, has been transformed (Cook in Kruger, 1982; Moll and Bossi, 1984). About 20% of the area of Mountain Fynbos has legal conservation status but for the lowlands this fraction is barely more than 1% (Greyling and Huntley, 1984).

Like the Cape, California and Chile have been markedly transformed in the lowlands (Aschmann and Bahre, 1977; Canas *et al.*, 1982; Mooney *et al.*, 1986). The uplands are well conserved in California, with around 50% of the state in public ownership (Mooney *et al.*, 1986), whereas little of Chile is formally conserved and the uplands are still used in a system of transhumance (Bahre 1979; Cañas *et al.*, 1982) as well as being used through the year for grazing, small-scale agriculture, and for wood-gathering (E. R. Fuentes, personal communication). In Australia, land within the 500 mm rainfall isohyet in the mediterranean regions is now mainly under pastures for cattle and sheep if not under crops (Moore, 1975). Beard (1984) estimated that up to 54% of the approximately 120 000 square kilometers the sclerophyllous shrublands of western Australia have been transformed. There are nevertheless large tracts of natural vegetation in national parks and forestry reserves (e.g. Hopper and Muir, 1984).

The uplands in California and South Africa are managed principally as watersheds and for nature conservation. Management is relatively intensive and, in the latter case at least, the practices specifically provide for the exclusion or control of invasive species (Macdonald and Richardson, 1986). Fire exclusion policies have given way to the managed use of fire to serve conservation objectives. Much effort has gone into afforestation with introduced species in the Cape, some of this having been fairly haphazard (e.g. Shaughnessy, 1986). There has been little or no afforestation in California, where instead there have been attempts to convert chaparral to relatively non-flammable herbaceous sward or to pastures, and introduced plants have been sown extensively for this purpose as well as for soil conservation (Hanes, 1977). The intensive exploitation of the matorral in Chile has been accompanied by little management for conservation. It is difficult to generalize for the complex picture in Australia, beyond to say that in such a sparsely populated region, distance is important so that natural areas remote from centres of development show less invasions than do those closer by (e.g. Bridgewater and Backshall, 1981; Muir, 1983).

These patterns are summarized schematically in Table 8.4. Each region is distinguished by patterns of land use that must in some degree have a differential effect on the establishment and spread of introduced species. The effect of people on remaining natural vegetation is apparently strongest in Chile. In California, the active propagation and spread of introduced species is presently perhaps confined to those for amenity in urban areas, though until recently a significant fraction of the wild landscape was subject to the introduction and spread of exotic plants for pasturage and watershed management (Hanes, 1977; Mooney *et al.*, 1986). The situation in Australia is dominated by the past energetic efforts to manipulate the pastures, and competing grazers and predators as described previously, as well as legacies of animal introductions. In places, plantation forests provide a permanent source of plant invasion. In South Africa, although afforestation with introduced trees provides a permanent source of propagules for invading species of *Pinus* and *Eucalyptus*, most of the extensive invasions are by plants introduced over the past century or more, but which are now little used.

In every case where the evidence is available, the degree of invasion of any given habitat is in some degree a function of distance from urban centres or centres of occupation (Frenkel, 1977; Muir, 1983; Macdonald *et al.*, 1988).

These differences in the histories of introductions affect our ability to provide ecological explanations for differences between the regions in the degree and nature of invasions found there. Because no adequate historical analyses are available, with exceptions in detailed cases such as that of Shaughnessy (1986) for trees and shrubs in the Cape Peninsula and for the rabbit in Australia (Myers, 1986), it is necessary to consider history separately for each species examined. It is equally important to consider the differences that may exist in the pattern and trend in human disturbance to the environment.

8.4 ENVIRONMENTAL CORRELATES WITH PATTERNS OF INVASION

8.4.1 Climate

Homoclimatic analysis reveals the geographical patterns of similarities in climates (e.g. Russel and Moore, 1976) and such similarity permits probabilistic

Table 8.4. A summary of historical factors likely to have influenced invasions of introduced plants in the four recently colonized mediterranean climate regions, collated from sources quoted in the text. The symbols '+' indicates an increase in the given factors, by up to three levels, and '–' a decrease

FACTOR	REGION			
	CAPE	AUSTRALIA	CALIFORNIA	CHILE
Length of human occupancy, years	1, 0–1, 5 million	40 000	12 000 (30 000?)	12 000 (30 000?)
Length of time since European settlement, years	330	150	215	450
Fire Regime				
Changes to natural regime due to early humans				
—Frequency	+ +	+ + +	+	+
—Size	—	———	?	———
—Season	?	?	?	?
Period of exposure to altered fire regime, years	150 000	40 000	12 000	12 000
Changing due to Europeans, from early human regime				
—Frequency	0/–	—	+/–	–
—Size	+/–	+ + +	+?	–
—Season	+ +	?	+ + +	?
Grazing Regime				
Change due to humans				
—Intensity	+ + (lowlands)	+ + +	+ + + (lowlands)	+ + +
Type of introduced herbivores				
—Domestic	0	+ + +	+ + +	+ + +
—Invasive	nil	+ + +	nil	nil
Period of exposure to altered regime prior to European settlement, years	1700	nil	nil	nil
Elimination of native large herbivores	+ + + (post settlement)	+ + (post settlement)	+ (lowlands)	+ (uplands)

Deforestation				
Historical	0/+	+ + +	+ + (lowlands)	+
Current	0	0	0	+ + +
Afforestation				
Extent	+ + +	+	0	+ +
Number of species	+ + +	+	0	+
Time since initation (years)	330	20–40	—	50
Sanddune stabilization	+ + +	+	+	0
Introduction of pasture forage plants	0	+ + +	+ + +	0
Cultivation				
Extent	+ + +	+	+ +	?
Period of time before European settlement, years	nil	nil	?	1000
Proportion of area now transformed	lowlands–68% montane–10%	mesic–c90% xeric–c10%	lowlands–c90% montane–c5%	?50%
Urban areas	+ +	+	+ + +	+

predictions of invasive success, i.e. species which are not pre-adapted to the new environment by evolution in similar climates in their source regions are unlikely to invade in the new environment (Specht, 1981; Groves, 1986b; Nix and Wapshere, 1986). Groves (1986b) has analysed the nature of mediterranean-type climates in relation to the establishment and spread of plants in new regions, emphasising the phenology of reproduction. In *Chrysanthemoides monilifera* invading Australia the subspecies are evidently confined to their appropriate homoclimes (Weiss, 1986). With a few exceptions, the Australian species of *Acacia* have remained within the Cape equivalent of their climatic and edaphic range in Australia (Milton 1981). For example, *Acacia pycnantha*, which is native to inland areas in southern Australia, has disappeared from the coastal Cape Flats while *A. cyclops* and *A. saligna* dominate the area: these three species were introduced at the same time (Shaughnessy, 1986). An obvious exception to this is *A. longifolia* which invades riparian habitats in both mediterranean and savanna habitats in South Africa (Macdonald, Jarman and Beeston, 1985) and spreads independently of climate.

But species often invade regions with climates differing in a greater of lesser degree from that of the source region. *Pinus radiata*, for example, thrives in mediterranean climates on various continents but also in climates different from that of its native habitats. It often regenerates and spreads unaided under such circumstances (Chilvers and Burdon, 1983; Hunter and Douglas, 1984; Richardson and Brown, 1986). The California poppy, *Escholtzia californica*, has naturalized and spread widely in Chile, extending into climatic regions that are markedly different from those at its source (Gulmon, 1977; Bahre, 1979; F. di Castri, personal communication). Mediterranean weeds spread in climates in Australia and California which often differ substantially from their native climates (Wapshere, 1984; Jackson, 1985). Similarly, *Hakea sericea*, introduced to South Africa from narrow provenances in south-eastern Australia (Neser, 1968; Kluge, 1983), spreads in habitats with diverse climates in the Cape (Fugler, 1982; see Russel and Moore, 1976 for climate similarities and contrasts). Competition from existing plants or predation by natural enemies is often important in confining the ranges of species in their source regions. For example, *Pinus halepensis* is limited in its distribution within its habitats in the Mediterranean Basin by competition from native grasses (Acherar, Lepart, and Debussche, 1984). Once released from these pressures, introduced species can sometimes invade areas with different climatic conditions to those in their native ranges. *P. halepensis* spreads wherever planted on more fertile soils in the Cape. Wapshere (1984) has emphasized this phenomenon for several herbaceous species in Australia. *Hypericum perforatum* invaded a wide range of habitats after its introduction to California. However, when insect herbivores from its natural range were introduced, *H. perforatum* became restricted to a limited subset of these habitats (Harper, 1977; see also Groves, this volume).

Nevertheless, climate, or at least site water balance, may explain the failure of certain species originating from regions with mediterranean climates to spread in some mediterranean region, when they are weeds in analogous regions elsewhere. *Acacia cyclops* and *A. longifolia*, which are important weeds in the Cape, are widely used in horticulture in California (Mathias and McLintock, 1963) yet have not naturalized and spread (Mooney *et al*., 1986). Shrubs in the California chaparral and the Chilean matorral are subject to pronounced water stress in summer, even when well established (Poole, Roberts and Miller 1981), much more so than their analogues in the Cape fynbos (Miller, Miller and Miller, 1983). *Acacia* spp. that are invasive in the Cape fynbos have relatively weak root systems which are slow to penetrate deeper layers of soil (Milton, 1981), and though they succeed in the Cape they are very likely to be excluded from chaparral habitats by summer drought.

8.4.2 Nutritional status of soils

The distribution of seasonal herbaceous species and of the various forms of evergreen sclerophylls in mediterranean climate regions is limited in the first instance by climate, principally the duration of the summer drought, and the availability of mineral nutrients (Miller, 1981; Specht and Moll, 1983).

There are pronounced differences between and within continents in the levels of mineral nutrients in soils of mediterranean regions, and this is reflected in patterns of vegetation (Specht and Moll, 1983). Difference in the success of invasion of contrasting habitats within the fynbos biome of the Cape by the various forms of introduced plants are correlated with the nutrient status of the soils. Trees and shrubs succeed on the highly leached soils of the Mountain Fynbos, whereas herbaceous forms tend to be the prevalent invaders on the more fertile sites of lowland habitats. In a sample of 14 reserves, the proportion of herbaceous species in the lowland habitats was significantly greater than in the montane habitats (Table 8.5; $F = 8.54$, $P > 0.05$, in a one-way analysis of variance). This trend is consistent with the observation that the introduced invasive flora of California, with uniformly richer soils than fynbos (di Castri, 1981), is nearly entirely dominated by herbaceous taxa (Mooney *et al*., 1986).

Specht (1963) and Heddle and Specht (1975) have demonstrated the limitations imposed by nutrient supply on invasions of heathland habitats by Mediterranean herbs. Fertilization allowed such species as *Aira caryophyllea* and *Vulpia myuros* to establish in communities from which they were otherwise absent. Annuals and many other seasonal herbs are unable to complete their lifecycles where rates of nutrient supply are low. Conversely, the evergreen shrubs of the nutrient-deficient ecosystems of the heathland and fynbos could be excluded from more fertile habitats by the toxic effects of high concentrations of nutrients in the soils or inherent limitations to nutrient uptake and use (Heddle and Specht 1975; Ozanne and Specht, 1981).

Although California grasslands in general are dominated by introduced

Table 8.5. Sizes of native and introduced floras in reserves in the fynbos regions with an analysis of the composition of the introduced floras in terms of life forms. Data collated by I. A. W. Macdonald and D. M. Richardson from lists supplied by the Forestry Branch of the Department of Environment Affairs and other sources

Vegetation type and reserve	Size of flora		Percentage of introduced species in given category		
	Native	Introduced	Trees and shrubs	Herbs	Other
Mountain fynbos					
Cape of Good Hope	1050	71	42	56	1
Fernkloof	773	31	48	48	3
Jonkershoek	1231	53	43	49	8
Table Mountain	1362	78	50	45	5
Vogegat	697	25	32	64	4
Zachariashoek	623	8	38	63	0
Lowland fynbos					
Cape Flats	165	34	6	91	3
Goukamma	356	15	27	67	7
Pella	379	18	28	72	0
Rondevlei	229	42	24	62	4
Coastal Renosterveld					
Bontebok National Park	446	46	46	50	4
Tygerberg	373	116	20	74	6
Strandveld					
Rocherpan	207	12	8	83	9
Fynbos–karoo transition					
Nieuwoudtville	276	19	5	84	11

species, those on serpentine soils are not (Mooney *et al.*, 1986). This is evidently because the introduced flora includes few species able to compete with the natives under the nutrient limitations which characterize these habitats. When nutrient supply is improved, as on the mounds formed by gophers *Thomomys bottae* some introduced species such as *Bromus mollis* are able to establish and reproduce successfully (Hobbs and Mooney, 1985).

8.5 BIOTIC CORRELATES

8.5.1 Biotic interactions in plant invasions

Diverse interactions between the introduced and the native species, or between introduced species themselves, are involved in the invasion by alien plants, but these interactions have seldom been analysed to the extent needed to explain the similarities and differences among regions in the invasions observed.

Dispersal of alien plants is mainly by wind, but animals may often be important. The dispersal of many species of herbs is epizoochorous (Baker, 1965). Thirty-two per cent of invasive woody species in South Africa have fruit dispersed by vertebrates, or with the potential for this dispersal (Knight, 1986). In the fynbos biome, disproportionately few species of aliens are dispersed by animals but native animals are implicated in the dispersal of some of the more important invasive trees (Kruger, Richardson and Van Wilgen, 1986). It is likely that potential invaders in some regions have failed for lack of suitable dispersal agents.

The role of herbivores other than domesticated grazers in governing the spread of introduced plants has been identified in several cases. Introduced herbs, together with the native plants, are eliminated from the understory of mature chaparral by rodents acting alone or in concert with the suppressing effect of the shrub canopies (Westman, 1983). This is evidently also the case in the matorral of Chile, at least where the community of shrubs is relatively undisturbed (Fuentes and Etchegaray, 1983; Fuentes *et al.*, 1984). On a smaller scale Bartholomew (1970) found that the distinctive flora in the bare zones around shrubs in the coastal shrublands of California, which included unpalatable aliens such as species of *Anagallis* and *Centaurea*, was maintained by preferential grazing by small mammals. The animals sheltered in the shrub habitats but avoided the open grassland, and so foraged in the transition zone. In the matorral of Chile, open patches between native shrubs are maintained as habitats for introduced herbs by rabbits *Oryctolagus cuniculus* which venture from the shelter of the shrubs, unlike the native herbivores which are inhibited by their predator avoidance behaviour (Fuentes and Simonetti, 1982; Fuentes and Etchegaray, 1983). The native predators, in turn, have not yet adapted to preying upon the introduced rabbit (Fuentes and Simonetti, 1982).

Small herbivores has been noted to prevent the establishment of transplants of *Pinus radiata* in Chile (Fuentes and Etchegary, 1983) and this may be the reason why the pine is unable to spread naturally in Chile (personal communication from F. di Castri; E. R. Fuentes; C. Gonzalez; and P. D. Valenzuela), though it does in South Africa and Australia (see earlier).

The otherwise analogous plant communities in each of the mediterranean climate regions differ from each other in important aspects of community structure. For example, the strand communities of California have a low proportion of annuals by comparison with those in Israel (Barbour *et al.*, 1981).

Overall, shrublands of the fynbos lack trees in habitats where they would normaly be encountered elsewhere. Such differences have been cited as evidence for vacant niches, which in turn correlate with marked invasion by species of the missing growth form (e.g. Campbell, Mackenzie, and Moll, 1979). In the Cape, a rich local flora of the "missing" form is available locally, for example in the riverine and montane forests of the Cape. The question as to what it is that excludes the native trees from the communities concerned but admits the introduced trees is pertinent. The native forest trees lack ectomycorrhizae, in contrast with invading pines and eucalypts (Lamont, 1983). Whatever prevented the evolution of such mutualisms in the natives does not appear to operate on the aliens, and this together with the absence of a drain on internal reserves by herbivores and pathogens on the latter, may explain their advantage.

8.5.2 Biotic interactions in the invasion by introduced mammals

The house mouse *Mus musculus* and the brown rat *Rattus rattus*, ubiquitously commensal with humans, sometimes invade undisturbed habitats in mediterranean climate regions. In the Cape fynbos the house mouse has been recorded in nature reserves close to or surrounded by urban areas, or on sites otherwise influenced by human occupation (Macdonald, Powrie and Siegfried, 1986). There is no evidence of extensive penetration of fynbos habitats by introduced species of small mammals. Intensive studies of small mammal communities in several vegetation formations, variously disturbed by grazing and fire, have revealed no introduced species (e.g. Bond, Ferguson and Forsyth, 1980; Rautenbach and Nel, 1980; Nel, Rautenbach, and Breytenbach, 1980; Willan and Bigalke, 1982, David and Jarvis, 1983). The same seems to hold in Chilean and California systems (e.g. Jaksic, Janez and Fuentes, 1981; Wirtz, 1982; Quinn, 1983). The exception is the presence of *Mus musculus* in disturbed coastal sage of California, where introduced herbs have invaded (Blaustein, 1981), and its general occurrence in Australian heathland and sclerophyll forest, where the brown rat is also often present (Fox, Quinn, and Breytenbach, 1985; Catling, 1986; Fox and Fox, 1986). It was recorded in seven of eight samples in Australian reserves, but was recorded in none of six similar samples in the Cape fynbos; in California it was found in each of four samples from coastal sage, but in none of six in chaparral (Barnett, How and Humphrey, 1978; Glanz and Meserve, 1982; Kruger and Bigalke, 1984; and references quoted above). In all these systems there is a marked faunal succession after fire, but it is only in the Australian communities that the mouse appears in the succession, and there it is present in equal or greater numbers than other species, almost throughout the succession (Fox *et al.*, 1985).

Fox and Fox (1986) maintained that it is the reduction in the numbers of native mammal species caused by disturbance, i.e. presumed alteration of the fire regime, that admits the house mouse to the heathland communities, but this argument would not seem to be valid for California and the Cape. The reason may lie in the structure of the small-mammal communities, and their relative proneness to disturbance. In Australia, the mouse has been recorded in small mammal communities with between one and eight indigenous species present (e.g. Fox *et al.*, 1985). There is no consistent association between richness and the presence of aliens in the 26 samples from the literature cited above. A more complete explanation of the different invasions by the house mouse is necessary, which must take account of the particular characteristics of the mouse and of the native animals it encounters (cf. Myers, 1986; Pimm, this volume).

Brown (1975) has argued that small mammals of like size, i.e. where the larger is no more than 1.3 times heavier than the smaller, are excluded from coexistence by competition. In chaparral, the house mouse is likely to encounter one or more of five other rodents of nearly identical size. In fynbos, there are two natives of about the same size. In Australian heathland, one mammal of like size, *Sminthopsis murina*, is largely scansorial and unlikely to interact with the house mouse. The other, *Pseudomys novaehollandiae*, tends to be successionally segregated from the mouse. Fox and Pople (1984) have manipulated small-mammal communities in the heathland by removing and adding *P. novaehollandiae* and demonstrated significant increases in the house mouse population where the native had been removed, and decreases where it had been introduced. They suggested that the house mouse tended to avoid *P. novaehollandiae*, rather than that there was direct competition for resources. The house mouse is evidently prevented from invading undisturbed California coastal sage in a similar way by interaction with *Microtus californicus* (Blaustein, 1980).

It is possible that small disruptions to the relatively simple small-mammal communities of the Australian heathland (see Fox *et al.*, 1985) would suffice to admit the house mouse, whereas the native communities in California and the Cape, where fire regimes have also been modified in various directions, are more robust in this respect. Therefore, the susceptibility to disturbance may vary with the structure and dynamics of the small-mammal communities, in terms of the array of size classes and the behaviour of members of the relevant size classes, and this, rather than disturbance as such, may determine whether or not the mouse invades (e.g. Fox and Fox, 1986). However, the direct or indirect effects of predators, much depleted in Australian ecosystems (Myers, 1986) but not in others, may also govern the success of the mouse. For example in South African ecosystems in general, Macdonald, Powrie and Siegfried (1986) found that the incidence of remains of introduced rodents in owl pellets was negatively associated with the size of the pool of avian predators, and not at all with native small-mammal diversity. However, in Chile, pedators occur in sufficient abundance without controlling the invasion of European rabbit (Fuentes and Simonetti, 1982).

There are many substantial differences between and within continents in the structure and organization of small-mammal communities and the food habits and patterns of habitat use by different species in mediterranean-type regions (e.g.

Glanz and Meserve 1982). These are likely to differentially affect interactions with introduced species and hence the invasion of the latter.

8.5.3 Linked invasions

Successive invasions by introduced species may be expected and is sometimes evident. For example, the cat and the fox have followed and to some extent depended on the spread of the rabbit in Australia. In large areas the rabbit forms the main prey item of both these predators (Myers, 1986). The new arboreal habitats arising from the spread of introduced trees and shrubs in the Cape have favoured the extension of the range of the European starling (Winterbottom and Liversidge, 1954), as well as the immigration of native birds previously absent from the fynbos region (Macdonald 1986; Macdonald, Richardson and Powrie, 1986). In this way, the trend is set for the progressive change of natural communities.

8.5.4 Disturbance regimes

The natural or endogenous (Westman, 1986) disturbance regimes differ markedly among the regions discussed here. Chaparral ecosystems, especially those in the Transverse Ranges, have high rates of erosion because of rapid crustal uplift. Extensive mass-wastage of soils on the chaparral slopes occurs at San Dimas in California; up to 6% of the surface area on sites under burnt chaparral can consist of soil slips (Rice and Foggin, 1971; Rice, Ziemer and Hankin, 1982). Average annual yield of sediments and bed-load at San Dimas amounts to about 20 to 300 tonnes per ha per year, which may increase five-fold to 35-fold after fire (Rice, 1974). By contrast, the rates of erosion from the equally steep catchments of the Cape fynbos are negligible (Van Wyk, 1982). Fire regimes also differ markedly among the regions. Chaparral ecosystems tend to have large fires at long intervals (40 years and more), fynbos ecosystems have smaller fires more frequently (6–40 years), and matorral has even smaller fires (Aschmann and Bahre, 1977; Kruger, 1983). Heathlands in Australia experience fires once in about 5 to 15 or 20 years (Walker, 1981), perhaps more frequently elsewhere (Kruger, 1983), and the fires may be very large where natural vegetation is extensive (e.g. Coaldrake, 1951).

The role of natural disturbance regimes in determining the patterns of invasion has seldom been addressed directly. Chaparral, subject to the highest rates of erosion, has been little invaded. When chaparral is burnt, introduced herbs temporarily invade the community, but in low numbers, whereas they are abundant after fire in matorral, as discussed earlier. If, however, the chaparral is disturbed mechanically, for example by road cuts, then it is readily invaded by such plant species as *Brassica nigra* and *Cytisus scoparius* (e.g. Horton and Kraebel, 1955). Endogenous disturbance appears therefore to play a minor role in admitting invaders to chaparral relative to that of human disturbance.

Many invasive plant species are tolerant of or pre-adapted to the current fire regimes of their new habitats. Species such as *Hakea sericea, Pinus pinaster*, and *P. radiata* are well adapted to the fire regimes prevailing in Cape fynbos and their population growth is controlled by the incidence of fire (Kruger, 1977; Richardson and Brown, 1986; Richardson, Van Wilgen and Mitchell, 1987). In some cases an introduced species may invade when the disturbance regime differs substantially from that prevailing in its natural range. An example is *Pinus halepensis* which has invaded an area of Cape fynbos where the mean intervals between fires has been 11.6 years (Richardson, 1988), whereas in its natural range in the Mediterranean Basin fires occur at average intervals of 25 years (LeHouerou, 1981). On the other hand, some species which do not invade may be excluded from the natural habitat by the fire regime. A study of four species of Hakea in the Cape indicated that the limited invasion by one, *H. suaveolens*, was because its life cycle was marked by long juvenile periods and low fecundity at ages when fires are likely to recur (Richardson *et al.*, 1987). *H. salicifolia*, equally precocious and fecund as *H. sericea*, one of the successful invaders, has failed to invade natural vegetation, possibly because the follicles of its fruit has walls too thin to shield the seed against fires in the fynbos. Nevertheless, successful invaders such as *P. pinaster* and *P. radiata* do not necessarily depend on fire for their entry into the vegetation (Richardson and Brown, 1986), nor would they be excluded from any of the mediterranean regions by the fire regimes prevailing there.

Modification of the fire regime has led to successful invasion in at least some plants. Baird (1977) has monitored the invasion of the understorey of sclerophyll woodland in King's Park, Western Australia by veld grass *Ehrharta calycina* and has shown how this alien grass increased in abundance with fires recurring at intervals of about six years, a frequency shorter than the natural regime, and then declined when fire was excluded (see also Groves, this volume). Invasion of woody communities by herbaceous plants may in itself modify the fire regime to the advantage of the invader. Zedler *et al.* (1983) have documented such feedback. They recorded how the artificial establishment of *Lolium multiflorum* in chaparral after a wildfire, followed by favourable rainfall resulted in another fire after two years. This caused a reduction in populations of native shrub species and a shift in species composition, together probably with greater persistence of introduced herbs of several species.

Disturbance to vegetation by grazing is often cited as a key factor in invasions of exotic plants. Specht (1972) argued that grazing by the native Australian fauna, such as the kangaroo and the wallaby, had minimal effects on Australian plant communities. Sheep, cattle and horses introduced during the 19th century rapidly altered the herbaceous stratum in regions with relatively fertile soils so that native perennial grasses tended to disappear, to be replaced by introduced annuals from the typical Mediterranean genera as well as Mediterranean and Cape geophytes (see also Mack, this volume). Moore (1975, and references therein) has demonstrated this direct effect of grazing by means of field experiments. Grazing effects on this scale have occurred in California and Chile but not in fynbos, as mentioned earlier. It has been speculated that, if anything, the relaxation of native herbivore pressure through overhunting has facilitated invasion by certain palatable plant species (Macdonald, 1984).

Introduced domestic goats and feral European rabbits in Chile have disrupted recruitment of native trees and shrubs, while simultaneously reducing their cover. Rabbits forage more widely for seedlings of woody plants as well as for native herbaceous species than do the native small mammals, while allowing introduced herbaceous species to persist in openings (see earlier). These subtle disturbances to native patterns of herbivory and seedling predation evidently reinforce strongly the simple effects of direct disturbance by alien browsers, and thus the spread of introduced herbaceous vegetation (Fuentes and Simmonetti, 1982; Fuentes and Etchegaray, 1983).

Mack (this volume) contends that the native tussock grasses lacked the morphological and tillering characteristics necessary to survive the new patterns of defoliation and trampling to which they were subject to with the arrival of domestic stock. Experimental work to evaluate the effects of grazing on the native tussocks in California has not been done and, since the tussocks have virtually disappeared, the reasons for the invasion must remain conjectural (Heady, 1977).

Disturbance of the habitat in a way that permits the entry of exotics may involve synergisms. In sclerophyll vegetation of low-nutrient habitats in Western Australia, Bridgewater and Backshall (1981) detected significant differences between the ratios of introduced to native plant species in two sets of samples, one close to and the other relatively remote from urban areas, in environments which were otherwise matched. They ascribed this to the combined effects of nutrient enrichment and the increased frequency of fire in the former case. They argued for a positive feedback between the incidence of nutrient-demanding herbs such as *Hypochoeris glabra* and *Anagallis arvensis*, patches of fertile soil, and the incidence of fire as a mechanism governing the patterns of invasions observed (see earlier). Lambert and Turner (1987) have found similar correlations in the vicinity of Sydney. Irrespective of the kind of disturbance, it may be the effects on nutrient supply patterns that are crucial in the invasions of such highly leached mediterranean ecosystems (see Hobbs, this volume).

8.6 CONCLUSIONS

The problem of invasive biotas is a pervasive element of land management in regions of mediterranean climates settled by Europeans during the last 400 years, and is likely to continue to grow. There are overall similarities in the invasions of similar environments in the four regions of mediterranean climate compared here, but there are many anomalies. The fact that introduced species vary in their invasive behaviour in different regions provides a series of on-going natural experiments in biological invasion. These offer opportunities to advance our understanding of the processes in general, as well a means to better management of mediterranean ecosystems.

Certain general problems could be usefully addressed at this stage. Attention to homoclime analysis would contribute rapidly to a basis of broad prediction. More important in the long run is the need to grasp fully the role of disturbance in governing invasions. Two aspects seem to be relevant. First, the consequences of a given disturbance depends on the properties of the ecosystem or community, as we have seen in the contrasting results of grazing in the Cape, compared with the other regions, and in the behaviour of the house mouse in Australia, compared with California and the Cape. The relative susceptibility of ecosystems to disturbance needs to be better understood. Second is the need to evaluate disturbance, not in terms of the elements of a given regime, such as the intensity, frequency, and timing of fire, but rather in terms of ecological effects. A uniform basis of comparison among the different kinds and degrees of disturbance is needed, and this is likely to be found in measuring changes in the availability of resources, and in the intensities of interactions among the members of a community (see, for example, Hobbs, this volume).

Patterns of invasion are obviously confounded by the different histories of each region, which have differentially affected the disturbance regimes as well as the availability of introduced organisms for invasion. The patterns of success or failure among the alien species on the different continents are broadly associated with differences in land use, as well as with the physical environmental factors and the nature of the communities encountered by the introduced species. Sometimes, selective introduction may have played a role. The reasons for success and failure have seldom been addressed directly.

With time, more introduced species will invade in each of the mediterranean regions considered here. Studies of the processes of invasion where this is occurring, and of the factors preventing it where invasion does not occur, would contribute substantially to our ability to manage these invasions.

ACKNOWLEDGEMENTS

This work was conducted as part of the conservation research programme of the Forestry Branch, and was partly supported by the South African National Programme for Environmental Sciences. We thank Eduardo Fuentes for comments on the manuscript.

REFERENCES

Acherar, M., Lepart, J., and Debussche, M. (1984). La colonisation des friches le pin d'Alep (*Pinus halepensis* Miller) en Languedoc méditerraneen. *Acta Oecol. Plant.*, **5**, 179–89;

Aschmann, H. (1977). Aboriginal use of fire. In: Mooney, M. A. and Conard, C. E. (Eds), *Proceedings of the Symposium on the Environmental Consequences of Fire and Fuel*

Management in Mediterranean Ecosystems, August 1–5, 1977, Palo Alto, California. USDA Forest Service General Technical Report No. 3. pp. 132–44.

Aschmann, H. and Bahre, C. (1977). Man's impact on the wild landscape. In: Mooney, H. A. (Ed.), *Convergent Evolution in Chile and California: Mediterranean Climate Ecosystems*, pp. 73–84. Dowden, Hutchinson and Ross, Stroudsburg, Pennsylvania.

Bahre, C. J. (1979). Destruction of the natural vegetation of north-central Chile. *Univ. Calif. Publ. Geogr.*, **23**, 1–117.

Baird, A. M. (1977). Regeneration after fire in King's Park, Perth, Western Australia. *J. Roy. Soc. W. Aust.*, **60**, 1–22.

Baker, H. G. (1965). Characteristics and modes of origin of weeds. In: Baker, H. G. and Stebbins, G. L. (Eds.), *The Genetics of Colonizing Species*, pp. 147–69. Academic Press, New York.

Baker, H. G. (1986). Patterns of plant invasion. In: Mooney, H. A. and Drake, J. A. (Eds.), Ecology of biological invasions of North America and Hawaii. Springer-Verlag, New York, pp. 44–57.

Barbour, M., and Major, J. (1977). *Terrestrial Vegetation of California*. Wiley, New York.

Barbour, M. G., Shmida, A., Johnson, A. F., and Holton, B. (1981). Comparison of coastal dune scrub in Israel and California: physiognomy, association patterns, species richness, phytogeography. *Israel J. Bot.*, **30**, 181–198.

Barnett, J. L., How, R. A., and Humphrey, S. W. F. (1987). The use of habitat components by small mammals in eastern Australia. *Aust. J. Ecol.*, **3**, 277–285.

Bartholomew, B. (1970). Bare zone between California shrub and grassland communities: the role of animals. *Science*, **170**, 1210–12.

Beard, J. S. (1984). Biogeography of the kwongan. In: Pate, J. S. and Beard, J. S. (Eds.), *Kwongan: Plant Life of the Sandplain. Biology of the South-West Australian Shrubland Ecosystem*, pp. 1–26. University of Western Australia Press, Nedlands.

Blaustein, A. R. (1980). Behavioural aspects of competition in a three-species rodent guild of coastal southern California. *Behav. Ecol. Sociobiol.*, **6**, 247–255.

Blaustein, A. R. (1981). Population fluctuations and extinctions of small rodents in coastal southern California. *Oecologia*, **48**, 71–8.

Bond, W., Ferguson, M., and Forsyth, G. (1980). Small mammals and habitat structure along altitudinal gradients in the southern Cape mountains. *S. Afr. J. Zool.*, **15**, 34–43.

Bond, W. J. and Slingsby, P. (1984). Collapse of an ant-plant mutualism: the argentine ant (*Iridomyrmex humilis*) and myrmecochorous Proteaceae. *Ecology*, **65**, 1031–37.

Bond, W. J., and Slingsby, P. (1984). Collapse of ant-plant mutualism: the Agrentine ant (*Iridomyrex humilis*) and myrmecochorous Proteaceae. *Ecology*, **65**, 1031–37.

Bray, W. (1986). Finding the earliest Americans. *Nature*, **321**, 726.

Bridgewater, P. B., and Backshall, D. J. (1981). Dynamics of some Western Australian ligneous formations with special reference to the invasion of exotic species. *Vegetatio*, **46**, 141–8.

Brooke, R. K., Lloyd, P. H., and De Villiers, A. L. (1986). Alien and translocated vertebrates in South Africa. In: Macdonald, I. A. W., Kruger, F. J., and Ferrar, A. A. (Eds), *The ecology and Management of Biological Invasions in Southern Africa*, pp. 63–74. Oxford University Press, Cape Town.

Brown, J. H. (1975). Geographical ecology of desert rodents. In: Cody, M. L., and Diamond, J. M. (Eds), *Ecology and Evolution of Communities*. Belknap, Cambridge, Massachussetts.

Campbell, B. M., Mackenzie, B., and Moll, E. J. (1979). Should there be more tree vegetation in the mediterranean climate region of South Africa. *J. S. Af. Bot.*, **45**, 543–57.

Cañas, R., Aguilar, C., Paladines, O., and Muñoz, G. (1982). Biomass production and utilization of natural pastures in the Chilean mediterranean ecosystems. In: Conrad, C. E., and Oechel, W. C. (Eds), *Proceedings of the Symposium on Dynamics and Management of Mediterranean Type Ecosystems (1981)*, pp. 34–41. San Diego, California, USDA Forest Service, General Technical Report PSW–58.

Catling, P. C. (1986). *Rattus lutreolus*, colonizer of heathland after fire in the absence of Pseudomys species. *Aust. Wildl. Res.*, **13**, 127–39.

Chilvers, G. A., and Burdon, J. J. (1983). Further studies on a native Australian eucalypt forest invaded by exotic pines. *Oecologia*, **59**, 239–245.

Coaldrake, J. E. (1951). *The Climate, Geology, Soils, and Plant Ecology of Portion of the County of Buckingham (Ninety-Mile Plain), South Australia.* CSIRO, Australia, Bulletin 266.

Cowling, R. M. (1983). Phytochorology and vegetation history in the southeastern Cape, South Africa. *J. Biogeogr.*, **10**, 393–419.

David, J. H. M., and Jarvis, J. U. M. (1983). Notes on two brief surveys of the small mammal fauna on the Rooiberg, Ladysmith, southern Cape Province. *S. Afr J. Zool*, **18**, 370–7.

Deacon, J. (1986). Human settlement in South Africa and archaeological evidence for alien plants and animals. In: Macdonald, I. A. W., Kruger, F. J., and Ferrar, A. A. (Eds.), *The Ecology and Management of Biological Invasions in Southern Africa*, pp. 3–19. Oxford University Press, Cape Town.

De Kock, A. E. An Argentine ant survey in South African fynbos. *J. Ent. Soc. S. Afr.* (in press).

Di Castri, F. (1981). Mediterranean-type shrublands of the world. In: Di Castri, F., Goodall, D. W., and Specht, R. L. (Eds.), *Ecosystems of the World II: Mediterranean-type Sclerophyllous Shrublands.* Elsevier, Amsterdam.

Donelly, D., and Giliomee, J. H. (1985). Community structure of epigaeic ants (*Hymenoptera: Formicidae*) in fynbos vegetation in the Jonkershoek Valley. *J. Ent. Soc. S. Afr.*, **48**, 247–57.

Fox, M. D., and Fox, B. J. (1986). The susceptibility of natural communities to invasion. In: Groves, R. H., and Burdon, J. J. (Eds.), *Ecology of Biological Invasions: An Australian Perspective*, pp. 57–66. Australian Academy of Science, Canberra.

Fox, B. J., and Pople, R. (1984). Experimental confirmation of interspecific competition between native and introduced mice. *Aust. J. Ecol.*, **9**, 323–34.

Fox, B. J., Quinn, R. D., and Breytenbach, G. J. (1985). A comparison of small-mammal succession following fire in shrublands of Australia, California, and South Africa. *Proc. Ecol. Soc. Aust.*, **14**, 179–97.

Frenkel, R. E. (1977). Ruderal vegetation along some California roadsides, *Univ. Calif. Publ. Geogr.*, **20**, 1–163.

Fuentes, E. R., and Simonetti, J. A. (1982). Plant pattering in the Chilean matorral: are the roles of native and exotic mammals different? In: Conrad, C. E., and Oechel, W. C. (Eds.), *Proceedings of the Symposium on Dynamics and Management of Mediterranean-Type Ecosystems, June 22–26, 1981*, pp. 227–33. San Diego, California. USDA Forest Service General Technical Report PSW–58.

Fuentes, E. R., and Etchegaray, J. (1983) Defoliation patterns in matorral ecosystems. In Kruger, F. J., Mitchell, D. T., and Jarvis, J. U. M. (Eds.), *Mediterranean-Type Ecosystems: the Role of Nutrients.*, pp. 528–42. Springer-Verlag, Berlin.

Fuentes, E. R., Otaiza, R. D., Alliende, M. C., Hoffman, A., and Poiani, (1984). Shrub clumps of the Chilean matorral vegetation: structure and possible maintenance mechanisms. *Oecologia*, **62**, 405–11.

Fugler, S. R. (1982). Infestations of three Australian Hakea species in South Africa and their control. S. *Afr. For. J.*, **120**, 63–8.

Glanz, W. (1977). Small mammals. In: Thrower, N. J. W., and Bradbury D. E. (Eds.), *Chile-California Mediterranean Scrub Atlas*, pp. 232–7. Dowden, Hutchinson and Ross, Stroudsburg, Pennsylvania.

Glanz, W. E., and Meserve, P. L. (1982). An ecological comparison of small mammal communities in California and Chile. In: Oechel, W. C., and Conrad C. E. (Eds.), *Proceedings of the Symposium on Dynamics and Management of Mediterranean-Type Ecosystems, June 22–26, 1981*, pp. 222–6. San Diego, California. USDA Forest Service General Technical Report PSW–58.

Greyling, T., and Huntley, B. J. (1984). Directory of southern African conservation areas. *South African National Scientific Programmes Report* **98**. CSIR, Pretoria.

Groves, R. H. (1986a). Plant invasion of Australia: an overview. In: Groves, R. H. and Burdon, J. J. (Eds.), *Ecology of Biological Invasions: An Australian Perspective*. pp. 137–47. Australian Academy of Science, Canberra.

Groves, R. H. (1986b). Invasion of mediterranean ecosystems by weeds. In: Hopkins, A. J. M., and Lamont, B. B. (Eds.), *Resilience in Mediterranean-type Ecosystems*, pp. 129–45. Junk. Dordrecht.

Groves, R. H., and Buruon, J. J. (Eds.) (1986). *Ecology of Biological Invasions: An Australian Perspective*. Australian Academy of Science, Canberra.

Gulmon, S. L. (1977). A comparative sutdy of the grassland of California and Chile. *Flora*, **166**, 261–78.

Gulmon, S. L. (1979). Competition and coexistence: three annual grass species. *Amer. Midl. Nat.*, **101**, 403–16.

Hanes, T. L. (1977). California chaparral. In: Barbour, M. G., and Major, J. (Eds.), *Terrestrial Vegetation of California*, pp. 417–69. Wiley, New York.

Harper, J. L. (1977). *Plant Population Biology*. Academic Press, London.

Heady, H. F. (1977). Valley grassland. In: Barbour, M. G., and Major, J. (Eds.), *Terrestrial Vegetation of California*, pp. 491–514. Wiley, New York.

Heddle, E. M., and Specht, R. L. (1975). Dark Island heath (Ninety-Mile Plain, South Australia) VII: the effects of fertilizers on composition and growth, 1950–72. *Aust. J. Bot.*, **23**, 151–64.

Hobbs, R. J., and Mooney, H. A. (1985). Community and population dynamics of serpentine grassland annuals in relation to gopher disturbance. *Oecologia*, **67**, 342–51.

Hopper, S. D., and Muir, B. G. (1984). Conservation of the kwongan. In: Pate, J. S., and Beard, J. S. (Eds.), *Kwongan: Plant Life of the Sandplain. Biology of the Southwest Australian Shrubland Ecosystem*, pp. 253–66. University of Western Australia Press, Nedlands.

Horton, J. S., and Kraebel, C. J. (1955). Development of vegetation after fire in the chamise chaparral of southern California. *Ecology*, **36**, 244–62.

Hunter, G. G., and Douglas, M. H. (1984). Spread of exotic conifers on South Island Rangelands. *N. Z. Jl. For.*, **29**, 78–96.

Jackson, L. E. (1985). Ecological origins of California's mediterranean grasses. *J. Biogeogr.*, **12**, 349–61.

Jaksic, F. M., Yanez, J. L., and Fuentes, E. R. (1981). Assessing a small mammal community in Central Chile. *J. Mammal.*, **62**, 391–6.

Keeley, S. C., and Johnson, A. W. (1977). A comparison of the pattern of herb and shrub growth in comparable sites in Chile and California. *Am. Midl. Nat.*, **97**, 120–32.

Keeley, S. C., Keeley, J. E., Hutchinson, S. M., and Johnson, A. W. (1981). Postfire-succession of the herbaceous flora in the southern California chaparral. *Ecology*, **62**, 1608–21.

Kershaw, A. P. (1986). Climate change and Aborginal burning in northwest Australia during the last two glacial/interglacial cycles. *Nature*, **322**, 47–9.

Kluge, R. L. (1983). The hakea fruit weavil, *Erytenna consputa*, and the biological control of Hakea serices in South Africa. Ph.D thesis, Rhodes University, Grahamstown.

Knight, R. S. (1986). A comparative analysis of fleshy fruit displays in alien and indigenous plants. In: Macdonald, I. A. W., Kruger, F. J. and Ferrar, A. A. (Eds), *The Ecology and Management of Biological Invasions in Southern Africa*, pp. 171–8. Oxford University Press, Cape Town.

Kruger, F. J. (1982). Use and management of mediterranean ecosystems in South Africa—current problems. In: Oechel, W. C., and Conrad, C. E. (Eds), *Proceedings of the Symposium on Dynamics and Management of Mediterranean-Type Ecosystems, June 22–26, 1981*, pp. 42–8. San Diego, California. USDA Forest Service General Technical Report PSW–58.

Kruger, F. J. (1977). Invasive woody plants in the Cape fynbos, with special reference to the biology and control of *Pinus pinaster*. In *Proceedings of the Second National Weeds Conference of South Africa*, pp. 47–55. Balkema, Cape Town.

Kruger, F. J. (1983). Plant community diversity and dynamics in relation to fire. In Kruger, F. J., Mitchell, D. T., and Jarvis, J. U. M. (Eds), *Mediterranean-Type Ecosystems: The Role of Nutrients*, pp. 447–72. Springer-Verlag, Berlin.

Kruger, F. J., and Bigalke, R. C. (1984). Fire in fynbos. In Booysen, P. de V., and Tainton, N. M. (Eds.), *Ecological Effects of Fire in South African Ecosystems*, pp. 67–114. Springer-Verlag, Berlin.

Kruger, F. J., and Taylor, H. C. (1979). Plant species diversity in Cape fynbos: gamma and delta diversity. *Vegetatio*, **41**, 85–93.

Kruger, F. J., Richardson, D. M., and Van Wilgen, B. W. (1986). Processes of invasion by alien plants. In Macdonald, I. A. W., Kruger, F. J., and Ferrar, A. A. (Eds), *The Ecology and Management of Biological Invasions in Southern Africa*, pp. 145–55. Oxford University Press, Cape Town.

Lambert, M. J., and Turner, J. (1987). Suburban development and change in vegetation nutritional status. *Aust. J. Ecol.*, **12**, 193–6.

Lamont, B. B. (1983). Strategies for maximizing nutrient uptake in two mediterranean ecosystems of low nutrient status. In: Kruger, F. J., Mitchell, D. T., and Jarvis, J. U. M. (Eds), *Mediterranean-Type Ecosystems: The Role of Nutrients*, pp. 246–73. Springer-Verlag, Berlin.

LeHouerou, H. N. (1981). Impact of man and his animals on the mediterranean vegetation. In: Di Castri, F., Goodall, D. W., and Specht, R. L. (Eds), *Ecosystems of the world 11: Mediterranean-Type Shrublands*, pp. 479–521. Elsevier, Amsterdam.

Long, L. J. (1981). *Introduced Birds of the World: the Worldwide History, Distribution and Influence of Birds Introduced to New Environments.* New York, Universe Books.

Macdonald, I. A. W. (1984). Is the fynbos biome especially susceptible to invasion by alien plants? A re-analysis of available data. *S. Afr. J. Sci.*, **80**, 369–377.

Macdonald, I. A. W., and Jarman, M. L. (Ed.), (1984). Invasive alien organisms in the terrestrial ecosystems of the fynbos biome, South Africa. *South African National Scientific Programmes Report 85*. CSIR, Pretoria.

Macdonald, I. A. W. (1986). Range expansion in the pied barbet and the spread of alien tree species in southern Africa. *Ostrich*, **57**, 75–94.

Macdonald, I. A. W., Jarman, M. L., and Beeston, P. (Eds), (1985). Management of invasive alien plants in the fynbos biome. *South African National Scientific Programmes Report 111*. CSIR, Pretoria.

Macdonald, I. A. W., and Richardson, D. M. (1986). Alien species in terrestrial ecosystems of the fynbos biome. In: Macdonald, I. A. W., Kruger, F. J., and Ferrar, A. A. (Eds), *The Ecology and Management of Biological Invasions in Southern Africa*, pp. 77–91. Oxford University Press, Cape Town.

Macdonald, I. A. W., Powrie, F. J. and Siegfried, W. R. (1986). The differential invasion of southern Africa's biomes and ecosystems by alien plants and animals. In: Macdonald, I. A. W., Kruger, F. J., and Ferrar, A. A. (Eds), *The Ecology and Management of Biological Invasions in Southern Africa*, pp. 209–25. Oxford University Press, Cape Town.

Macdonald, I. A. W., Richardson, D. M., and Powrie, F. J. (1986). Range expansion of the hadeda ibis *Bostrychia hagedash* in southern Africa. *S. Afr. J. Zool.*, **21**, 331–42.

Macdonald, I. A. W., Graber, D. M., DeBenedetti, S., Groves, R. H., and Fuentes, E. R. (1988). Introduced species in nature reserves in mediterranean-type climate regions of the world. *Biological Conservation*, **44**, 37–66.

Majer, J. D. (1978). Preliminary survey of the epigaeic invertebrate fauna with particular reference to ants, in areas of different landuse at Dwellingup, Western Australia. *For. Ecol. Manage.*, **1**, 321–34.

Mathias, M. E., and McClintock, E. (1963). A checklist of woody ornamental plants of California. *University of California Division of Agricultural Sciences Experimental Station Extension Service Manual* **32**.

McNaughton, S. J. (1968). Structure and function in California grasslands. *Ecology*, **49**, 962–72.

Miller, P. C. (1981). Similarities and limitations of resource utilization in mediterranean type ecosystems. In: Miller, P. C. (Ed.), *Resource Use By Chaparral and Matorral: A Comparison of Vegetation Function in Two Mediterranean-Type Ecosystems*, pp. 369–407. Springer-Verlag, New York.

Miller, P. C., Miller, J. M., and Miller, P. M. (1983). Seasonal progression of plant water relations in fynbos in the Western Cape Province, South Africa. *Oecologia*, **56**, 392–96.

Milton, S. J. (1981). Australian acacias in the south western Cape: pre-adaption and success. In: Neser, S. and Cairns. A. L. P. (Eds), *Proceedings of the Third National Weeds Conference of South Africa*, pp. 68–78. Balkema, Cape Town.

Moll, E. J., and Bossi, L. (1984). Assessment of the extent of the natural vegetation of the fynbos biome of South Africa. *S. Afr. J. Sci.*, **80**, 355–8.

Montgomery, K. R., and Strid, T. W., (1976). Regeneration of introduced species of *Cistus* (Cistaceae) after fire in southern California. *Madrono*, **23**, 417–27.

Mooney, H. A., and Parsons, D. J. (1973). Structure and function of the California chaparral—an example from San Dimas. In: Di Castri, F. and Mooney, H. A. (Eds), *Mediterranean-Type Ecosystems: Origin and Structure*, pp. 83–113. Springer-Verlag, Berlin.

Mooney, H. A., Hamburg, S. P., and Drake, J. A. (1986). The invasion of plants and animals into California. In: Mooney, H. A., and Drake, J. A. (Eds), *Ecology of Biological Invasions of North America and Hawaii*, pp. 250–72. Springer-Verlag, New York.

Mooney, H. A., Dunn, E. L., Shropshire, F., and Song, L. (1972). Land-use history of California and Chile as related to the structure of the sclerophyll scrub vegetations. *Madroño*, **21**, 305–19.

Moore, R. M. (1975). *Australian Grasslands*, pp. 455. Australian National University Press, Canberra.

Muir, B. G. (1983). *Weeds in National Parks of Western Australia.* Paper presented at Weeds Conference, Adelaide 1–3 November 1983. 7 pp.

Myers, K. (1986). Introduced vertebrates in Australia with emphasis on the mammals. In: Groves, R. H., and Burdon, J. J. (Eds), *Ecology of Biological Invasions: An Australian Perspective*, pp. 120–36. Australian Academy of Science, Canberra.

Naveh, Z. (1967). Mediterranean ecosystems and vegetation types in California and Israel. *Ecology*, **48**, 445–59.

Naveh, Z. (1975). The evolutionary significance of fire in the mediterranean region. *Vegetatio*, **29**, 199–208.

Nel, J. A. J., Rautenbach, I. L., and Breytenbach, G. J. (1980). Mammals of the Kammanassie Mountains, Southern Cape Province. *S. Afr. J. Zool.*, **15**, 255–61.
Neser, S. (1968). Studies on some potential useful insects enemies of the needle-brush (*Hakea* spp.—Proteaceae). Ph.D. thesis, Australian National University, Canberra.
Nix, H. A., and Wapshere, A. J. (1986). Biogeographic origins of invading species. In: Groves, R. H., and Burdon, J. J. (Eds), *Ecology of Biological Invasions: An Australian Perspective*, P. 155, Australian Academy of Science, Canberra.
Ozanne, P. G., and Specht, R. L. (1981). Mineral nutrition of heathlands: phosphorus toxicity. In: Specht, R. L. (Ed.). *Ecosystems of the World 9B: Heathlands and Related Shrublands: Analytical Studies*, pp. 209–13. Elsevier, Amsterdam.
Poole, D. K., Roberts, S. W., and Miller, P. C. (1981). Water utilization. In: Miller, P.C. (Ed.), *Resource Use by Chaparral and Matorral: a Comparison of Vegetation Function in Two Mediterranean-Type Ecosystems*, pp. 123–49. Springer-Verlag, New York.
Quinn, R. D. (1983). *Short-term effects of habitat management on small vertebrates in chaparral.* Cal-Neva Wildlife Transactions. pp. 55–66.
Rautenbach, I. L., and Nel, J. A. J. (1980). Mammal diversity and ecology in the Cedarberg Wilderness Area, Cape Province. *Ann. Tvl. Mus.*, **32**, 101–24.
Raven, P. H. (1977). The California flora. In Barbour, M. G., and Major, J. (Eds), *Terrestrial Vegetation of California*, pp. 109–31. Wiley, New York.
Raven P. H., and Thomas, J. H. (1970). *Iris pseudacorus* in western North America. *Madroño*, **20**, 390–91.
Rice, R. M. (1974). The hydrology of chaparral watersheds. *Proceedings of the Symposium on Living with the Chaparral*, pp. 27–34. Riverside, California. Sierra Club, California Division of Forestry, U. S. Forest Service.
Rice, R. M., and Foggin, G. T. (1971). Effects of high intensity storms on soil slippage on mountainous watersheds in southern California. *Water Res. Res.*, **7**, 1485–96.
Rice, R. M., Ziemer, R. R., and Hankin, S. C. (1982). Slope stability effect of fuel management strategies—inferences from Monte Carlo simulations. In: Oechel, W. C., and Conrad, C. E. (Eds), *Proceedings of the Symposium on Dynamics and Management of Mediterranean-Type Ecosystems, June 22–26, 1981*, pp. 365–71. San Diego, California. USDA Forest Service Genetal Technical Report PSW–58.
Richardson, D. M. (1988). Age structure and regeneration after fire in a self-sown *Pinus halepensis* forest on the Cape Peninsula, South Africa. *S. Afr. J. Bot.*, **54**, 140–44.
Richardson, D. M., and Brown, P. J. (1986). Invasion of mesic mountain fynbos by *Pinus radiata*. *S. Afr. J. Bot.*, **52**, 529–36.
Richardson, D. M., Van Wilgen, B. W., and Mitchell, D. T. (1987). Aspects of the reproductive ecology of four australian *Hakea* species (Proteaceae) in South Africa. *Oecologia*, **71**, 345–54.
Rundel, P. W. (1981). The matorral zone of central Chile. In: Di Castri, F., Goodall, D. W., and Specht, R. L. (Eds.), *Ecosystems of the World 11: Mediterranean-Type Shrublands*, pp. 175–201. Elsevier, Amsterdam:
Russel, J. S., and Moore, A. W. (1976). Classification of climate by pattern analysis with Australasian and Southern African data as an example. *Agric. Meteorol.* **16**, 45–70.
Rutherford, M. C., and Westfall, R. H. (1986). Biomes of southern Africa: an objective categorization. *Mem. Bot. Surv. S. A.*, **54**, 1–98.
Scholtz, A. (1986). *Palynological and palaeobotanical studies in the southern Cape.* M. A. thesis, University of Stellenbosch.
Shaughnessy, G. L. (1986). A case study of some woody plant introductions to the Cape Town area. In: Macdonald, I. A. W., Kruger, F. J., and Ferrar, A. A. (Eds), *The Ecology and Management of Biological Invasions in Southern Africa*, pp. 37–46. Oxford University Press, Cape Town.
Specht, R. L. (1963). Dark Island heath (Ninety-Mile Plain, South Australia) VII: the

effects of fertilizers on composition and growth, 1950–1960. *Aust. J. Bot.*, **11**, 67–94.

Specht, R. L. (1972). *The Vegetation of South Australia*, pp. 328 Government Printer, Adelaide.

Specht, R. L. (1981). Major vegetation formations in Australia. In: Keast, A. (Ed.), *Ecological Biogeography of Australia*, pp. 165–297. Junk. The Hague.

Specht, R. L., and Moll, E. J. (1983). Mediterranean-type heathlands and sclerophyllous shrublands of the world: an overview. In: Kruger, F. J., Mitchell, D. T., and Jarvis, J. U. M. (Eds), *Mediterranean-Type Ecosystems: the Role of Nutrients*, pp. 41–65. Springer-Verlag, Berlin.

Van der Sommen, F. J. (1986). The use of spatial and temporal survivorship curves in predicting invasion by *Pinus radiata*. In: Groves, R. H., and Burdon, J. J. (Eds), *Ecology of Biological Invasions: an Australian Perspective*, pp. 159. Australian Academy of Science, Canberra.

Van Wyk, D. B. (1982). Influence of prescribed burning on nutrient budgets of mountain fynbos catchments in the south western Cape, Republic of South Africa. In: Conrad, C. E., and Oechel, W. C. (Eds.), *Proceedings of the Symposium on Dynamics and Management of Mediterranean Type Ecosystems, June 22–26, 1981*, pp. 390–6. San Diego, California. USDA Forest Service General Technical Report PSW–58.

Vlok, J. A. The effect of alien annuals on indigenous therophytes in lowland fynbos. South African Journal of Botany (submitted).

Walker, J. (1981). Fuel dynamics in Australian vegetation. In: Gill, A. A. M., Groves, R. H., and Noble, I, R. (Eds), *Fire and the Australian biota*, pp. 101–27. Australian Academy of Science, Canberra.

Wapshere, A. J. (1984). The invasiveness and importance in Australia of weeds from mediterranean regions of the old world. In: Dell, B. (Ed.), *1984 MEDECOS IV: Proceeding of the Fourth International Conference on Mediterranean Ecosystems*, pp. 157–8. Botany Department, University of Western Australia, Nedlands.

Weiss, P. W. (1986). Invasion of coastal plant communities by *Chrysanthemoides monilifera*. In: Groves, R. H., and Burdon, J. J. (Eds), *Ecology of Biological Invasions: an Australian Perspective*, p. 162. Australian Academy of Science, Canberra.

Weiss, P. W., and Milton, S. J. (1984). *Chrysanthemoides monilifera* and *Acacia longifolia* in Australia and South Africa. In: Bell, D. T. (Ed.), MEDECOS IV: *Proceedings of the Fourth International Conference on Mediterranean Ecosystems*, pp. 159–60. Botany Department, University of Western Australia, Nedlands.

Wells, M. J., Balsinhas A. A., Joffe, H., Engelbrecht, V. M., Harding, G., and Stirton C. H. (1986). A catalogue of problem plants in southern Africa incorporating the national weed list of South Africa. *Mem. Bot. Surv. S. Africa*, **53**. pp. 1–549.

Westman, W. E. (1983). Plant community structure—spatial partitioning of resources. In Kruger, F. J., Mitchell, D. T., and Jarvis, J. U. M. (Eds), *Mediterranean-Type Ecosystems: the Role of Nutrients*. pp. 417–45. Springer-Verlag, Berlin.

Westman, W. E. (1986). Resilience: concepts and measures. In Dell, B., Hopkins, A. J. M., and Lamont, B. B. (Eds.), *Resilience in Mediterranean-Type Ecosystems*, pp. 5–19. Junk, Dordrecht.

Willan, K., and Bigalke, R. C. (1982). The effects of fire regime on small mammals in south western Cape Montane Fynbos (Cape Macchia). *Proceedings of Symposium on Dynamics and Management of Mediterranean-Type Ecosystems, June 22–26 1981*, pp. 207–12. San Diego, California. USDA Forest Service General Technical Report PSW–58.

Winterbottom, J. M., and Liversedge, R. (1954). The European starling in the south west Cape. *Ostrich*, **25**, 89–96.

Wirtz, W. O. (1982). Postfire community structure of birds and rodents in southern

California chaparral. In: Conrad, C. E., and Oichel, W. C. (Eds.), *Proceedings of Symposium on Dynamics and Management of Mediterranean-Type Ecosystems, June 22–26 1981*, pp. 241–6. San Diego, California. USDA Forest Service General Technical Report PW–58.

Zedler, P. H., Gautier, C. H., and McMaster, G. S. (1983). Vegetation change in response to extreme events: the effect of a short interval between fires in California chaparral and coastal scrub. *Ecology*, **64**, 809–18.

Biological Invasions: a Global Perspective
Edited by J. A. Drake *et al.*
© 1989 SCOPE. Published by John Wiley & Sons Ltd

CHAPTER 9

Wildlife Conservation and the Invasion of Nature Reserves by Introduced Species: a Global Perspective

IAN. A. W. MACDONALD, LLOYD L. LOOPE, MICHAEL B. USHER AND O. HAMANN

9.1 INTRODUCTION

In this chapter we address some of the ways in which introduced species can affect nature conservation, the current extent of invasions of nature reserves by these species and some of the observed effects of these invasions. We then give examples of successes and failures in past efforts at controlling these invasions. Finally we draw some conclusions related to the prospects of controlling them in the future.

The World Conservation Strategy gives three main objectives for nature conservation; the maintenance of essential life support systems, the maintenance of natural diversity and the sustained utilization of species and ecosystems. In terms of conventional ecological terminology the first two objectives can be paraphrased as maintaining ecosystem function and ecosystem structure (Watt, 1947).

This chapter is restricted to wildlife conservation (i.e. the *in situ* conservation of natural communities) as implemented within a framework of proclaimed nature reserves. For the purposes of this chapter a 'nature reserve' is any area set aside primarily for the conservation of its native biota and these can range from urban reserves of a few hectares to national parks thousands of square kilometres in extent.

Whenever possible, reserves have been classified according to the biome and realm system of Udvardy (1975) and the names used throughout the chapter follow this system. The area of natural vegetation in each of these biomes has not yet been surveyed. Rough approximations of these areas derived from two surveys using different 'biome' classification systems, (Whittaker and Likens, 1975; World Resources Institute, 1986) together with the number and area of national parks and equivalent reserves that had been proclaimed up to 1982 (after Harrison *el al.*, 1984) are given in Table 9.1.

The percentage of the world's non-marine area currently included in nature reserves (*Sensu* Harrison *et al.*, 1984) lies somewhere between 2.7 and 3.4%,

Table 9.1. The area of natural ecosystems estimated as remaining in the continental and insular biomes of the world and the total number and area of reserves in these biomes

Biome (after Udvardy, 1975)	Area of Natural Vegetation (million of km^2) (after World Resources Institute, 1980)	(after Whittaker and Likens, 1975)	No. of Reserves	Area of Reserves (from Darrison *et al.*, 1984) (thousand of km^2)
Tropical humid forests	15.6	17.0	355	475.2
Subtropical and temperate rainforests or woodlands	1.5	8.5	201	113.3
Temperate broadleaf forests or woodlands	12.5	12.0	270	144.1
Temperate needleleaf forests or woodlands	11.8	12.0	114	278.3
Evergreen sclerophyllous forests, scrub or woodlands	42	Included in temperate forest	426	82.7
Tropical dry or deciduous forests or woodlands	10.2	7.5	481	703.7
Tropical grasslands and savanna	27.4	15.0	31	87.7
Temperate grasslands	?	9.0	05	32.8
Warm deserts and semi deserts	24.5	42.0	67	603.3
Cold winter deserts		42.0	48	76.1
Tundra communities and barren arctic communities	7.4	8.0	31	1012.4
Mixed mountain and highland systems with complex zonation	Included above under constituent biome	Included above under constituent biome	318	283.7
Mixed island systems			74	15.8
River and lake systems	Not included	2.0	10	4.3
Swamp and marsh	Not included	2.0	Included above under constituent biome	
[Cultivated lands]	[17.5]	[14.0]	[not normally present within reserves.]	
Total land area	132.5	149.0	2511	3973.2

depending on which estimate of the total area one accepts (Table 9.1). By 1983 some 13% of the total ice-free land area had been transformed by cultivation (Table 9,1; World Resources Institute, 1986). Eight of the 30 global vegetation types had already been reduced by between 25% and 45% of their original areas (World Resources Institute, 1986). As the human population increases the conversion of natural vegetation through cultivation will intensify. The remaining areas of natural vegetation outside nature reserves will become progressively more modified. Increasingly, the conservation of the world's wild genetic resources will come to depend on that small percentage of its area included in nature reserves.

This chapter concentrates on invasive introduced species which affect the maintenance of ecosystem structure (i.e. species composition, genetic diversity) and ecosystem function (e.g. nutrient cycling, hydrology, soil erosion, decomposition) within nature reserves. An introduced species is defined as one which only occurs in the reserve as a result of the intentional or accidental movement of the species by humans from its natural distribution range. To be classified as 'invasive' the introduced species must be capable of establishing self-sustaining populations in areas of natural or semi-natural vegetation (i.e. untransformed ecosystems) within the reserve.

Any significant alteration of ecosystem function by an introduced species is considered deleterious. This is not simply a subjective value judgement as it is a fundamental goal of nature reserves that they should serve as unaltered natural baselines against which can be measured the effects of anthropogenic changes in adjacent unprotected areas.

If an invasion alters the relative abundances of a reserve's native species this cannot always be taken to be deleterious. Thus, although introduced species most probably have important effects on nature reserves through changing their native species composition, such cases do not form the main subject matter of this chapter. There are, however, two categories of structural change to ecosystems that can be relatively easily quantified and which are universally held to be in conflict with the goals of nature conservation. The first category is the reduction in species richness of a community through local extinction. The second is the adulteration of gene pools through the hybridization of introduced and native species. However, before examining these effects, the extent of the invasion of nature reserves by introduced species in some of the world's biomes is first described.

9.2 THE EXTENT OF THE INVASION OF NATURE RESERVES BY INTRODUCED SPECIES

The measurement of the actual extent of invasions of nature reserves has been carried out only for a handful of introduced species in a few of the world's reserves. In almost all cases the measures of extent differ between reserves and between different types of introduced organisms. In the absense of a large, standardized data base, three simpler measures have been used: the proportion of reserves in a biome reporting problems with introduced species (e.g. Machlis and Tichnell, 1985), the number of such species recorded from the reserve (Loope, in press; Macdonald, in press; Macdonald *et al.*, 1986) and the extent of resource allocation to control operations within a reserve (Macdonald, in press). Data on the number of invasive introduced species known to be present, expressed both as an absolute number and as a percentage of the total number of species (native and invasive), are given in Table 9.2.

The number of invasive introduced plant species present in a reserve is an order of magnitude greater than the number of such species in any vertebrate class. Introduced invertebrates are however known to reach numbers similar to those of vascular plants in the national parks of Hawaii (Loope, in press). If the number of introduced species is expressed as a percentage of the total species present, the differences between plants and vertebrate classes are not so great, with fish often showing the highest percentage. The major difference between biomes is that the reserves in mixed island systems have higher mean absolute and proportional numbers of introduced species than do reserves in any of the continental biomes.

If there is any trend between continental biomes it is for the more xeric environments to have smaller absolute numbers of introduced species, in particular of introduced vascular plant species. This is most obvious in the warm deserts and semi deserts but is also apparent in the tropical dry forests or woodlands. There are no invasive introduced plants and vertebrates in the Barren Arctic Deserts of Antarctica (Usher, unpublished). The trend then is for biomes with extreme abiotic conditions i.e. very hot, cold or dry, to have fewer invasive introduced species.

In their analyses of the numbers of introduced species in 41 southern African nature reserves, Macdonald *et al.* (1986) found that the only reserve characteristic which gave rise to significantly different numbers of introduced vascular plants was the annual number of visitors to the reserve. This relationships was explored more fully in both the southern African data set (slightly modified to include more recent data) and one for 21 continental reserves in the USA (Loope, in press) with annual visitor numbers for 1985 (Statistical Office, Denver Service Center, 1986). The relationships between the numbers of introduced and native species per reserve were also investigated by regression analyses of untransformed and logarithmically transformed variables.

The significant results obtained in these regression analyses are presented in Table 9.3. Some of the bivariate scatter plots showing the linear regressions are presented in Figures 9.1 and 9.2.

The correlation between the number of introduced species and visitor numbers was significant for 11 of the 13 reserve groupings tested here. There were also positive correlations between numbers of introduced and native species in nine of the 13 groupings of reserves.

Table 9.2. Data on the extent of invasions of nature reserves in different biomes. The number of reserves for which data were available is given in brackets; the three numbers represent the minimum, mean and maximum for that sample respectively. No information was available for biomes listed in Table 9.1 but omitted from this table

Biome	No. of Invasive Spp. per Reserve				Invasive Introduced Spp. as Percentage of Total Species			
	Vascular plants	Freshwater fish	Birds	Mammals	Vascular plants	Freshwater fish	Birds	Mammals
Subtropical and temperate rainforests or woodlands	20-139-260 (3)				3-15-22 (3)			
Temperate broadleaf forests or woodlands	23-157-249 (5)				7-13-17 (5)			
Temperate needleleaf forests or woodlands	27-115-222 (3)				4-13-21 (3)			
Evergreen scleropnyllous forests, scrub or woodlands	69-88-113 (5)	1-3-6 (4)	1-3-5 (6)	2-4-8 (5)	6-11-28 (5)	37-56-86 (4)	1-3-6 (6)	5-13-26 (5)
Tropical dry or deciduous forests or woodlands	12-60-113 (5)	0-1-1 (3)	0-1-3 (3)	1-2-4 (3)	3-5-7 (5)	0-17-50 (3)	0-0-1 (3)	1-2-5 (3)
Warm deserts and semi-deserts	7-20-33 (3)	0-2-6 (5)	1-3-6 (5)	0-1-3 (5)	4-5-6 (3)	0-30-100 (5)	1-1-2 (5)	0-3-6 (5)
Mixed mountain and highland systems with complex zonation	67-98-120 (4)				7-10-12 (4)			
Mixed island systems	240-358-520 (3)		1-9-17 (2)	9-9-10 (2)	31-47-64 (3)		3-20-53 (2)	71-81-90 (2)

Table 9.3. Significant results in the regression analyses of the number of alien plant species in continental nature reserves of North America and southern Africa. (P = significance of the regression. Where $P > 0.05$ the regression was considered not to be significant (N.S.) % = percentage of variance accounted for by regression)

Group of reserves	No. of reserves in sample	No. of alien species on log visitors/year	Log no. of alien spp. on log visitors/year	No. of alien species on no. of native spp.	Log no. of alien spp. on log no. native spp.
All American reserves	21	$P < 0.001$ % = 43	$P < 0.001$ % = 45	N.S.	$P < 0.002$ % = 41
Temperate broadleaf forests or woodlands	6	$P < 0.004$ % = 90	$P < 0.004$ % = 70	$P < 0.04$ % = 72	$P < 0.006$ % = 88
Warm deserts and semideserts	4	$P < 0.002$ % = 60	N.S.	N.S.	N.S.
Mixed mountain systems	5	N.S.	N.S.	N.S.	N.S.
All Southern African reserves	41	$P < 0.001$ % = 40	$P < 0.001$ % = 26	$P < 0.001$ % = 59	$P < 0.001$ % = 44
Tropical dry forest or woodlands	15	$P < 0.015$ % = 38	N.S.	$P < 0.001$ % = 77	$P < 0.001$ % = 71

Evergreen sclerophyllous forests, scrub or woodland	15	$P < 0.001$	% = 71	$P < 0.001$	% = 64	$P < 0.01$	% = 41	N.S.	
Warm deserts and semideserts	8	N.S.		N.S.		N.S.		N.S.	
All reserves	62	$P < 0.001$	% = 53	$P < 0.001$	% = 56	$P < 0.01$	% = 38	$P < 0.001$	% = 54
Tropical dry forest or woodlands	16	$P < 0.002$	% = 52	$P < 0.02$	% = 37	$P < 0.002$	% = 51	$P < 0.001$	% = 65
Evergreen sclerophyll vegetation	16	$P < 0.001$	% = 74	$P < 0.001$	% = 67	$P < 0.04$	% = 29	N.S.	
Warm deserts and semideserts	12	$P < 0.003$	% = 60	$P < 0.02$	% = 43	$P < 0.001$	% = 86	$P < 0.001$	% = 80
Mixed mountain systems	8	$P < 0.002$	% = 81	$P < 0.001$	% = 87	$P < 0.04$	% = 65	$P < 0.008$	% = 72

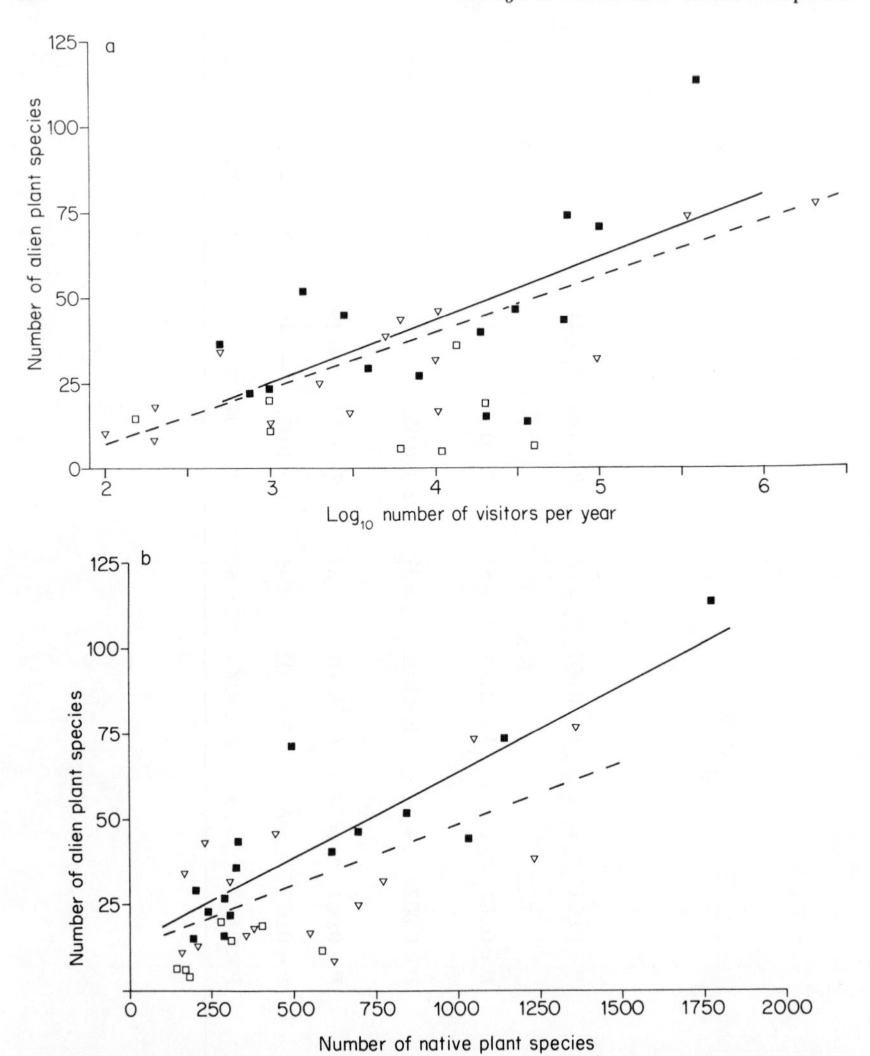

Figure 9.1. The number of introduced vascular plant species (y) recorded from southern African nature reserves, plotted against (A) the log of the number of vi-itors per annum to the reserve (x) and (B) the number of native vascular plant species recorded from the reserve (z).

Key: Reserves in Tropical Dry Forest or Woodland biome

(■; ———— = regressions $y = 18{,}41x - 30.93$

$y = 0{,}05z + 13{,}52$)

Reserves in Evergreen Sclerophyllous Forest, Scrub or Woodland biome

(▽; — — — — = regressions $y = 14{,}55x - 21{,}55$

$y = 0{,}035z + 12{,}57$)

Reserves in Warm Desert or Semidesert biome

(□)

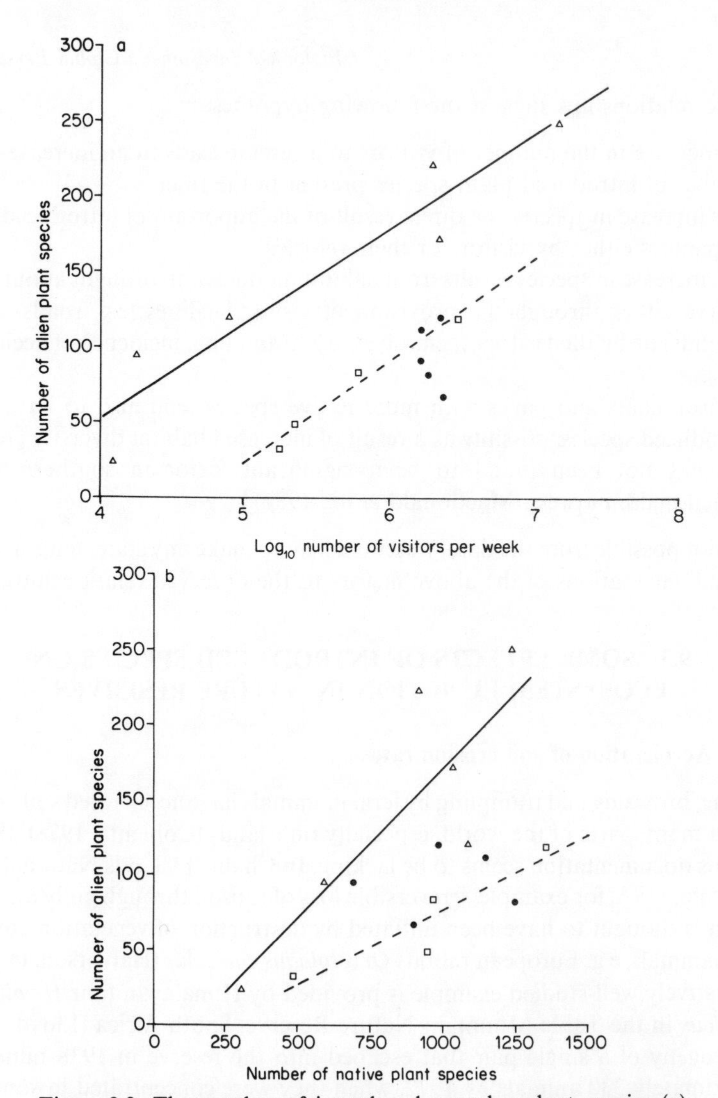

Figure 9.2. The number of introduced vascular plant species (y) recorded from North American nature reserves, plotted against (A) the log of the number of visitors per annum to the reserve (x) and (B) the number of native vascular plant species recorded from the reserve (z).

Key: Reserves in Temperate Broadleaf Forest or Woodland biome
(△;——— = regressions $y = 63{,}17x - 200{,}86$
$y = 0{,}20z - 29{,}32$)

Reserves in Warm Desert or Semidesert biome
(□;– – – = regressions $y = 67{,}66x - 315{,}74$
$y = 0{,}098z - 21{,}60$)

Reserves in Mixed Mountain or Highland Systems with Complex Zonation biome
(●)

These relationships suggest the following hypotheses:

1. An increase in the number of visitors to a reserve leads to an increase in the number of introduced plant species present in the reserve.
2. This increase in species is a direct result of the important of introduced plant propagules either by visitors or their vehicles.
3. This increase in species results from habitat modification brought about in the reserve either through the provision of visitor facilities (e.g. roads, campgrounds) or by the visitors themselves (e.g. trampling, incidence of accidental fires).
4. Environments and parks with more native species tend also to have more introduced species, possibly as a result of increased habitat diversity [reserve size has not been found to be a significant factor in southern Africa (Macdonald, in press; Macdonald *et al.*, 1986)].

It is not possible from the data presented here to make any statements as to the relative contributions of the above factors to the observed relationships.

9.3 SOME EFFECTS OF INTRODUCED SPECIES ON ECOSYSTEM FUNCTION IN NATURE RESERVES

9.3.1. Acceleration of soil erosion rates

Grazing, browsing and trampling by feral mammals have accelerated soil erosion rates in many parts of the world, especially on islands (Coblentz, 1978), though rigorous documentation seems to be lacking. In Channel Islands National Park, California, USA, for example, irreversible loss of topsoil through gully and sheet erosion is thought to have been initiated by destruction of vegetation cover by feral mammals, e.g. European rabbits *Oryctolagus cuniculus* (Halvorson, in press).

A relatively well-studied example is provided by Himalayan tahr *Hemitragus jemlahicus* in the Table Mountain Nature Reserve, South Africa (Lloyd, 1975). The progeny of a single pair that escaped into the reserve in 1938 numbered approximately 330 animals by 1972, when they were concentrated in some 300 hectares of optimal habitat. Using exclosures situated in the main concentration areas, the tahr were estimated to be removing through browsing or reducing by trampling the annual above-ground net plant production by 5300 kg dry matter (DM) ha^{-1} (73% of that recorded inside the enclosures). All the concentration areas showed accelerated soil erosion, with a loss of 8 mm of soil being measured over an 11 month period in one area (Lloyd, 1975).

9.3.2 Alteration of other geomorphological processes

Any changes in the geomorphological processes which shape the landscapes and habitats of nature reserves will affect the long-term conservation of their constituent ecosystems. Introduced species have already been demonstrated to have affected some of the more dynamic of these processes, namely soil erosion, sand dune formation and changes in riverbank configuration.

The European grass *Ammophila arenaria* has been planted world-wide for sand-fixation. It has become an aggressive invader along the coast of western North America, in northern California and Oregon, and is believed to have changed the topography of dunes in this area. Before its introduction, foredunes were low, rose gradually, and were accompanied by an inland dune system parallel to the prevailing on-shore winds (perpendicular to the coast). Most dune systems now have steep foredunes dominated by *A. arenaria* with an inland system of dunes and swales oriented parallel to the coast (Barbour and Johnson, 1977). Introduced plants have been implicated in similar alterations of dune configurations in Africa and Australia (Macdonald and Richardson, 1986).

Stands of invasive *Casuarina equisetifolia* on the west coast of Everglades National Park, Florida, have been found to result in sand accumulation, leading to development of steeper shorelines that can not be surmounted by sea turtles *Caretta caretta* searching for nesting sites (Klukas and Truesdell, 1969).

A study by Graf (1978) on the Green River in Canyonlands National Park, Utah, documented the effects of the invasive tree *Tamarix ramosissima* on the fluvial geomorphology of the area. Before the introduction of tamarisk, the native plants (primarily *Distichlis stricta* and *Salix exigua*) occupying the river floodplain were repeatedly swept away by floods, but once introduced and established, tamarisk was able to withstand inundation, continue to grow, stabilize the underlying surface, and induce sedimentation. Dense tamarisk thickets along the Green River have trapped and stabilized sediment, causing a reduction in channel width of 13–57%, increasing overbank flooding, and stabilizing and altering dimensions of islands and bars.

An interesting example of how an introduced species has caused a range of geomorphological effects is provided by *Spartina* species in the estuaries of England. *Spartina alterniflora* appears to have been introduced to Southampton Water, southern England, accidentally, probably in shipping ballast, from the east coast of North America. Prior to 1870 it hybridized with the native species, *Spartina maritima*, to produce the sterile hybrid, *Spartina* × *townsendii*. *S. alterniflora* has not been invasive and, although it has persisted, it is now considered to be 'very rare' (Hubbard, 1968). To some extent the sterile hybrid has been invasive, spreading throughout the southern coastal areas of England. However, a highly invasive fertile amphidiploid, *Spartina anglica*, has been derived from the sterile hybrid by a doubling of the chromosomes, and it is the new species which has invaded many coastal mud-flats around the shores of Great Britain.

The effects of *s* × *townsendii* and *S. anglica* on geomorphological processes are documented in Doody (1984). Hubbard (1968) recorded that these species, especially the latter, stabilized the soft coastal mud of tidal mud-flats by means of their extensive system of roots and rhizomes. The filtering action of the culms and leaves collects silt and debris, thus raising the level of the mud-flats. Due to this, *S. anglica* has been widely planted to protect foreshores from erosion and to reclaim previously useless land. Such accretion had occurred, for example, in the North Solent National Nature Reserve, but the *S. anglica* stands are now experiencing an unexplained dieback. Degeneration within the reserve is so severe that it is accompanied by destruction of the terminal cliffs of the marsh along an exposed face and the Nature Conservancy Council is having to devise means of encouraging longshore drift to build a protective shingle barrier (C.R. Tubbs, personal communication). The *Spartina* species have, during the first half century or so of their invasion, been associated with a geomorphological process of accelerated accretion; subsequently their degeneration is threatening the plant and animal communities that have previously been protected from marine erosion in many areas.

9.3.3 Alteration of biogeochemical cycling

Singer *et al.* (1984) found that feral pigs *Sus scrofa* rooting in deciduous forests of the Great Smoky Mountains USA accelerated leaching of Ca, P, Zn, Cu, and Mg from leaf litter and soil. Nitrate concentrations were higher in soil, soil water, and stream water from rooted areas suggesting alterations in nitrogen mineralization processes. Rooting was not found to increase sediment yield in this study, however.

The invasive introduced plants *Myrica faya* and *Leucaena leucocephala* in Hawaii Volcanoes National Park actively fix nitrogen in association with symbiotic bacteria, resulting in much elevated nitrogen levels on otherwise nitrogen-poor sites on young lava flows. Since most introduced plant invaders in Hawaii are most successful on more fertile sites, the invasion of these nitrogen-fixers may facilitate further invasions (Vitousek, 1986).

At Channel Islands National Park, California, the introduced annual iceplant *Mesembryanthemum crystallinum* accumulates salt which enters the soil profile during decomposition. The species excludes other vegetation by shading and by increasing the salt content in the soil above the osmotic tolerance of potential competitors in former grassland ecosystems (Halvorson, in press). [The mechanism was documented for mainland California by Vivrette and Muller (1977)].

9.3.4 Alteration of hydrological cycles

Invasive introduced plants can primarily affect hydrological cycles by utilizing more or less of the annual precipitation than was used by the native vegetation they replace. Several examples of these changes are known to have occurred in nature reserves.

The invasion of nature reserves in the jarrah *Eucalyptus marginata* forests of southwestern Australia by the introduced soil-borne fungus *Phytophthora cinnamomi* (see Von Broembsen, this volume) has given rise to major hydrological changes. The extent of the invasion is considerable, with 10% of the total area of 1.5×10^6 hectares covered by this forest type being severely diseased by 1972 (Shea, 1981). Once it has invaded the forest the fungus causes the death of all the individuals of the deep-rooted dominant tree species, *E marginata*. The majority of the other species in the shrub and understorey layer of the forest are also killed (Shea, 1981). Following dieback, transpiration losses are severely reduced, the soil water table rises and catchment discharge increases. In areas where soil salinity levels are high this change in soil hydrology results in salination of the runoff water (Shea *et al.*, 1975).

Another example of introduced species giving rise to hydrological changes in a mediterranean-type climate region comes from the fynbos biome of South Africa, Numerous nature reserves in the mountain catchments are being invaded by introduced trees and shrubs, such as *Hakea sericea*, *Pinus pinaster*, *Acacia longifolia* and *Acacia mearnsii* (Macdonald *et al.*, 1985), which reduce stream flow from the affected mountain catchments (Macdonald and Richardson, 1986; Versfeld and van Wilgen, 1986).

In parks and reserves of southwestern USA (including Death Valley National Monument and Big Bend National Park), evapotranspiration of invasive *Tamarix* spp. lowers water tables and dries up springs (Vitousek, 1986). One of the indirect consequences of this is habitat degradation for, and population decline of, desert bighorn sheet *Ovis canadensis* (US National Park Service, 1981). In the Namib Naukluft and Skeleton Coast Parks of South West Africa/Namibia, invasions of the rivercourses by introduced plants such as *Datura innoxia*, *Nicotiana glauca* and *Prosopis* spp. are considered to be having an analogous effect on the subsurface hydrology of these desert rivers (Brown *et al.*, 1985; Macdonald and Nott, 1987).

Another example of the potential of an invasive species to alter hydrological cycles is provided by a mammal, the coypu *Myocastor coypus*. In moving towards a policy for the total eradication of this species in Great Britain, the case for Drainage Authorities was summarized by Morton *et al.*, (1978) as 'The damage caused... was not quantifiable but involved boring into the bank of watercourses, railway embankments, culverts and foundations of pumping stations'. Although many of these effects were not observed on nature reserves the potential of this introduced species to upset the hydrological balance in a countryside of relatively small nature reserves interspersed in an agricultural environment is considerable.

9.3.5 Alteration of fire regimes

The fire regime is a critical component of the environment in all except the wettest (e.g. tropical humid forests and tundra) and driest terrestrial biomes (e.g. deserts). Where introduced species alter the fire regime this can have profound implications for ecosystem structure and function.

In both the New World and Australia the invasion of introduced grasses has affected fire regimes (Clark, 1956; Parsons, 1972; Christensen and Burrows, 1986). In all cases the change is thought to have been to a regime of more frequent fires. Invasion by *Andropogon* species and other introduced grasses has greatly increased the quantity and continuity of fine fuels and consequently the extent and intensity of wildfires in Hawaii Volcanoes National Park (Smith, 1985). The introduced annual cheatgrass *Bromus tectorum* has a similar effect over wide areas of the western USA. Unlike most perennial grasses of western North America, which reach maturity slowly and often do not readily burn until the major fire season is past, *B. tectorum* matures in early June and dries 1 to 2 weeks after maturing. At maturity, it has an exceptionally low moisture content (Klemmedson and Smith, 1964). Following fire, stands typically return quickly to domination by cheatgrass. The invasion has radically altered fire regimes in several reserves, e.g. in Capital Reef, Zion, Canyonlands, and Arches National Parks, in Utah. Similarly, in three reserves in the mediterranean-type climate region of California the almost total replacement of the indigenous ground layer by introduced European grasses is thught to have made spring and early summer fires more frequent (Macdonald *et al.*, 1988).

In African savanna reserves fire is generally considered to have limited the invasion of introduced plants rather than to have been affected by their invasion. (Macdonald and Frame, 1988). An exception to this is the composite climber *Chromolaena* (*Eupatorium*) *odorata*. This species has extensively invaded reserves, such as the Hluhluwe-Umfolozi Game Reserve, in the moister eastern savannas of South Africa (Macdonald, in press). The plant is extremely flammable and it invades savanna–forest ecotones which normally act as natural fire breaks. Following invasion by *C. odorata*, fires burn from the savanna into the forest margins killing the fire-sensitive forest tree species and thus reducing the size of forest patches (Macdonald, 1983).

In the fynbos reserves of South Africa the replacement of the indigenous sclerophyll vegetation by dense thickets of introduced trees such as *Hakea sericea* and *Acacia saligna* is thought to have decreased the frequency and mean intensity of fire by altering fuel characteristics (Van Wilgen and Richardson, 1985). However, when dense stands of these species are felled and then burned, high intensity fires result. These fires reduce both the diversity of plant species, particularly those which have underground structures which normally survive fires, and the density of termites (Breytenbach, 1986b).

These altered fire regimes often lead to recruitment failures of native species. However, invasions can have more direct effects on the recruitment of native species.

9.3.6 Prevention of recruitment of native species

The prevention of recruitment is often the causal process in long-term changes in the gross structural, and hence the functional, aspects of a reserve's ecosystems.

An example of an introduced plant preventing the recruitment of indigenous plant species is provided by one of the most serious plant invasions in the British Isles. In the mixed oak *Quercus petraea* and holly *Ilex aquifolium* woodland reserves in southwest Ireland, the introduced species, *Rhododendron ponticum*, is thought to inhibit woodland regeneration both by casting a dense shade and by forming an impenetrable litter layer (Usher, 1987). Cross (1982) quoted data for a mixed oak and holly stand, with 9% of incident daylight beneath the canopy, having seedlings of four tree species as well as a ground flora of seven fern and herbaceous plant species and thirteen bryophyte species. In an equivalent stand invaded by rhododendron, the light intensity was only 2% of daylight; there were no tree seedlings and only two species of fern and herbaceous plant and four species of bryophytes.

In one of the few quantitative studies ever made of the impacts of introduced plants in a nature reserve, Thomas (1980) showed that two species of vines were, in combination, preventing the recruitment of native forest species on Theodore Roosevelt Island, Washington DC. Japanese honeysuckle *Lonicera japonica* inhibited the reproduction of dominant forest trees such as *Ulmus americana*, *Prunus serotina* and *Liriodendron tulipifera*. English ivy *Hedera helix*, by contrast, primarily inhibited the recruitment of herbaceous species. Both species suppress and eventually kill established forest trees through shading. By inhibiting the subsequent recruitment of native species they were literally 'destroying the forests of this low lying island' (Thomas, 1980).

On Juan Fernandez Island the introduced plant *Aristotelia chilensis* is gradually replacing the endemic species *Myrceugenia fernandeziana* and *Drimys confertifolia*. Similarly, the native *Ugni selkirkii* is being replaced by the invasive *U. molinae*. Sanders, *et al.* (1982) conclude that these and other native plant species, 'especially many of the rosette Compositae, are disappearing, seemingly because seedlings are not becoming established to replace the senescing plants. Most of these oddities appear to be adapted to specific habitats and do not compete well for growing space as the vegetation around them changes.'

Alien animals can have equally significant impacts on the recruitment of native plants. In the Pinnacles National Monument, California, feral pigs are considered to be preventing the recruitment of oak, *Quercus* species (Macdonald *et al.*, 1988). Similarly European rabbits are thought to be inhibiting the recruitment of perennial herbs and shrubs in La Campana National Park, Chile (Macdonald *et al.*, 1988).

In the fynbos reserves of South Africa wherever the introduced Argentine ant *Iridomyrmex humilis* invades, it displaces the native ant community (Macdonald and Richardson, 1986). Many of the native plant species are myrmecochorous and the Argentine ant fails to carry out the seed removal and seed burial function of the native ant species. This leads to an almost complete recruitment failure in myrmecochorous plant species (Bond and Slingsby, 1984; Bond and Breytenbach, 1985).

In island reserves the introduction of animals has often given rise to recruitment failures. An example of an introduced animal apparently preventing the recruitment of an island plant is provided by the house mouse *Mus musculus* on the south Atlantic Gough Island. Seed predation by the mouse is thought to be preventing the recruitment of the locally important tree *Phylica arborea* (Breytenbach, 1986a). Feral goats *Capra hircus* are probably the single most important introduced herbivore in terms of their effect on the recruitment of native plants on island nature reserves (Holdgate, 1967; Coblentz, 1978). In the Galapagos Islands the recruitment of trees such as *Scalesia pedunculata*, *Zanthoxylum fagara* and *Acnistus ellipticus* has been totally prevented by goat browsing (Hoeck, 1984).

Island nature reserves also provide examples where introduced animals have prevented the recruitment of native animal species. In the Galapagos Islands, the introduced black rat *Rattus rattus* has had a devastating effect on the Pinzon Island race of the giant tortoise *Geochelone elephantopus ephippium*. According to Hoeck (1984) black rats were established on this island around 1880 and each year since then they 'have killed almost every tortoise hatchling that emerges from its nest. MacFarland *et al.* (1974) calculated that in a period of 10 years between 7000 and 19 000 hatchlings were produced, but only a single one-year-old tortoise was found on the entire island. On other islands in the Galapagos (Santa Cruz, San Cristobal, Isabela and Santiago), feral pigs effectively reduce the recruitment of tortoises, and of land and marine iguanas, by regularly digging up and eating the eggs. Furthermore, on a number of sandy beaches feral pigs dig up every nest of the green turtle *Chelonia mydas* and eat the eggs.

An example of an introduced plant having the potential to prevent the recruitment of a native animal species is provided by *Lantana camara* on the Galapagos island of Floreana. The dense thickets formed by this species are spreading towards the largest remaining colony of the threatened darkrumped petrel *Pterodroma phaeopygia*. If this invasion is not halted, it is considered likely (Cruz *et al.*, 1986) that the thickets will prevent the petrels from flying in and out of their nest burrows.

On the island of Mauritius a wide range of introduced species in combination prevent the recruitment of native plant species. Introduced monkeys *Macaca fascicularis* cause extensive damage to native trees such as *Erythrospermum monticolum* and *Mimusops petiolaris* by selectively pulling off branches, flowers and unripe fruits. The few seeds of these native trees that survive to germinate are subjected to heavy browsing at the seedling stage by deer *Cervus timorensis* and feral pigs and intense competition from introduced plants such as *Ligustrum robustum* and *Psidium cattleyanum* which form dense thickets. In addition, the established individuals of 26 native tree species are being ringbacked by the deer (Strahm, 1986).

9.3.7 Concluding comment on effects on ecosystem function

This is by no means an exhaustive listing of the perturbations in ecosystem function that can be brought about by introduced species in nature reserves. If one looks at any one of the myriad different ecosystem functions already known, one can normally find an example of how an introduced species is affecting this function. However, few of these ecological impacts have been studied and the listing simply documents some of those which have. These examples serve merely to indicate the wide range and, often, the insidious nature of such effects.

9.4 SOME EFFECTS OF ALIEN SPECIES ON ECOSYSTEM STRUCTURE IN NATURE RESERVES

9.4.1 Acceleration of local and global extinction rates

Although it would be desirable to use data from recent extinctions to investigate the role that introduced species play in extinction, this is generally not possible for two reasons. Firstly, it is very difficult to 'prove' that a species has in fact become extinct, e.g. the case of the Australian noisy scrub-bird *Atrichornis clamosus* which was incorrectly thought to have been extinct for 70 years (Whittell, 1943; Webster, 1982). Secondly, very few of the species that have become extinct were studied prior to their extinction, so that the factors responsible for their demise are almost invariably unknown or have been inferred from incomplete information. Thus, in the case of *A. clamosus*, it was originally deduced that its 'extinction' had been brought about by deforestation by humans and predation by feral cats *Felis catus* (Whittell, 1943). It was only after detailed investigations of the remnant population that it was possible to ascribe the decline to alterations in fire regime (Smith and Robinson, 1976) and, more specifically, to reductions in invertebrate prey following too frequent burning (Blakers *et al.*, 1984). Currently *A. clamosus* is not considered threatened by any introduced species (Conservation Monitoring Centre, personal communication). Thus, although recent extinctions will be covered they will not be analysed quantitatively.

Instead, data on the species currently considered to be threatened with extinction have been analysed to indicate the potential role introduced invasions might play in future extinctions. These data have been extracted from the computerized Red Data Book data bases of the International Union for the Conservation of Nature's Conservation Monitoring Centre in Cambridge, UK Only data on vertebrates have been used (Table 9.4) as these are the only taxonomic groups for which global coverage is considered to be adequate to show real trends as distinct from artefacts of incomplete knowledge (N. Phillips, personal communication).

The proportion of threatened species, all species with any Red Data Book status other than 'O' (out of danger), that are known to be threatened in any way by introduced species is shown to vary markedly between taxonomic groups, between biogeographic realms and between island and mainland species. Islands were defined as all landmasses smaller than New Guinea.

Using chi-squared tests with Yates' correction with 1 d.f., the number of

Table 9.4. The percentage of threatened terrestrial vertebrate species affected by introductions in the continental landmasses of the different biogeographic realms and on the world's islands. The total number of threatened species in the realm is given in brackets

Taxonomic Group	Percentage of threatened species affected by introductions																			
	Continental areas within biogeographic realms																All mainland areas		All insular areas	
	Eurasia		N. America		Africa		Indo–Malaya		Oceania		Australia		Antarctica		S. America					
	%	(*n*)	%	(*n*)	%	(*n*)	%	(*n*)	%	(*n*)	%	(*n*)	%	(*n*)	%	(*n*)	%	(*n*)	%	(*n*)
Mammals	16.7	(42)	3.4	(29)	8.0	(100)	12.7	(55)	0.0	(8)	64.4	(45)	0.0	(0)	10.0	(60)	19.4	(283)	11.5	(61)
Birds	4.2	(24)	13.3	(15)	2.5	(118)	0.0	(30)	0.0	(1)	27.3	(11)	0.0	(0)	4.2	(71)	5.2	(250)	38.2	(144)
Reptiles	5.9	(17)	16.7	(24)	25.0	(16)	4.3	(23)	14.3	(7)	22.2	(9)	0.0	(0)	14.3	(28)	15.5	(84)	32.9	(76)
Amphibians	0.0	(8)	6.3	(16)	0.0	(3)	0.0	(0)	0.0	(0)	0.0	(2)	0.0	(0)	0.0	(1)	3.3	(30)	30.8	(13)
Total for all groups considered	9.9	(91)	9.5	(84)	6.3	(237)	7.4	(108)	6.3	(16)	50.7	(67)	0.0	(0)	8.1	(160)	12.7	(647)	31.0	(294)

threatened species having introductions as a threat as compared to those having only other threats were compared for all islands and mainlands. The frequency of introduction-threatened species was significantly higher on islands for birds ($\chi^2 = 67{,}4$; $P < 0{,}001$), reptiles ($\chi^2 = 5{,}76$; $P < 0{,}03$) and amphibians ($\chi^2 = 4{,}24$; $P < 0{,}05$) but not for mammals ($\chi^2 = 1{,}65$; $P > 0{,}1$). The percentage of threatened mammals threatened by introductions was actually lower on islands than on mainlands (Table 9.4) but this was almost entirely due to 29 of Australia's threatened mammal species having introduced species as a major factor in their ecology. When this Australian frequency was compared with that of the sum of all other continental areas it was found to be significantly higher ($\chi^2 = 78{,}2$; $P < 0{,}001$). Over all groups, the Australian fauna tends to be more severely affected by introduced invasions than the fauna of the other continents (Table 9.4).

Of the 941 vertebrate species currently thought to be in danger of extinction 18,4% are known to be threatened in some way by introduced species. This figure is probably an underestimate as many of the threatened species have not been intensively studied, and the effects of introduced species are not always obvious.

It is not possible from these data to show how significant introductions are to the fauna as a whole, as distinct from that component of the fauna that is known to be threatened. Collar and Stuart's (1985) study of the threatened birds of Africa and related islands allowed Stuart and Collar (in press) to estimate the threat introductions are posing to the fauna as a whole. For the Africotropical portion of Africa they estimate the total avifauna to comprise 1481 species. Of these 96 (6.5%) are considered threatened and 91 (6.1%) 'near-threatened'. Only two of the 'near-threatened' species are known to be affected by introduced animals and one of the threatened species is believed to be so affected, constituting less than 0.2% of the Africotropical continental avifauna. By contrast, on the oceanic islands 35 of the 49 threatened species and three of the 13 'near-threatened' are known or believed to be affected by introduced animals. Although the native breeding avifauna of these islands is not given by Stuart and Collar (in press) a total of 285 species was obtained from published lists of the individual island faunas (Macdonald and Jones, unpublished). This means that approximately 13% of the insular avifauna is considered threatened by introductions.

Analysing the total avifauna of Australia, the continent showing the highest proportion of introduction-threatened species, there are 656 regularly occurring species (Blakers *et al.*, 1984) of which only three are considered threatened by introduced species (Table 9.4). This is less than 0.5% of the total avifauna. By contrast, 12.9% of Australia's mammalian fauna of 224 species (Pianka and Schall, 1984) are considered threatened by introductions.

It therefore appears that, with the exception of Australian mammals, it is only on islands that introductions are currently playing a significant role in the extinction of vertebrate faunas.

9.4.1.1 Extinction of oceanic island birds

Although only a small percentage of the world's land and freshwater avifauna occurs on oceanic islands '93% of the 93 species and 83 subspecies of birds that have become extinct since 1600 A. D. have been island forms' (King, 1985). After habitat destruction, predation by introduced predators ranks as the second most important cause of these extinctions; rats (*Rattus* spp.) are implicated in 54% of these (King, 1985).

This is hardly surprising as rats have been introduced to 82% of the world's major islands and island groups (Atkinson, 1985). Of 53 island bird species known to have been preyed on by the brown rat *R. norvegicus* four are known to have been locally eliminated by this predation. The comparable statistics for *R. rattus* are 39 species of which 12 have become extinct either as a species or as a subspecies. The Pacific rat *R. exulans* was generally introduced prior to the period of scientific observation and only 15 species are known to have been preyed on by it and none of these have become extinct as a result (Atkinson, 1985). However, *R. exulans* had been introduced to islands such as those of the Hawaiian Archipelago in the prehistoric period during which most of the endemic avifauna became extinct (Olson and James, 1984). The contribution of *R. exulans* predation to these extinctions has not yet been assessed although it is considered likely to have played a role (Olson and James, 1982).

Two examples of the extinction of oceanic island birds by introduced mammalian predators will be given. The endemic Socorro dove *Zenaida graysoni* has apparently become extinct since the introduction of cats to Isla Socorro in about 1958, with two other ground-foraging endemic bird species becoming less common during the same period. Although data are not available to test whether it was predation by feral cats that caused these declines, this is considered likely by the researchers involved (Jehl and Parkes, 1982). Similarly, Curry (1986) argues that the local extinction of the Floreana mockingbird *Nesomimus trifasciatus* from Floreana Island in the Galapagos was the result of nest predation by introduced *R. rattus.* The principle that island species that have evolved in the absence of predators are more likely to become extinct when predators are introduced (e.g. Clark, 1981) is born out by both the above studies: On Isla Socorro a mainland dove of the genus *Zenaida* became common during the period when the island species became extinct (Jehl and Parkes, 1982). Similarly, in the Galapagos, mockingbirds of the genus *Nesomimus* have been less severely affected by introduced rats on islands which originally held native rat species, than on islands, such as Floreana, which had never had any rats present (Curry, 1986). Atkinson (1985) has argued that rats have had most impact on temperate islands because the avifaunas of tropical islands had evolved predator defence mechanisms in response to the presence of land crabs.

It is not only rats and cats that have brought about extinctions of birds on islands. On the island of Guam the brown tree snake *Boiga irregularis*, which was introduced in 1947, suddenly expanded its population in the late 1970s and by early 1985 had severely reduced several of the island's endemic bird species and subspecies. The Guam broadbill *Myiagra freycinetti*, and rufous-fronted fantail *Rhipidura rufifrons uraniae* had been reduced to a few individuals while the Guam bridled white-eye *Zosterops conspicillata conspicillata* was already extinct. It was reported that experimental work carried out on the island 'just about fixes the entire blame upon the snake' (Marshall, 1985). However, there is a chance that introduced diseases might also have played a role in these declines.

In the Hawaiian Islands introduced diseases are considered to be important: Hawaiian honeycreepers evolved without exposure to avian malaria. The mosquito *Culex quinquefasciatus* was introduced into the islands in the 1800s. The avian malaria protozoan parasite *Plasmodium relictum* arrived in the early 1900s. The result is thought to have been extinction of some honeycreeper populations and reduction of others (van Riper *et al.*, 1982).

9.4.1.2 *Other island extinctions*

Although avian extinctions on islands are relatively well studied this does not mean that they are the faunal group most threatened by extinctions. For example, of 11 native rodent species, or island populations of species, known from the Galapagos Islands, eight are now extinct. *Rattus rattus* is implicated in five or six of these extinctions. Native rodents only persist on the two islands from which *R. rattus* is absent (Clark, 1984). Predation by mongooses in the Virgin Islands has caused, in addition to the near extinction of the ground-nesting quail dove *Geotrygon mystacea*, the local extirpation of a ground lizard *Ameiva polops*, and the destruction of many eggs and hatchlings of hawksbill turtles *Eretmochelys imbricata* on beaches (Nellis and Everard, 1983; Philobosin and Ruibal, 1971; Small, 1982).

Mauritius provides a classic example of the acceleration in extinction rates in a tropical insular fauna following human colonization and the associated introduction of plant and animal species. Prior to colonization in the 17th century there were at least 23 taxa of endemic landbirds, 12 reptiles (including giant land tortoises, skinks, geckos and snakes) and two fruit bats. Currently only nine endemic landbirds, four geckos, one skink and one fruit bat survive on the mainland of Mauritius (Cheke, in press). Most of the island's surface (95%) has been transformed and the remnants of the natural ecosystems have been highly modified by a wide range of plant and animal introductions (Strahm, 1986; Cheke, in press).

Honegger (1981) lists two amphibian and 28 reptile taxa known to have become extinct since 1600 AD. The reptiles were all island forms and introduced species are implicated in the extinction of at least eight of them and one of the two amphibians. For most of the other taxa no information is available on their extinction, or they were extirpated by humans.

Invertebrate groups have also been severely affected by introductions. On the island of Moorea, French Polynesia, the carnivorous snail *Euglandina rosea* was introduced in 1977 and by 1982 had already caused the endemic snail *Partula aurantia* to become extinct. On the basis of its rate of spread and its observed ability to eliminate native snails, Clarke *et al.*, (1984) predicted the extinction of all but one of the remaining six species of *Partula* on Moorea by 1987. *E. rosea* has also brought about the extinction of the Hawaiian endemic snail *Achatinella mustelina* (Hadfield and Mountain, 1980).

A similar example is the ant *Wasmannia auropunctata* which now occurs on four of the islands comprising the Galapagos Islands National Park (Santa Cruz, Floreana, San Cristobal, and Isabela). It is believed to have been introduced in the early 1900s to Santa Cruz Island, where at least 17 of 28 ant taxa present, including four probably endemic species, are negatively affected. *Wasmannia* was found to reduce population densities of, or eliminate altogether, three species of arachnids as well as reducing abundance and species diversity of flying and arboricolous insects (Lubin, 1984).

It should be emphasized that it is not only the native fauna of islands that has been depleted by introductions. The flora has also been affected. For example, feral goats have damaged the native flora of Hawaii Volcanoes and Haleakala National Parks for many years (Stone and Loope, 1987). At Haleakala, browsing of feral mammals (goats and cattle *Bos taurus*) has been directly responsible for extinction of at least three flowering plant species and local elimination and depletion of many others (Medeiros *et al.*, 1986). At Hawaii Volcanoes Park, construction of a goat-proof exclosure in 1971 resulted in the appearance of a previously undescribed leguminous plant species, *Canavalia kauensis*, which is believed to have survived nearly two centuries of goat browsing only as stored seeds in the soil (Mueller-Dombois, 1981).

9.4.1.3 Extinctions in continental ecosystems

Although no example is known of the global extinction of a continental species purely as a result of the invasion of introduced organisms, there are several cases where a species has either been locally eliminated or brought to the brink of global extinction by introductions. Not surprisingly several of these come from the 'island' continent, Australia (see Table 9.4). At least 19 introduced mammal species have successfully invaded large areas within Australia (Myers, 1986). By contrast only three introduced rodents and the feral cat can be considered to have widespread populations in southern Africa and these are generally at low densities away from transformed areas (Brooke *et al.*, 1986; Macdonald *et al.*, 1986).

Foxes *Vulpes vulpes* were introduced to Australia about 1870 for hunting; feral populations of domestic cats have probably existed there for longer. Kinnear *et al.* (1984) posed the question 'Has the cat and the fox affected our [Australian] native fauna?' Despite an element of controversy in interpreting a mass of observational data, their studies of the rock wallaby *Petrogale lateralis* indicate that these invasive introduced predators have a marked impact on these native marsupials (see Usher, this volume).

Further circumstantial evidence for the importance of introduced predators comes from the moister southwest of Western Australia. There the fox is now the major predator of medium-sized indigenous mammals (Christensen, 1980). The introduction of the fox in about 1915 was followed by the disappearance from much of its former distribution of the woylie *Bettongia penicillata*. Also, recent declines in abundance and distribution of medium-sized indigenous mammals in the jarrah forests have been correlated with an increase in fox abundance following the reduction of poison baiting programmes for rabbits in the mid-1970s (King *et al.*, 1981; Christensen, 1983). Ironically this adverse effect on indigenous mammal species was apparently brought about by the improvement in myxomatosis biocontrol of rabbits following the successful release of this disease's vector, the European rabbit flea *Spilopsyllus cuniculi* (Christensen, 1983; King and Wheeler, 1984). The subsequent increases of some of the affected mammal species in localized areas has thrown doubts on Christensen's (1980) initial interpretation of the fox's central role in the decline. However, as Christensen (1980, 1983) makes abundantly clear, the population dynamics of these native species are controlled by a complex interaction of fire, habitat and predation. There seems little doubt that introduced predation pressure is now one of the more important factors, and one which in time could lead to local and, possibly, to global extinction of some of these species. Already two species of rat-kangaroos, *Bettongia gaimardi* and *B. lesueur*, have become locally extinct on the Australian mainland: *B. gaimardi* is now only found on the fox-free island of Tasmania and *B. lesueur* became extinct in mainland Western Australia in the 1930s after the area's colonization by foxes (King *et al.*, 1981).

It is not only mammals that have been affected by fox predation. On the Murray River in South Australia the fox has been found to destroy 93% of nests of two species of aquatic tortoise *Chelodina longicollis* and *Emydura macquarii* (Thompson, 1983). This predation is thought likely to lead to a significant decline in the populations of both species unless fox predation is decreased.

From Australia too comes one of the few observations of an introduced invasive herbivore giving rise to local extinctions of plant species in a continental situation. The intense grazing pressure exerted by feral buffalo *Bubalis bubalis* in the Northern Territory is considered likely to eliminate certain native plant species from the swamp vegetation, e.g. the grasses *Hymenachne acutigluma* and *Phragmites karka* (Williams and Ridpath, 1982).

There is only one well-documented case where invasions by introduced plants pose a major threat to native plant species diversity in a continental situation. In the fynbos and karoo biomes of South Africa there are 1808 threatened plant species. Analyses of the threatened species from two fynbos areas show that just over 50% are threatened by the spread of introduced trees and shrubs, mainly species of *Acacia, Hakea* and *Pinus* (Hall and Ashton, 1983; Hall and Veldhuis,

1985). This threat is present in nature reserves, as has been documented for the Cape of Good Hope Nature Reserve by Taylor (1977).

9.4.1.4 Continental nature reserves—the significance of introduced plant pathogens

One group of invasions that has brought about extinctions on continents is plant pathogens. The invasion of southwestern Australia by the fungus *Phytophthora cinnamomi* is likely to have such an effect given the large number of localized endemic plant species found in the affected area (Rye, 1982). But it is from North America that there are definite examples.

Chestnut blight, caused by a fungal pathogen *Cryphonectria parasitica*, within 50 years eliminated the American chestnut *Castanea dentata* as a dominant tree species over much of the eastern USA (see Von Broembsen, this volume), including Great Smoky Mountains and Shenandoah National Parks. One apparent side-effect of the demise of the American chestnut is the loss of at least five native microlepidopterans, including the chestnut borer *Synanthedon castaneae* (Pyle *et al.* 1981). In addition to this, Quimby (1982) has postulated that the increase in oak wilt disease *Ceratocystis fagacearum* affecting native oaks in the northeastern USA resulted from changes in the density and distribution of the susceptible red oak *Quercus rubra* following the removal of *C. dentata* by chestnut blight. These 'ripple effects' of introduced species generally go unnoticed.

Dutch elm disease, caused by the fungal pathogen *Ophiostoma ulmi* (see Von Broembsen, this volume), primarily affected the native American elm *Ulmus americana*, an important successional tree species throughout eastern North America (Braun, 1950). American elm has been reduced but not eliminated in native stands; but the disease continues to run its course. Some indication of the extent of the dieback in a nature reserve comes from Theodore Roosevelt Island, where in 1972 there were 187 healthy *U. americana* canopy trees, 32 classed as dying, 105 standing dead individuals, 46 trees which had died and fallen and 94 which had been cut as part of a Dutch elm disease control programme in 1963 (Thomas, 1980). Although it is not possible to attribute all the deaths to *O. ulmi*, it is apparent that the proportion of the recent population known to have been infected (94 out of 464) is high. In addition it is inferred by Thomas (1980) that the individuals classified as dying were mainly doing so as a result of *O. ulmi* infestations.

9.4.1.5 Continental nature reserves—the significance of introduced insect herbivores

Introduced insect predators of temperate forest trees have also had major impacts on nature reserves, in some cases leading to the near extinction of affected species. *Lymantria dispar*, the gypsy moth, introduced to the northeastern USA

(Massachusetts) from Europe in 1869, has a very broad host tolerance (it prefers *Quercus* spp., but attacks most broad-leaved tree species and even conifers) and has been spreading rapidly during the past decade. It has recently caused heavy defoliation in Shenandoah National Park, Virginia, and is expected to spread throughout much of the USA. Mortality mostly affects less robust, subdominant trees (Campbell and Sloan, 1977; Marshall, 1981).

Balsam woolly aphid *Adelges piceae*, native to Europe, was introduced into Maine, USA, about 1900 and has now spread throughout *Abies* forests of eastern North America. It was first detected in the mountains near Great Smoky Mountains National Park in the 1950s. It currently threatens nearly to eliminate Fraser fir *Abies fraseri*, a dominant in the high-elevation spruce/fir zone of Great Smoky Mountains National Park and elsewhere in the southern Appalachian mountains. Interactive effects between pollutant-induced stress of Fraser fir and aphid-caused mortality are suspected but untested. Total mortality of fir overstorey and partial mortality of undestorey fir is expected within the next few years. In addition, the understorey flora adapted to cool, moist, low-light environments will lose much of its habitat (Eager, 1984). In high-elevation stands that have been invaded in the last three to seven years, fuel loading has been found to be two to four times that in stands not yet attacked by aphids, creating a short-term fire hazard. Afterwards, decomposition results in fuel loads similar to those in uninvaded stands (Nicholas and White, 1985).

9.4.1.6 Future risks, arising through 'trophic cascades'

One of the important points that emerges from several of the above examples is that an introduced species might indirectly bring about the extinction of a species which is itself unaffected by the introduced organism. This arises through the interrelationships that exist between organisms at different trophic levels in an ecosystem, and the sequence of such extinctions has been termed a 'trophic cascade'.

A possible example of this phenomenon is the extinction of insectivorous Hawaiian bird species as a result of reductions in native lepidopteran populations following the introduction of insect parasites. These lepidopterans formed a critical food resource for the birds involved and it has been convincingly argued that their reduction could have been an important contributory factor in several avian extinctions (Howarth, 1983).

Two further examples come from recent studies of the effects of the Argentine ant *Iridomyrmex humilis*. The first of these is the possible future extinction of fynbos plant species as a result of the displacement of native ants formerly responsible for these plants' seed dispersal (Bond and Slingsby, 1984). The second is the ant's possible future role in the extinction of endemic Hawaiian plant species. The Hawaiian biota lacks endemic ants and introduced ants have proved to be devastating to the low elevation insect fauna of these islands. Argentine ants have recently become established within a limited area (184 hectares) at high elevations in Haleakala National Park. Comparisons of areas with and without *I. humilis* have shown that these ants severely deplete the native ground-dwelling arthropod faunas of high-elevation shrubland, including major pollinators of native shrubs and herbaceous plants (Medeiros *et al.*, 1986). The Haleakala silversword *Argyroxiphium macrocephalum* and some of its near relatives, *Dubautia* spp., have been shown to be self-incompatible, requiring pollinators for seed set. These locally endemic species reproduce only by seed and would decline rapidly if pollination failed.

9.4.2 Genetic effects of introduced species

Another deleterious impact of introduced species on ecosystem structure is that of hybridization. The following examples serve to illustrate the threat posed by invasions in this regard.

The mallard duck *Anas platyrhynchos* is a highly successful species, has substantial invasive capacity, and has expanded its range into many areas with habitats modified by human activities. It has a remarkable degree of reproductive and genetic compatibility with congeners with which it has only recently come into contact—including *Anas poecilorhyncha superciliosa* of New Zealand and Australia, *Anas rubripes* of eastern North America, and *Anas diazi* in southwestern North America. Nevertheless, the maximum recorded frequencies of hybrids (12.9% for *A. rubripes* × *A. platyrhynchos* in Massachusetts, USA) suggest that hybrids have lower fitness than parent forms. A high embryonic death rate has been noted for hybrids in New Zealand, reducing genetic genetic swamping, but probably leading to depression of population size of the native taxon (Cade, 1983).

Another widespread and successful introduction, *Felis catus*, has also been responsible for hybridization with native congeners. In southern Africa there is concern for the long-term survival of the African wild cat *F. lybica* due to its observed ability to hybridize with *F. catus* (Brown *et al.*, 1985; Brooke *et al.*, in press). Similarly, there is increasing evidence from many areas in Scotland, including nature reserves especially in the Grampian Region, for the presence of *F. catus* genes in populations of the Scottish wild cat *F. sylvestris grampia* (D. French, personal communication).

The hybridization of native and introduced fish species has been pinpointed as potentially being the most important impact of fish introductions in southern Africa (Bruton and van As, 1986). This mostly arises from interbasin transfers of formerly allopatric southern African species. This problem has already arisen in North America and is currently affecting several taxa within nature reserves, For example, in Rocky Mountain National Park, Colorado the greenback cutthroat trout *Salmo clarki stomias* narrowly escaped extinction in the recent past due to overharvesting, competition, and interbreeding with introduced trout. A restoration programme is underway, involving removal of non-native fish (using a toxicant) above barriers to upstream movement and restocking with pure genetic stocks of *S. c. stomias* (Stevens and Rosenlund, 1986).

An invertebrate example is provided by the introduction of African honeybees *Apis mellifera adansonii* into the fynbos biome where the Cape honeybee *A. m. capensis* occurs. Fortunately, although hybridization occurs, the native subspecies is able to withstand this human-aided invasion with only minimal genetic pollution (Tribe, 1983).

An example of how an introduced species can indirectly result in the hybridization of a taxonomically unrelated native species is provided by the introduction of herbivorous fish to Madagascan freshwaters and the subsequent threat to the genetic integrity of the Alaotra grebe *Tachybaptus rufolavatus*. The little grebe *T ruficollis*, although native, has recently increased in abundance throughout Madagascar, apparently as a consequence of the widespread introduction of introduced *Tilapia* species to lakes and pools. The cosmopolitan little grebe has invaded the waters, mainly Lake Alaotra, formerly occupied by this localized island species and is now hybridizing freely with it. As a direct result of this hybridization Collar and Stuart (1985) conclude that 'nothing can be done to prevent the extermination of the Alaotra grebe in the wild'.

The hybridization of the introduced marsh grass *Spartina arterniflora* with the native *S. maritima* in England, as detailed previously, provides, a classic example of this phenomenon in plants. Another less well-known example is the hybridization of the introduced *Tamarix ramosissima* with the native *T. usneoides* on the Swakop River in South West Africa/Namibia (Palgrave, 1977). The hybridization between introduced and native *Rubbs* species in South Africa provides a further example (Spies and du Plessis, 1985).

The irreversible nature of these genetic effects makes them of particular significance to nature conservation. Another aspect which is possibly significant is that several hybrids have exhibited a high invasive capacity (see Dean *et al*., 1986). Plant examples of this are *Spartina anglica, Lantana camara* (Spies, 1984) and the *Psidium guajava* complex (Macdonald and Jarman, 1985). The Africanized honeybee, an *Apis mellifera* subspecific hybrid, has invaded vast areas in the neotropics (Michener, 1975). These 'new' genetic combinations have often proved difficult to control, particularly in respect to the application of classical biocontrol measures (Kluge *et al*., 1986). This aspect has considerable relevance to the possible conservation implications of genetic engineering (Colwell *et al*., 1985).

The widespread occurrence of hybridization following introductions in certain groups, e.g. the genus *Anas* and some freshwater fishes, must inevitably bring up the taxonomic question of species limits. Although taxonomic identity is not a prime concern of conservationists, genetic variability is. Perhaps the primary aim of conserving ecosystem structure is the retention of the maximum genetic variation in the ecosystem. Hence the loss of species and subspecies are both detrimental. In addition population 'bottlenecks', which often result from the effects of introductions (see Usher, this volume), are undesirable since, in general, genetic variation is lost during these events.

9.5 CONTROL OF ALIENS IN NATURE RESERVES

The preceding sections have attempted to outline some of the problems introduced organisms pose to wildlife conservation. In this section the possible solutions to these problems are investigated. Where information is available these solutions are illustrated using actual examples of past control programmes. Once again the emphasis is on programmes carried out within nature reserves.

9.5.1 Controlling introduced plant invasions

In a questionnaire survey of 299 southern African reserves it was found that, on average, only 0.4 invasive introduced plant species were known to have been eradicated from each reserve. This statistic should be compared with the average of 8.2 species recorded as having invaded each reserve and the average of 2.1 species for which control operations were known to have been initiated. The proportion of invasive introduced plants eliminated is considered to be even lower than these statistics indicate as the total number of introduced plants reported per reserve is considered to be a gross underestimate (possibly by a factor of 2 or 3) since herbaceous introduced plants tend to be ignored by reserve managers (Macdonald, in press).

In the Cape of Good Hope Nature Reserve monitoring of fixed plots has shown that two decades of control operations have eliminated only two of the 19 introduced tree and shrub species present in the reserve. Both species were initially rare and only marginally invasive within the reserve. The frequency of occurrence of the really invasive species such as *Acacia cyclops* and *Acacia saligna* has not declined significantly over the period despite intensive control operations. However, the density of all species has been reduced and species such as *Pinus pinaster*, in which the plants are easily killed and which do not accumulate large soil-stored seed banks, have been substantially reduced in frequency of occurrence. Relatively rare species, such as *Eucalyptus lehmannii*, for which cost-effective techniques for killing established plants have not been developed, have not been reduced in frequency (Taylor *et al.*, 1985; Macdonald *et al.*, in press).

Usher (1987) investigated the control of *Rhododendron ponticum* in nature reserves in Great Britain. Data from 19 National Nature Reserves indicated that no mechanical or chemical means of control—cutting, hand-pulling of regrowth, spraying—was successful. These examples from nature reserves confirm Robinson's (1980) conclusion that rhododendron invasions of forestry plantations could only be contained, not controlled or eradicated. Rhododendron has an extremely small and widely distributed seed that germinates prolifically in disturbed sites (Cross, 1981). Indeed, control operations disturb the litter and fermentation layers of forest soils and, paradoxically, make the environment even more suitable for regeneration of rhododendron seedlings (Usher, 1987).

Myrica faya in Hawaii Volcanoes National Park provides an example of a particularly discouraging failure to control a highly invasive and damaging introduced plant. One individual was noted in the park in 1961. By 1978, 609 hectares were mapped as supporting *M. faya* infestations of varying densities. In 1985, 12 000 hectares were infested—indicating a 20-fold increase in 8 years in spite of attempted control by park managers (Whiteaker and Gardner, 1986). In this case success was thwarted by the biology of the plant (abundant seed production, dispersal by birds, rapid growth to maturity), occurrence of seed sources adjacent to park boundaries and probably by periodic lapses in control efforts.

Another major failure at Hawaii Volcanoes National Park involved fountain grass *Pennisetum setaceum.* Over a 5-year period, US $ 250 000 were spent in efforts to control this invader through hand-pulling. Tunison (in press) determined that long-term control using this method would require at least US $ 100 000 per year and the project had to be abandoned. Success was here thwarted by the biology of the plant (abundant seed production, dispersal by wind, inconspicuousness of young non-flowering individuals, huge area of potential habitat), occurrence of seed sources adjacent to park boundaries, and probably by hesitance in control efforts in the early stages of the invasion.

In the Galapagos National Park repeated attempts to control guava *Psidium guajava* have all failed. A proportion of the plants treated, using a variety of herbicides, has always managed to survive, often suckering from the roots after extensive time periods [An analogous situation has been reported for this species in the Hluhluwe-Umfolozi Game Reserve, South Africa (Macdonald, 1983).] Until an effective control technique has been developed this species will continue to spread; already it occupies extensive areas on four of the islands and on Santa Cruz it has been suggested to be the main reason for the decrease of the endemic *Miconia robinsonia.* Recent control efforts have centred on a 14 hectare area on Santa Cruz Island and this is being kept clear of both *P. guajava* and the shrub *Cinchona succirubra.* However, the long-term solution to the Park's problems lies in controlling these species in adjacent farming areas; the seeds of *P. guajava* are dispersed into the park from these areas by cattle and birds while *C. succirubra* has light wind-dispersed seeds (Hamann, 1984).

Success in controlling or eradicating introduced plants in nature reserves is not common, but it should also be documented (see Macdonald, in press). The giant hogweed *Heracleum mantegazzianum* has been successfully controlled in the Aberlady Bay Local Nature Reserve in the east of Scotland. As individuals established within the reserve they were killed by cutting and spraying with herbicides as pinpoint applications (Usher, 1973). On the 40 hectare Masthead Island Reserve off Queensland, Australia, the prickly pear *Opuntia strica* has been effectively controlled using herbicides (Messersmith, 1986). In Organ Pipe Cactus National Monument, Arizona, two species of *Tamarix* and *Nicotiana glauca* have been recorded and removed wherever found and have so far not become a problem (Loope *et al.*, 1988). At Haleakala National Park, *Ulex europaeus, Pinus radiata, Pinus pinaster*, and *Pinus patula* have been kept in check along the northwest park boundary through periodic removal (Loope *et al.*, in press). *Melaleuca quinquenervia*, a tree invading marshes of southern Florida, has been successfully kept out of Everglades National Park for 15 years (in spite of established stands nearby) by periodically searching for and removing all individuals established within the park (La Rosa, in press). Managers at Hawaii Volcanoes National Park have been and continue to be successful in controlling 37 invasive plant species that are confined to small areas within park boundaries, forestalling their range expansion and attainment of levels at which they are uncontrollable (Taylor, in press). It is not possible to say how much they would have spread without this control.

9.5.2 Controlling introduced animals in nature reserves

Although control of certain introduced animal species has proved to be impossible, control of animal invaders appears to have been more successful than control of plant invaders in nature reserves. The following examples indicate the type of programmes that have been carried out and which of these have proved successful.

Between 1920 and 1970 approximately 70 000 goats were removed from Hawaii Volcanoes National Park, yet the population had not been noticeably reduced. A long term plan was developed that included the construction and maintenance of a goat-proof boundary fence and internal drift fences, frequent organized goat hunts and drives, long term monitoring of vegetation recovery and goat population levels, and measures to gain public support. Between 1971 and 1975 13 000 goats were eliminated; from 1976 to 1979, 1600 goats were eliminated. By 1980 most goat habitat had been fenced and less than 100 goats remained in managed areas (Stone and Loope, 1987).

Feral burros *Equus asinus*, originally released by prospectors and miners in the southwestern USA in the 19th century, have caused extensive damage to vegetation of numerous parks and reserves in that area. Serious efforts at removal, begun in the 1960s and early 1970s were thwarted by organizations and individuals who opposed any killing of burros on ‘humanitarian’ grounds. This obstacle was finally overcome in such reserves as Grand Canyon National Park and Death Valley National Monument, after 10 years of assessment, study and public participation. In Death Valley, the project included boundary fencing; 5724 burros were removed by live-trapping during 1983–1986, leaving only about 230 burros as of March 1986 (Loope *et al.*, 1988).

Rabbits were eradicated from the 260 hectare Santa Barbara Island in Channel

Islands National Park through intensive shooting during 1979–1981. By 1985, native shrubs were conspicuously recovering (Halvorson, in press). At Haleakala National Park, live trapping is being used successfully to control mongooses and feral cats near burrows of dark-rumped petrels *Pterodroma phaeophgia*. Petrel reproductive success rose from 35% in 1979 to 66% in 1980 and 1981 when predation was eliminated, apparently due to trapping (Simons, 1983). Coblentz and Coblentz (1985) have reported success in dramatically reducing mongoose populations in Virgin Islands National Park by repeated intensive trapping.

The importance of limiting reinvasion during the control programme was highlighted on Marcus Island, which forms part of the Langebaan Lagoon National Park in South Africa. This 11 hectare offshore island was joined to the mainland by a causeway in 1976 and within two years had been invaded by at least eight mammal species, including three viverrids and three canids. Control operations failed to reduce the frequency of sightings of introduced predators or the associated mortality rate in island birds until a predator-proof wall across the causeway was completed in April 1982. After this date trapping and shooting operations finally led to the elimination of introduced mammalian predators (Cooper *et al.*, 1985). However, even where reinvasion is not possible, as on the subantarctic Marion Island, where the island is big and cover for the introduced mammal abundant, the elimination of a species such as *Felis catus* has, to date, proven to be impossible (Cooper and Brooke, 1986).

When reserve managers are faced with introduced animals which prove difficult to control, e.g. the cats on Marion Island, they often consider the introduction of some form of biological control agent (see Usher, this volume). Although these measures are often relatively inexpensive, the possibilities for deleterious side effects need to be very carefully evaluated prior to their introduction (Howarth, 1983).

9.6 SUMMARY AND CONCLUSIONS

There are several principles affecting management which emerge from this study. Firstly, it must be emphasized that in most cases what we are observing is simply the short-term effects of a few hundred years or a few decades of rapid species movement into nature reserves. There is no *a priori* reason for assuming that invasions will not continue to occur. In fact, the indications are that plant introductions are likely to occur with increasing frequency unless improved preventative measures are implemented; i.e. all reserves can look forward to receiving propagules of all the world's worst weeds. The prospects for faunal invasions, excluding invertebrates, are not quite so bleak. Most of these have tended to be deliberate introductions and a heightened awareness of the problem should limit further such introductions in the future. The current situation where reserve managers are more aware of introduced animals as a problem than of introduced plants (e.g. Machlis and Tichnell, 1985) will need to be reversed in the future. Also, reserve managers are often unaware of plant and animal pathogens, and the impact of introduced pathogens in reserves remains largely unknown.

Given the fragmentation of the world's major biomes into relatively small 'quasi-insular' reserves with all the attendant changes in faunal composition, altered microclimates, altered fire regimes, and increased proximity to transformed areas, we can actually predict that the circumstances conducive to the invasion of introduced species will become more widespread in the future, not less widespread. It is, however, unlikely that 'island' continental reserves will rapidly come to resemble oceanic islands in their extreme vulnerability to introduction-induced extinctions. This vulnerability apparently stems from a history of evolution in isolation, something which to a large extent cannot manifest itself in continental situations.

There is a need for reserve managers to identify and subsequently eliminate or ameliorate changes in disturbance regimes which are beyond the evolutionary experience of the native. biota. These anthropogenic changes in disturbance regimes have often been implicated in the invasion of nature reserves by introduced species. An example is the invasion of Kosciusko National Park, Australia, by a number of introduced species as a result of an artificial fire regime (Medhurst and Good, 1984). In some cases these changes in disturbance regimes within the park arise from land use practices outside the park's boundaries; e.g. the altered flooding regime in Everglades National Park, Florida, USA which favours the establishment of introduced plant species (Morehead, 1984). In these cases the conservation agency has to work in cooperation with external agencies in an attempt to ameliorate the changes.

Where more than one serious invasion is occurring within a reserve, and where resource limitations prevent all invasions being combated simultaneously, it is important that reserve managers establish their priorities correctly. In particular there is a need to identify keystone introduced species which themselves give rise to ecosystem-level impacts (see Sections 9.3 and 9.4.1.6). An example of this comes from the well-studied Haleakala and Hawaii Volcanoes National Parks. In both parks the feral pig is recognized as constituting a keystone introduced species, since pigs are the single major factor contributing to the spread of many introduced plants, not only by creating open habitats through digging, but also by transporting progagules in their hair and faeces. They also create depressions that collect standing water in which the bird malaria vector *Culex quinquefasciatus* breeds (Stone and Loope, 1987).

In almost all cases the successful elimination or control of invasive introduced species has only been possible where management has been initiated during the early stages of the invasion. Wherever the populations of the invading plant or animal have been allowed to build up within the reserve's boundaries prior to the initiation of a control programme, control has subsequently proven to be either extremely costly or ineffective. Accordingly the creation of an 'early warning' monitoring system to detect invasions in their earliest stages is essential for all reserves. Similarly, the critical importance of follow-up control operations must be recognized and budgeted for from the outset. It is during follow-up operations that the last few remaining individuals are killed or kept under control. Lapses in vigilance at this stage have often undermined all that has been achieved in initially successful large-scale campaigns, e.g. the gypsy moth campaign in Massachusetts (Marshall, 1981).

In controlling both introduced plants and animals the critical factor is in all cases the balance between the rate of removal of established individuals from within the reserve and their recruitment within the reserve. In highly mobile species, for instance plants with windblown seeds and many animals, it is often necessary to create impermeable barriers along the reserve boundaries or to initiate control in areas adjacent to the reserve in order to limit this recruitment. Where introduced organisms are dispersed into the reserve by riverflow it is important to conduct control on a 'whole catchment' basis (Macdonald, in press). Where introduced plant species are recruiting from buried seed banks it is important that the dynamics of this seedbank are understood at the outset of the control programme (Milton and Hall, 1981; Holmes *et al.*, 1987).

Finally, in all control campaigns the possible deleterious side effects of alternative control techniques should be predicted, assessed and included in the cost/benefit analysis of the techniques. However, short-term deleterious effects should not be allowed to preclude the adoption of techniques which will in the long-term give rise to benefits through the effective eradication of the introduced species. Thus the killing of large numbers of non-target animals during the successful eradication of muskrat in Scotland was a short-term disadvantage which was outweighed by the long-term advantage of eradicating this species (Matthews, 1952; Usher, 1987). Conservation agencies will have to counter the public's emotional arguments against campaigns if they are to discharge their conservation duties effectively.

ACKNOWLEDGEMENTS

The authors would like to thank the other participants in the SCOPE Working Group on Introduced Species in Nature Reserves for their contributions to the ideas and data presented in this paper. The IUCN's Conservation Monitoring Centre at Cambridge and, in particular, Nick Phillips are acknowledged for producing the data used in Table 9.4. The senior author's research is funded by the Nature Conservation Research Committee of the CSIR. SCOPE is thanked for making possible a trip by the senior author to consult with the co-authors of this chapter. Richard Knight is thanked for carrying out the regression analyses and Pauline Solomon and Susan Macdonald for word processing successive drafts of the manuscript.

REFERENCES

Atkinson, I. A. E. (1985). The spread of commensal species of *Rattus* to oceanic islands and their effects on island avifaunas. *ICBP Technical Publication*, **3**, 35–81.

Barbour, M. G., and Johnson, A. F. (1977), Beach and dune. In: Barbour M. G., and Major J. (Eds), *Terrestrial Vegetation of California*, pp. 223–61. John Wiley and Sons, New York.

Blakers, M., Davies, S. J. J. F., and Reilly, P. N. (1984). *The Atlas of Australian Birds*. Melbourne University Press, Melbourne. 738 pp.

Bond, W. J., and Breytenbach, G. J. (1985). Ants, rodents and seed predation in Proteaceae. *S. Afr. J. Zool.*, **20**, 150–4.

Bond, W. J., and Slingsby, P. (1984). Collapse of an ant-plant mutualism: the Argentine ant *Iridomyrmex humilis*, and myrmecochorous Proteaceae. *Ecology*, **65**, 1031–7.

Braun, E. L. (1950). *Deciduous Forests of Eastern North America*. Blakiston, Philadelphia, Pa. 596 pp.

Breytenbach, G. J. (1986a). Dispersal: The case of the missing ant and the alien mouse. *S. Afr. J. Bot.*, **52**, 463–6.

Breytenbach, G. J. (1986b). Impacts of alien organisms on terrestrial communities with emphasis on communities of the south-western Cape. In: Macdonald, I. A. W., Kruger, F. J. , and Ferrar, A. A. (Eds), *The Ecology and Management of Biological Invasions in southern Africa*, pp. 229–38. Oxford University Press, Cape Town.

Brooke, R. K., Lloyd, P. H., and de Villiers, A. L. (1986). Alien and translocated terrestrial vertebrates in South Africa. In: Macdonald, I. A. W, Kruger, F. J., and Ferrar, A. A. (Eds.), *The Ecology and Management of Biological Invasions in southern Africa*, pp. 63–74. Oxford University Press, Cape Town.

Brown, C. J., Macdonald, I. A. W., and Brown, S. E. (Ed.) (1985). Invasive alien organisms in South West Africa/Namibia. *S. Afr. Natn. Sci. Progr. Rept.*, **119**, 1–74.

Bruton, M. N., and van As, J. G. (1986). Faunal invasions of aquatic ecosystems in southern Africa, with suggestions for their management. In: Macdonald, I. A. W., Kruger, F. J., and Ferrar, A. A. (Eds), *The Ecology and Management of Biological Invasions in southern Africa*, pp. 47–61. Oxford University Press, Cape Town.

Cade, T. J. (1983). Hybridization and gene exchange among birds in relation to conservation. In: Schonewald-Cox, C. M., Chambers, S. M., McBryde, B., and Thomas, W. L. (Ed.), *Genetics and Conservation: a Reference for Managing Wild Animal and Plant Populations*, pp. 288–309. Benjamin/Cummings, London.

Campbell, R. N., and Sloan, R. J. (1977). Forest Stand Responses to Defoliation by the Gypsy Moth. *Forest Science Monograph* 19.

Cheke, A. S. (in press). A review of the ecological history of the Mascarene Islands, with particular reference to the extinctions and introductions of land vertebrates. In: Diamond, A. W. (Ed.), *Studies of the Mascarene Avifauna*. Cambridge University Press, Cambridge.

Christensen, P. E. S. (1980). The biology of *Bettongia penicillata* Gray, 1987 and *Macropus eugenii* (Desmarest, 1817) in relation to fire. *Forests Dept. of Western Australia. Bulletin*, **91**, 1–90.

Christensen, P. E. S. (1983). A sad day for native fauna. *Forest Focus*, **23**, 3–12.

Christensen, P. E., and Burrows, N. D. (1986). Fire: an old tool with a new use. In: Groves, R. H., and Burdon, J. J. (Eds), *Ecology of Biological Invasions: an Australian perspective*, pp. 97–105. Australian Academy of Science, Canberra.

Clark, A. H. (1956). The impact of exotic invasion on the remaining new world mid-latitude grasslands. In: *Man's Role in Changing the Face of the Earth*, pp. 737–62. Chicago University Press, Chicago.

Clark, D. A. (1981). Foraging patterns of black rats across a desert-montane forest gradient in the Galapagos Islands. *Biotropica*, **13**, 182–94.

Clark, D. A. (1984). Native land mammals. In: Perry, R. (Ed.), *Key Environments. Galapagos*, pp. 225–32. Pergamon Press, Oxford.

Clarke, B., Murray, J., and Johnson, M. S. (1984). The extinction of endemic species by a program of biological control. *Pacific Sci.*, **38**, 97–104.

Coblentz, B. E. (1978). The effects of feral goats (*Capra hircus*) on island ecosystems. *Biol. Conserv.*, **13**, 279–86.

Coblentz, B. E., and Coblentz, B. A. (1985). Control of the Indian mongoose *Herpestes auropunctatus* on St. John, U. S. Virgin Islands. *Biol. Conserv.*, **33**, 281–8.

Collar, N. J., and Stuart, S. N. (1985). *Threatened birds of Africa and related islands.* ICBP and IUCN, Cambridge, UK. 761 pp.

Colwell, R. K., Norse, E. A., Pimentel, D., Sharples, F. E., and Simberloff, D. (1985). Genetic engineering in agriculture. *Science*, **229**, 111–2.

Cooper, J., and Brooke, R. K. (1986). Alien plants and animals on South African continental and oceanic islands: species richness, ecological impacts and management. In: Macdonald, I. A. W., Kruger, F. J., and Ferrar, A. A. (Eds), *The Ecology and Management of Biological Invasions in southern Africa*, pp. 133–42. Oxford University Press, Cape Town.

Cooper, J., Hockey, P. A. R., and Brooke, R. K. (1985). Introduced mammals on South and South West African Islands: History, effects on birds and control. *Proc. Symp. Birds and Man. Jhbg*, pp. 179–203. Witwatersrand Bird Club, Johannesburg.

Cross, J. R. (1981). The establishment of *Rhododendron ponticum* in the Killarney oakwoods, S. W. Ireland, *J. Ecol.*, **69**, 807–24.

Cross, J. R. (1982). The invasion and impact of *Rhododendron ponticum* in native Irish vegetation. *J. Life Sci., Royal Dublin Society*, **3**, 209–20.

Cruz, F., Cruz, J., and Lawesson, J. E. (1986). *Lantana camara L.*, a threat to native plants and animals. *Noticias de Galapagos*, **43**, 10–11.

Curry, R. L. (1986). Whatever happened to the Floreana Mockingbird? *Noticias de Galapagos*, **43**, 13–15.

Dean, S. J., Holmes, P. M., and Weiss, P. W. (1986). Seed biology of invasive alien plants in South Africa and South West Africa/Namibia. In: Macdonald, I. A. W., Kruger, F. J., and Ferrar, A. A. (Eds), *The Ecology and Management of Biological Invasions in southern Africa*, pp. 157–78. Oxford University Press, Cape Town.

Doody, P. (Ed.) (1984). *Spartina anglica* in Great Britain. *Focus on Nature Conservation*, **5**, 1–72.

Eager, C. C. (1984). Review of the biology and ecology of the balsam woolly aphid in southern Appalachian spruce-fir forests. In: White, P. S. (Ed.), *The Southern Appalachian Spruce-Fir Ecosystem: Its Biology and Threats.* U. S. Dept. of the Interior, National Park Service, Research/Resources Management Report SER–71. 268 pp.

Graf, W. L. (1978). Fluvial adjustments to the spread of tamarisk in the Colorado Plateau region. *Geol. Soc. Amer. Bull.*, **89**, 1491–501.

Hadfield, M. G., and Mountain, B. S. (1980). A field study of a vanishing species, *Achatinella mustelina* (Gastropoda, Pulmonata), in the Waianae Mountains of Oahu. *Pacific Science*, **34**, 345–58.

Hall, A. V., and Ashton, E. R. (1983). *Threatened plants of the Cape Peninsula.* Threatened Plants Res. Group, University of Cape Town, Cape Town. 26 pp.

Hall, A. V., and Veldhuis, H. A. (1985). South African Red Data Book: Plants—Fynbos and Karoo Biomes. *S. Afr. Natn. Sci. Progr. Rep.*, **117**, 1–160.

Halvorson, W. (in press). Alien plants at Channel Islands National Park. In: Stone, C. P., Tunison, T., and Smith, C. W. (Eds), *Proceedings of Symposium on Control of*

Introduced Plants in the Hawaiian Islands. University of Hawaii Press, Honolulu, Hawaii.

Hamann, O. (1984). Changes and threats to the vegetation. In: Perry, R. (Ed.), *Key Environments. Galapagos*, pp. 115–32. Pergamon Press, Oxford.

Harrison, J., Miller, K. R., and McNeely, J. (1984). The world coverage of protected areas: development goals and environmental needs. In: McNeely, J., and Miller, K. R. (Eds), *National Parks, Conservation and Development*, pp. 24–31. Smithsonian Institution Press, Washington, DC.

Hoeck, H. N. (1984). Introduced fauna. In: Perry, R. (Ed.), *Key Environments. Galapagos*, pp. 233–45. Pergamon Press, Oxford.

Holdgate, M. W. (1967). The influence of alien species on the ecosystems of temperate oceanic islands. In: *Towards a New Relationship of Man and Nature in Temperate Lands*. IUCN Publication New Series, **9**, 151–76.

Holmes, P. M., Macdonald, I. A. W., and Juritz, J. (1987). Effects of clearing treatment on seed banks of the alien invasive *Acacia saligna* and *Acacia cyclops* in the southern and southwestern Cape, South Africa. *J. Appl. Ecol.*, **24**, 1045–51.

Honegger, R. E. (1981). List of amphibians and reptiles either known or thought to have become extinct since 1600. *Biol. Conserv.*, **19**, 141–58.

Howarth, F. G. (1983). Classical biocontrol: panacea or Pandora's box. *Proc. Hawaiian Ent. Soc.*, **24**, 239–44.

Hubbard, C. E. (1968). *Grasses: a Guide to their Structure, Identification, Uses, and Distribution in the British Isles*. Penguin Books, Harmondsworth, Middlesex.

Jehl, J. R., and Parkes, K. C. (1982). The status of the avifauna of the Revillagigedo Islands, Mexico. *Wilson Bull.*, **94**, 1–19.

King, D. R., Oliver, A. J., and Mead, R. J. (1981). *Bettongia* and fluoroacetate: a role for 1080 in fauna management. *Aust. Wildl. Res.*, **8**, 529–36.

King, D. R., and Wheeler, S. H. (1984). Myxomatosis in Western Australia. *Journal of Agriculture—Western Australia*, **25**, 9–11.

King, W. B. (1985). Island birds: will the future repeat the past? *ICBP Technical Publication*, **3**, 3–15.

Kinnear, J., Onus, M., and Bromilow, B. (1984). Foxes, feral cats and rock wallabys. *Swans*, **14**, 3–8.

Klemmedson, J. O., and Smith, J. G. (1964). Cheatgrass (*Bromus tectorum* L.). *Bot. Review*, **30**, 226–62.

Kluge, R. L., Zimmerman, H. G., Cilliers, C. J., and Harding, G. B. (1986). Integrated control for invasive alien weeds. In: Macdonald, I. A. W., Kruger, F. J., and Ferrar, A. A. (Eds), *The Ecology and Management of Biological Invasions in southern Africa*, pp. 295–303. Oxford University Press, Cape Town.

Klukas, R. W., and Truesdell, W. G. (1969). *The Australian Pine problem in Everglades National Park. Part 1: The problem and some possible solutions. Part 2: Management plan for exotic plant eradication* (*Casuarina equisetifolia*). Everglades National Park, Homestead, Florida. 22 pp.

La Rosa, A. M. (in press). Alien plant management and research in Everglades National Park. In: Stone, C. P., Tunison, T., and Smith, C. W. (Eds), *Proceedings of Symposium on Control of Introduced Plants in the Hawaiian Islands*. University of Hawaii Press, Honolulu, Hawaii.

Lloyd, P. H. (1975). A study of the Himalayan Thar (*Hemitragus jemlahicus*) and its potential effects on the ecology of the Table Mountain Range. *Unpub. Report*. Cape Dept. of Nature and Environmental Cons. 80 pp.

Loope, L. L. (in press). Haleakala National Park and the 'Island Syndrome'. *Proceedings of the Science in the National Parks Conference, Fort Collins*, July, 1986. 30 pp.

Loope, L. L., Sanchez, P. G., Tarr, P. W., Loope, W. L., and Anderson, R. L. (1988). Biological invasions of arid land nature reserves. *Biol. Conserv.*, **44**, 95–118.

Loope, L. L., Nagata, R. J., and Medeiros, A. C. (in press). Alien plants in Haleakala National Park. In: Stone, C. P., Tunison, T., and Smith, C. W. (Eds), *Proceedings of Symposium on Control of Introduced Plants in the Hawaiian Islands.* University of Hawaii Press, Honolulu, Hawa.

Lubin, Y. D. (1984). Changes in the native fauna of the Galapagos islands following invasion by the little red fire ant, *Wasmannia auropunctata. Biol. J. Linn. Soc.*, **21**, 229–42.

Macdonald, I. A. W. (1983). Alien trees, shrubs and creepers invading indigenous vegetation in the Hluhluwe-Umfolozi Game Reserves Complex in Natal. *Bothalia*, **14**, 949–59.

Macdonald, I. A. W. (in press). Invasive alien plants and their control in southern African nature reserves. *Proceedings of the 'Science in the National Parks' Conference, Fort Collins*, July, 1986.

Macdonald, I. A. W., Clark, D. L., and Taylor, H. C. (in press). The history and effects of alien plant control in the Cape of Good Hope Nature Reserve, 1941–1987. *S. Afr. J. Bot.*

Macdonald, I. A. W., and Frame, G. W. (1988). The invasion of introduced species into nature reserves in tropical savannas and dry woodlands. *Biol. Conserv.*, **44**, 67–93.

Macdonald, I. A. W., Graber, D. M., DeBenedetti, S., Groves, R. H., and Fuentes, E. R. (1988). Introduced species in nature reserves in Mediterranean-type climatic regions of the world. *Biol. Conserv.*, **44**, 37–66.

Macdonald, I. A. W., and Jarman, M. L. (Eds) (1985). Invasive alien plant in the terrestrial ecosystems of Natal, South Africa. *S. Afr. Natn. Sci. Progr. Rept.*, **118**, 1–88.

Macdonald, I. A. W., Jarman, M. L., and Beeston, P. (Eds) (1985). Management of invasive alien plants in the fynbos biome. *S. Afr. Natn. Sci. Progr. Rept.*, **111**, 1–140.

Macdonald, I. A. W., and Nott, T. B. (1987). Invasive alien organisms in central South West Africa/Namibia: results of a reconnaisance survey conducted in November, 1984. *Madoqua*, **15**, 21–34.

Macdonald, I. A. W., Powrie, F. J., and Siegfried, W. R. (1986). The differential invasion of southern Africa's biomes and ecosystems by alien plants and animals. In: Macdonald, I. A. W., Kruger, F. J., and Ferrar, A. A. (Eds), *The Ecology and Management of Biological Invasions in southern Africa*, pp. 209–25. Oxford University Press, Cape Town.

Macdonald, I. A. W., and Richardson, D. M. (1986). Alien species in terrestrial ecosystems of the fynbos biome. In: Macdonald, I. A. W., Kruger, F. J., and Ferrar, A. A. (Eds), *The Ecology and Management of Biological Invasions in southern Africa*, pp. 77–91. Oxford University Press, Cape Town.

MacFarland, C. G., Villa, J., and Toro, B. (1974). The Galapagos Giant Tortoises (*Geochelone elephantopus*). Part 1: Status of the surviving populations. *Biol. Conserv.*, **6**, 118–33.

Machlis, G. E., and Tichnell, D. L. (1985). *The state of the world's parks.* Westview Press, Boulder, Colorado. 147 pp.

Marshall, E. (1981). The summer of the gypsy moth. *Science*, **213**, 991–3.

Marshall, J. T. (1985). Guam: A problem in avian conservation. *Wilson Bull.*, **97**, 259–62.

Maythews, L. H. (1952). *British Mammals.* Collins, London.

Medeiros, A. C., Loope, L. L., and Cole, F. R. (1986). Distribution of ants and their effects on endemic biota of Haleakala and Hawaii Volcanoes National Parks: a preliminary assessment. In: Smith, C. W., and Stone, C. P. (Eds), *Proceedings of 6th Conference in Natural Sciences, Hawaii Volcanoes National Park*, pp. 39–51. Coop. Natl. Park Resources Studies Unit, University of Hawaii, Honolulu.

Medeiros, A. C., Loope, L. L., and Holt, R. A. (in press). *Status of native flowering plant species on the south slope of Haleakala, East Maui, Hawaii.* Coop. Natl. Park Resources Studies Unit, University of Hawaii, Honolulu. Tech. Rept.

Medhurst, G., and Good, R. (1984). Fire and pest species: a case study of Kosciusko National Park. In McNeely, J., and Miller, K. R. (Eds), *National Parks, Conservation and Development*, pp. 296–300. Smithsonian Institution Press, Washington DC.

Messersmith, J. (1986). Eradication of prickly pear (*Opuntia stricta*) on Masthead Island. *Aust. Ranger Bull.* **3**, 14–15.

Michener, C. D. (1975). The Brazilian Bee Problem. *Ann. Rev. Entomol.*, **20**, 399–416.

Milton, S. J., and Hall, A. V. (1981). Reproductive biology of Australian Acacias in the South-Western Cape Province, South Africa. *Trans. Roy. Soc. S. Afr.*, **44**, 465–87.

Morehead, J. M. (1984). Attempts to modify significant deterioration of a park's natural resources: Everglades National Park. In: McNeely, J., and Miller, K. R. (Eds), *National Parks, Conservation and Development*, pp. 496–502. Smithsonian Institution Press, Washington DC.

Morton, J., Calver, J., Jefferies, D. J., Norris, J. H. M., Roberts, K. E., and Southern, H. N. (1978). *Coypu: Report of the Coypu Strategy Group.* London, HMSO (Ministry of Agriculture, Fisheries and Food).

Mueller-Dombois, D. (1981). Vegetation dynamics in a coastal grassland of Hawaii. *Vegetatio*, **46**, 131–40.

Myers, K. (1986). Alien vertebrates in Australia, with emphasis on the mammals. In: Groves, R. H., and Burdon, J. J. (Eds), *Ecology of Biological Invasions: an Australian Perspective*, pp. 120–36. Australian Academy of Science, Canberra.

Nellis, D. W., and Everard, C. O. R. (1983). The biology of the mongoose in the Caribbean. *Studies on the Fauna of Curacao and Other Caribbean Islands*, **64**, 1–162.

Nicholas, N. S., and White, P. S. (1985). *The effect of balsam woolly aphid infestation on fuel levels in spruce-fir forests of Great Smoky Mountains National Park.* US Dept. of the Interior, National Park Service, Research/Resources Management Report SER–71. 268 pp.

Olson, S. L., and James, H. F. (1982). Prodromus of the fossil avifauna of the Hawaiian Islands. *Smithsonian Contributions to Zoology*, **365**, 1–59.

Olson, S. L., and James, H. F. (1984). The role of Polynesians in the extinction of the avifauna of the Hawaiian Islands. In: Martin, P. S., and Klein, R. G. (Eds), *Quaternary extinctions. A prehistoric revolution*, pp. 768–80. University of Arizona Press, Tucson.

Palgrave, K. C. (1977). *Trees of Southern Africa.* Struik, Cape Town.

Parsons, J. J. (1972). Spread of African pasture grasses to the American tropics. *J. Range Manage.*, **25**, 12–17.

Philobosin, R., and Ruibal, R. (1971). Conservation of the lizard *Ameiva polops* in the Virgin Islands. *Herpetologia*, **27**, 450–4.

Pianka, E., and Schall, J. (1984). Species densities of Australian vertebrates. In: Archer, M., and Clayton, G. (Eds), *Vertebrate Zoogeography and Evolution in Australasia*, pp. 119–24. Hesperian Press, Carlisle, WA.

Pyle, R., Bentzien, M, and Opler, P. (1981). Insect conservation. *Ann. Rev. Entomol.*, **26**, 233–58.

Quimby, P. C. (1982). Impact of diseases on plant populations. In: Charudattan, R., and Walker, H. L. (Eds), *Biological Control of Weeds with Plant Pathogens*, pp. 47–60. John Wiley, New York.

Robinson, J. D. (1980). *Rhododendron ponticum*—a weed of woodlands and forest plantations seriously affecting management. *Proceedings of the Weed Control in Forestry Conference.* pp. 89–96.

Rye, B. L. (1982). *Geographically Restricted Plants of Southern Western Australia.* Department of Fisheries and Wildlife, Western Australia Report No. 49, pp. 1–63.

Sanders, R. W., Stuessy, T. F., and Marticorena, C. (1982). Recent changes in the flora of the Juan Fernandez Islands, Chile. *Taxon*, **31**, 284–9.

Severns, M. (1980). Land molluscs of Kipahulu Valley below 2000 feet. In: Smith, C. W. (Ed.), *Resources Base Inventory of Kipahulu Valley below 2000 feet*. University of Hawaii at Manoa, Coop. Natl. Park Studies Unit, Honolulu, Hawaii. 175 pp.

Shea, S. R. (1981). Multiple use management in a Mediterranean ecosystem—the Jarrah Forest, a case study. In: Conrad, C. E., and Oechel, W. C. (Eds), *Proceedings of symposium on Dynamics and Management of Mediterranean-Type Ecosystems. San Diego*, pp. 49–55. General Techn. Rept. PSW–58. Pacific Southwest Forest and Range Experiment Station, Berkeley, California.

Shea, S. R., Hatch, A. B., Havel, J. J., and Ritson, P. (1975). The effects of changes in forest structure and composition on water quality and vield from the northern Jarrah Forest. In: Kikkana, J., and Nix, H. A. (Eds), *Managing terrestrial ecosystems*. Proceedings of the Ecological Society of Australia **9**.

Simons, T. R. (1983). *Biology and conservation of the endangered Hawaiian dark-rumped petrel* (*Pterodroma phaeopygia sandwichensis*). Univ. Wash. Coop. Natl Park Resource Studies Unit, Tech. Rept 83–2. Seattle, Washington.

Singer, F. J., Swank, W. T., and Clebsch, E. E. C. (1984). Effects of wild pig rooting in a deciduous forest. *J. Wildl. Management*, **48**, 464–73.

Small, V. (1982). *Sea turtles nesting at Virgin Islands National Park and Buck Island Reef National Monument. 1980 and 1981*. USDI, National Park Service, Research/Resource Management Report, SER–61, Virgin Islands National Park.

Smith, C. W. (1985). The impact of alien plants on Hawaii's native biota. In: Stone, C. P., and Scott, J. M. (Eds), *Hawaii's Terrestrial Ecosystems: Preservation and Management*, pp. 180–250. University of Hawaii Press, Honolulu, Hawaii.

Smith, G. T., and Robinson, F. N. (1976). The Noisy scrub-bird: an interim report. *Emu*, **76**, 37–42.

Spies, J. J. (1984). A cytotaxonomic study of *Lantana camara* (Verbenaceae) from South Africa. *S. Afr. J. Bot.*, **3**, 231–50.

Spies, J. J., and Du Plessis, H. (1985). The genus *Rubus* in South Africa. 1. Chromosome numbers and geographical distribution of species. *Bothalia*, **15**, 591–6.

Statistical Office, Denver Service Center. (1986). *National Park Statistical Abstract 1985*. National Park Service, United States Dept. of Interior, Denver. 62 pp.

Stevens, D. R., and Rosenlund, B. D. (1986). Greenback cutthroat trout restoration in Rocky Mountain National Park, (Abstract). In: *Program and Abstracts, Conference on Science in the National Parks, Colorado State University, Fort Collins, Colorado, July 13–18, 1986*. p. 53.

Stevens, R. B. (1974). *Plant disease*. The Ronald Press Company, New York.

Stone, C. P., and Loope, L. L. (1987). Reducing impacts of alien animals on native biota in Hawaii.: what is being done, what needs doing, and the role of national parks. *Environ. Conserv.*, **14**, 245–58.

Strahm, W. (1986). *Case Study on Plant Conservation in Mauritius*. Forestry Department, FAO, Rome.

Stuart, S. N., and Collar, N. J. (in press). Birds at risk in Africa and related islands: the causes of their rarity and decline. *Proc. VI Pan-Afr. Orn. Congr.*

Taylor, D. (in press). A manager's guide for controlling weeds in natural areas. In: Stone, C. P., Tunison, T., and Smith, C. W. (Eds), *Proceedings of Symposium on Control of Introduced Plants in the Hawaiian Islands*. University of Hawaii Press. Honolulu, Hawaii.

Taylor, H. C. (1977). Aspects of the ecology of the Cape of Good Hope Nature Reserve in relation to fire and conservation. In: Mooney, H. A., and Conrad, C. E. (Eds), *Proceedings of the Symposium on the Environmental Consequences of Fire and Fuel*

Management in Mediterranean Ecosystems. USDA. For. Serv. Gen. Tech. Rept WO–3. pp. 483–7.
Taylor, H. C., Macdonald, S. A., and Macdonald, I. A. W. (1985). Invasive alien woody plants in the Cape of Good Hope Nature Reserve. 2. Results of a second survey in 1976–1980. *S. Afr. J. Bot.*, **51**, 21–9.
Thomas, L. K. (1980). The impact of three exotic plant species on a Potomac Island. *National Park Service Scientific Monograph Series*, **13**, 1–179.
Thompson, M. B. (1983). Populations of the Murray River Tortoise, *Emydura* (Chelodina): the effect of egg predation by the red fox, *Vulpes vulpes*. *Aust. Wildl. Res.*, **10**, 363–71.
Tribe, G. D. (1983). What is the Cape Bee? *S. Afr. Bee J.*, **55**, 77–87.
Tunison, T. (in press). *Pennisetum setaceum* control in Hawaii Volcanoes National Park: effort, economics, and feasibility. In: Stone, C. P., Tunison, T., and Smith, C. W. (Eds), *Proceedings of Symposium on Control of Introduced Plants in the Hawaiian Islands*. University of Hawaii Press, Honolulu, Hawaii.
Udvardy, M. D. F. (1975). *A classification of the Biogeographical Provinces of the World*. IUCN Occasional paper No. 18, Morges, Switzerland. 46 pp.
US National Park Service. (1981). *Proposed Natural and Cultural Resources Management Plan and Draft Environmental Impact Statement, Death Valley National Monument*. U. S. Dept. of the Interior. 234 pp.
Usher, M. B. (1973). *Biological Management and Conservation*. Chapman and Hall, London.
Usher, M. B. (1987). Invasibility and wildlife conservation: invasive species on nature reserves. *Phil. Trans. R. Soc. Lond., B*, **314**, 695–710.
van Riper, C., van Riper, S. G., Goff, M. L., and Laird, M. (1982). *The impact of malaria on birds in Hawaii Volcanoes National Park*. University of Hawaii at Manoa, Coop. Natl Park Studies Unit, Tech. Rept. 47.
Van Wilgen, B. W., and Richardson, D. M. (1985). The effects of alien shrub invasions on vegetation structure and fire behaviour in South African Fynbos shrublands: a simulation study. *J. Appl. Ecol.*, **22**, 1–11.
Versfeld, D. B., and van Wilgen, B. W. (1986). Impact of woody aliens on ecosystem properties. In: Macdonald, I. A. W., Kruger, F. J., and Ferrar, A. A. (Eds), *The Ecology and Management of Biological Invasions in southern Africa*, pp. 239—46. Oxford University Press, Cape Town.
Vitousek, P. (1986). Biological invasions and ecosystem properties: can species make a difference? In: Mooney, H. A., and Drake, J. A. (Eds), *Ecology of Biological Invasions of North America and Hawaii*, pp. 163–76. Springer-Verlag, New York.
Vivrette, N. J., and Muller, C. H. (1977). Mechanism of invasion and dominance of coastal grassland by *Mesembryanthemum crystallinum*. *Ecol. Monogr.*, **47**, 301–18.
Watt, A. S. (1947). Pattern and process in the plant community. *J. Ecol.*, **35**, 1–22.
Webster, H. O. (1962). Rediscovery of the noisy scrub-bird. *West. Aust. Nat.*, **8**, 81–4.
Whiteaker, L. D., and Gardner, D. E. (1986). *The distribution of Myrica faya Ait. in the state of Hawaii*. University of Hawaii at Manoa, Coop. Natl Park Studies Unit, Tech. Rept 55. 31 pp.
Whittaker, R. H., and Likens, G. E. (1975). The biosphere and man. In: Lieth, H., and Whittaker, R. H. (Eds), *Primary Productivity of the Biosphere*, pp. 305–28. Springer-Verlag, New York.
Whittell, H. M. (1943). The Noisy Scrub-bird (*Atrichornis clamosus*). *Emu*, **42**, 217–34.
Williams, C. K., and Ridpath, M. G. (1982). Rates of herbage ingestion and turnover of water and sodium in feral swamp buffalo, *Bubalis bubalis*, in relation to primary

production in a Cyperaceous Swamp in monsoonal Northern Australia. *Aust. Wildl. Res.*, **9**, 397–408.
World Resources Institute (1986). *World Resources 1986*. International Institute for Environment and Development.

Biological Invasions: a Global Perspective
Edited by J. A. Drake *et al.*
© 1989 SCOPE. Published by John Wiley & Sons Ltd

CHAPTER 10

Characteristics of Invaded Islands, with Special Reference to Hawaii

LLOYD L. LOOPE AND DIETER MUELLER-DOMBOIS

10.1 INTRODUCTION

Islands have long fascinated biologists and as a tool have much more yet to contribute in the search for general principles in ecology and evolution (Williamson, 1981). A better understanding of how island ecosystems differ from continental ecosystems may help to clarify the nature of various ecosystem properties. This chapter attempts, within the limits imposed by our current state of knowledge, to answer the question: Why are islands so vulnerable to biological invasions? Because few locations on earth have had such pervasive biological invasions as the Hawaiian Islands, an archipelago that is relatively well studied and the one with which we are most familiar, we will focus most of our discussion there.

10.2 CHARACTERISTICS OF ISLANDS: AN OVERVIEW

Reviews by MacArthur and Wilson (1967), Carlquist (1965, 1974) and Williamson (1981) provide good general coverage to this topic, and it will only be touched upon here. Islands are by definition relatively small and surrounded by water, but otherwise exhibit remarkable individuality. Most islands are geologically young in relation to continents, but the Seychelles are partly underlain by Precambrian granites. Table 10.1 compares selected parameters for various island groups. The islands included in Table 10.1 exhibit substantial topographic and environmental diversity, but many islands of the world are low lying atolls. Islands near continents tend to have more native species and lower endemism than more isolated islands. Older islands, especially those with great environmental diversity, exhibit higher endemism than young islands. Heavily glaciated islands, such as the Faeroes, contain relatively few endemic species.

In relation to the Hawaiian Islands, most archipelagoes of the world are less isolated from continents, geologically younger, have less topographic and climatic diversity, possess lower endemism and relatively few examples of evolutionary adaptive radiation, have had and continue to receive less human

Table 10.1. A comparison of selected archipelagoes of the world

	Hawaiian Islands	Galapagos Islands	Canary Islands	Juan Fernandez Islands	New Zealand
Location	19°–22°N 155°–157°W	0°–2°N 90°–92°W	28°–29°N 14°–18°W	33°S 79°–81°W	33°–47°S 167°–178°W
Distance from nearest continent (km)	3200	800	115	665	2000
Age of oldest islands in group (in millions of years (m.y.))	ca. 70	ca. 3	ca. 80	ca. 6	200 +
Total area of islands (km^2)	16,500	7900	7300	140	268,000
Highest elevation (m)	4206	1707	3718	1650	3765
Range of mean annual precipitation (mm)	200–13,000	< 750	50–1000	1000 +	350–8000
No. of endemic angiosperm genera	31	8	19	10	39
No. of native angiosperm spp.	970–1400	434	ca. 1700	146	1996
% endemism in angiosperm flora	91–96 +	51	28	66	81
No. of invasive introduced spp.	800	240	700	many	ca. 500
References	Table 2	Porter, 1984	Kunkel, 1976	Perry, 1984 Stuessy *et al.*, 1984	Kuschel, 1975

Table 10.1. (*cont*)

	Tristan da Cunha group	Granitic Seychelle Islands	Faeroe Islands	Channel Islands California
Location	37°–40°S 11°–13°W	4°–5°S 55°–56°E	62°N 7°W	33°–34°N 119°–120°W
Distance from nearest continent (km)	3000	1600	300	21
Age of oldest islands in group (in millions of years (m.y.))	20	650	50–60	ca. 20
Total area of islands (km^2)	160	260	1400	900
Highest elevation (m)	2060	914	882	753
Range of mean annual precipitation	ca. 1000	1250–2540	ca. 1500	250–400
No. of endemic angiosperm genera	0	9	0	1
No. of native angiosperm spp.	41	222	310	621
% endemism in angiosperm flora	37	31	0	22
No. of invasive introduced spp.	97	165	30	227
References	Moore, 1979 Williamson, 1981	Stoddart, 1984	Rutherford, 1982	Wallace, 1985

influence, and have suffered fewer extinctions. Most of them have probably been affected by biological invasions to a lesser extent than the Hawaiian Islands, although definitive data are lacking.

10.3 CHARACTERISTICS AND INVASIONS OF THE HAWAIIAN ISLANDS

The Hawaiian Archipelago, the most isolated island group of comparable size and topographic diversity on earth, is located over 3000 km from the nearest continent. The Hawaiian Islands consist of eight major high islands (Figure 10.1). These islands are part of a much longer island chain which was produced by the northwestward movement of the Pacific Plate over a hot spot in the earth's mantle over a 70 million year period (Figure 10.2, Macdonald *et al.*, 1983). Islands of the chain extending to the northwest are eroded and submerged remnants of what were once high islands. The area of the present high islands totals 16 500 km^2, with the largest and youngest island (Hawaii) comprising 63% of the total area. Their elevations range from sea level to over 4000 m. Average annual precipitation, strongly influenced by topography in relation to the northeast trade winds, varies from 200 mm to over 10 000 mm.

The Hawaiian biota started to evolve as much as 70 million years ago in nearly

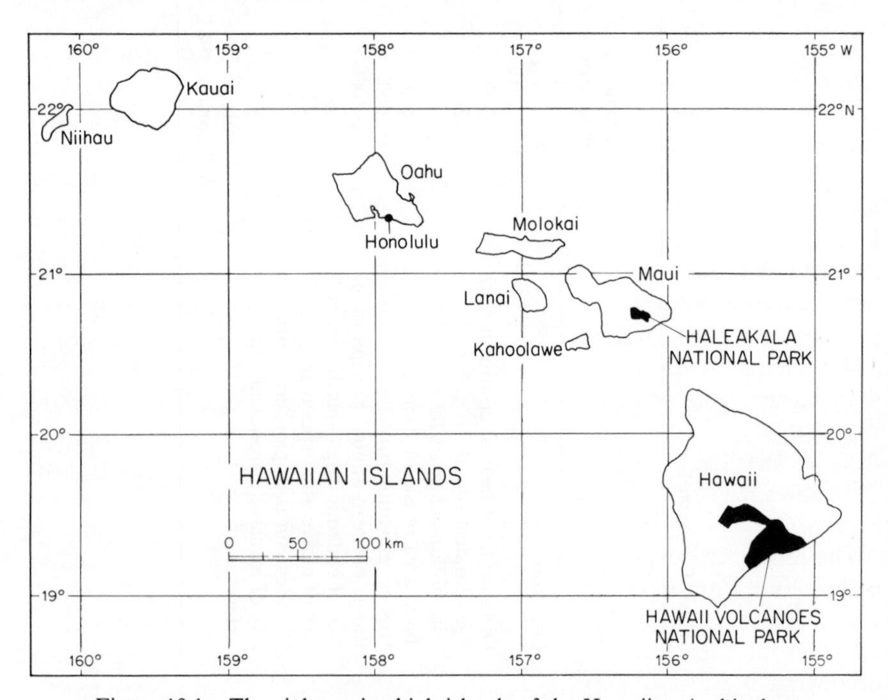

Figure 10.1. The eight major high islands of the Hawaiian Archipelago

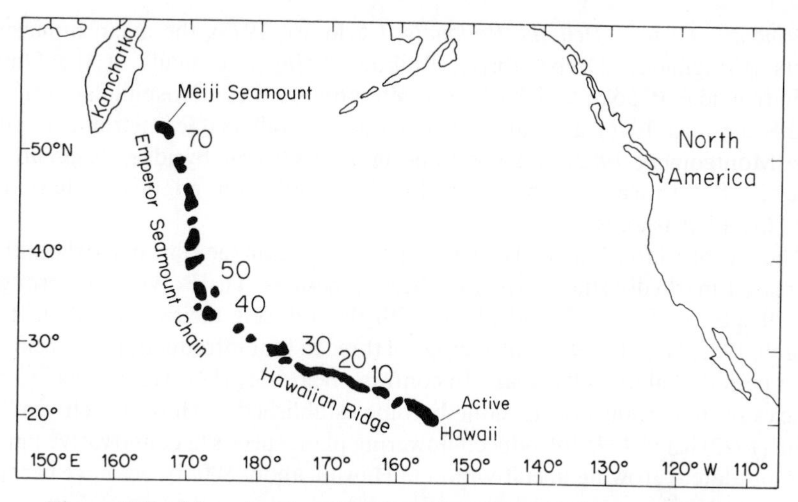

Figure 10.2. Map showing the Hawaiian–Emperor island chain with approximate ages given in millions of years. The islands were formed by northwestward movement across the active 'hot spot' now near the eastern side of the island of Hawaii. (Adapted from Macdonald *et al.*, 1983.)

total isolation—with persisting colonization through long-distance dispersal for major taxonomic groups occurring at very infrequent intervals. In only a few very efficiently dispersed groups (e.g. diatoms, McMillan and Rushforth, 1985) have dispersal and establishment not been major factors limiting the development of the Hawaiian biota. In general, the percentage of endemism is very high. The known terrestrial endemic Hawaiian vascular flora and macroscopic metazoan fauna are thought to have evolved from fewer than 1000 colonizing ancestors. A flora of 1000–1500 species of flowering plants (St John, 1973; Wagner, in Gentry, 1986) evolved from about 272 colonizing ancestors (Fosberg, 1948); 168 pteridophyte taxa developed from about 135 original immigrants (Fosberg, 1948); a native arthropod biota of 6000–10 000 species evolved from about 300–400 ancestral immigrant species (Gagné and Christensen, 1985; Hardy, 1983); a native mollusc biota of *ca* 1000 species evolved from as few as 22–24 long-distance immigrants, probably carried by birds (Zimmerman, 1948); and about 100 known species of endemic land birds (including species known only as fossils) evolved from as few as 20 ancestors (Olson and James, 1982).

The differential dispersal and establishment of various taxonomic groups as colonizers of Hawaii has allowed adaptive radiation (Carlquist, 1974). Beginning with a single ancestral population, certain groups have undergone spectacular adaptive radiation, resulting in a diverse assemblage of closely related species occupying a wide range of habitats. The silversword alliance (Asteraceae: *Argyroxiphium–Dubautia–Wilkesia*, Carr, 1985), lobeliads (Lobeliaceae:

Clermontia–Cyanea–Delissia–Rollandia, Carlquist, 1970), the avian honeycreepers (Fringillidae: Drepanidinae, Raikow, 1976), tree snails of the family Achatinellidae (Cooke and Kondo, 1960), pomace flies (Drosphilidae, Carson and Kaneshiro, 1976), a group of predacious caterpillars (Geometridae: *Eupithecia*, Montgomery, 1982), and wood-boring beetles (Cerambycidae: *Plagithmysus*, Gressitt, 1978) are among the groups that have undergone adaptive radiation in the Hawaiian Islands.

The rate of establishment of new immigrant species in the Hawaiian Islands has increased markedly since arrival of the Polynesians. The Polynesians brought 40–50 species of animals and plants with them during their own colonization (Kirch, 1982; Nagata, 1985), an average of three to four introductions per century for a period of about 1400 years. In contrast, Beardsley (1979) found that 15–20 species of immigrant insects alone become established in Hawaii each year. St John (1973) listed 4275 introduced flowering plant species (a conservative figure, now outdated) growing in cultivation in Hawaii; about 800 species have become invasive (Wagner, personal communication).

The Hawaiian Islands have had more bird species become established than any other area on earth (Long, 1981). Moulton and Pimm (1986) found that 63% of the documented intentional or unintentional introductions of bird species to individual Hawaiian Islands have resulted in establishment. For mammals and reptiles the success rate was assessed at 93% and 91% respectively (Moulton and Pimm, 1986). Table 10.2 provides the most currently available assessment of the number of native and introduced species in the various taxonomic groups of Hawaiian biota.

10.4 WHY DO INVADERS OF ISLAND ECOSYSTEMS HAVE SUCH A HIGH RATE OF SUCCESS?

Darwin (1859) attributed the vulnerability of island biota to invasions to the fact that only small numbers of indigenous species occur on islands. Much biological work has been done on islands since. Although the broad patterns are evident, the details are not yet fully clear. We can class probable reasons into four general categories as follows.

10.4.1 Evolution of island organisms in isolation

The Hawaiian biota and its ecosystems have evolved with disturbance from several sources including volcanism, windthrow, and landslides, but without grazing and trampling of ungulates and with a reduction in frequency and intensity of fire relative to most continental systems (Mueller-Dombois, 1981). Disturbance has been such an important evolutionary force in continental situations that opportunistic species have evolved that are adapted to persistence, dispersal and colonization of unoccupied areas (Pickett, 1976). The Hawaiian

Table 10.2. Statistics for various taxonomic groups in the Hawaiian Islands

Taxonomic Group	No. of Native Species	No./% of Endemic Species	No. of Invasive Species	Reference
Flowering plants	1442	1394 (97%)	326	St. John, 1973
	970	883 (91%)	800	W. Wagner in Gentry, 1986 and pers. comm.
Ferns and allies	143	105 (73%)	21	Lamoureux pers. comm.
Hepaticae (liverworts)	168	ca. 112 (67%)	2	Miller, 1956. Yoshida and Smith, 1976
Musci (mosses)	233	112 (48%)	3	Hoe, 1979
Lichens	678	268 (38%)	0	Magnusson, 1956. Smith pers. comm.
Resident birds	57	44 (77%)	38	Pyle 1983
Mammals (on land)	1	0 (0%)	18	Tomich, 1986
Reptiles (on land)	0	0	13	McKeown, 1978
Amphibians	0	0	4	
Freshwater fish	6	6 (100%)	19	Maciolek, 1984
Arthropods	6000–10,000	98%	ca. 2000	Gagné and Christensen, 1985 Hardy, 1983. Howarth, 1985
Molluscs	ca. 1060	99%	9	Zimmerman, 1984. Howarth, 1985

biota is well adapted locally to disturbances related to volcanism (e.g. Smathers and Mueller-Dombois, 1974; Winner and Mooney, 1985). Grazing and trampling of ungulates are assumed to comprise a major evolutionary force, at least in certain savanna and grassland ecosystems of the world (e.g. McNaughton, 1985), but were entirely lacking in Hawaii before 1778 AD. In most terrestrial environments of the world, fire has been the most pervasive disturbance. In areas long influenced by frequent fire, the entire flora has become closely attuned to this type of disturbance (Naveh, 1975). In any given fire-influenced area, the plants and animals have evolved a broad spectrum of adaptations to deal with this aspect of their environment (Gill, 1981; Keeley and Zedler, 1981).

Fire does not appear to have played an important role in most native ecosystems of Hawaii (Mueller-Dombois, 1981), since few native plants of the Hawaiian Islands possess adaptations to fire. Lightning is relatively uncommon on islands because their small land mass is not conducive to convective buildup of thunderheads. Many native Hawaiian ecosystems may have lacked adequate fuel to carry fires that may have been ignited by lightning or vulcanism. Humans were not present as an agent of ignition until the 4th century AD, but subsequently Polynesian colonizers undoubtedly used fire in clearing for agriculture (Kirch, 1982). Fires in modern Hawaii, carried mainly by introduced grasses, are generally highly destructive to native plant species. Opportunistic invasive plant species, on the other hand, spread rapidly following fire or other disturbance, notably on mineral soil exposed by pig-digging (Spatz and Mueller-Dombois, 1975).

Island biotas have proved vulnerable to many other types of changes in conditions after the arrival of man. Unquestionably, the lack of such ecologically important groups as ants, rodents, mammalian carnivores, and herbivorous ungulates in the Hawaiian Islands and other isolated island groups has increased the vulnerability of endemic species when members of these groups are introduced. Certain introduced ants are voracious predators and have obliterated most of Hawaii's endemic lowland arthropod fauna (Zimmerman, 1978). Rodents are selective feeders, eat great quantities of seeds, and may prevent reproduction of certain plant species that have not evolved mechanisms to protect even a fraction of their seeds from rodents (e.g. Clark, 1981). Rodents may also be devastating to the reproductive success of endemic birds (Atkinson, 1977, 1985). Predators such as the mongoose (*Herpestes auropunctatus*) are especially destructive to birds with conservative reproductive strategies such as the Hawaiian dark-rumped petrel (*Pterodroma phaeopygia*), a long-lived species that has low natural mortality of its single, well cared for, chicks (Simons, 1983). Flightlessness in birds is a condition that has evolved independently on many islands and usually proves to be a fatal condition when predators are introduced (James and Olson, 1983). Evolution in the absence of large mammalian herbivores, which consume large quantities of vegetation and cause perpetual disturbance through trampling and digging (in the case of pigs), has resulted in high vulnerability to damage by introduced ungulates. Hawaiian plants are relatively nonpoisonous and free from many other characteristics that deter mammalian herbivores (Carlquist, 1970, 1974), although they retain physical and chemical defenses against insect herbivory. Evolution in the absence of exposure to avian malaria has resulted in high susceptibility of the surviving Hawaiian land bird fauna to the mosquito-borne protozoan, *Plasmodium relictum* (van Riper *et al.*, 1986).

Loss of coevolved organisms may further hasten the demise of island species. Some of the many extinct birds of the Hawaiian Islands (Olson and James, 1982) may have been important in dispersal and/or scarification of seeds. From East Maui, there is evidence that necessary scarification of seeds of the native shrub *Styphelia tameiameiae* is provided by passing through the digestive tract of introduced pheasants (*Phasianus colchicus*) (A. C. Medeiros, personal communication). Successful alien plant species in Hawaii are often dependent upon introduced dispersal vectors as well.

Pollination is an essential process for which many plant species are dependent upon insects, birds, or other animals. Even when appropriate pollinators are present, seed set is often limited by a suboptimal quantity of pollinators (Bierzychudek, 1981). At least some island plants, including the silversword (*Argyroxiphium–Dubautia*) alliance of the Hawaiian Islands, have evolved obligate outcrossing mechanisms (Carr *et al.*, 1986) and are thus vulnerable to extinction of pollinators. Pollinators of the Haleakala silversword (*Argyroxiphium sandwicense macrocephalum*) may indeed be vulnerable to elimination by the Argentine ant (*Iridomyrmex humilis*) (Medeiros *et al.*, 1986).

10.4.2 Modification of island environments by humans

The Hawaiian Islands have been heavily exploited by the colonizing Polynesians (Kirch, 1982; Olson and James, 1982) and much more so by continental man after 1778 (Gagné, 1975; Daws, 1968, Cooper and Daws, 1985). The small land area in relation to its potential (augmented by products from the sea) for supporting a self-sustaining large human population contributed to the Polynesian impacts. In Polynesia in general, dispersal of human populations through voyaging during the past several thousand years was apparently triggered by large population pressure (Jennings, 1979). Throughout the Hawaiian Islands, most land below 600 m with even moderately good soils was cultivated by the Hawaiians in the 13th–18th centuries (Kirch, 1982). The pre-contact Hawaiians also eliminated over half the species of endemic birds of the islands through habitat alteration and hunting (Olson and James, 1982). Archeological reconstruction of the prehistory of the Hawaiian island of Kahoolawe may illustrate an extreme case of a broad pattern. Now an uninhabited and barren wasteland used as a bombing range by the US military, Kahoolawe is unique in that the archeological evidence for settlement has been preserved intact (Kirch, 1985). Colonization of Kahoolawe

(117 km^2 area) is inferred by Hommon (1980) to have taken place about 1000 AD, with initial exploitation of the coastal zone. The inland central plateau zone (up to 450 m above sea level), initially a 'dryland forest or parkland', was occupied and extensively used for agriculture (sweet potatoes, etc.) from 1400 to 1550 AD. Hommon (1980) proposed that agricultural clearance and burning led to 'island-wide degradation', leading to a rapid decline in human population during 1550–1650 AD. The island area was completely abandoned after about 1750 AD. At the time of Western contact, the island's population was estimated at about 60 persons, dwelling entirely along the coast. Although environmental degradation was much less on other Hawaiian Islands, there is relatively good archeological evidence on western Hawaii for a human population decline, presumably due to attainment of a population level exceeding carrying capacity and accompanying resource deterioration, prior to Western contact (Kirch, 1985).

Human-related prehistoric extinctions have now been documented throughout the world, both on islands (e.g. Olson, 1975, 1977; McCoy, 1979) and continents (e.g. Martin, 1984). Massive deforestation has been demonstrated in prehistoric Mesoamerica (Turner and Harrison, 1981). Effects of prehistoric man in altering island environments may have been more consistently severe largely because of their smaller size and lack of alternative land for exploitation.

Westernization has led to unprecedented resource exploitation and degradation throughout the oceanic world (e.g. Greenway, 1958; Holdgate and Wace, 1961; Millot, 1972; Lee, 1974; Wace, 1976; Coblentz, 1978; Melville, 1979; Rauh, 1979; van der Werff, 1979; Cronk, 1980, 1986; Perry, 1984; Gade, 1985). For example, the portion of New Zealand covered by forest has been reduced from 68% to 14% by human activity over the past 200 years (Kuschel, 1975). Again, continental environments, though subjected to the same types of devastating damage that have occurred on islands, have proved relatively resilient because at least until recently, possibility existed for exploiting an area and moving elsewhere.

10.4.3 Invasibility of island ecosystems

Island biotas typically have low species numbers in certain groups in relation to their relative proportions in continental areas of the world. 'Disharmony' is a term often used in the context of island biology to denote a taxonomic balance that differs from continental norms. For example, the Hawaiian Islands have only four native species of orchids, one of the largest angiosperm families worldwide, particularly in warm and moist climatic regions. The native biota also lacks representatives of the following taxa important on continents: gymnosperms, Aceraceae, Araceae, Betulaceae, Bignoniaceae, Cunoniaceae, Fagaceae, *Ficus*, *Piper*, reptiles and amphibians. Mammals are lacking except for a single species of bat. Two-thirds of the world's insect orders have no representatives in Hawaii. On the other hand, the Hawaiian Drosophilidae comprise about 600 species, a large percentage of the total species worldwide in this family (Carson and Templeton, 1984).

Simberloff has justifiably questioned whether the taxonomically disharmonic nature of the Hawaiian entomofauna has been responsible for the large number of invading insect species there. He shows (Simberloff, 1986, Table 1.1) that the insect orders in Hawaii with the largest numbers of natives (Coleoptera, Lepidoptera, Hymenoptera) are also those with large numbers of introductions.

In some instances, nevertheless, some of the same factors that have promoted adaptive radiation over evolutionary time appear to promote vulnerability to invasion in modern time. The absence or near absence of a taxonomic group or ecological guild on an island will often result in minimum resistance to invading continental species of that group until saturation is approached. An example is the successful purposeful introduction of 19 species of freshwater fish in Hawaiian waters, which were occupied prior to 1800 only by six species of native gobiids, freshwater-tolerant fish derived from ancestors that spent their entire life cycles in the sea (Heere, 1940; Maciolek, 1984). Absence of native mammalian herbivores and ants has undoubtedly facilitated the highly successful invasion of those groups. On the other hand, invasive species are not necessarily responding in most cases to 'vacant niches waiting to be filled.' A common pattern is creation and exploitation of new niches through multiple invasion. For example, the extremely successful invading tree *Myrica faya* is able to establish and thrive on nutrient-poor young volcanic substrates on the island of Hawaii; it is aided by an introduced microbial nitrogen-fixing symbiont (*Frankia*) and introduced birds that disperse its seeds (Walker *et al.*, 1986). Feralization of pigs in the Hawaiian Islands has been facilitated by invasion of introduced earthworms and of such introduced plants as *Psidium cattleianum* and *Passiflora mollissima* (Stone and Loope, 1987). In turn, massive invasion of forest understories by numerous introduced plant species is made possible by exposure of mineral soil by pig digging.

The total number of species per unit area is smaller on islands than in continental situations. MacArthur (1972) states the principle that 'no island has nearly the number of species it would have if it were part of the mainland.' The central equation of MacArthur and Wilson's (1967) theory of island biogeography, presenting island species number as a balance between immigration rate and extinction rate, might seem to shed light on island vulnerability to invasions, with accelerated immigration driving up extinction rates or increased extinction rates facilitating establishment of immigrants. Although this theory has stimulated much thought and numerous investigations regarding island ecology, it would seem to have little to do with an isolated oceanic island system such as Hawaii where pre-Polynesian immigration rates were exceeded by rates of evolution of new species (Williamson, 1981). Now that immigration rates have been accelerated from one introduction per 50 000 years to 20 or more introductions per year, there is little indication that species numbers *per se* have much immediate relevance to forcing species extinctions or retarding establishment of further immigrants. However, individual introductions (e.g. the ant *Pheidole megacephala*, the feral goat, the feral pig, etc.) can be crucial in this regard.

Comparisons of bird communities on oceanic islands with those of comparable mainland habitats suggests to some workers that as a result of lower species numbers on islands, competition is often reduced and niches are broader (Crowell, 1962; MacArthur *et al.*, 1972). Williamson (1981) questions whether niche expansion on islands has really been demonstrated, but concludes from his review of island ecology that 'isolated islands have a distinctive biota, and the number of species per unit area is less than on an equivalent area of the mainland, or of less isolated islands.' Mountainspring and Scott (1985), working in relatively undisturbed upland habitats of Hawaii, Maui and Kauai, found only sporadic interspecific competition between native and introduced bird species and virtually none among native species or among introduced species. Competition among native Hawaiian species would be expected to have been drastically reduced by massive extinctions in the past 1500 years related to habitat destruction or disturbance (Olson and James, 1982; Kirch, 1982; Gagné, 1975) and by severely negative effects of avian pox and malaria, introduced within the past century (van Riper *et al.*, 1986). There appears to be at least some evidence that upland Hawaii is not presently saturated with bird species, making it highly vulnerable to continued invasion.

On the other hand, Moulton and Pimm (1983) inferred substantial competitive interaction in the Hawaiian Islands among introduced bird species below 600 m elevation. Since most inadvertent introductions occur at low elevations in areas of high human population density, the most effective barrier against further bird invasions may be the relative saturation of lowland habitats with introduced bird species.

Likewise, invasion of low and middle elevation areas in the Hawaiian Islands by about 40 species of alien ants has produced alien ant communities that appear relatively resistant to further invasions (Huddleston and Fluker, 1968; Fluker and Beardsley, 1970). Immigration or introduction of numerous alien parasitoid hymenopterans and ants not only adversely effects native biota (Howarth, 1985), but retards establishment of intentional biocontrol introductions (Howarth, 1983) and depresses populations of introduced pest species (Wong *et al.*, 1984).

Disharmony and low species numbers appear to contribute in at least a minor way to vulnerability of islands to invasions.

10.4.4 Reduced aggressiveness and vulnerability to extinction of island biotas

Numerous workers have noted that native island species have reduced 'aggressiveness' or increased vulnerability to extinction even under optimal environmental conditions. A theory first proposed by E. O. Wilson (1961) in describing this phenomenon for the Melanesian ant fauna is the concept of the taxon cycle. Later, Greenslade (1968) and Ricklefs and Cox (1972, 1978) applied the theory to the avifauna of the Solomon Islands and West Indies respectively. The taxon cycle concept involves the increasing habitat specialization and increasing vulnerability to extinction that a taxonomic group undergoes in the progressive invasion of an archipelago. In Solomon Island birds, for example, recent colonizers (Stage I) are expanding and are present in coastal and cultivated habitats throughout the island group (Greenslade, 1968). In Stage II, there is a fragmentary distribution in tropical rainforests, leading to subspeciation and some local extinction. In Stage III, involving speciation, populations become highly fragmented as a result of habitat shift to montane forests with range contraction and extensive extinction of local populations. Extinctions are due both to invasion by later arriving forms and to over-specialization in small habitats.

Williamson (1981) has reviewed the taxon cycle concept favorably, concluding with the following statement: 'Only a few cases of the taxon cycle have been described, but no studies have been published showing it to be inapplicable. How general a phenomenon the cycle is remains to be determined.' Kruckeberg and Rabinowitz (1985) have, on the other hand, criticized the taxon cycle concept as unfalsifiable. Carlquist (1974) cites the work of McDowall (1969) as suggesting that the older an island endemic is, the more prone it is to extinction. McDowall showed that the higher the taxonomic level of endemism for New Zealand land birds (e.g. endemic family versus endemic subspecies) the more chance of their being extinct or endangered with extinction.

Application of a similar approach to McDowall's to the Hawaiian angiosperm flora and bird fauna yields similar results (Tables 10.3 and 10.4). Among Hawaiian flowering plants (Table 10.3), a taxon in an endemic Hawaiian genus is 1.3 times as likely to be extinct or endangered as one endemic at the species level, 4 times as

Table 10.3. Relationship between vulnerability of taxa in the Hawaiian flora and the taxonomic level of endemism. Computations made based on data in St. John (1973) for numbers of taxa in flora and levels of endemism and in Fosberg and Herbst (1975) for extinction or level of endangerment

	Taxa Endemic at Level of:			
	Genus	Species	Variety	Not endemic
No. of taxa in flora	872	1562	33	43
No. (%) of extinct taxa	135(15%)	133(9%)	0	0
No. (%) of 'endangered' taxa	311(36%)	480(31%)	4(12%)	2(5%)
Total extinct or endangered taxa	446(51%)	613(40%)	4(12%)	2(5%)
No. of 'rare' taxa	25	63	0	1
No. taxa of 'uncertain' status	235	278	4	0
Total extinct or vulnerable taxa	706(81%)	956(61%)	8(24%)	3(7%)

likely as one endemic at an infraspecific level, and 10 times as likely as a native taxon not endemic at any level. The same pattern holds true if taxa classed as 'rare' and 'of uncertain status' (many of which are actually extinct or endangered) by Fosberg and Herbst (1975) are included. In the historically known Hawaiian bird fauna (Table 10.4), no taxon endemic at a level below species is extinct and no nonendemic taxon is classified as 'endangered,' whereas for endemics at the species, genus, and subfamily levels, 32% are extinct and 50% endangered. Such analyses seem to give at least some support to the taxon cycle theory.

The genetic basis of the phenomenon of reduced aggressiveness with progressive island evolution is virtually unexplored, but the extensive work with Hawaiian drosophilids by H. L. Carson and colleagues provides a basis for educated speculation. To begin with, we must emphasize that Carson's work shows no evidence for genetic impoverishment of local drosophilid populations in Hawaii. Carson (1981, pp. 471–2) summarized studies of genetic variation in Hawaiian *Drosophila* as follows:

> The genetic data presented here provide no evidence that there are unique properties to the genetic variation systems of insular species. Thus the levels of genetic variation found within a series of endemic and introduced species of *Drosophila* are basically similar to their continental counterparts. Perhaps the most important point is that island species, even some with quite small total populations, are capable of carrying as much genetic variability in a local population as are species with very large populations.

In spite of comparable variability of local populations, termed 'the growing point of evolutionary change,' insular and continental species differ in their gross population size (Carson, 1981). Continental species have many local populations; island populations have few. Carson continues (p. 474):

> Accordingly the total genetic variance carried in a continental species should be far greater than that found in insular ones. A continental deme will be able to draw variability from adjacent demes, given a capacity for gene flow between demes. Any deme is limited in the amount of genetic variability it can carry. Continental demes... have the advantage of being able to be enriched continually by gene flow. This probably represents an important difference between island and continental populations. The isolated nature of most island demes may be conducive to the evolution of restrictive specializations, whereas continental conditions are capable of giving rise to the genetic basis of a generalism, wherein the organism is homeostatic. This difference may underlie the observed failure among island organisms to evolve aggressive weedy organisms that have genotypes adapted to general purposes.

Table 10.4. Relationship between vulnerability of taxa in the historically known Hawaiian resident bird fauna and the taxonomic level of endemism. Computations made based on taxonomic status and status on Federal List of Endangered Species given in Pyle (1983).

	No. of Species Endemic at Level of:				
	Subfamily	Genus	Species	Subspecies	Not endemic
Phaethontidae					
Non-endangered	—	—	—	—	1
Sulidae					
Non-endangered	—	—	—	—	3
Fregatidae					
Non-endangered	—	—	—	—	1
Ardeidae					
Non-endangered	—	—	—	—	1
Anatidae					
Endangered	—	1	2	—	—
Acciptridae					
Endangered	—	—	1	—	—
Rallidae					
Extinct	—	—	2	—	—
Endangered	—	—	—	2	—
Recurvirostridae					
Endangered	—	—	—	1	—
Laridae					
Non-endangered	—	—	—	—	3
Strigidae					
Non-endangered	—	—	—	1	—
Corvidae					
Endangered	—	—	1	—	—
Muscicapidae					
Endangered	—	—	3	—	—
Non-endangered	—	—	1	—	—
Melaphagidae					
Extinct	—	4	—	—	—
Endangered	—	1	—	—	—
Fringillidae					
Extinct	8	—	—	—	—
Endangered	13	—	—	—	—
Non-endangered	7	—	—	—	—
Totals					
Extinct	8	4	2	0	0
Endangered	13	2	7	3	0
Non-endangered	7	0	1	1	9

Recent investigations of allozyme variation in several island groups that have undergone extensive adaptive radiation shed additional light on the genetic structure of island taxa. Lowrey and Crawford (1985) examined allozyme divergence in Hawaiian *Tetramolopium*, a morphologically and ecologically diverse group in the Asteraceae occupying habitats from sea level to above 3000 m. They found that 'the mean genetic identity for pairwise comparison of 19 populations from seven species is 0.95, a very high value normally obtained for conspecific plant populations.' This type of pattern has been found in Hawaiian *Bidens* (also Asteraceae) by Helenurm and Ganders (1985) as well as in such diverse island groups as Hawaiian *Drosophila* (Diptera: Drosophilidae) (Carson and Kaneshiro, 1976), Galapagos finches (Fringillidae: Geospizinae) (Yang and Patton, 1981), and the snail genus *Partula* in Moorea of French Polynesia (Johnson *et al*., 1977).

Another genetic factor, although a highly controversial one (e.g. Charlesworth *et al*., 1982), that could lead to increased vulnerability of island species through evolutionary time is the repetition (and compounding) of founder events in the genetic history of many island species. The Hawaiian Islands have an ancient insular biota which has established by long-distance dispersal (Fosberg, 1948; Carlquist, 1974; Gressitt, 1978; Carson and Templeton, 1984) and undergone adaptive radiation on an archipelago much older than the currently existing islands. Whereas the oldest current high island, Kauai, has no dated rocks more than 6 million years old, geological evidence suggests an age of 30–40 million years for the now largely eroded older islands in the Leeward Hawaiian chain and up to 70 million years for the islands in the adjacent Emperor Seamount Chain (Macdonald *et al*., 1983). Biological evidence from DNA and protein 'clocks' suggests that the ancestral forms of Drosophilinae and Drepanidinae arrived in the island chain prior to the existence of the current high islands (Sibley and Ahlquist, 1982; Beverley and Wilson, 1985). The rigor of natural selection in such an evolving insular system may be relaxed by the large number of genetic bottlenecks (founder events) many groups have undergone in island-hopping.

Carson (1981), referring to the founding of populations of Hawaiian *Drosophila* on geologically new islands (often by a single gravid female) states: 'At each of these events, drastic effects on the genetics of the new species resulting from founders would be expected. Species that inhabit a moderately old archipelago such as Hawaii have populations that both ancestrally and currently have been and are being rent by forces of chance to which no specific adaptational response is possible.' Natural selection operates, of course, on these populations and the island forms achieve a high level of adaptation to their new local environment—but in many instances may not be so well adapted as the 'general purpose genotypes' of invasive introduced species.

In summary, there are suggestive patterns and some interesting genetic evidence possibly related to the relative lack of aggressiveness and tenacity of some island species. However, further work is needed before definitive statements can be made. Granted that some island species exhibit reduced aggressiveness and increased vulnerability to extinction, we wish to emphasize that by no means all island species suffer such restriction. For example, the native fern *Dicranop-*

teris linearis aggressively colonizes disturbed areas, in some cases displacing aggressive non-native woody plants. *Metrosideros polymorpha*, the dominant tree in many Hawaiian forests, exhibits a number of ecotypes, some of which are successful colonizers of recent lava flows, others are better adapted to older, more mesic soils, and still others to bogs (Stemmermann, 1983). Aggressive non-native plants notably lack competitive superiority over natives in extreme habitats such as on new lava flows (Smathers and Mueller-Dombois, 1974), on soils with aluminum toxicity (Gerrish and Mueller-Dombois, 1980), and in montane bogs (Canfield, 1986).

10.5 ISLAND INVASIONS AND CONSERVATION

The inference that conservation of island ecosystems is a hopeless task because of their vulnerability to invasions has been often used as a rationalization for either wholesale destruction of island biotas or benign neglect of their protection. For example, an early influential figure in Hawaiian botany and forestry, H. L. Lyon, noted (1909, 1918, 1919) the widespread dieback of native *Metrosideros* rainforest on northern East Maui (termed the 'Maui forest disease') and concluded that Hawaiian native forests could not be maintained or restored and that a new forest flora must be built from introduced species in order to save the Islands' watershed (Holt, 1983). Lyon's interpretations led to massive introductions of introduced trees to Maui with limited positive results and much subsequent damage from aggressive invasive species. Similarly, Burgan and Nelson (1972) and Petteys *et al.* (1975) interpreted large-scale *Metrosideros* dieback on the island of Hawaii as a progressive disease-induced decline which would ultimately lead to the demise of most *Metrosideros*-dominated native forests. An obvious implication of this work was that these native forests should be exploited commercially before their natural demise took place. However, work by Mueller-Dombois and his co-workers (summarized in Mueller-Dombois *et al.*, 1981; Mueller-Dombois, 1985) has largely contradicted interpretations of earlier workers on the demise of *Metrosideros* forests, and has shown that in most instances forest dieback is followed by vigorous regeneration (e.g. Jacobi *et al.*, 1983). Mueller-Dombois (1983) interprets *Metrosideros* dieback in Hawaii as a recurring natural phenomenon that may be related to similar phenomena in New Zealand, Australia, and New Guinea.

Studies of vegetation response within fenced exclosures in Hawaii indicate that at least partial recovery of native vegetation occurs in most instances after the influence of alien ungulates is removed (Loope and Scowcroft, 1985). The synthesis volume of work in Hawaii as part of the International Biological Program's Island Ecosystems project (Mueller-Dombois *et al.*, 1981) took a positive view of the ability of island organisms to survive, given reasonable human assistance. Increasingly active management by the US National Park Service in Hawaii's national parks in the past two decades has yielded some success (Stone and Loope, 1987). With active management, primarily involving control of feral ungulates and some of the more aggressive plant invaders, chances appear good for preservation of the still largely intact systems at high elevations and on specialized sustrates.

10.6 SUMMARY AND CONCLUSIONS

Isolated oceanic islands were predisposed to certain types of human-related invasions because of long isolation from the continual challenge of some of the selective forces that shape continental organisms—including such forces as virulent diseases, browsing and trampling of herbivorous mammals, ant predation, and frequent and intense fire. In spite of their very limited resources, islands (particularly tropical ones) have fascinated and attracted humans, resulting in severe direct and indirect human impacts. The structure of island ecosystems, generally comprised of relatively few species in comparison to comparable mainland habitats, and certain genetic properties of island taxa may also contribute significantly to their vulnerability to invasion. Much remains to be understood. Island biology has played a major role in the development of evolutionary theory. As native island species continue to be lost to the onslaught of invaders, accompanied by the further loss of dependent coevolved species, opportunities are lost for important studies that can continue to contribute to the mainstream of biological theory. Well planned, scientifically based efforts at active management of strategically selected island ecosystems appear highly worthwhile. At the very least, they will greatly prolong the time that these systems will be available for study.

REFERENCES

Atkinson, I. A. E. (1977). A reassessment of the factors, particularly *Rattus rattus* L., that influenced the decline of endemic forest birds in the Hawaiian Islands. *Pacific Sci.*, **31**, 109–33.

Atkinson, I. A. E. (1985). The spread of commensal species of *Rattus* to oceanic islands and their effects on island avifaunas. In: Moors, P. J. (Ed.), *Conservation of Island Birds*, pp. 35–81. Intl. Council for Bird Preservation Tech. Publ. No. 3.

Beardsley, J. W. (1979). New immigrant insects in Hawaii: 1962 through 1976. *Proc. Hawaiian Entomol. Soc.*, **23**, 35–44.

Beverley, S. H., and Wilson, A. C. (1985). Ancient origin for Hawaiian Drosophilinae inferred from protein comparisons. *Proc. Nat. Acad. Sci. U.S.A.*, **82**, 4753–7.

Bierzychudek, P. (1981). Pollinator limitation of plant reproductive effort. *Amer. Nat.*, **117**, 838–40.

Burgan, R. E., and Nelson, R. E. (1972). *Decline of Ohia Lehua Forests in Hawaii.* USDA, Forest Service, General Tech. Rept. PSW-3. Pacific Southwest Forest and Range Expt Sta., Berkeley, California.

Canfield, J. (1986). The role of edaphic factors and plant water relations in plant distribution in the bog/wet forest complex of the Alakai Swamp, Kauai, Hawaii. Ph.D. dissertation, Department of Botany, University of Hawaii, Honolulu.

Carlquist, S. (1965). *Island Life: A Natural History of the Islands of the World*, Natural History Press, Garden City, New York. 451 pp.
Carlquist, S. (1970). *Hawaii: A Natural History*. Natural History Press, Garden City, New York. 467 pp.
Carlquist, S. (1974). *Island Biology*. Columbia University Press, New York. 660 pp.
Carr, G. D. (1985). Monograph of the Hawaiian Madiinae (Asteraceae): *Argyroxiphium, Dubautia*, and *Wilkesia*. *Allertonia*, **4**(1), 1–123.
Carr, G. D., Powell, E. A., and Kyhos, D. W. (1986). Self-incompatibility in the Hawaiian Madiinae (Compositae): an exception to Baker's rule. *Evolution*, **40**(2), 430–4.
Carson, H. L. (1981). Microevolution in insular ecosystems. In: Mueller-Dombois, D., Bridges, K. W., and Carson, H. L. (Eds), *Island Ecosystems: Biological Organization in Selected Hawaiian Communities*, pp. 471–82. Hutchinson-Ross. 583 pp.
Carson, H. L., and Kaneshiro, K. Y. (1976). *Drosophila* of Hawaii: systematics and ecological genetics. *Ann. Rev. Ecol. Syst.*, **7**, 311–46.
Carson, H. L., and Templeton, A. R. (1984). Genetic revolutions in relation to speciation phenomena: the founding of new populations. *Ann. Rev. Ecol. Syst.*, **15**, 97–131.
Charlesworth, B., Lande, R., and Slatkin, M. (1982). A neo-Darwinian commentary on macroevolution. *Evolution*, **36**, 474–98.
Clark, D. A. (1981). Foraging patterns of black rats across a desert-montane forest gradient in the Galapagos Islands. *Biotropica*, **13**(3), 182–94.
Coblentz, B. E. (1978). The effects of feral goats (*Capra hircus*) on island ecosystems. *Biol. Conserv.*, **13**, 279–86.
Cooke, C. M., and Kondo, Y. (1960). *Revision of Tornatellinidae and Achatinellidae* (*Gastropoda, Pulmonata*). B.P. Bishop Mus. Bull. 221. Honolulu, Hawaii. 303 pp.
Cooper, G., and Daws, G. (1985). *Land and Power in Hawaii*. Benchmark Books, Honolulu, Hawaii. 518 pp.
Cronk, Q. C. B. (1980). Extinction and survival in the endemic vascular flora of Ascension Island. *Biol. Conserv.*, **17**, 207–19.
Cronk, Q. C. B. (1986). The decline of the St. Helena ebony *Trochetiopsis melanoxylon*. *Biol. Conserv.*, **35**, 159–72.
Crowell, K. L. (1962). Reduced interspecific competition among the birds of Bermuda. *Ecology*, **43**, 75–88.
Darwin, C. R. (1859). *On the Origin of Species by means of Natural Selection, or, The Preservation of Favoured Races in the Struggle for Life*. John Murray, London.
Daws, G. (1968). *Shoal of Time: A History of the Hawaiian Islands*. University Press of Hawaii, Honolulu. 494 pp.
Fluker, S. S., and Beardsley, J. W. (1970). Sympatric associations of three ants: *Iridomyrmex humilis, Pheidole megacephala*, and *Anoplolepis longipes* in Hawaii. *Ann. Entomol. Soc. Amer.*, **63**(5), 1290–6.
Fosberg, F. E. (1948). Derivation of the flora of the Hawaiian Islands. In: Zimmerman, E. C. (Ed.), *Insects of Hawaii, Vol. 1. Introduction*, pp. 107–19. University of Hawaii Press, Honolulu, Hawaii, 206 pp.
Fosberg, F. R., and Herbst, D. (1975). Rare and endangered species of Hawaiian vascular plants. *Allertonia*, **1**, 1–72.
Gade, D. W. (1985). Man and nature in Rodrigues: tragedy of an island common. *Environ. Conserv.*, **12**(3), 207–16.
Gagné, W. C. (1975). Hawaii's tragic dismemberment. *Defenders*, **50**(6), 461–9.
Gagné, W. C., Christensen, C. C. (1985). Conservation status of native terrestrial invertebrates in Hawaii, In: Stone, C. P., and Scott, J. M. (Eds), *Hawaii's Terrestrial Ecosystems: Preservation and Management*, pp. 105–26. Coop. Natl Park Resources Studies Unit, University of Hawaii, Honolulu, Hawaii. 584 pp.

Gentry, A. H. (1986). Endemism in tropical versus temperate plant communities. In: Soulé, M. E. (Ed.), *Conservation Biology: the Science of Scarcity and Diversity*, pp. 153–81. Sinauer Assoc., Sunderland, Mass. 584 pp.

Gerrish, G., and Mueller-Dombois, D. (1980). Behavior of native and non-native plants in two tropical rain forests on Oahu, Hawaiian Islands. *Phytocoenologia*, **8**, 237–95.

Gill, A. M. (1981). Fire adaptive traits of vascular plants. In: Mooney, H. A., Bonnicksen, T. M., Christensen, N. L., Lotan, J. E., and Reiners, W. A. (Eds), *Proceeding of Conference on Fire Regimes and Ecosystem Properties*, pp. 208–30. US Forest Service, Gen. Tech. Rept WO-26. 594 pp.

Greenslade, P. J. M. (1968). Island patterns in the Solomon Islands bird fauna. *Evolution*, **22**, 751–61.

Greenway, J. C. (1958). *Extinct and Vanishing Birds of the World*. American Committee for International Wildlife Protection, Special Publication No. 13.

Gressitt, J. L. (1978). Evolution of the endemic Hawaiian Cerambycid beetles. *Pacific Insects*, **18**, 137–67.

Hardy, D. E. (1983). Insects. In: Armstrong, R. W. (Ed.), *Atlas of Hawaii*, pp. 80–2. University Press of Hawaii. 238 pp.

Heere, W. C. T. (1940). Distribution of fresh-water fishes in the Indo-Pacific. *Sci. Month.*, **51**, 165–8.

Helenurm, K., and Ganders, F. R. (1985). Adaptive radiation and genetic differentiation in Hawaiian *Bidens*. *Evolution*, **39**(4), 753–65.

Hoe, W. J. (1979). Phytogeographical relationships of Hawaiian mosses. Ph.D. dissertation. Department of Botany, University of Hawaii, Honolulu. 357 pp.

Holdgate, M. W., and Wace, N. W. (1961). The influence of man on the floras and faunas of southern islands. *Polar Record*, **10**, 475–93.

Holt, R. A. (1983). The Maui forest trouble: a literature review and proposal for research. Hawaii Bot. Sci. Pap. No. 42. University of Hawaii, Honolulu.

Hommon, R. J. (1980). *Kahoolawe: Final Report for the Archaeological Survey*. Report prepared for US Navy. Honolulu, Hawaii.

Howarth, F. G. (1983). Classical biocontrol: panacea of Pandora's box. *Proc. Hawaiian Entomol. Soc.*, **24**(2&3), 239–44.

Howarth, F. G. (1985). Impacts of alien land arthropods and mollusks on native plants and animals in Hawaii. In: Stone, C. P., and Scott, J. M. (Eds), *Hawaii's Terrestrial Ecosystems: Preservation and Management*, pp. 149–79. Coop. Natl. Park Resources Studies Unit, University of Hawaii, Honolulu, Hawaii, 584 pp.

Huddleston, E. W., and Fluker, S. S. (1968). Distribution of ant species in Hawaii. *Proc. Hawaiian Entomol. Soc.*, **20**(1), 45–69.

Jacobi, J. D., and Gerrish, G., and Mueller-Dombois, D. (1983). Ohia dieback in Hawaii: vegetation changes in permanent plots. *Pacific Sci.*, **37**(4), 327–38.

James, H. F., and Olson, S. L. (1983). Flightless birds. *Nat. Hist.*, **92**(9), 30–40.

Jennings, J. D. (Ed.) (1979). *The Prehistory of Polynesia*. Harvard University Press, Cambridge, Mass.

Johnson, M. S., Clarke, B., and Murray, J. (1977). Genetic variation and reproductive isolation in *Partula*. *Evolution*, **31**, 116–26.

Keeley, J. E., and Zedler, P. H. (1981). Reproduction of chaparral shrubs after fire: a comparison of the sprouting and seedling strategies. *Amer. Midl. Natur.*, **99**, 142–61.

Kirch, P. V. (1982). The impact of the prehistoric Polynesians on the Hawaiian ecosystem. *Pacific Sci.*, **36**(1), 1–14.

Kirch, P. V. (1985). *Feathered Gods and Fishhooks: an Introduction to Hawaiian Archaeology and Prehistory*. University of Hawaii Press, Honolulu, Hawaii. 349 pp.

Kruckeberg, A. R., and Rabinowitz, D. (1985). Biological aspects of endemism in higher plants. *Ann. Rev. Ecol. Syst.*, **16**, 447–79.

Kunkel, G. (Ed.) (1976). *Biogeography and Ecology in the Canary Islands.* Monographie Biologicae, Vol. 30. Junk, The Hague, 511 pp.

Kuschel, G. (Ed.) (1975). *Biogeography and Ecology in New Zealand.* Monographie Biologicae, Vol. 27. Junk, The Hague, 687 pp.

Lee, M. A. B. (1974). Distribution of native and invader plant species on the island of Guam. *Biotropica*, **6**, 158–64.

Long, J. L. (1981). *Introduced Birds of the World.* A. H. and A. W. Reed, Sydney, Australia.

Loope, L. L., and Scowcroft, P. G. (1985). Vegetation response within exclosures in Hawaii: a review. In: Stone, C. P., and Scott, J. M. (Eds), *Hawaii's Terrestrial Ecosystems: Preservation and Management*, pp. 377–402. Coop. Natl Park Resources Studies Unit, University of Hawaii, Honolulu, Hawaii. 584 pp.

Lowrey, T. K., and Crawford, D. J. (1983). Allozyme divergence and evolution of *Tetramolopium* (Compositae: Astereae) in the Hawaiian Islands. *System. Bot.*, **10**(1), 64–72.

Lyon, H. L. (1909). The forest disease on Maui. *Hawaiian Planter's Record*, **1**, 151–9.

Lyon, H. L. (1918). The forests of Hawaii. *Hawaiian Planter's Record*, **20**, 276–81.

Lyon, H. L. (1919). Some observations on the forest problems of Hawaii. *Hawaiian Planter's Record*, **21**, 289–300.

MacArthur, R. H. (1972). *Geographical Ecology: Patterns in the Distribution of Species.* Harper & Row, New York, 269 pp.

MacArthur, R. H., Diamond, J. M., and Karr, J. R. (1972). Density compensation in island faunas. *Ecology*, **53**, 332–42.

MacArthur, R. H., and Wilson, E. O. (1967). *The Theory of Island Biogeography.* Princeton University Press, Princeton, NJ.

McCoy, P. C. (1979). Easter Island. In: Jennings, J. D. (Ed.), *The Prehistory of Polynesia*, pp. 135–66. Harvard University Press, Cambridge, Mass. 369 pp.

Macdonald, G. A., Abbot, A. T., and Peterson, F. L. (1983). *Volcanoes in the Sea: the Geology of Hawaii.* University of Hawaii Press, Honolulu. 517 pp.

McDowall, R. M. (1969). Extinction and endemism in New Zealand land birds. *Tuatara*, **17**, 1–12.

Maciolek, J. A. (1984). Exotic fishes in Hawaii and other islands of Oceania. In: Courtenay, W. R., and Stauffer, J. R. (Eds), *Distribution, Biology and Management of Exotic Fishes*, pp. 131–61. Johns Hopkins University Press, Baltimore, Maryland.

McKeown, S. (1978). *Hawaiian Reptiles and Amphibians.* The Oriental Publ. Co., Honolulu, Hawaii.

McMillan, M., and Rushforth, S. R. (1985). The diatom flora of a steam vent of Kilauea Crater, island of Hawaii. *Pacific Sci.*, **39**(3), 294–301.

McNaughton, S. J. (1985). Ecology of a grazing ecosystem: the Serengeti. *Ecol. Monogr.*, **55**(3), 259–94.

Magnusson, A. H. (1956). A catalogue of the Hawaiian lichens. *Arch. Bot. (Stockholm)*, **3**(10), 223–402.

Martin, P. S. (1984). Prehistoric overkill: the global model. In: Martin, P. S., and Klein, R. G. (Eds), *Quaternary Extinctions: A Prehistoric Revolution*, pp. 354–403. University of Arizona Press, Tucson, Arizona. 892 pp.

Mayr, E. (1963). *Animal Species and Evolution.* The Belknap Press of Harvard University Press, Cambridge, Mass. 797 pp.

Medeiros, A. C., Loope, L. L., and Cole, F. R. (1986). Distribution of ants and their effects on endemic biota of Haleakala and Hawaii Volcanoes National Parks: a preliminary assessment. In: Smith, C. W., and Stone, C. P. (Eds), *Proceedings of 6th Conference in Natural Sciences*, pp. 39–51. *Hawaii Volcanoes National Park.* Coop. Natl Park Resources Studies Unit, Department of Botany, University of Hawaii, Honolulu.

Melville, R. (1979). Endangered island floras. In: Bramwell, D. (Ed.), *Plants and Islands*, pp. 361–78. Academic Press. New York. 459 pp.

Miller, H. A. (1956). A phytogeographical study of Hawaiian Hepaticae. Ph.D. dissertation, Stanford University, Stanford, California. 123 pp.

Millot, J. (1972). In conclusion. In: Battistini, R., and Richard-Vindard, G. (Eds.), *Biogeography and Ecology of Madagascar*. Monographie Biologicae, Vol. 21. Junk, The Hague.

Montgomery, S. L. (1982). Biogeography of the moth genus *Eupithecia* in Oceania and the evolution of ambush predation in Hawaiian caterpillars (Lepidoptera: Geometridae). *Entomologia Generalis*, **8**(1), 27–34. Stuttgart and New York.

Moore, D. M. (1979). Origins of temperate island floras. In: Bramwell, D. (Ed.), *Plants and Islands*, pp. 69–85. Academic Press. New York. 459 pp.

Moulton, M. P., and Pimm, S. L. (1983). The introduced Hawaiian avifauna: biogeographic evidence for competition. *Amer. Nat.*, **121**, 669–90.

Moulton, M. P., and Pimm, S. L. (1986). Species introductions to Hawaii. In: Mooney, H. A., and Drake, J. (Eds), *Ecology of Biological Invasions of North America and Hawaii*, pp. 231–49. Springer-Verlag, New York.

Mountainspring, S., and Scott, J. M. (1985). Interspecific competition among Hawaiian forest birds. *Ecol. Monogr.*, **55**(2), 219–39.

Mueller-Dombois, D. (1981). Fire in tropical ecosystems. In: Mooney, H. A., Bonnicksen, T. M., Christensen, N. L., Lotan, J. E., and Reiners, W. A. (Eds), *Proceedings of Conference on Fire Regimes and Ecosystem Properties*, pp. 137–76. US Forest Service, Gen. Tech. Rept WO-26. 594 pp. Stroudsberg, Pennsylvania and Woode Hole, Massachusetts.

Mueller-Dombois, D. (1983). Canopy dieback and successional processes in Pacific forests. *Pacific Sci.*, **37**(4), 317–25.

Mueller-Dombois, D. (1985). Ohia dieback in Hawaii: 1984 synthesis and evaluation. *Pacific Sci.*, **39**, 150–70.

Mueller-Dombois, D., Bridges, K. W., and Carson, H. L. (Eds) (1981). *Island Ecosystems: Biological Organization in Selected Hawaiian Communities*. Hutchinson-Ross. 583 pp.

Nagata, K. M. (1985). Early plant introductions in Hawaii. *Hawaiian J. Hist.*, **19**, 35–61.

Naveh, Z. (1975). The evolutionary significance of fire in the Mediterranean region. *Vegetatio*, **29**, 199–208.

Olson, S. L. (1975). Paleornithology of St Helena Island, South Atlantic Ocean. *Smithsonian Contributions Zoology*, **23**, 1–43.

Olson, S. L. (1977). Additional notes on subfossil remains from Ascension Island. *Ibis*, **119**, 37–43.

Olson, S. L., and James, H. F. (1982). Prodromus of the fossil avifauna of the Hawaiian Islands. *Smithsonian Contributions to Zoology*, **365**, 1–59.

Perry, R. (1984). Juan Fernandez Islands: a unique botanical heritage. *Environ. Conserv.*, **11**(1), 72–6.

Petteys, E. Q. P., Burgan, R. E., and Nelson, R. E. (1975). *Ohia Forest Decline: its Spread and Severity in Hawaii*. Pac. SW For. Range Expt Sta. Res. Pap. PSW-105. USDA, Forest Service, Berkeley, California.

Pickett, S. T. A. (1976). Succession: an evolutionary interpretation. *Amer. Nat.*, **110**, 107–19.

Porter, D. M. (1984). Relationships of the Galapagos flora. *Biol. J. Linn. Soc.* (*London*), **21**, 243–51.

Pyle, R. L. (1983). Checklist of the birds of Hawaii. *Elepaio*, **44**, 47–58.

Raikow, R. J. (1976). The origin and evolution of the Hawaiian honeycreepers (Drepanididae). *The Living Bird*, **15**, 95–117.

Rauh, W. (1979). Problems of biological conservation in Madagascar. In: Bramwell, D. (Ed.), *Plants and Islands*, pp. 405–22. Academic Press, New York. 459 pp.
Ricklefs, R. F., and Cox, G. W. (1972). Taxon cycles in the West Indian avifauna. *Amer. Nat.*, **106**, 195–219.
Ricklefs, R. F., and Cox, G. W. (1978). Stage of taxon cycle, habitat distribution, and population density in the avifauna of the West Indies. *Amer. Nat.*, **112**, 875–95.
Rutherford, G. (Ed.) (1982). *The Physical Environment of the Faeroe Islands.* Monographic Biologicae, Vol. 46. Junk, The Hague. 148 pp.
St John, H. (1973). *List and Summary of the Flowering Plants in the Hawaiian Islands.* Pacific Tropical Bot. Garden, Memoir No. 1. Lawai, Kauai, Hawaii. 519 pp.
Sibley, C. G., and Ahlquist, J. E. (1982). The relationships of the Hawaiian honeycreepers (Drepaninini) as indicated by DNA-DNA hybridization. *Auk*, **99**, 130–40.
Simberloff, D. I. (1986). Introduced insects: a biogeographic and systematic perspective. In: Mooney, H. A., Drake, J. A. (Eds), *Ecology of Biological Invasions of North America and Hawaii*, pp. 3–26. Springer-Verlag, New York.
Simons, T. R. (1983). *Biology and Conservation of the Endangered Hawaiian Dark-rumped Petrel* (*Pterodroma phaeopygia sandwichensis*). University of Washington Coop. Natl Park Resources Studies Unit, Seattle. 311 pp.
Smathers, G. A., and Mueller-Dombois, D. (1974). *Invasion and Recovery of Vegetation after a Volcanic Eruption in Hawaii.* National Park Service Sci. Monogr. Series No. 5. 129 pp.
Spatz, G., and Mueller-Dombois, D. (1975). Succession pattern after pig digging in grassland communities on Mauna Loa, Hawaii. *Phytocoenologia*, **3**, 346–73.
Stemmermann, L. (1983). Ecological studies of Hawaiian *Metrosideros* in a successional context. *Pacific Sci.*, **37**(4), 361–73.
Stoddart, D. R. (1984). *Biogeography and Ecology of the Seychelles Islands.* Monographie Biologicae, Vol. 55. Junk, The Hague. 691 pp.
Stone, C. P., and Loope, L. L. (1987). Reducing negative effects of introduced animals on native biota in Hawaii: what is being done, what needs doing, and the role of national parks. *Environ. Conserv.*, **14**, 245–258.
Stuessy, T. F., Foland, K. A., Sutter, J. F., Sanders, R. W., and Silva, M. O. (1984). Botanical and geological significance of potassium-argon dates from the Juan Fernandez Islands. *Science*, **225**, 49–51.
Tomich, P. Q. (1986). *Mammals in Hawaii: A Synopsis and Notational Bibliography*, 2nd Edn, B.P. Bishop Museum Spec. Publ. 76. 375 pp.
Turner, B. L., and Harrison, P. D. (1981). Prehistoric raised-field agriculture in the Maya lowlands. *Science*, **213**, 399–405.
van der Werff, H. (1979). Conservation and vegetation of the Galapagos Islands. In Bramwell, D. (Ed.), *Plants and Islands*, pp. 391–404. Academic Press. New York. 459 pp.
van Riper, C., van Riper, S. G., Goff, M. L., and Laird, M. (1986). The epizootiology and ecological significance of malaria in Hawaiian landbirds. *Ecol. Monogr.*, **56**(4), 327–44.
Wace, N. M. (1976). Man and nature in the Tristan da Cunha islands. *IUCN Monograph*, **6**, 1–114.
Walker, L. R., Vitousek, P. M., Whiteaker, L. D., and Mueller-Dombois, D. (1986). The effect of an introduced nitrogen-fixer (*Myrica faya*) on primary succession on volcanic cinder. (Abstract). In: Smith, C. W., and Stone, C. P. (Eds), *Proceedings of 6th Conference in Natural Sciences, Hawaii Volcanoes National Park*, p. 98. Coop. Natl Park Resources Studies Unit, Department of Botany, University Hawaii at Manoa.
Wallace, G. D. (1985). *Vascular Plants of the Channel Islands of Southern California and Guadalupe Island, Baja California Mexico.* Natural History Museum of Los Angeles County, Contr. in Science, No. 365, pp. 1–136.

Williamson, M. (1981). *Island Populations.* Oxford University Press, Oxford. 286 pp.

Wilson, E. O. (1961). The nature of the taxon cycle in the Melanesian ant fauna. *Amer. Nat.*, **95**, 169–93.

Winner, W. E., and Mooney, H. A. (1985). Ecology of SO2 resistance. V. Effects of volcanic SO2 on native Hawaiian plants. *Oecologia* (*Berlin*), **66**, 387–93.

Wong, T. Y. Y., McInnis, D. O., Nishimoto, J. I., Ota, A. K., and Chang, V. C. S. (1984). Predation of the Mediterranean fruit fly (Diptera: Tephritidae) by the Argentine ant (Hymenoptera: Formicidae) in Hawaii. *J. Econ. Entomol.*, **77**, 1454–8.

Yang, S. H., and Patton, J. L. (1981). Genic variability and differentiation in Galapagos finches. *Auk*, **98**, 230–42.

Yoshida, L., and Smith, C. W. (1976). Two recent thalloid liverworts introduced to Hawaii. *Bull. Pacific Tropical Bot. Garden*, **6**, 18–20.

Zimmerman, E. C. (1948). *Insects of Hawaii, Vol. 1. Introduction.* University of Hawaii Press, Honolulu. 206 pp.

Zimmerman, E. C. (1978). *Insects of Hawaii, Vol. 9. Microlepidoptera, Parts I and II.* University of Hawaii Press, Honolulu. 1903 pp.

Biological Invasions: a Global Perspective
Edited by J. A. Drake *et al.*
© 1989 SCOPE. Published by John Wiley & Sons Ltd

CHAPTER 11

Ecosystem-level Processes and the Consequences of Biological Invasions

P. S. RAMAKRISHNAN and PETER M. VITOUSEK

11.1 INTRODUCTION

Numerous studies demonstrate that biological invasions by exotic species can alter the population dynamics and community structure of native ecosystems (Elton, 1958; Mooney and Drake, 1986). However, considerably less information is available for the ecosystem-level consequences of biological invasions. In this paper, we define ecosystem-level changes as those that alter the fluxes of water and energy or the cycling and loss of material. Commonly studied characteristics such as primary productivity, decomposition, secondary production, mineral nutrient availability, and hydrological balances would be included in this definition, as would the type and frequency of dsiturbance. Changes in these properties can alter the conditions of life for all of the organisms in an ecosystem.

Our emphasis on ecoystem-level characteristics of invasions is based in part on the need to evaluate the potential impacts of invasions on ecosystems from the point of view of managing or mitigating those effects. However, we believe that this emphasis also addresses fundamental concerns in ecosystem-level ecology. For example, there is a considerable debate on the functional significance of individual species in ecosystems. We believe that a demonstration of widespread ecosystem-level consequences of biological invasions would constitute an explicit demonstration that species make a difference on the ecosystem level, and would further suggest ways in which to integrate the often disparate approaches of population biology and ecosystem-level ecology.

Limited research on ecosystem-level consequences of biological invasions has been carried out, and some of the generalizations that can be drawn from that work are summarized by Vitousek (1986). Exotic animals, especially mammals, clearly alter ecosystems in many areas (see Batcheler, 1983; Singer *et al.*, 1984). Determining the ecosystem-level consequences of invasions by plants is more difficult, although some clear examples of major effects are well documented (cf. Mueller-Dombois, 1973; Thomas, 1981; Neill, 1983; Halvorson, in press). In part, this distinction between the ecosystem-level effects of invading plants and animals probably reflects a real difference—most exotic animals have a larger effect on native ecosystems than do most exotic plants (Vitousek, 1986).

One reason for this difference may be that invading plants most often occupy disturbed habitats, especially in sites altered by humans. It can be difficult to separate the ecosystem-level effects of biological invasions from those of the disturbance that created the invaded habitats. The consequences of biological invasions in disturbed sites and secondary succession represent at once the most difficult and the most important area of research. It is here that the effects of invasions are most easily confounded with those of the massive, prolonged, or novel disturbances which often form the invaded habitat, here that economic effects on humans are most important, and here that changes in ecosystem-level characteristics are most rapid even in the absence of biological invasions.

In this chapter, we will briefly review the ecosystem-level consequences of biological invasions into intact native ecosystems and into primary succession. We will then develop a detailed case study of the effects of biological invasion into secondary successions initiated by shifting cultivation.

11.2 INVASIONS OF INTACT ECOSYSTEMS

The generalization that biological invasions are primarily successful in disturbed habitats (Allan, 1936; Egler, 1942) is well supported as it applies to plants, but exotic animals frequently invade intact native ecosystems and cause significant changes in ecosystem-level properties of the areas invaded. Pigs probably provide the best example. They invade intact forest and savanna ecosystems in 12 states within the *United States* (Wood and Barrett, 1979), including the most diverse forest area in North America, the Great Smoky Mountains National Park in North Carolina and Tennessee. Wherever they are abundant, they alter soil structure, soil nutrient availability, and even nutrient losses in streamwater (Singer *et al.*, 1984; Vitousek, 1986).

Many other examples of the effects of exotic animals have been documented. Goats remove vegetative cover and cause increased soil erosion on many oceanic islands (see Stone *et al.*, in press). Even exotic insects are likely to be important. The ant *Pheidole megacephala* totally changed the lowland invertebrate fauna (including many of the pollinators and organisms on the decomposer food chain) of Hawaii about 100 years ago, and the Argentinian ant (*Iridomyrmex humilis*) is now doing so at higher elevations (Medeiros *et al.*, 1986). Similar processes are important in the South African fynbos (Bond and Slingsby, 1984). Ecosystem-level effects of these invasions have not yet been studied.

Invasions by exotic plants into intact native ecosystems are less common, but they do occur. For example, Eurasian phreatophytes of the genus *Tamarix* invade both natural water courses and reservoir margins in the arid southwestern United States (Robinson, 1969). Their transpiration is much more rapid than that of native communities, and they can convert marshes which support surface water during part of the year into wholly dry areas. In one case, removal of

Tamarix led to regeneration of a marsh (Neill, 1983). The floating aquatic plant *Salvinia molesta* can change parts of tropical river systems into thick masses of live and dead plant material. Exotic grasses in seasonal montane forests or shrublands can increase fire frequency or intensity, thereby altering ecosystems in a way that favors increased dominance by the grasses (Parsons, 1972; Smith, 1985; Christensen and Burrows, 1986). Finally, exotic trees can invade certain native shrublands and woodlands (Chilvers and Burdon, 1983, Macdonald *et al*., this volume), thereby altering rates of productivity and nutrient circulation.

Exotic plants can also interact with exotic animals in ways that facilitate invasions of disturbed habitats. Invasions of seemingly intact mature ecosystems by exotic plants often take place in close association with the activities of exotic animals (Merlin and Juvik, in press). For example, guavas (Psidium *guajava* and *Psidium cattleianum*) are common and serious invaders of oceanic islands worldwide. Their dispersal and success are closely tied to the movements and soil-disturbing activities of cattle and pigs. Similarly, feral pigs consume and disperse the exotic vine *Passiflora mollisima* in Hawaii, and then deposit the seeds within mounds of organic fertilizer in seedbeds cleared by the pigs' rooting activity. *Passiflora mollisima* then apparently affects patterns of mineral cycling in montane Hawaiian rainforests (Scowcroft, 1986).

Overall, detailed studies of the population biology and plant–animal interactions of that subset of invaders able to colonize intact native ecosystems would be most useful. To the extent that these species are able to alter ecosystem-level characteristics, they are likely to have impacts disproportionate to their numbers.

11.3 INVASIONS OF PRIMARY SUCCESSION

Primary succession involves the development of ecosystems on areas that are free of the influence of previous biotic communities (Clements, 1916). Primary succession can be initiated by glacial recession, eolian activity, river meanders, or vocanic activity; some human activities (such as strip mining) reproduce it quite closely.

The effects of biological invasions upon primary succession are relatively little-studied, in large part because primary succession does not occur over large areas at present. It is slower and often much simpler (fewer species in the early stages, more predictable patterns of soil-plant interaction) than secondary succession, however (Vitousek and Walker, 1987), so it may lend itself better to studies of the effects of biological invasions. For example, nitrogen availability is generally low early in primary succession (Walker and Syers, 1976): Robertson and Vitousek, 1981); Vitousek *et al*., 1983). The exotic symbiotic nitrogen fixer *Myrica faya* is now invading young volcanic substrates in Hawaii and substantially altering nitrogen availability there (Vitousek *et al*., 1987); it is likely to alter the course of soil development substantially. Australian species of *Acacia* may have a similar effect in South Africa (MacDonald *et al*., this volume).

Casuarina equisetiifolia invades beaches in south Florida and in many oceanic islands. Its effects on nitrogen availability have been not studied, but it significantly alters the form of shorelines and patterns of beach erosion (Macdonald *et al.*, this volume, LaRosa, in press).

11.4 INVASIONS OF SECONDARY SUCCESSION

Exotic species are prominent in disturbed sites throughout the world, on continents as well as islands and in the tropics as well as the temperate zone. The association between disturbed sites and invaders may be due to greater invasiveness on the part of successional species, or it may simply reflect the association between humans and disturbance which gives early successional species greater opportunities to invade. In either case, it is often difficult or impossible to determine how invaders have altered secondary succession because it is difficult to know what secondary succession was like before humans altered disturbance regimes and brought in exotic species.

Nevertheless, there are several documented cases in which biological invaders have altered ecosystem processes during secondary succession. The phenology of exotic *Andropogon* in some areas of Hawaii does not match the seasonal distribution of rainfall, and in consequence boggy conditions develop in invaded sites (Mueller-Dombois, 1973). The ice-plant *Mesembryanthemum crystallinum* invades degraded pastures in California and Australia (Vivrette and Muller, 1977; Kloot, 1983); once established it redistributes salt from throughout the rooting zone onto the soil surface, thereby interfering with the growth of other species and eventually increasing soil erosion (Halvorson, in press). In fact, invaders may alter ecosystem processes in secondary succession wherever they can obtain resources the natives cannot, or wherever they differ substantially from the natives in resource use efficiency (Vitousek, 1986)—but it is difficult to document the effects of many invasions for the reasons outlined above.

One type of disturbance which lends itself well to studies of the effects of invasion on secondary succession is shifting cultivation. It is practiced over large areas of the tropics, and in many of these areas it has been the major form of disturbance for millenia. The consequences of biological invasions in these relatively well-defined successional systems are therefore more understandable than is true in systems with novel disturbance regimes.

11.5 INVASIONS AND SHIFTING CULTIVATION—A CASE STUDY

In the humid tropics of Asia, Africa, and Latin America, shifting agriculture is a major land use and an important, frequent perturbation to forest ecosystems. Shifting agriculture involves clearing and burning forest vegetation, cropping for one to several years, and then abandoning the land for several to many years before another cycle of clearing and cultivation is initiated (Nye and Greenland,

1960; Walters, 1971; Ruthenberg, 1976; Ramakrishnan, 1985a). Today, exotic species along with natives are important components of early successional communities in the fallow phase of shifting cultivation in many areas. This section analyzes the consequences of invasion by exotic species, using the shifting agriculture systems in north eastern India as a case study.

11.5.1 Shifting agriculture and secondary succession

Slash and burn agriculture, locally known as 'jhum,' is extensively practiced by the tribal populations of the hill areas in the northeastern region of India. This practice consists of cutting down the vegetation at various stages of development on the hill slopes, allowing the slash to dry for a few months, burning it, and then cropping for 1 or 2 years. The average size of a plot varies from 1 to 2.5 hectares. The fallow period before the land is again cultivated is now short, around 4 or 5 years, but in the traditional system when population pressure was not so great the cycle was as long as 20–30 years (Ramakrishnan *et al.*, 1981). Longer cycles are occasionally encountered at present in more remote areas.

The soil of the region is highly leached because of average annual rainfall of 200 cm or more. The climate is seasonal, supporting a mixed subtropical humid forest at lower elevations (Singh and Ramakrishnan, 1982); high elevation sites have cooler climates with subtropical montane forests (Boojh and Ramakrishnan, 1983). Most of the rainfall occurs during the monsoon period, which is followed by a dry winter and a brief warm summer.

When a forest is cleared for cultivation, not only is its original vegetation destroyed, but the site is subject to continuing perturbations owing to fire, the introduction of crop species, weeding, hoeing, and other disturbances to the soil attendant on harvesting. These result in a progressive reduction in species diversity. Hence, early successional stages following the abandonment of cultivation contain few species, and the number of species increases gradually as secondary succession proceeds.

The pattern of secondary succession and the rapidity with which a forest community develops depend upon the degree of destruction of the prefarming vegetation and of its propagules in the soil. During the first few years of succession, when weeds of cultivated sites predominate, there is considerable variation in community composition according to the length of the agricultural cycle, weeding intensity, and the availability of seeds. For example, at least four different types of weed-dominated communities were observed at lower elevations of Meghalaya in northeast India (Toky and Ramakrishnan, 1983). Either the exotic species *Eupatorium odoratum* or *Mikania micrantha*, or the natives *Imperata cylindrica* or *Saccharum spontaneum* predominate under short agricultural cycles. At higher elevations of Meghalaya, however, the important exotic weed is *Eupatorium adenophorum*, which occurs with many natives such as *Imperata cylindrica* and *Pteridium aquilinum* (Mishra and Ramakrishnan, 1983c).

If the agricultural cycle is a short one of 4 to 5 years and is imposed continuously at the same site, succession is arrested at the weed stage (Toky and Ramakrishnan, 1983); Saxena and Ramakrishnan, 1984a). If the cycle is longer, however, vegetation changes rapidly from weed colonizers to bamboo (*Dendrocalamus hamiltonii*) and other shade-intolerant tree species at lower elevations. Bamboo itself is eliminated after about 25 years. A 50-year-old forest at the lower-elevation study site had broad-leaved tree species such as *Schima wallichii, Castanopsis indica, Shorea robusta, Millusa roxburghiana*, and *Artocarpus chaplasha* (Singh and Ramakrishnan, 1982). At higher elevations of Meghalaya the weeds are replaced by rapidly regenerating pine trees (*Pinus kesiya*) along with broad-leaved trees such as *Schima wallichii* (Mishra and Ramakrishnan, 1983c). A mature forest in the region would be a mixed broad-leaved forest dominated by species of *Quercus* and *Castanopsis*, such as now exists only in sacred groves (Boojh and Ramakrishnan, 1983).

11.5.2 Weed potential under shifting agriculture

As observed elsewhere, the species that form early successional communities are commonly found as weeds in agricultural plotts (Woodwell and Whittaker, 1968;

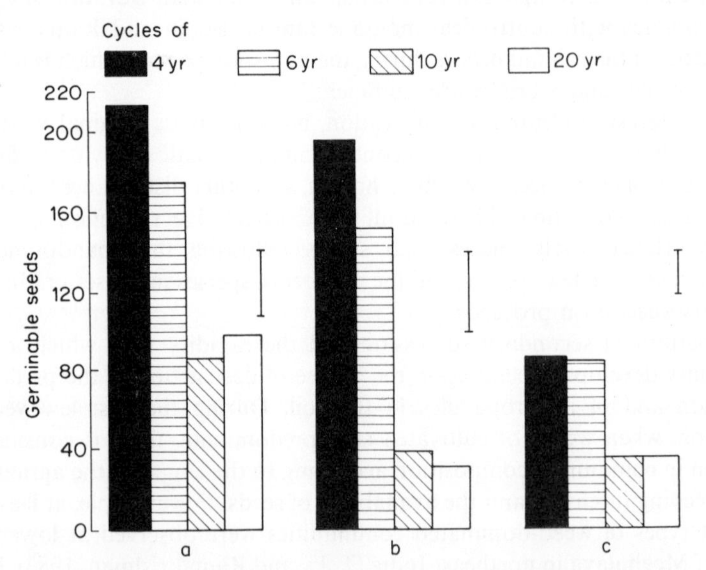

Figure 11.1. Germinable soil seed population of herbaceous species under different jhum cycles. Vertical lines represent least significant difference ($P = 0.01$): (A) just before slashing and burning; (B) uncropped sites; (C) cropped sites (from Saxena and Ramakrishnan, 1984a). Reproduced by permission of Blackwell Scientific Publications Ltd.

Perozzi and Bazzaz, 1978). Continuous imposition of short agricultural cycles has facilitated the establishment of herbaceous weeds characterized by high reproductive potentials, in contrast to later successional species such as *Oplismenus compositus, Centotheca lappacea*, and *Oryza granulata* with low reproductive potential.

The high reproductive potential of the early successional species leads to a high rate of propagule production during the fallow phase following cultivation. Also, short fallow cycles can only support a low intensity burn following land clearing, and short cycles may also be associated with comparatively lower losses of viability in buried seed banks. Overall, these processes cause significant differences in the germinable soil seed population under different agricultural cycles (Figure 11.1).

11.6 ADAPTIVE STRATEGIES OF NATIVES VERSUS EXOTICS: SYSTEM-LEVEL CONSEQUENCES

Several interacting factors contribute to invasion by exotic species and to the success of the exotics. Obviously, climatic and edaphic similarity between the original habitat of the exotics and their new habitat is one important factor (Holdgate, 1986). Humid tropical areas of Asia and Africa with highly leached soils are similar to the Latin American home of species such as *Eupatorium* spp. and *Mikania micrantha*, enabling them to invade and colonize appropriate sites on these two continents (Bennett and Rao, 1970; Nye and Greenland, 1960; Zinke *et al.*, 1978; Toky and Ramakrishnan, 1983; Mishra and Ramakrishnan, 1983c).

Under a shifting agricultural system, the initial perturbation by clearing and burning is further accentuated by cropping operations such as weeding and crop harvest (Toky and Ramakrishnan, 1981a; Mishra and Ramakrishnan, 1981). Tremendous losses of sediment and nutrients from steep slopes occur through runoff and infiltration water during the cropping phase (Toky and Ramakrishnan, 1981b; Mishra and Ramakrishnan, 1983a). Rapid uptake and reproduction which make efficient use of nutrients from these highly heterogeneous soils are critical to the establishment of successional vegetation, especially under the intense rain of the monsoon climate (Ramakrishnan, 1985b).

The exotic species which occupy sites after slash and burn have two sets of strategies which differ from those of early successional natives: (i) establishment largely through seeds (non-sprouting regeneration strategy) versus establishment largely through below-ground vegetative organs (sprouting strategy), and (ii) C_3 versus C_4 photosynthetic pathways.

11.6.1 Allocation and establishment

The non-sprouting regeneration strategy has been suggested to be successful under frequent perturbations through fire, whereas sprouting may be effective only under certain fire regimes (Keeley, 1981; Gill, 1975). Similar patterns have also been observed for shifting agriculture.

Exotics such as *Eupatorium* spp. have a high seed production potential (Kushwaha *et al.*, 1981; Ramakrishnan and Mishra, 1981) with a maximum seed production in a 3-year-old fallow. The survival of a species such as *Eupatorium odoratum* is favored by its rapid growth rate, rapid net assimilation, and allocation of many resources to the shoot system (Table 11.1). On the other hand, the slow growth of natives such as *Imperata cylindrica* and *Thysanolaena maxima* may result from initial allocation of energy and nutrients to below-ground parts, later followed by translocation of these resources for shoot growth during the early stages of regrowth (Figure 11.2).

Because a larger allocation of resources is made to below-ground organs in the sprouting species such as *Imperata cylindrica* and *Thysanolaena maxima*, they have less to allocate to sexual reproduction compared to a non-sprouting species (Keeley and Keeley, 1977; Saxena and Ramakrishnan, 1983). Thus, *Thysanolaena maxima* had much lower allocation of biomass and most nutrients to reproduction than *Eupatorium odoratum.* Both, however, allocated proportionately more nitrogen and phosphorus to reproduction in comparison with their allocation of energy. *Imperata cylindrica* did not flower in the first post-fire year. Clipping and burning promoted flowering (Kushwaha *et al.*, 1983), an observation also made by Schlippe (1956) for the effects of fire on this species. Gill (1975) reported similar observations on a number of successional shrubs in Australia.

While the success of exotic species such as *Eupatorium odoratum* and *E. adenophorum* depends upon heavy seed production (Kushwaha *et al.*, 1981; Ramakrishnan *et al.*, 1981), *Mikania micrantha* is an exotic that combines an effective seed-based reproduction with clonal propagation. *Mikania micrantha* also illustrates elegantly the importance of frequent perturbation such as fire (Swamy, 1986). This species responded to burning with increased birth and death

Table 11.1. Mean values (± standard error of mean) of growth functions for four different species

	Growth function		
Species	Relative growth rate ($mg\ mg^{-1}\ day^{-1}$)	Net assimilation rate ($mg(cm^2)^{-1}\ day^{-1}$)	Leaf area ratio ($cm^2\ mg^{-1}$)
Eupatorium odoratum	0.036 ± 0.015	0.302 ± 0.115	0.122 ± 0.031
Grewia elastica	0.008 ± 0.003	0.065 ± 0.014	0.121 ± 0.040
Imperata cylindrica	0.012 ± 0.004	0.133 ± 0.043	0.089 ± 0.016
Thysanolaena maxima	0.010 ± 0.004	0.108 ± 0.020	0.093 ± 0.018

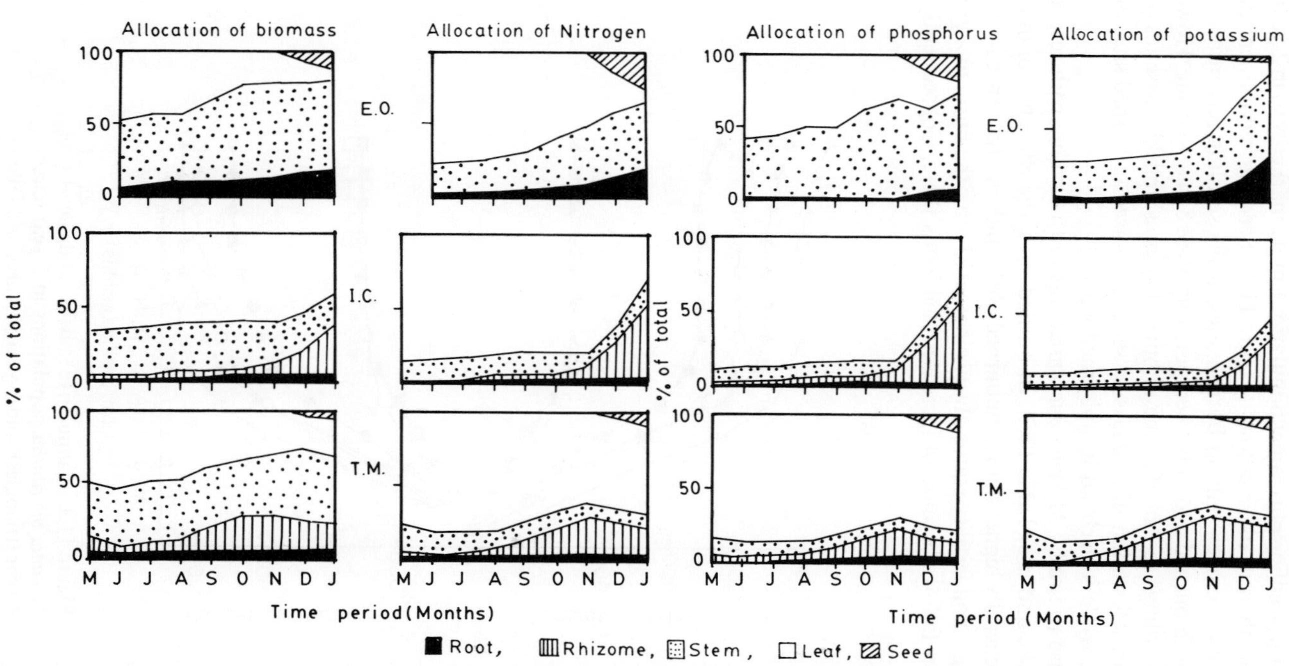

Figure 11.2. Allocation of biomass and nutrients to various compartments (expressed as a percentage of the total pool) during the growth of different species. E. O., *Eupatorium odoratum*; I.C., *Imperata cylindrica*; T.M., *Thysanolaena maxima*. (From Saxena and Ramakrishnan, 1983, reproduced by permission of the National Research Council of Canada)

rates of the total (seedling plus ramet population), more markedly so in 2-year-old and 4-year-old than in older fallows (Figure 11.3). Seedling recruitment in the younger fallows occurred only in burned sites, while seedling recruitment was observed in both burned and unburned portions of an 8-year-old site. However, none of the seedlings recruited in the unburned sites established rosettes in the following year. Allocation of biomass to seeds increased in burned sites of the older fallows, while it declined in unburned sites.

The response to fire by *Mikania micrantha* was most striking with respect to nutrient uptake and use (Table 11.2). In burned sites, the efficiency of nutrient uptake was generally higher than in unburned plots, and this difference increased with fallow age. Soil fertility status always increased after the burn, to a greater extent in older fallows because of increased fuel loads there. Consequently, the

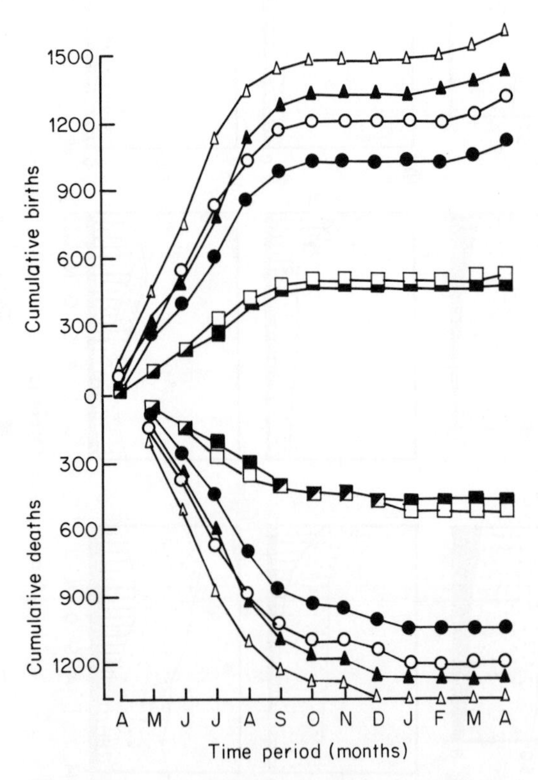

Figure 11.3. Cumulative births and deaths of *Mikania micrantha* populations in 2-year (circles), 4-year (triangles), and 8-year (squares) old fallows, after shifting agriculture. Burnt sites, closed symbols; unburnt sites, open symbols (from Swamy, 1986)

Table 11.2. Elemental uptake (mg nutrient absorbed per g root biomass) and use (mg dry matter produced per mg nutrient absorbed) efficiencies of *Mikania micrantha* in burnt and unburnt successional fallow plots after shifting agriculture. Values in parentheses are for nutrient use efficiencies (from Swamy, 1986)

Elemental efficiencies	Fallow age	Burnt	Unburnt
Nitrogen	2	108.1 (78.1)	85.2 (98.3)
	4	120.7 (75.6)	92.7 (100.4)
	8	128.0 (69.3)	45.1 (175.0)
LSD ($P = 0.05$)		10.6 (7.6)	7.3 (7.6)
Phosphorus			
	2	11.5 (699.3)	9.5 (895.7)
	4	13.6 (672.1)	9.3 (1005.8)
	8	14.5 (583.4)	4.4 (2029.9)
LSD ($P = 0.05$)		1.3 (24.9)	1.2 (157.0)
Potassium	2	133.8 (60.3)	114.6 (73.7)
	4	154.8 (56.5)	120.9 (77.1)
	8	166.1 (50.7)	81.2 (97.9)
LSD ($P = 0.05$)		14.6 (6.0)	11.0 (5.2)

LSD = least significant difference.

nutrient use efficiency in burned plots was generally lower than in unburned plots. Fire is an important process at various stages in the life cycle of this species, and the species itself is closely adapted to a ruderal environment subject to frequent perturbation (Swamy, 1986).

11.6.2 C_3/C_4 strategy and microdistribution

The C_3 species with large biomass contributions to secondary successional communities are all exotics, including the *Eupatorium* species and *Mikania micrantha*. On the otrher hand, all of the important native species, including *Imperata cyclindrica*, *Saccharum* spp., *Panicum* spp., and *Thysanolaena maxima*, have the C_4 photosynthetic pathway. Several less abundant C_3 species occur among the natives.

The biomass of C_3 species is higher than that of C_4 species in early succession (Saxena and Ramakrishnan, 1984b). The rapid growth of a species such as *Eupatorium odoratum* (Figure 11.4) is achieved partly through a larger light interception surface, and partly through its allocation of resources to the shoot system. Similar results were obtained for *Mikania micrantha* (Swamy, 1986). These results demonstrate that C_4 species are not always more productive than C_3 species (Caldwell, 1977); the productive potential of plants depends on factors in addition to photosynthetic rate per unit of leaf area (Black *et al.*, 1969; Black, 1971).

Continuous imposition of short agricultural cycles of 4 to 6 years results in early successional communities monopolized by exotic C_3 species (Saxena and Ramakrishnan, 1984b). In contrast, long agricultural cycles impoverish the soil seed bank owing to a more intense fire effect (Saxena and Ramakrishnan, 1984a),

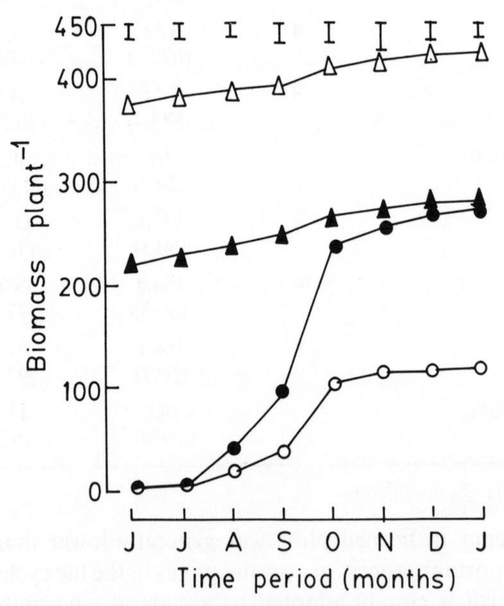

Figure 11.4. Growth curves of *Eupatorium odoratum* in different fallow fields. Recent fallow, after the burn and uncropped (●); soon after cropping (○); 2-year old (▲); 6-year old (△). Vertical bars represent LSD ($P = 0.05$) (from Saxena and Ramakrishnan, 1984a, reproduced by permission of Blackwell Scientific Publications Ltd.)

thereby affecting the seed-propagated C_3 species more than the vegetatively propagated C_4 species.

An important consequence of the different photosynthetic pathways is a temporal separation in peak photosynthetic production between the two groups, a difference that can contribute to their successful coexistence in some sites. Dry matter production by C_4 species occurred more efficiently during the early part of the growing season (April–June), when solar radiation was more intense and temperatures were higher. On the other hand, biomass accumulation by C_3 species was more pronounced during the latter part of the growing season (October–December) when solar, radiation and temperature were both less (Saxena and Ramakrishnan, 1981, 1984b), a situation when C_3 photosynthesis should be superior to C_4 (Ehleringer and Bjorkman, 1977).

Nitrogen is volatilized from the soil during fires, causing a decline in soil nitrogen status, and nitrification in the soil is inhibited in proportion to fire intensity (Saxena and Ramakrishnan, 1986). The high nutrient use efficiency of C_4 plants (Figure 11.5) may then allow them to grow more effectively during the early part of the growing season when nitrogen status is low. The input of detritus causes increased nutrient availability and nitrification later in the first growing season following burning, possibly favouring the growth of less nutrient efficient C_3 species (Ramakrishnan and Toky, 1981; Saxena and Ramakrishnan, 1986).

Under conditions of successive short agricultural cycles, nitrogen accumulation during the fallow period is inadequate to replace the amounts removed in harvest, volatilized during fires, or leached (Mishra and Ramakrishnan, 1984). Moreover, on steep slopes the loss of nutrients through leaching and erosion (Toky and Ramakrishnan, 1981b); Mishra and Ramakrishnan, 1983b) results in a heterogeneous distribution of soil nutrients after shifting cultivation (Ramakrishnan and Toky, 1981; Mishra and Ramakrishnan, 1983b). Under these conditions, the coexistence of C_3 and C_4 species is facilitated because C_3 species can colonize nutrient rich microsites while C_4 species occupy nutrient poor microsites (Saxena and Ramakrishnan, 1983, 1984b).

11.6.3 Nutrient conservation by exotic species

Exotic species contriburte a major fraction of the total biomass of early successional communities, and hence they may play a significant role in reducing losses of nutrients in run-off after cropped plots are abandoned (Table 11.3) (Toky and Ramakrishnan, 1981b; Ramakrishnan, 1984). The extent to which this function of exotic weeds would be replaced by natives if the former were not available is a moot point. However, with their C_3 photosynthetic pathway the exotics do have a systematically higher nutrient uptake (Swamy, 1986), and therefore they accumulate larger quantities of nutrients in plant tissue (Table 11.4). Under the present shortened agricultural cycles, it is likely that nutrient losses would be accelerated in the absence of exotic species.

The manner in which an exotic species can affect nutrient retention is

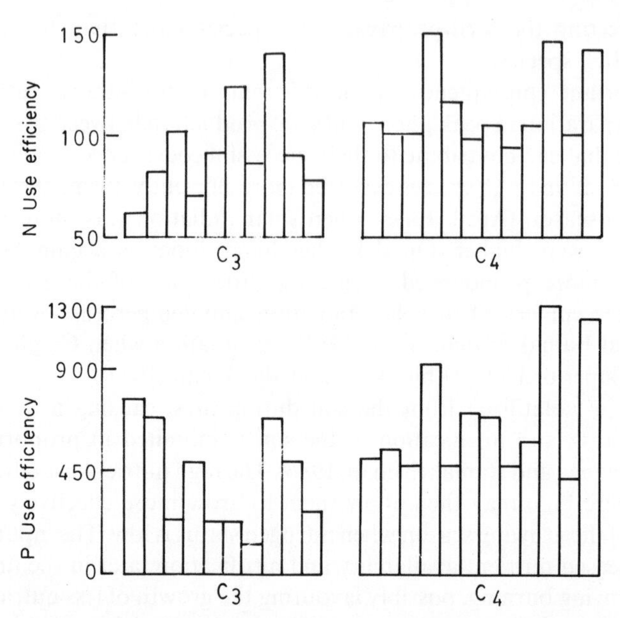

Figure 11.5. Nutrient use efficiency (expressed as mg dry matter production per mg nutrient absorbed) of different species coming after slash and burn. Different columns from left to right are: C_3 species: *Ageratum conyzoides*, *Boweria articularis*, *Cassia tora*, *Costus* sp., *Crossocephalum crepidioides*, *Erigeron linifolius*, *Eupatorium odoratum*, *Grewia elastica*, *Mimosa pudica*, *Mollugo stricta*; C_4 species: *Braclinaria distachya*, *Digitaria adscendens*, *Euphorbia hirta*, *Imperata cylindrica*, *Manisuris granularis*, *Panicum khasianum*, *P. maximum*, *Paspalidium punctatum*, *Rott boclia goalparer*, *Saccharum arundinaceum*, *Setaria palmifolia*, *Thysanoleana maxima*. Vertical bars represent standard error of the mean. (From Saxena and Ramakrishnan, 1984b.)

exemplified by the effect of *Mikania micrantha* on potassium cycling. Potassium is accumulated rapidly by vegetation up to 4 years into the fallow period—the enrichment quotient for potassium was the highest observed (Table 11.5). The bamboo species *Dendrocalamus hamiltonii* apparently functions in a similar way during the 10–25 year interval of longer fallows (Toky and Ramakrishnan, 1981b, 1983).

11.6.4 Weed extinction

As the shifting cultivation cycle lengthens, exotic weeds are suppressed during the natural progress of succession. Demographic analyses of invaders such as *Eupatorium odoratum* (Kushwaha *et al.*, 1981), *E. adenophorum* (Ramakrishnan

Table 11.3. Nutrient losses (kg ha^{-1} year^{-1}) through run-off and percolation water under a 10-year shifting agriculture cycle and a 5-year-old weed fallow in northeast India (from Toky and Ramakrishnan, 1981b)

	10-year cycle jhum plot		5-year fallow	
	Run-off	Percolation	Run-off	Percolation
Low elevation jhum:				
Nitrate nitrogen	4.2	10.7	0.8	1.1
Available phosphate	1.3	0.1	0.1	0.02
Potassium	91.2	21.2	0.9	0.5
High elevation jhum:				
Nitrate nitrogen	1.7	0.5	1.0	0.9
Available phosphate	0.9	0.1	ND	ND
Potassium	80.1	25.8	19.6	ND

ND = not detectable.

Table 11.4. Mean concentration of three nutrients ($\pm$ standard error of the mean) during the growing season in leaf tissue of C_3 versus C_4 species. (From Saxena and Ramakrishnan, 1983, reproduced by permission of the National Research Council of Canada)

Species	Nutrient concentration (%)		
	Nitrogen	Phosphorus	Potassium
Eupatorium odoratum	2.92 ± 0.12	0.28 ± 0.06	3.60 ± 0.32
Grewia elastica	1.70 ± 0.14	0.21 ± 0.04	1.32 ± 0.24
Imperata cylindrica	0.85 ± 0.05	0.11 ± 0.02	1.46 ± 0.24
Thysanolaena maxima	1.40 ± 0.09	0.19 ± 0.04	1.54 ± 0.20

Table 11.5. Enrichment quotients (element held in vegetation element uptake) for *Mikania micrantha* and the vegetation in successional fallows developed after shifting agriculture in northeast India. Values in parentheses are for total vegetation (from Swamy, 1986)

Years	N	P	K	Ca	Mg
0–1	0.85	0.97	1.00	0.85	0.88
	(1.14)	(1.25)	(1.32)	(1.08)	(1.17)
1–2	1.03	1.12	1.18	1.03	0.98
	(1.42)	(1.74)	(1.60)	(1.33)	(1.29)
2–4	1.19	1.30	1.61	1.22	1.07
	(2.12)	(2.67)	(2.74)	(1.90)	(2.41)
4–8	1.85	1.94	2.55	2.44	1.46
	(2.63)	(5.62)	(4.01)	(2.64)	(3.20)
8–12	1.93	2.19	2.81	2.47	1.52
	(3.08)	(4.64)	(6.15)	(3.58)	(3.34)

and Mishra, 1981), and *Mikania micrantha* (Swamy 1986) over a successional gradient suggest that they are eliminated after 5–6 years largely because of reduced light availability under the developing shrub and tree strata of the forest community (Kushwaha and Ramakrishnan, 1982). However, patterns of enregy and nutrient allocation could also contribute to their elimination. Whereas all individuals of *Eupatorium odoratum* in 1-year-old and 3-year-old fallows were fertile, only 50% of the plants in a 5-year-old fallow produced seeds and none did so in 10-year-old fallows.

11.7 CONCLUSION

We have summarized patterns of invasion, colonization, and extinction of exotic weeds following shifting agriculture in northeastern India. This example illustrates very well both the potential significance of biological invasions into secondary succession and the analytical difficulties inherent in demonstrating those effects.

Short agricultural cycles of 4–6 years are an important factor allowing the successful invasions that have largely occurred subsequent to the Second World War. Two species of *Eupatorium* and *Mikania micrantha* are now among the most aggressive weeds of the subcontinent. These species differ from most of the natives important early in secondary succession by having the C_3 photosynthetic pathway, reproduction primarily by seeds, and high nuitrient uptake. Thus, it is probable that invasion by these species has markedly altered myriad ecosystem properties, including productivity and nutrient budgets, in this succession. However, the success of these biological invaders is inextricably linked with a novel distrubance regime (short agricultural cycles), and it is difficult to separate the effects of the invaders *per se* those of that disturbance regime.

ACKNOWLEDGEMENTS

We thank F. di Castri, H. A. Mooney, D. Simberloff, and especially L. F. Huenneke for critical reviews of an earlier draft of this manuscript. PMV's contribution was supported by NSF Grant BSR-8415821 to Stanford University.

REFERENCES

Allan, H. H. (1986). Indigene versus alien in the New Zealand plant world. *Ecology*, **17**, 187–93.

Batcheler, C. L. (1983). The possum and rata-kamahi dieback in New Zealand: a review *Pacific Sci.*, **37**, 426.

Bennett, F. D., and Rao, V. P. (1978). Distribution of an introduced weed *Eupatorium odoratum* L. (compositae) in Asia and Africa and possibilities for its biological control. *Pest Abstracts and News Summaries, Section C*, **14**, 277–81.

Black, C. C. (1971). Ecological implications of dividing plants into groups with distinct photosynthetic production capabilities. *Adv. Ecol. Res.*, **7**, 87–144.

Black, C. C., Chen, T. M., and Brown, R. H. (1969). Biochemical basis for plant competition. *Weed Sci.*, **17**, 338–44.

Bond, W. J., and Slingsby, P. (1984). Collapse of an ant-plant mutualism: the Argentine ant *Iridomyrmex humilis* and *Myrmecochorous proteaceae*. *Ecology*, **65**, 1031–7.

Boojh, R. and Ramakrishnan, P. S. (1983). Sacred groves and their role in environmental conservation. In: *Strategies for Environmental management*, pp. 6–8. Souvenier Vol. Dept Sci. & Environ., Govt UP, Lucknow.

Caldwell, M. M., White, R. S., Moore, R. T. and Camp, L. B. (1977). Carbon balance, productivity and water use of cold winter desert shrub communities dominated by C_3 and C_4 species. *Oeclogia*, **29**, 275–300.

Chilvers, G. A., and Burdon, J. J. (1983). Further studies on a native Australian eucalypt forest invaded by exotic pines. *Oecologia*, **59**, 239–45.

Christensen, P. E., and Burrows, N. D. (1986). Fire: an old tool with a new use. In: Groves, R. H., and Birdon J. J. (Eds.), *Ecology of Biological Invasions: An Australian Perspective*, pp. 97–105. Australian Academy of Science, Canberra.

Clements, F. E. (1916). *Plant Succession: Analysis of Development of Vegetation*. Carnegie Inst. Wash. Publ., 242. 512 pp.

Egler, F. E. (1942). Indigene versus alien in the development of arid Hawaiian vegetation. *Ecology*, **23**, 14–23.

Ehleringer, J. R., and Bjorkman, O. (1977). Quantum yields for CO_2 uptake in C_3 and C_4 plants: dependence on temperature, CO_2 and O_2 concentration. *Plant Physiol.*, **59**, 86–92.

Elton, C. S. (1958). *The Ecology of Invasions of Animals and Plants*. Methuen, London. 181 pp.

Gill, M. (1975). Fire and the Australian flora: a review. *Austr. Forestry*, **38**, 4–25.

Halvorson, W. C. Alien plants at Channel Islands National Park. In: Stone, C. R., Smith C. W., and Tunison, J. T. (Eds.), *Alien Plant Invasions in Hawaii: Management and Research in Near-native Ecosystems*, Cooperative National Park Resources Study Unit. University of Hawaii, Honolulu (in press).

Holdgate, M. W. (1986). Summary and conclusions: characteristics and consequences of biological invasions. *Phil. Trans. R. Soc. Lond.* (in press).

Keeley, J. E. (1981). Reproductive cycles and fire regimes. In: Mooney, H. A., Bonnicksen, T. M., Christensen, N. L, Lotan, J. E., and Reiners, W. A. (Eds.), *Fire Regimes and Ecosystem Properties*, pp. 237–77. USDA For. Ser. Gen. Tech Rep.

Keeley, J. E., and Keeley, S. E. (1977). Energy allocation patterns of a sprouting and a non sprouting species of *Arctostaphylos* in the California chaparral. *Am. Midl. Nat.*, **98**, 1–10.

Kloot, P. M. (1983). The role of the common iceplant (*Mesembryanthemum crystallinum*) in the deterioration of medic pastures. *Aust. J. Ecol.*, **8**, 301–6.

Kushwaha, S. P. S., and Ramakrishnan, P. S. (1982). Observations on growth of *Eupatorium odoratum* L. and *Imperata cylindrica* (L.) Beauv. var major under different light and moisture regimes. *Proc. Indian Natn. Sci. Acad.*, **B48**, 689–93.

Kushwaha, S. P. S., Ramakrishnan, P. S., and Tripathi, R. S. (1981). Population dynamics of *Eupatorium odoratum* in successional environments following slash and burn agriculture. *J. Ecol.*, **18**, 529–35.

Kushwaha, S. P. S., Ramakrishnan, P. S., and Tripathy, R. S. (1983). Population dynamics of *Imperata cylindrica* (L.) Beauv. var. major related to slash and burn agriculture (jhum) in north eastern India. *Proc. Indian Acad. Sci. (Plant Sci.)*, **92**, 313–21.

LaRosa, A. M. Alien plant management in Everglades National Park: a historical perspective. In: Stone, C. P., Smith, C. W., and Tunison, J. T. (Eds.), Cooperative National Park Resources Study Unit University of Hawaii, Honolulu, (in press).

Medeiros, A. C., Loope, L. L., and Cole, F. R. (1986). Distribution of ants and their effects on endemic biota of Haleakala and Hawaii Volcanoes National Parks: a preliminary assessment. pp 39–51. In: Smith, C. W., and Stone, C. P. (Eds.), *Proceedings of Sixth Conference in Natural Sciences, Hawaii Volcanoes National Park*, Cooperative National Park Resources Study Unit, University of Hawaii, Honolulu.

Merlin, M. D., and Juvik, J. O. Relationships between native and alien plants on islands with and without wile ungulates. In: Stone, C. P., Smith, C. W., and Tunison, J. T. (Eds.), *Alien Plant Invasions in Hawaii: Management and Research in Near-native Ecosystems*, Resources Study Unit, University of Hawaii, Honolulu, (in press).

Mishra, B. K., and Ramakrishanan, P. S. (1981). The economic yield and energy efficiency of hill agro-ecosystem at higher elevations of Meghalaya in north-eastern India. *Acta Oecol./Oecol. Applic.*, **2**, 369–89.

Mishra, B. K., and Ramakrishnan, P. S. (1983a). Slash and burn agriculture at higher elevations in north-eastern India. I. Sediment. water and nutrient losses. *Agric. Ecosys. Environ.*, *9*, 69–82.

Mishra, B. K., and Ramakrishnan, P. S. (1983b). Slash and burn agriculture at higher elevations in north-eastern India. II. Soil fertility changes. Agric. Ecosys. Environ., **9**, 83–96.

Mishra, B. K., and Ramakrishnan, P. S. (1983c). Secondary succession subsequent to slash and burn agriculture at higher elevations of north-east India. I. Species diversity, biomass and litter production, *Acta Oecol./Oecol. Applic.*, **4**, 95–107.

Mishra, B. K., and Ramakrishnan, P. S. (1984). Nitrogen budget under rotational bush fallow agriculture (jhum) at higher elevations of Meghalaya in north-eastern India. *Plant Soil*, **81**, 37–46.

Mooney, H. A., and Drake, J. A. (Eds) (1986). *Biological Invasions of North America and Hawaii*. Springer-Verlag, New York.

Mueller-Dombois, D. (1973). A non-adapted vegetation interferes with water removal in a tropical rain forest area in Hawaii. *Trop Ecol. 14*, 1–18.

Neill, W. M. (1983). The tamarisk invasion of desert riparian areas. Educational Bulletin 83–4, Desert Protective Council, Spring Valley, California, 4. pp.

Nye, P. H., and Greenland, D. J. (1960). *The Soil under Shifting cultivation.* Tech. Comm. No. 51, Commonwealth Bureau of Soils, Harpenden. 156 pp.

Parsons, J. J. (1972). Spread of African pasture grasses to the American tropies. *J. Range Manage.*, **25**, 12–17.

Perozzi, R. E., and Bazzaz, F. A. (1978). The response of an early successional community to shortened growing season. *Oikos*, **31**, 89–93.

Ramakrishnan, P. S. (1984). The science behind rotational bush fallow agriculture systems (jhum). *Proc. Indian Acad. Sci.* (*Plant Sci.*), **93**, 379–400.

Ramakrishnan, P. S. (1985a). Tribal man in the humid tropics of the north-east. *Man in India*, *65*, 1–32.

Ramakrishnan, P. S. (1985b). *Research on Humid Tropical Forests* Regional Meeting, Natl MAB Comm. Central and South Asian Countries, India, New Delhi. 39 pp.

Ramakrishnan, P. S., and Mishra B. K. (1981). Population dynamics of *Eupatorium adenophorum* Spreng during secondary succession after slash and burn agriculture (jhum) in north-eastern India. *Weed Res.*, **22**, 77–84.

Ramakrishnan, P. S., and Toky, O. P. (1981). Soil nutrient status of hill agro-ecosystems and recovery pattern after slash and burn agriculture (jhum) in north-eastern India. *Plant Soil*, **60**, 41–64.

Ramakrishnan, P. S., Toky, O. P., Mishra, B. K., and Saxena, K. G. (1981). Slash and burn agriculture in north-eastern India. In: Mooney, H. A., Bonnicksen, T. M., Christensen, N. L., Lotan, J. E., and Reiners, W. A. (Eds.), *Fire Regimes and Ecosystem Properties*, pp. 570–86. USDA Forest Serv. Gen. Gen. Tech. Rep. WO–26, Honolulu, Hawaii.

Robertson, G. P., and Vitousek, P. M. (1981). Nitrification potentials in primary and secondary succession. *Ecology*, **62**, 376–86.

Robinson, R. W. (1969). *Introduction, Spread, and Aerial Extent of Saltcedar (Tamarix) in the Western States*, US Dept Interior Geol. Survey Prof. Paper 491–A. 12 pp.
Ruthenberg, H. (1976). *Farming System in the Tropics*, 2nd Edn. Clarendon Press, London, 366 pp.
Saxena, K. G., and Ramakrishanan, P. S. (1981). Growth strategy and allocation pattern of *Eupatorium odoratum* and *Imperata cylindrica* at different fertility levels of the soil. *Proc. Indian Natn. Sci. Acad.*, **B47**, 861–6.
Saxena, K. G., and Ramakrishnan, P. S. (1983). Growth and allocation strategies of some perennial weeds of slash and burn agriculture (jhum) in north eastern India. *Can. J. Bot.*, **61**, 1300–6.
Saxena, K. G., and Ramakrishnan, P. S. (1984a). Herbaceous vegetation development and weed potential in slash and burn agriculture (jhum) in north-eastern India. *Weed Res.*, **24**, 135–42.
Saxena, K. G., and Ramakrishnan, P. S. (1984b). C_3/C_4 species distribution among successional herbs following slash and burn in north-eastern India. *Acta Oecol./Oecol. Plant.*, **3**, 335–46.
Saxena, K. G., and Ramakrishnan, P. S. (1986). Nitrification during slash and burn agriculture (jhum) in north-eastern India. *Acta Oecol./Oecol. Plant.*, **7**, 319–31.
Schlippe, P. D. (1956). *Shifting Cultivation in Africa: The Zande System of Agriculture*, Routledge and Kegan Paul, London. 304 pp.
Scowcroft, P. G. (1986). Fine litterfall and leaf decomposition in a montane koa-ohia forest. In: Smith, C. W., and Stone, C. P. (Eds.), *Proceedings of Sixth Conference in Natural Sciences, Hawaii Volcanoes National Park* pp. 66–82. Cooperative National Park Resources Study Unit, University of Hawaii, Honolulu.
Singer, F. J., Swank, W. T., and Clebsch, E. E. C. (1984). Effects of wild pig rooting in a deciduous forest. *J. Wildlife Manage.*, **48**, 464–73.
Singh, J., and Ramakrishnan, P. S. (1982). Structure and function of a sub-tropical humid forest of Meghalaya. I. Vegetation, biomass and its nutrients. *Proc. Indian Acd. Sci. (Plant Sci.)*, **97**, 241–53.
Smith, C. W. (1985). Impact of alien plants on Hawaii native biota. In: Stone, C. P., and Scottt, J. M. (Eds.), *Hawaii Terrestrial Ecosystems; Preservation and Management*, pp. 180–250. Cooperative National Parks Study Unit, University of Hawaii, Honolulu.
Stone, C. P., Higashino, P. K., Tunison, J. T., Cuddihy, L. W., Anderson, S. J., Jacobi, J. D., Ohashi, T. J., and Loope, L. L. Success of alien plants after feral goat and pig removal. In: Stone, C. P., Smith, C. W., and Tunison, J. T. (Eds.), *Alien Plant Invasions in Hawaii: Management and Research in Near-native Ecosystems.* Cooperative National Park Resources Study Unit, University of Hawaii, Honolulu, (in press).
Swamy, P. S. (1986). Ecophysiological and Demographic Studies on Weeds of Successional Environments after Slash and Burn Agriculture in North-eastern India. Ph.D. Thesis, North-eastern Hill University, Shillong. 270 pp.
Thomas, K. J. (1981). The role of aquatic weeds in changing the pattern of ecosystems in Kerala. *Environ. Conserv.*, **8**, 63–6.
Toky, O. P., and Ramakrishnan, P. S. (1981a). Cropping and yields in agricultural systems of the north-eastern hill region of India. *Agro-Ecosystems*, **7**, 11–25.
Toky, O. P., and Ramakrishnan, P. S. (1981b). Run-off and infiltration losses related to shifting agriculture in north-eastern India. *Environ. Conserv.*, **8**, 313–21.
Toky, O. P., and Ramakrishnan, P. S. (1933). Secondary succession following slash and burn agriculture in north-eastern India. I. Biomass, litterfall and productivity. *J. Ecol.*, **71**, 737–45.
Vitousek, P. M. (1986). Biological invasions and ecosystem properties: can species make a difference? In: Mooney, H. A., and Drake, J. A. (Eds.), pp. 163–76. *Biological Invasions of North America and Hawaii*, Springer-Verlag, New York.

Vitousek, P. M. Effects of alien plants on native ecosystems. In: Stone, C. P., Smith, C. W., and Tunison, J. T. (Eds.), *Alien Plant Invasions in Hawaii: Management and Research in Near-native Ecosystems*, Cooperative National Park Resources Study Unit, University of Hawaii, Honolulu. (in press).

Vitousek, P. M. and Walker, L. R. (1987). Colonization, succession, and stability: ecosystem level interactions. In: Gray, A., Crawley, M., and Edwards, P. J. (Eds.), *Colonization, Succession and Stability*, Blackwell Scientific, Oxford.

Vitousek, P. M., Van Cleve, K., Balakrishnan, N., and Mueller-Dombois, D. (1983). Soil development and nitrogen turnover in montane, rainforest soils on Hawaii. *Biotropica*, **15**, 268–74.

Vitousek, P. M., Walker, L. R., Whittacker, L. D., Mueller-Dombois, D., and Matson, P. A. (1987). Biological invasion by *Myrica faga* alters ecosystem development in Hawaii, *Science*, **238**, 802–4.

Vivrette, N. J., and Muller, C. H. (1977). Mechanism of invasion and dominance of coastal grassland by *Mesembryanthemum crystallinum*. *Ecol. Monogr.*, **47**, 301–18.

Walker, T. W. and Syers, J. K. (1976). The fate of phosphorus during pedogenesis. *Geoderma*, **14**, 1–19.

Walters, H. (1971). *Ecology of Tropical and Sub-tropical Vegetation*. Oliver and Boyd, Edinburgh. 539 pp.

Wood, G. W., and Barrett, R. H. (1979). Status of wild pigs in the United States. *Wildlife Soc. Bull.*, **7**, 237–46.

Woodwell, G. M., and Whittaker, R. H. (1968). Primary production in terrestrial communities. Amer. Zool., **8**, 318–32.

Zinke, P. J., Sabhasri, S., and Kunstadter, P. (1978). Soil fertility aspects of the 'lua' forest fallow system of shifting cultivation. In: Kunstadter, P., Chapman, E. C. and Sabhasri, S. (Eds.), *Formers of the Forest*, pp. 134–59, East-West Centre, Honolulu, Hawaii.

Biological Invasions: a Global Perspective
Edited by J. A. Drake *et al.*
© 1989 SCOPE. Published by John Wiley & Sons Ltd

CHAPTER 12

Attributes of Invaders and the Invading Process: Terrestrial and Vascular Plants

IAN R. NOBLE

12.1 INTRODUCTION

In a summary paper on the consequences of biological invasions Holdgate (1986) concluded 'what is clear ... is that many species reach potential new habitats, but that establishment depends critically on habitat features and that success may have counterintuitive attributes'. Here I examine whether, by looking at the attributes of invaders and the invasion process, we can find ways of improving our understanding so that we may better predict which species are potential invasives.

Newsome and Noble (1986) described four major types of invasions. The first dealt with new habitats associated with human settlements. Human activities have radically changed most ecosystems and the spread of new agricultural and pastoral techniques and areas of habitation has created a range of new environments in many regions. We must expect that a group of fortuitously pre-adapted species will invade and dominate these environments. Much of the world's literature on invasions deals with this commensal flora and fauna and I do not want to emphasize this material here. Instead I will concentrate on the invasion of localities where little prior alteration has occurred.

A second type of invasion is associated with the filling of vacant niches (Lawton, 1984). I have not been able to find evidence that such vacancies occur in plant communities, or even of how they should be recognized (e.g. Weiss and Noble, 1984a), and thus they appear to be of little importance.

The remaining invasions occur in habitats where native species are displaced from relatively undisturbed communities. Newsome and Noble (1986) recognized two types here. In one the invading species has distinct competitive superiority over an ecologically similar native species and thus might be expected to become a permanent feature of the flora (without necessarily leading to the extinction of native species). The other occurs where the invasive species have characteristics that allow them to survive under only some conditions and not under other conditions (such as extreme events). The continued success of these risk-taking invaders is dependent on either the prevention of the extreme events by human actions (e.g. the prevention of fires in some native forests in Australia and the invasion by *Pinus radiata*), the existence of local refuges, or on continual re-invasion. It is these last two type of invasions that I will emphasize here.

12.2 THE IDEAL INVADER

There are many descriptions of the attributes of the ideal invader. Most reflect our concern with the invasion of agricultural crops and must be treated with caution if we are to retain the emphasis on natural communities that is the theme of this volume.

The best known description of invasive plants is Baker's (1965) description of the 'ideal weed'. In summary, an ideal weed is a plastic perennial which will germinate in a wide range of physical conditions, grow quickly, flower early, is self-compatible, produces many seeds which disperse widely, reproduces vegetatively and is a good competitor. However, as Baker points out, no one species is likely to possess all these characters; nor does a species need all these features to be a successful invader. Conversely, the possession of a single, or indeed several, characters from the list does not mean the species will be a successful invader.

Baker's list is over 20 years old and we might ask whether it can be revised to add more recent information about eco-physiological properties such as photosynthetic pathways. For example, C_4 grasses from Africa have been successful in displacing native grasses elsewhere (Baruch *et al.*, 1985). Fourteen out of the first 18 species listed in *The World's Worst Weeds* (Holm *et al.*, 1977) are C_4 species compared with only three out of 15 of the world's major crop plants (Harlan, 1975). C_4 species appear 17 times more frequently than expected in some lists of weeds (Elmore and Paul, 1983). However, we should be cautious about interpreting C_4 species as potentially invasive in all environments and especially in temperate regions (Pearcy and Ehleringer, 1984). Each characteristic must be assessed in relation to the environment subject to invasion. The prominence of C_4 species in weed lists may reflect the compilers' concerns with tropical crops where weed control practices tend to be less effective or rigorous.

We must also be cautious of drawing conclusions about the interactions of plant populations based on a few eco-physiological characteristics. For example, the South African Shrub *Chrysanthemoides monilifera* is displacing *Acacia longifolia* from coastal dunes in Australia. Counter to expectations, pot experiments show that *C. monilifera* has a lower per unit leaf area assimilation rate and lesser drought tolerance than *A. longifolia* (Weiss and Noble, 1984b) However, when grown in competition, seedlings of *C. monilifera* out-grow those of *A. longifolia* through a more effective arrangement of their photosynthetic tissue. This study and others such as that of Patterson *et al.* (1984) gives emphasis to the comments by Mooney and Chiariello (1984) and Ehleringer *et al.* (1986) on the necessity to move eco-physiological studies away from the single leaf and towards the whole plant in interaction with its environment. Only in this way will an integration of physiological and population biology be achieved.

Lists such as Baker's (1965) are of value as checklists of potential warning sings, but they have little predictive value about the likelihood that a particular species will be a problem in particular environment.

12.3 PLANT STRATEGIES

There has long been interest amongst ecologists in describing ecological strategies. King (1966) has summarized some of the schemes that have been applied to 'weeds'. One approach (e.g. Thellung, 1912, quoted in King, 1966) is based on the relationship between the weed and humans and emphasizes the method of introduction. Such schemes have little value as a predictive tool. By contrast, the classification by Korsmo (1930) is based on method of reproduction and is a precursor to some more recent classifications (e.g. Purdie and Slatyer, 1976).

The relationship between invasive properties and the *r*–*K* continuum of MacArthur and Wilson (1967) has been described by many. In fact, Baker's list of properties of the ideal weed are similar to the properties of a typical *r* strategy species. Southwood (1976) classified 'pest' species on this continuum and concluded that both *r* and *K* pests exist. Further, knowledge of an organism's position on the continuum provided not only a convenient framework on which to discuss 'pests' but also an indication as to which control strategies might be successful. However, since both *r* and *K* pests occur, the classification has little value in predicting whether a species is a potential invader.

One of the most general, and most widely applied, descriptions of ecological strategies is an extension of the McArthur and Wilson (1967) *r*–*K* continuum by Grime (1974, 1979). He argued that stress and disturbance are to two dominant factors in the environment of a plant and a species' reaction to these factors can be classified along three axes representing its degree of tolerance to stress, to competition and its tendency towards ruderal behaviour. Grime, suggested a methodology by which a competitive index and RGR_{max} (maximum relative growth rate) can be used to position species on a triangular graphical representation of the strategy. The position of a species on the triangle tends to convey some information about other properties that the species might have. Similarly, particular life forms tend to cluster together; thus the life form of a species gives some indication about the strategies the species might exhibit. Grime (1979) has pointed out that many colonists are ruderals by his classification. Although this point has been confirmed by many others it provides little predictive value. Just as Southwood (1976) recognized both *r* and *K* pests, some invaders have C (competitive) or S (stress tolerant) properties.

Milton (1979) classified the invasive Australian acacia species on South West Cape of South Africa on to Grime's triangle and concluded that the acacia were

C-strategists with some tendency towards S. The classification was based on their observed fast growth rates and tall erect habit compared with the native fynbos species. The acacias were able to displace the largely S-strategy native species of the fynbos by spreading along the less stressful water-courses and gradually reducing the degree of stress by nitrogen fixation and soil modification. They, therefore, created an environment that was less suited to the fynbos species and when subsequent disturbances (usually fires) occurred, the fynbos species did poorly in competition with the acacias and other invasive species.

In this example, Grime's strategies were a useful framework on which to describe the observed invasion of the fynbos by acacias. However, I doubt that they would have given much direct insight into the potential for acacias to invade if the classification had been known before the introduction of acacias into the fynbos. It is true that this is an unfair test of the overall utility of Grime's scheme. Grime (1982) defended his scheme against over-specialized application and he (1985) asserted that many ecologists were nervously 'putting from the tee' in avoiding generalizations. But, we must also avoid the pitfall of staying with schemes which are overly generalized for the task in hand (Grubb, 1985).

Thus, I have rejected two approaches to finding a general description of invasive organisms—namely the checklist of properties which are likely to contribute to the invasive success of an organism and the use of a framework based on broadly defined plant strategies. Is there a better classification of terrestrial plants which may help us recognize potentially invasive species?

12.4 FUNCTIONAL GROUPS

In the past few years there has been increasing interest in the concept of 'functional groups' in plant ecology. The ieda has sometimes been expressed indirectly (MacArthur, 1972, quoted in May 1986; Mooney *et al.*, 1977; Noble and Slatyer, 1980; Jain, 1983; Grime, 1985) or in different terminology such as the functional types of Huston and Smith (1987). The goal of a classification based on functional groups is to describe a set of physiological, reproductive and life history characteristics where variation in each characteristic has specific, ecologically predictive (rather than descriptive) value. It is argued that related groups of plants can be recognized because there are a limited number of sets of physiological, morphological and life history options that are feasible in dealing with the trade offs necessary to cope with the multiple requirements of survival and reproduction.

Botkin (1975) was apparently the first to use the term (although without a precise definition) when he defined the purpose of a functional grouping of organisms to be 'to reduce the analysis of ecosystems to tractable problems from the mathematical point of view, while still allowing consideration of the important population interactions'. If the functional groups concept is to be useful, the groups should be defined by the minimal set of characters necessary to distinguish between groups, otherwise the classification will become unwieldy. However, the set of characters should not be reduced to the point where the classification is over-generalized for the task in hand. Thus if we are to have functional group classification appropriate to dealing with invasive organisms, we need to establish the important combinations of population processes that determine the likelihood of successful invasion, i.e. an 'invasive syndrome'. Only with these insights will we be able to choose the set of characteristics which will define the functional groups.

Here I use a simple model of the process of invasion which gives some insights into the important population processes. Initially, I omit the process of dispersal and assume that some individuals of the invading species have reached the site. There are several reasons for this ommission. Most species with a potential for true long distance dispersal (i.e. unaided by humans) have already spread around the globe. Humans are now the dominant vector of plant dispersal and any attempt to classify the properties of a plant that increased its susceptibility to long distance dispersal is more likely to be based on the psychology of higher primates than the biology of plants. Short distance dispersal is an important property in determining the rate (e.g. Nip-van der Voort *et al*., 1979) and likelihood of spread (Forcella, 1985; Davis and Mooney, 1985) and I will discuss this below.

I also assume that the species is invading an area where the physical environment is similar to its native habitat. This assumption is supported both by general reviews of invasive organisms (e.g. Groves, 1986) and by specific case studies (e.g. Forcella and Wood, 1984; Milton, 1979). Kruger *et al*. (1986) use South African examples to show that pre-adaptation to the new habitat increases the chances of successful establishment, but does not guarantee it.

Thus, if we assume that the species is present, but rare, on a recently invaded site, the subsequent population dynamics can be described by the stage grouped Leslie matrix equation in Table 12.1 (cf. Sarukhan and Gadgil, 1974). Here, S is the number of seeds successfully incorporated in the seed pool or in potential establishment sites (i.e. where dispersal is complete), J is the number of successfully established but not yet reproductively mature individuals and A is the number of reproductively mature individuals. In the matrix, s is the rate of seed survival over a year (or growing season) and e the rate of establishment; j is the rate of juvenile survival, m is the rate of maturation of juveniles and a is the

Table 12.1. Stage grouped Leslie matrix of the simplified invasion process. The symbols are defined in the text

Current state: seed, juvenile, adult; rows (Next): Seed, Juvenile, Adult

$$\begin{bmatrix} s & 0 & r \\ e & j & 0 \\ 0 & m & a \end{bmatrix} \begin{bmatrix} S \\ J \\ A \end{bmatrix}_{t} = \begin{bmatrix} S \\ J \\ A \end{bmatrix}_{t+1}$$

survival rate (inverse of longevity) of the adults. The reproductive output of the adults, measured as seeds incorporated in the seed pool is r and it can be considered to be made up of two terms, f, the flowering (or fruiting) effort and, v, which is a measure of all reproductive losses between flowering and the seed pool.

If the population is assumed to be in approximate equilibrium in its native habitat (i.e. its intrinsic rate of increase is approximately 1.0 and a stable age structure exists), some simple matrix algebra shows that the following relationship holds:

$$fvem/[(1-s)(1-j)(1-a)] = 1$$

Of these parameters, s, a and f are largely genetically determined, physiological properties of the species and are unlikely to change dramatically if the species invades a site with a similar physical environment. The survival of the juveniles is also largely genetically fixed, unless they have a prolonged juvenile stage such as the lignotuberous seedling stage of the eucalypts. The remaining parameters, v, e, and m, obviously have some genetic component, but are much more affected by the biotic and abiotic environment. The probability of survival from flowering to seed pool (v) is largely determined by seed predation, while the probability of establishing (e) and of reaching maturity (m) are determined to a great degree by competition for limiting resources, i.e. site dependent factors. Thus, if on dispersing to a new site of similar abiotic conditions, any one of v, e or m changes, f is unlikely to be appropriate to maintain approximately equilibrium populations. In particular, if any one of v, e or m increases then an f which was appropriate for the native conditions will now be excessive and population growth rapid. If this is coupled with even limited powers of local dispersal, the species may be a successful invader.

This simple model, therefore, gives us some insight into the properties of the species which may contribute to making it a successful invader. The first clue is a large flowering or fruiting effort (f) since this implies that in the native habitat there must be heavy losses between this point and the establishment of a replacement adult. This would probably have been a more direct warning device than Grime's strategies to the early settlers of South Africa when they introduced the prolifically flowering and fruiting Australian acacias. It should also have been a warning to the soil conservation services along the east coast of Australia when they duplicated the mistake by introducing *Chrysanthemoides monilifera* from South Africa as dune stabilizers. Table 12.2 summarizes the results of Weiss and Milton (1984) which show a dramatic increase in v for both species in their introduced habitats leading to the establishment of extraordinarily large seed pools. In terms of direct competition (i.e. e or m) there is evidence (described above) that *C. monilifera* can out-compete *A. longifolia*, but this effect is apparently minor in comparison with the changes in v and the resultant increase in the size of the seed pools.

A change in e, or m, is an important factor in the expansion of both the range

Table 12.2. Reproductive performance of *Chrysanthemoides monilifera* (native to South Africa) and *Acacia longifolia* (native to Australia). All units are number per m^2. (Adapted from Weiss and Milton, 1984.)

	Chrysanthemoides monilifera	*Acacia longifolia*
South Africa		
Fruits	3800	?
Ripe seeds	2200	2900
Soil seeds	2300	7600
Viable soil seeds	50	7400
Australia		
Fruits	6700	600
Ripe seeds	4500	400
Soil seeds	2500	10
Viable soil seeds	2000	10

and density of some native species. An example is the increase in density and apparent expansion in range of native, woody weed species under grazing by domestic herbivores. For example, *E. mitchellii* is a 3 to 9 m tall shrub found in eastern Australia which has dramatically increased in density in what was previously an open grassy woodland (Hodgkinson and Beeston, 1982). It is a long-lived species with a life span of 50 to 100 years and once established is resistant to grazing and resprouts quickly after fire. It establishes only rarely (about three times this century in most regions) after a sequence of rainfall events which stimulate flowering, seed set and germination. The seedlings are slow growing and poor competitors with grasses. *E. mitchellii* appears to have gained a double advantage under current land management practices. First, in areas where grazing has reduced grass densities, higher than normal establishment rates (e) occur. Secondly, pastoral management has suppressed the fires that would normally have followed such a run of good seasons with the result that many more of the fire-sensitive juvenile plants reach maturity (i.e. increased m).

Pulsed flowering on establishment, as shown by *E. mitchellii*, is another warning sign of a potential invader. This strategy effectively increase the value of f at any one recruitment event and usually implies that quite specific conditions are required to produce the mortality of fruit, seed and plants that leads to stable populations. A change in the biotic or abiotic environment may either increase the frequency with which the pulses occur or reduce the effectiveness of the mortality filters.

The term f is relatively insensitive to seed and adult longevities (s and a) amongst perennials and even a change from an annual to a biennial life history can change f by a factor of only about 2. Competitive effects from other species which either increase or decrease the longevity of an established plant will usually result in a relatively small change in the term $(1 - a)$ especially in perennial plants. Thus the perenniality of the plant is little evidence of its invasive potential in stable environments. This last qualification is extremely important since the ability of the species to persist through adverse environmental conditions is as much funciton of its ability to disperse in time as its ability to disperse in space (Comins and Noble, 1985). Also, long-lived seed pools (i.e. high s) will often lead to problems of species eradication, but this is a separate issue from the problem of invasion itself.

Dispersal can be incorporated in the model. Net dispersal to or from the site will enter as an additive term in r but in most cases will be small with respect to local seed production and hence unimportant. Dispersal powers are important where new sites are being invaded or where site variability is such that the probability of local extinction is significant (Platt, 1975; Gleadow, 1982; Hobbs and Mooney, 1985; Marks and Mohler, 1985; McClanahan, 1986). Thus, good, short-distance dispersal abilities increase the possibility that a species will be a successful invader and species which are not reliant on special vectors (e.g. specific animals) have a greater chance of retaining those powers in the new environment. However, the problem of trade offs must be considered or else we will start to redefine the 'ideal weed'. For example, Morse and Schmitt (1985) have shown that in the wind-dispersed seeds of *Asclepias syrica*, heavy seeds disperse shorter distances but have greater germination success and faster initial seedling growth. Similarly Stanton (1985) has shown that increases in seed size in wild radish (*Raphanus raphanistrum*) give disproportionate increases in total reproductive output from the plants arising from the seeds. Therefore, the trade offs between seed numbers, sizes and dispersal powers must always be considered.

I have discussed invasion using a simple model of population dynamics. The model has several omissions. It deals with only a limited number of life stages, although the addition of extra stages makes little difference since the effect is to split the individual terms in this model into several. It deals with only the initial stages of invasion (i.e. before the invader starts to affect the environment itself), but this is justified since this is the critical stage if we are trying to prevent an invasion. The model also assumes stable population age structures, but simulations show that these will develop quickly even when the population establishes from a single propagule. More importantly, it is deterministic and therefore under-estimates the importance of the s and a terms as described above.

In summary, the model directs our attention to several factors in determining the likely invasive potential of a terrestrial plant. First, high population numbers at any life stage in the native environment followed by high mortalities should be seen as a warning. These high numbers do not have to be produced every year, but can occur as pulses. Secondly, adult or seed longevity are not reliable indicators of ivasive potential except in so far as they will allow the species to persist in variable environments over periods which are unsuitable for establishment. Thirdly, the model emphasizes that it is the invaded environment, as much as the properties of the invading species itself, that determine invasive success.

I started this section by asking whether we might distinguish functional groups of plants that give insight into whether a species is a potential invader. There appear to be particular population parameters which may form a basis for defining groups but a direct test of this approach has not been tried. Some estimate of the feasibility of deriving such groups can be gained by looking at the invasive flora of a particular environment to see if groups of species with correlated sets of properties can be recognized and if the sets of properties appear to be consistent with particular strategies of invasion and survival. Newsome and Noble (1986) found that such groups could be recognized in the noxious weed species listed by Parsons (1973) for Victoria, Australia.

The evidence in this paper and several other reviews (e.g. Crawley, 1986; Lawton, 1986) emphasize the importance of the invaded environment in determining the success of an invader. Even if we assume that the abiotic properties of the invaded environment are similar to the originating environment, biotic differences will still occur. Thus the groups may have to be formulated with respect to particular environments (see Jain, 1983) and it is unlikely that we will be able to make *a priori* predictions of whether a particular species will be a successful invader. Further development of the functional groups approach may help us identify those species which present the greatest risk, however, we need additional insight into the factors determining community composition structure.

Models of community structure and composition exist (see the review by Williamson, this volume, and May, 1986). The emphasis of these models is on broad questions of how many species we might expect in a community and questions of size distributions and trophic interactions. Even if answers were available they would be of limited help in determining whether a particular species might successfully invade a community. Many of the models of community structure are based on assumptions of equilibrium conditions, but an invader need be successful for only a limited time to become 'pest' or 'weed'. The models also deal with whether species with particular broadly defined properties may be successfully inserted in a community, but we have few guidelines on whether the properties of a species will change in a new biotic environment.

Clearly we need better understanding of the way in which important characteristics of a species will change in a new environment. Crawley (1986) was pessimistic about the possibility of doing this and concluded that we would be unlikely to achieve genuine prediction until we develop models of species interactions in which competition coefficients are context specific. Austin and his colleagues (Austin and Austin, 1980; Austin 1982; Austin *et al.*, 1985) have pursued the question of predicting the performance of plant species in multi-species mixtures along environmental gradients based on their performance in isolation. Their results indicate that the performance of a species in a mixture can be predicted from a measure of its performance in isolation in comparison with similar measures of performance in isolation of the other species in the mixture. This is promising, but the predictions are dependent on the position on the gradient, i.e. they are sensitive to the environmental context; thus the performance of a potential invader can be assessed only by trials carried out in the target environment.

The above conclusions are not encouraging and we have yet to take into account the other factors affecting the success of a new introduction to a community, including the effect of chance in community assembly (Lawton, 1986), of herbivores (Crawley, 1983) and of iteractions with other non-plant species (mutualisms). Nevertheless, our best approach appears to be to look for 'syndromes' of invasiveness. Lists of ideal properties and broadly based plant strategies do not seem to be sufficient. The functional groups approach does appear to hold some promise in assisting in recognizing species with the syndrome.

12.5 SUMMARY AND CONCLUSIONS

I conclude with the following points.

1. Highly generalized classifications or lists of preferred characteristics are of little help in recognizing potentially invasive organisms.
2. Groups of species with correlated sets of ecological and physiological characteristics (functional groups) do seem to occur. These groups probably represent particular trade offs in the allocation of plant resources.
3. The absence of an inherent long distance (e.g. trans-oceanic) dispersal mechanism is not a hindrance to invasion as humans are now the main vector.
4. Short distance dispersal mechanisms will increase both the probability and rate of invasion.
5. Species with high reproductive output (even if massively unsuccessful) in their native habitat have a high invasive potential.
6. The properties of the invaded habitat are a critical determinant of the likely success of any invader.
7. Thus, any system to assist in the recognition of potentially invasive species will require both studies of the species in its native habitat to assess its ecological characteristics and studies in the target habitat to assess the changes likely to arise in those characteristics.

REFERENCES

Austin, M. P. (1982). Use of a relative physiological performance value in the prediction of performance in multispecies mixtures from monoculture performance. *J. Ecol.*, **70**, 559–70.

Austin, M. P., and Austin B. O. (1980). Behaviour of experimental plant communities along a nutrient gradient. *J. Ecol.*, **68**, 891–918.

Austin, M. P., Groves, R. H., Fresco, L. M. F., and Kaye, P. E. (1985). Relative growth of six thistle species along a nutrient gradient with multispecies competition. *J. Ecol.*, **73**, 667–84.

Baker, H. G. (1965). Characteristics and modes of origin of weeds. In: Baker, H. G., and Stebbins, C. L. (Eds), *The Genetics of Colonizing Species*, pp. 147–69. Academic Press, New York.

Baruch, Z., Ludlow, M. M., and Davis, R. (1985). Photosynthetic responses of native and introduced C4 grasses from Venezuelan savannas. *Oecologia.*, **67**, 388–93.

Botkin, D. B. (1975). Functional groups of organisms in model ecosystems. In: Levin, S. A. (Ed.), *Ecosystem Analysis and Prediction*, pp. 98–102. Society for Industrial and Applied Mathematics, Philadelphia.

Comins, H. N., and Noble, I. R. (1985). Dispersal, variability, and transient niches: species coexistence in a uniformly variable environment. *Amer. Nat.*, **126**, 706–23.

Crawley, M. J. (1983). *Herbivory: The Dynamics of Animal Plant Interactions.* Blackwell Scientific Publications, Oxford.

Crawley, M. J. (1987). What makes a community invasible? In: Gray, A. J., Crawley, M. J., and Edwards, P. J. (Eds), *Colonization, Successions and Stability*, pp. 429–53. Blackwell Scientific Publications, Oxford.

Davis, S. D., and Mooney, H. A. (1985). Comparative water relations of adjacent California shrub and grassland communities. *Oecologia*, **66**, 522–9.

Ehleringer, J. R., Pearcy R. W., and Mooney, H. A. (1986). Recommendations of the workshop on the future development of plant physiological ecology. *Bull. Ecol. Soc. Amer.*, **67**, 48–58.

Elmore, C. D., and Paul, R. N. (1983). Composite list of C4 weeds. *Weed Sci.*, **31**, 686–92.

Forcella, F. (1985). Final distribution is related to rate of spread in alien weeds. *Weed Res.*, **25**(3), 181–92.

Forcella, F., and Wood, J. T. (1984). Colonization potentials of alien weeds are related to their native distributions—implications for plant quarantine. *J. Aust. Inst. Agric. Sci.*, **50**, 35–41.

Gleadow, R. M. (1982). Invasion by *Pittosporum undulatum* of the forests of central Victoria. 2. Dispersal, germination and establishment. *Aust. J. Bot.*, **30**(2), 185–98.

Grime, J. P. (1974). Vegetation classification by reference to strategies. *Nature*, **250**, 26–31.

Grime, J. P. (1979). *Plant Strategies and Vegetation Processes.* Wiley, Chichester.

Grime, J. P. (1982). The concept of strategies: use and abuse. *J. Ecol.*, **70**, 863–5.

Grime, J. P. (1985). Towards a functional description of vegetation. In: White, J. (Ed.), *The Population Structure of Vegetation*, pp. 501–54. Dr W. Junk, Dordrecht.

Groves, R. H. (1986). Plant invasions of Australia: an overview. In: Groves, R. H., and Burdon, J. J. (Eds), *Ecology of Biological Invasions: An Australian Perspective*, pp. 137–49. Australian Academy of Science, Canberra.

Grubb, P. J. (1985). Plant populations and vegetation in relation to habitat, disturbance and competition: problems of generalization. In: White, J. (Ed.), *The Population Structure of Vegetation*, pp. 595–620. Dr W. Junk, Dordrecht.

Harlan, J. (1975). *Crops and Man.* American Society of Agronomy, Madison, Wisconsin.

Hobbs, R. J., and Mooney, H. A. (1985). Community and population dynamics of serpentine grassland annuals in relation to gopher disturbance. *Oecologia*, **67**, 343–51.

Hodgkinson, K. C., and Beeston, G. R. (1982). The biology of Australian weeds 10. Eremophila mitchelli Benth. *J. Aust. Inst. Agric. Sci.*, **48**(4), 200–8.

Holdgate, M. W. (1986). Summary and conclusions: characteristics and consequences of biological invasions. *Phil. Trans. R. Soc. Lond. B* (in press).

Holm, L. G., Plucknett, D. L. Pancho, J. V., and Herberger, J. P. (1977). *The World's Worst Weeds: Distribution and Biology.* University Press, Hawaii.

Huston, M., and Smith, T. (1987). Plant succession: Life history and competition. *Amer. Nat.*, **130**, 168–98.

Jain, S. (1983). Genetic characteristics of populations. In: Mooney, H. A., and Godron, M. (Eds), *Disturbance and Ecosystems: Components of Response*, pp. 240–58. Springer-Verlag, Berlin.

King, L. J. (1986). *Weeds of the World: Biology and Control.* Interscience Publishers, New York.

Korsmo, E. (1930). Unkrauter in Ackerbau der Neuzeit. *Biologische und Practische Untersuchungen.* J. Springer, Berlin.

Kruger, F. J., Richardson, D. M., and Van Wilgen, B. W. (in press). Processes of invasion by alien plants. *South African Synthesis Volume.*

Lawton, J. H. (1984). Non-competitive populations, non-convergent communities, and vacant niches: The herbivores of bracken. In: Strong, D. R., Simberloff, D., Abele, L. G., and Thistle, A. B. (Eds), *Ecology Communities: conceptual Issues and the Evidence.* Princeton University Press, Princeton.

Lawton, R. H. (1986). Are there assembly rules for successional communities? In: Gray, A. J., Grawley, M. J., and Edwards, P. J. (Eds), *Colonization, Succession and Stability*, pp. 225–44. Blackwell Scientific Publishers, Oxford.

MacArthur, R. H. (1972). Coexistence of species. In: Behnke, J. (Ed.), *Challenging Biological Problems*, pp. 253–9. Oxford University Press, Oxford.

MacArthur, R. H., and Wilson, E. O. (1967). *The Theory of Island Biogeography.* Princeton University Press, Princeton.

McClanahan, T. R. (1986). The effect of a seed source on primary succession in a forest ecosystem. *Vegetatio*, **65**, 175–8.

Marks, P. L., and Mohler, C. L. (1985). Succession after elimination of buried seeds from a recently plowed field. *Bull. Torrey Bot. Club*, **112**, 376–82.

May, R. M. (1986). The search for patterns in the balance of nature: Advances and retreats. *Ecology*, **67**, 1115–26.

Milton, S. J. (1979). Australian acacias in the S. W. Cape: pre-adaptation, predation and success. In: Neser, S., and Cairns, A. L. P. (Eds), *Proceedings of Third National Weeds Conference of South Africa*, pp. 69–78. A. A. Balkema, Cape Town.

Mooney, H. A., and Chiariello, N. R. (1984). The study of plant function—the plant as a balanced system. In: Dirzo, R., and Sarukhan, J. (Eds), *Perspectives on Plant Population Ecology*, pp. 305–23. Sinauer Associates, MA, USA.

Mooney, H. A., Kummerow, J., Johnson, A. W., Parsons D. J., Keeley, S., Hoffman, A., Hays, R. I., Gilberto, J., and Chu, C., (1977). The producers—their resources and adaptive responses. In: Mooney, H. A. (Ed.), *Convergent Evolution in Chile and California: Mediterranean Climate Ecosystems*, pp. 85–143. Dowden, Hutchinson and Ross, Stroudsburg.

Morse, D. H., and Schmitt, J. (1985). Propagule size, dispersal ability, and seedling performance in *Asclepias syriaca. Oecologia*, **67**, 327–9.

Newsome, A. E., and Noble, I. R. (1986). Ecological and physiological characters of invading species. In: Groves, R. H., and Burden, J. J. (Eds), *Ecology of Biological Invasions: An Australian Perspective*, pp. 1–20. Australian Academy of Science, Canberra.

Nip-van der Voort, J., Hengeveld, R., and Haeck, J. (1979). Immigration rates of plant species in three Dutch polders. *J. Biogeog.*, **5**, 301–8.

Noble, I. R., and Slatyer, R. O. (1980). The use of vital attributes to predict successional changes in plant communities subject to recurrent disturbances. *Vegetatio*, **43**, 5–21.

Parsons, W. T. (1973). *Noxious Weeds of Victoria.* Inkata Press, Melbourne.

Patterson, D. T., Flint, E. P., and Beyers, J. L. (1984). Effects of CO_2 enrichment on competition between a C4 weed and a C3 crop. *Weed Sci.*, **32**, 101–5.

Pearcy, R. W., and Ehleringer, J. (1984). Comparative eco-physiology of C_3 and C_4 plants. *Plant Cell Environ.*, **7**, 1–13.

Platt, W. J. (1975). The colonization and formation of equilibrium plant species associations on badger disturbances in a tall-grass prarie. *Ecol. Monogr.*, **45**, 285–305.

Purdie, R. W., and Slatyer, R. O. (1976). Vegetation succession after fire in sclerophyll woodland communities in south-eastern Australia. *Aust. J. Ecol.*, **3**, 223–36.

Sarukhan, J., and Gadgil, M. (1974). Studies on plant demography: *Ranunculus repens* L., *R., bulbosus* L., and *R. acris* L.: III. A mathematical model incorporating multiple modes of reproduction. *J. Ecol.*, **62**, 921–36.

Southwood, T. R. E. (1976). The relevance of population dynamic theory to pest status. In: Cherrett, J. M., and Sagar, G. R. (Eds), *Origins of Pest., Parasite, Disease and Weed Problems.* Symposium of the British Ecological Society 18, pp. 35–54. Blackwell Scientific Publications, Oxford.

Stanton, M. L. (1985). Seed size and emergence time within a stand of wild radish (*Raphanus raphanistrum* L.): the establishment of a fitness hierarchy. *Oecologia*, **67**, 524–31.

Thellung, A. (1912). La flore adventice de Montpellier. *Mém. Soc. Nat. Sci., Cherbourg*, **38**, 57–728.

Weiss, P. W., and Milton, S. (1984). *Chrysanthemoides monilifera* and *Acacia longifolia* in Australia and South Africa. In: Dell, B. (Ed.), *Proceedings of 4th International Conference on Mediterranean Ecosystems*, pp. 159–60. Botany Department, University of Western Australia.

Weiss, P. W., and Noble, I. R. (1984a). Status of coastal dune communities invaded by *Chrysanthemoides monilifera. Aust. J. Ecol.*, **9**, 93–8.

Weiss, P. W., and Noble, I. R. (1984b). Interactions between seedlings of *Chrysanthemoides monilifera. Aust. J. Ecol.*, **9**, 107–15.

Biological Invasions: a Global Perspective
Edited by J. A. Drake *et al.*
© 1989 SCOPE. Published by John Wiley & Sons Ltd

CHAPTER 13

Attributes of Invaders and the Invading Processes: Vertebrates

PAUL R. EHRLICH

13.1 INTRODUCTION

Population biologists have long wondered why some animals are extremely successful invaders (colonists), while close relatives are not. Are there special evolutionary attributes of species that tend to make some better than others at entering new communities and thriving in them? Are there any general rules that would permit an ecologist to predict which of an array of exotic species might be a potential invader (or perhaps more importantly, an economically damaging invader)? Or does the success of invaders normally depend on an interaction between attributes of a potential invader and features of the communities to be invaded? Or does chance normally dominate invasion processes? These questions are examined in this chapter from the viewpoint of vertebrates; it is based in part on earlier work on the attributes of animal invaders (Ehrlich, 1986).

13.2 WHAT IS AN INVADER?

I use the two terms 'invading species' and 'colonizing species' interchangeably for an organism that easily crosses barriers (with or without the help of *Homo sapiens*), rapidly establishes itself on the other side, and then expands its numbers and range relatively quickly in its new habitat. Most of the really successful animal invaders are ones that, for a variety of reasons, are able to cross major barriers because of their relationship with *Homo sapiens* (Elton, 1958). Thus the coyote has been able to expand its range because areas of human disturbance which are impassable barriers for wolves are relatively hospitable to *Canis latrans*. The principal exceptions to the need for a 'human connection' are creatures such as tardigrades that are passively wind dispersed and have become ubiquitous. No larger animals, however, approached ubiquity until humanity did.

13.3 PAIRED COMPARISONS

One approach to the problem of why some species are very successful invaders and others are not is to compare closely related species with very different records. The examples of successful versus unsuccessful sets of vertebrate species that I compared previously (Ehrlich, 1986) were: *Homo sapiens vs* other great apes; coyotes *vs* wolves; black and Norway rats *vs* other *Rattus*; house sparrow *vs* tree sparrow; blue-spotted grouper *vs* other groupers; and *Lutjanus kasimira vs* other snappers.

These cases clearly are dissimilar. At one extreme there are organisms such as the rats and coyotes that have invaded without any *deliberate* aid from humanity. In the middle are the sparrows, which were purposely introduced, but without systematic thought as to consequences (many avian species have been introduced around the world by colonists from Western nations who wanted familiar birds around them). At the other extreme are the reef fishes, which were transplanted by biologists with considerable knowledge of the organisms to be moved and the communities into which they were being introduced.

It seems obvious that the characteristics of successful invaders in these three groups might be very different. An additional category of invaders for which I did no paired comparisons are vertebrates introduced as biological control agents. It may, indeed, be stretching the definition of 'invader' to include them. After all, they are introduced specifically because it is believed, sometimes after very careful investigation, that an especially suitable environment—an empty niche—is awaiting. Ordinarily that most basic of resources, food, is present in superabundance.

Nonetheless, the fate of biocontrol agents can be instructive, and their introduction does not seem so different in principle from the introduction of, say, game birds or game fishes. For example, ferrets (*Mustela putorious furo*—a domestic version of the European polecat, *M. putorius*), stoats (*Mustela erminea*), and weasels (*Mustela nivalis*) were all introduced in large numbers into New Zealand in the second half of the last century. It was hoped that these energetic predators would end that island nation's plague of rabbits, but they did not (rabbits were eventually brought under control in the 1950s by spreading poisoned bait from airplanes and night-shooting). Interestingly, the mustelids were differentially successful; today stoats are common everywhere, ferrets are locally abundant, and weasels are generally rare (King, 1984).

Thus three organisms deliberately introduced into a land with few competitors and abundant prey had little influence on the size of the prey populations, and one of them was relatively unsuccessful. All three species are native and sympatric in Europe and the stoats and weasels have extremely broad distributions, suggesting that physical conditions in New Zealand are unlikely to be more favorable to stoats than to weasels. Instead, the differences seem to be due primarily to different diets (King, 1984). The stoat ordinarily feeds on small rodents but does attack animals, such as hares, that are much larger than itself (Nowak and Paradiso, 1983). It dined in New Zealand on birds and to a lesser extent rabbits, possums, and rats. The large ferret (about eight times as heavy as the stoat) was originally bred, in part, to drive rabbits from their burrows. When rabbits were brought under control by other means, ferret populations declined.

The weasel is somewhat smaller than the stoat and has similar habits. The most likely explanation for the relative failure of the weasel is found in the abundance of prey of different sizes. In England its most common prey are voles (*Microtus, Arvicola, Clethrionomys*), which are much more abundant in woodlands than are their ecological equivalents, feral house mice, in New Zealand (King and Moors, 1979). Even though the three mustelid species were imported with the goal of controlling an abundant pest, their relative success seems to have largely been determined by their diets. These appear to be sufficiently different, however, so that all three species are able to coexist. These biocontrol introductions thus follow a pattern reminiscent of that seen in the lizard and bird faunas of the Caribbean and Hawaiian Islands respectively. There, closely related species with diets (as judged from body size or bill morphology) sufficiently different have coexisted, whereas the presence of close competitors has repeatedly led to extinctions (Roughgarden, 1986; Moulton and Pimm, 1986). Such patterns, of course, were first brought to our attention by Lack (1947).

13.4 ARE SOME SPECIES JUST NATURALLY GOOD COLONIZERS?

Evidence that intrinsic characteristics of vertebrate species can predispose them to be successful invaders comes from instances of some species *repeatedly* being successful invaders while close relatives *repeatedly* are less successful. For example, the house sparrow (*Passer domesticus*) occupied the entire United States in a little over 50 years after it was first successfully introduced. It had the help of additional releases by *Homo sapiens* and, possibly, an ability to disperse via railroad freight trains (Robbins, 1973). Other introductions have led to extensive ranges in South America, southern Africa, Australia, and New Zealand. The closely related tree sparrow (*Passer montanus*) has a range in Eurasia almost as large as that of the house sparrow. It was successfully introduced into North America at St Louis in 1870, but until the Second World War it was still largely confined to the St Louis area. Since then the tree sparrow has spread into central Illinois (Barlow, 1973), but it has not colonized with anything like the vigor of its close relative. Indeed, in some places it appears to have suffered some competitive displacement by *P. domesticus*. It is slightly smaller than the house sparrow, and in areas of eastern Asia where the house sparrow does not occur, *P. montanus* becomes a much closer commensal of humanity (Sumners-Smith, 1973). The tree sparrow was introduced into Australia at about the same time as the house sparrow, but also has a much more restricted range there as well. It did not successfully colonize New Zealand, a set of islands that in general have proven exceedingly vulnerable to occupation by exotics, but it has colonized some islands in the western Pacific, while failing to 'take' in Bermuda.

A similar case to that of the two sparrows is that of the rock dove or domestic pigeon (*Columba livia*) and its congener, the woodpigeon or ring dove (*Columba palumbus*). The former has spread from a presumed primitive distribution along the coasts of southern Europe, south Asia, and North Africa to occupy most of the northern hemisphere and substantial portions of the southern. The woodpigeon has a similar but larger original homeland, extending much further inland in Eurasia than did the rock dove. The woodpigeon reaches high densities feeding on crops, and is considered a pest; it thrives near human habitation. Yet it has hardly spread at all except by gradual extension of its original distribution. Two attempts to introduce the woodpigeon into North America, including one of 30 birds in New York, failed.

The extraordinarily successful invaders are the black or roof rat, *Rattus* (*Rattus*) *rattus*, and the Norway or brown rat, *R.* (*R.*) *norvegicus*. Both species are found throughout the world generally in association with human beings. The Polynesian rat, *R.* (*R.*) *exulans* is widely distributed in Southeast Asia, having spread through the islands of the Pacific by accompanying prehistoric human migrations (Wodzicki and Taylor, 1984). But more than 40 other species of the subgenus *Rattus*, a few of which are commensals with *Homo sapiens* in limited areas, have not become as widespread as even the Polynesian rat, let alone ubiquitous.

Gray and red wolves (*Canis lupus* and the closely related *C. rufus*) have undergone a dramatic decline in North America as a result of depredation by human beings (Nowak and Paradiso, 1983). With the exception of a gray wolf population in Minnesota and scattered reds in the coastal swamps of Louisiana and Texas, wolves are essentially gone from these coterminous United States. Their congener, the coyote (*Canis latrans*), with which they hybridize easily even though the species have had separate evolutionary histories since the late Pliocene, has proven to be a successful invader. The range of the coyote expanded dramatically while those of its relatives were shrinking.

As a final example, *Homo sapiens* has been the most successful invader of all, while its closest living relatives, *Pan troglodytes* and *Gorilla gorilla*, have been pushed almost to the brink of extinction.

13.5 CHARACTERISTICS OF GOOD INVADERS

What sort of biogeographic, ecological, genetic, and physiological attributes might one expect *a priori* to characterize successful and unsuccessful invaders? Table 13.1 (modified from Ehrlich, 1986) indicates some possible distinctions extracted from the ecological literature.

Most of these characteristics are elements of what can be summarized as 'broad

Table 13.1. Possible concomitants of invasion potential

Successful invaders	Unsuccessful invaders
Large native range	Small native range
Abundant in original range	Rare in original range
Vagile	Sedentary
Broad diet	Relatively restricted diet
Short generation times	Long generation times
Able to shift between *r* and *K* strategy (fishes)	Unable to shift
Much genetic variability	Little genetic variability
Gregarious	Solitary
Female able to colonize alone	Female unable to colonize alone
Larger than most relatives	Smaller than most relatives
Associated with *H. sapiens*	Not associated with *H. sapiens*
Able to function in a wide range of physical conditions	Only able to function in a narrow range of physical conditions

ecological amplitude.' Comparisons of closely related vertebrate species do not always support these *a priori* distinctions. At one extreme, there is no obvious reason why the tree sparrow does not colonize as easily as the house sparrow. I am tempted to speculate that *P. domesticus* simply outcompetes *P. montanus* in areas of human disturbance. Outside their native communities in Eurasia those may be the only areas invasible by either.

Almost as obscure is the reason for the relative lack of success of the Polynesian rat (known as the kiore in New Zealand) compared with black and Norway rates. The kiore has become a 'house rat' in Vietnam, Laos, Kampuchea, Malaya, Thailand, and Burma, and in most of these and the Philippines and Indonesia it is a serious pest of rice in the field and in storage. After it reached New Zealand, the kiore apparently played a role in the decimation of the flightless bird fauna there and is still a serious predator of birds' nests, native invertebrates, lizards, and so on. So on one hand the kiore seems to be quite a successful invader. On the other hand, with the arrival of black and Norway rats on New Zealand, the kiore declined, and it is now largely restricted to small, offshore islands. And the kiore has never spread northwestward through Eurasia or over the Australian continent, and has never, like *R. rattus* and *R. norvegicus*, invaded Africa or the western hemisphere. The kiore *is* smaller than the other two, but otherwise it seems to be their equal in invasion potential—and the house mouse is smaller still, and much more widely distributed. It is possible that poor competitive ability of the kiore (relative to the two ubiquitous rats) *on ships* has limited its distribution by denying it a means of dispersal.

The difference between invading coyotes and declining wolves is, in contrast, relatively clear. It seems to be a function of diet and social structure. Coyotes eat small game and carrion, wolves hunt large mammals cooperatively. Large areas disturbed by humanity are barriers to wolves, especially since their social behavior makes them easy prey for hunters. Coyotes can cross the worst of these areas and thrive in many of them.

The reason that *H. sapiens* is the most successful colonizing primate is certainly its extreme employment of cultural evolution, which permitted the development of artifacts and survival techniques (clothes, shelters, tools, weapons) that allowed expansion into previously inaccessible habitats.

One characteristic of organisms that are good invaders is an ability to tolerate a wide range of physical conditions. Classic examples are the mosquito fish, a poecilid (*Gambusia affinis*), which because of its use as a biological control agent now may be the most widely distributed species of freshwater fish, and the Mozambique tilapia, *Tilapia* (*Oreochromis*) *mossambicus*. *Gambusia* can survive in water as cold as 6° C and as hot as 35° C, extremely low oxygen concentrations, and salinities as high as twice that of seawater, given time to acclimate (Ahuja, 1964). *Tilapia* can tolerate 13–37° C and a wide range of oxygen saturations and salinities.

Both of these fishes also have reproductive strategies that appear to help make them successful colonizers (Bruton, 1986). *Tilapia mossambicus* is able to switch from a precocial to an altricial strategy when it finds itself in an unstable environment, relatively free of competitors, and to a more precocial strategy when competition is severe. *Gambusia affinis* is a live-bearer that produces a few well-quarded young and preys on the young of competitors. It seems possible, though, that physiological tolerances are more important, since other *Tilapia* are able to adjust their reproductive strategies and other *Gambusia* species share the live-bearing reproductive strategy of *G. affinis*. What we do not know is the degree to which the 'popularity' of *T. mossambicus* and *G. affinis* with introducers has affected their success relative to that of congeners.

Different innate responses to physical factors of the environment appear to explain the contrasting successes in North America of the European starling (*Sturnus vulgaris*) and the closely related Southeast Asian crested myna (*Acridotheres cristatellus*). While the former took the continent by storm, the latter, introduced into Vancouver in 1897, established itself but did not spread significantly. Studies in Vancouver, where the two species are sympatric, indicated that a lack of attentiveness to incubating eggs, and their resultant low temperature, led to low hatching rates. Furthermore, the mynas fed their young more low-protein vegetable food than did the starling, producing lower growth rates. In both cases the mynas retained habits appropriate to their tropical homelands, and relatively inappropriate to British Columbia (Johnson and Cowan, 1974). There must be other differences between these two species, however, since the myna has been a failure as a colonist virtually everywhere. For example, it took at least three attempts to establish it in the tropical Philippines where it has spread only in the vicinity of Manila (Long, 1981). It also has not managed to spread southward from Vancouver. Perhaps the myna simply is unable to build a large enough population to provide sufficient dispersers; perhaps it can only persist in areas of high human disturbance.

Being native to a relatively stressful environment may well be one characteristic of a good invader. Many temperate zone or arid zone vertebrate species (or species whose natural ranges include such areas) successfully invade rather benign, moist, tropical habitats. For example, numerous naturalized members of Hawaii's tropical lowland bird fauna came from more rigorous climes. At least among birds there are very few examples of species indigenous to benign environments successfully invading stressful habitats (Long, 1981).

13.6 SIZE OF THE INTRODUCTION; FREQUENCY OF ATTEMPTS

Given intrinsic attributes that make an organism a potential invader, it would seem that the size of the introduction should be an important factor in the success of a given invasion. Larger invading groups should bring with them a larger ration of genetic variability, and should be less subject to stochastic extinction before there is a chance for the population to increase in size. What data there are on purposeful introductions, however, do not universally support this view. For instance, Himalayan tahr, the only vertebrate species to invade relatively undisturbed ecosystems in South Africa, did so by building a population based on a single pair. In that same nation the European starling population was based on an introduction of 18 individuals, and Indian mynas on perhaps 20–30. In contrast, an introduction of about 200 rooks (*Corvus frugilegus*) failed, even though there are few South African crows, and the Indian house crow has managed to establish itself in Durban on the basis of much smaller groups of invaders (Brooke *et al.*, 1986).

An analysis of avian introductions in Australian showed those of foreign birds to be statistically more likely to succeed if propagule number was large, while there was no significance of propagule size when Australia birds were translocated (Newsome and Noble, 1986). Overall one suspects that invasions will be more likely to succeed if the number of individuals introduced is large, everything else being equal. But it is a weak statistical association, and everything else never equal.

Everything else is never equal because random factors, environmental and demographic stochasticity but possibly also genetic stochasticity, must play a substantial role in the success of invasions. This is evident from repeated failures of 'good invaders' to become established. A pattern of frequent failure has been best documented for birds (Long, 1981); even European starlings and common mynas did not succeed in establishing themselves after each introduction. Eight pairs of house sparrows, one of the most successful colonizers, were released in New York in 1851, but failed to become established. Another 50 did not manage to establish a population when introduced in 1852. The first successful release was another 50 in 1853 (Long, 1981). The European starling did not become established in two attempts in New York (1872, 1873), and 35 pairs or more released in a Portland, Oregon park around 1890 did not lead to colonization. Success came immediately, however, from the introduction in New York of two flocks of some 60 (1890) and 40 (1891) birds. On the other hand, attempts in 1875, 1889 and 1892 in Quebec all failed, even though the advancing front of starlings dispersing from the New York introductions swept through Quebec by 1920. There were several unsuccessful introductions of rooks into New Zealand before it finally 'took'. The ring-necked pheasant (*Phasianus colchicus*) is one of the most successful temperate-zone invaders, and yet its introductions have only succeeded about half the time (Long, 1981). Repeated early attempts to establish the species in the United States failed in areas where the birds are now plentiful.

Lack of repeated colonization attempts may be the principal reason that the woodpigeon has not spread as far as has the rock dove. Yet other birds, especially columbids such as the Senegal turtledove (*Streptopelia senegalenis*), spotted turtledove (*S. chinensis*), and the peaceful dove (*Geopelia striata*) seem to succeed whenever they are released. The frequency of failure for most species is probably considerably higher than indicated in the literature, however, since records of successes are much more likely to get into print than those of cases where establishment did not follow introduction.

13.7 INVASIBILITY

Some large regions may have communities that are generally less invasible by vertebrates than others. South Africa, for example, seems to be relatively resistance, Brooke *et al.* (1986) suggest that this is because the tip of Africa is a temperate extension of a large, tropical continent with an impressive suite of native predators and pathogens that evolved in the tropics and moved south. In contrast, New Zealand was 'easy'—with no native mammalian predators and very few predatory birds. European rabbits, a plague in New Zealand, have only been successful on offshore islands of South Africa. Rooks failed to invade South Africa, but succeeded in 'taking' in New Zealand and spreading slowly in spite of control measures there (Long, 1981). Mammals seem to have been less successful at invading Eurasia than other continents, possibly because it has a more diverse mammalian fauna than the New World (de Vos *et al.*, 1956).

Indeed, when the reasons for resistance of areas to invasion are discussed, competition is the one that seems most frequently mentioned. Support for this view is seen in the impressive *lack* of success that exotic birds have had in New Zealand forests not disturbed by humanity (Diamond and Veitch, 1981; Clout and Gaze, 1984)). On the islands of New Zealand's Hauraki Gulf, which are virtually devoid of mammalian predators and have no introduced browsing mammals, six exotic passerines (five from Europe and one recently self-introduced *Zosterops* from Australia) are essentially absent from climax native forest (Diamond and Veitch, 1981). On one offshore island, Cuvier, there were once cats, wild goats, and domestic stock, and four European bird species and the *Zosterops* were breeding in the forest. In 1959 the cats and goats were removed, and the stock were fenced. In 5 years a dense forest understory had regenerated; the four European exotics had disappeared, and the population of *Zosterops* had declined precipitously. Diamond and Veitch conclude that exotics are excluded from intact climax forest bird communities, and can only invade browsed forests where the native avian community has been decimated and the forest structure has been altered.

Native passerines in Hawaii are almost entirely confined to fragments of relatively undisturbed habitat, mostly above 900 m (Berger, 1981; Scott *et al.*, 1986). A large number of introduced bird species appear to be competing to some degree in the disturbed lowlands, and may be approaching a dynamic immigration—extinction equilibrium (Moulton and Pimm, 1983). The chances of two invading species both persisting is, as discussed earlier, in part an inverse function of their similarity in body size or bill morphology, which are considered an indirect measure of competition. Whether undisturbed Hawaiian bird communities, like those of New Zealand, were originally invasion resistant has not been determined, but seems a reasonable surmise since there are residual lowland forests dominated by native birds.

Thirty-three species of mammals have been successfully introduced into New Zealand, which had no endemic mammals except bats. None of these exotics is confined to areas of severe human disturbance, and several are abundant in native forests (Gibb and Flux, 1973). In contrast, New Guinea has a rich native mammalian fauna—some 200 species—and few successful invaders (Diamond and Case, 1986). Even the black rat is largely restricted to settlements, and the Norway rat is scarce, in striking contrast to the New Zealand situation.

Similarly, the two reptile species that have been successfully introduced to Australia are largely commensal with *Homo sapiens*. Both are geckos (*Hemidactylus frenatus* and *Leptodactylus lugubris*) that live mostly on insects attracted to house lights (Myers, 1986). The only successful amphibian species, is *Bufo marinus*. This organism is an extremely successful invader in many areas, and is not restricted to areas of human disturbance. Its ultimate distribution is expected to be determined by climatic barriers (Floyd and Easteal, 1986).

Detailed work by Jonathan Roughgarden and his colleagues (e.g. Roughgarden, 1986) has shown that the success of *Anolis* lizards invading Caribbean islands is largely a function of the anole community already established. Invasion of islands already containing an anole species of the same size is difficult or impossible, even though the lizard fauna of those islands is depauperate—being depauperate is not in itself enough to make a fauna invasible.

The degree to which intact freshwater fish communities are invasible is unclear (e.g. Moyle, 1985). There seems, however, little doubt that human-altered systems are more likely to support invaders than undisturbed lakes, rivers, and streams.

13.8 WHAT CONTROLS THE SUCCESS OF INVASIONS?

So far I have focused primarily on broad patterns, but how is the fate of any given introduction determined? It is evident that the success of any given introduction depends on an interaction between the characteristics of the invading species, the communities already established in the recipient area, and that area's physical environment. But, unfortunately, neither invaders nor environments are static. Consider some of the variables on the invader side of a vertebrate introduction:

1. Number of individuals.
2. Sex ratio.
3. Physiological status—are individuals mature, pregnant, at an appropriate state of the breeding cycle, healthy, acclimated, etc.
4. Genetic composition—including ecotype of origin and amount of genetic variability present.
5. Behavioral status—experience of individual(s), social relationships within group.

On the physical/biological environment side of the equation the variables include:

1. Season.
2. Weather.
3. Size and structure of populations of resource organisms.
4. Size and structure of populations of competitors.
5. Size and structure of populations of predators.
6. Size and structure of populations of parasites and pathogens.

Many of these factors are discussed by Bump (1963) in connection with the introduction of grouse into North America. Twelve attempts to establish capercaillie (*Tetrao urogallus*) and 10 to establish black grouse (*Tetrao tetrix*) all failed. The failures occurred even though climatic conditions were very similar in donor and recipient areas, all of the 40 genera of plants known to supply them with buds, leaves, seeds, and berries as food in Europe were present in the recipient area, and the numbers released in most cases should (by standards of successful introductions) have been adequate.

Grouse tend to show cycles of abundance, and Bump attempted to determine if either the state of the cycle of the introduced birds or of native potential competitors could explain the failures. There were, however, no obvious patterns. For example, in the 10 cases of failed capercaillie introductions for which state of cycle of the invaders could be estimated, in two the capercaillie were from populations close to their peak, two on the upswing, and six at the bottom. Seven of the black grouse introductions were made from populations probably near peak abundance, and one on the upswing. The ruffed grouse (*Bonasa umbellus*) is a close North American relative of the two European species. The ruffed grouse were near the peak of their cycle for 8/12 of the capercaillie and 6/10 of the black grouse liberations.

What, then, could be responsible for the failures? It seems unlikely that these species have their mating behavior in some way deleteriously modified by the process of capture, transport, and release (most were wild-trapped). Capercaillie were successfully reintroduced into Scotland, where they had been exterminated, using 49 individuals wild-trapped in Sweden. Poor condition of individuals after transport could very well be a factor in the lack of success in North America, but the data are inadequate to evaluate this. Bump's basic conclusion is that success in introductions only occurs when all or nearly all the crucial factors are 'in productive conjunction.' This could be modified into an 'invasion law of the minimum'—the success of an introduction is likely to depend on whatever factor (or interaction of factors) is in the least auspicious state at the place and time of the attempt. It is tempting to speculate that grouse are very vulnerable to predators, especially if they are not in prime condition. Ruffed grouse are subject to the depredations of hunters and predators such as foxes, lynxes, feral dogs and cats, goshawks, and great-horned owls and a variety of nest robbers. In the scramble to become established and familiar with new territory, many or most of the introduced birds may become victims. Unfortunately comparisons of predator populations in the North American and Scottish release sites are not available.

There is also a general pattern of lower invasibility in the Tetraoninae (grouse) compared to members of other subfamilies of the Phasianidae, Phasianinae (pheasants and partridges), Odontophorinae (quail), and Numidinae (guineafowl) Long, 1981). The chukar (*Alectoris chukar*; Phasianinae), for example, has been introduced with great success to many arid, mountainous parts of western North America. So in this case, as in so many others, there would appear to be subtle interactions between the properties of the invaders and the state of the ecosystem invaded. These interactions may make the probability of a grouse introduction succeeding in broadly 'suitable' habitat always smaller than the probability of success of a ring-necked pheasant (or house sparrow or Norway rat) introduction.

13.9 CONCLUSIONS

Ecologists can make some powerful and wide-ranging predictions about invasions. For instance, vertebrates will generally be more successful invaders than herbivorous insects, since the latter tend to be monophagous or oligophagous and can only colonize places in which suitable plants are already established. Ecologists (on the basis of observed patterns of success and knowledge of their biology) can predict that most organisms will not be successful invaders, and that among those that are invasive, most colonization 'attempts' will fail (due primarily to environmental and demographic stochasticity). Indeed, most of the *a priori* assumptions that I listed, representing an approximation of a consensus of trained ecologists, seem to be valid in many cases. And the rule that human disturbance will almost always pave the way for vertebrate invasion seems quite robust.

On the other hand, ecologists cannot accurately predict the results of a single invasion or introduction event. Even the arrival of a flock of starlings or house sparrows, or a pregnant Norway rat, in an apparently hospitable area and at a favorable time, is no guarantee of successful establishment.

The inability to make the latter kinds of prediction should not be considered a failure of ecology as a science—it does not necessarily represent a lack of adequate theory or a failure of the discipline (Ehrlich, 1986). Physicists, after all, cannot predict which of two identical radioactive nuclei will decay first or which of a series of nearly identical missiles launched from the same silo will come closest to the target. Moreover the task of ecologists in making predictions about potential invaders and invasions is much more complex. They are attempting to predict the fate of diverse, often little-known organisms launched in varying numbers at diverse, complex, usually barely-studied environments.

In spite of this, ecologists can say a great deal about both the probability of invasion success of different organisms in different environments (and the possible consequences of that success)—even if they can not yet generate probability distributions such as those associated with nuclear decay. Clearly we should strive to develop better predictive tools, including mathematical models for the behavior of invaders in some groups (and most successful models will almost certainly be group-specific). But, considering the enormous complexity of the problem, what can already be predicted is far from trivial.

ACKNOWLEDGEMENTS

I am grateful to James H. Brown, Department of Ecology and Evolutionary Biology, University of Arizona, A. H. Ehrlich, R. W. Holm, H. A. Mooney, D. D. Murphy, J. Roughgarden, P. M. Vitousek, and B. A. Wilcox of the Department of Biological Sciences, Stanford University, and Stuart L. Pimm, Graduate Program in Ecology, University of Tennessee, for helpful comments on the manuscript. This work was supported in part by a series of grants from the National Science Foundation (the most recent DEB82-069611), and a grant from the Koret Foundation of San Francisco.

REFERENCES

Ahuja, S. K. (1964). Salinity tolerance of *Gambusia affinis*. *Ind. J. Exp. Biol.*, **2**, 9–11.

Barlow, J. C. (1973). Status of the North American population of the European Tree Sparrow. *Ornithol. Monogr.*, **14**, 10–23.

Berger, A. J. (1981). *Hawaiian Birdlife*, 2nd Edn. University Press of Hawaii, Honolulu.

Brooke, R. K., Lloyd, P. H., and de Villiers, A. L. (1986). Alien and translocated terrestrial vertebrates in South Africa.

Bruton, M. N. (1986). The life history styles of invasive fishes in southern Africa. In: Macdonald, I. A. W., Kruger, F. J., and Ferrar, A. A. (Eds), *The Ecology and Management of Biological Invasions in Southern Africa*, Oxford University Press, Cape Town.

Bump, G. (1963). History and analysis of tetraonid introductions into North America. *J. Wildl. Manage.*, **27**, 855–67.

Clout, M. N., and Gaze, P. D. (1984). Effects of plantation forestry on birds in New Zealand. *J. Appl. Ecol.* **21**, 795–816.

Diamond, J., and Case, T. J. (1986). Overview: introductions, extinctions, exterminations, and invasions. In: Diamond, J., and Case, T. (Eds), *Community Ecology*. Harper and Row, New York.

Diamond, J. M., and Veitch, C. R. (1981). Exatinctions and introductions in the New Zealand avifauna: cause and effect? *Science*, **211**, 499–501.

Ehrlich, P. R. (1986). Which animal will invade? In: Mooney, H. A., and Drake, J. A. (Eds), *Ecology of Biological Invasions of North America and Hawaii*. Springer-Verlag, New York.

Elton, C. S. (1958). *The Ecology of Invasions by Animals and Plants*, Methuen, London.

Floyd, R. B., and Easteal, S. (1986). The giant toad (*Bufo marinus*): introduction and spread in Australia. In: Groves, R. H., and Burdon, J. J. (Eds), *Ecology of Biological Invasions*, p. 151. Australian Academy of Science Canberra.

Gibb, J. A., and Flux, J. E. C. (1973). Mammals. In: Williams, G. R. (Ed.), *The Natural History of New Zealand*, pp. 334-71. Reed, Wellington.

Johnson, S. R., and Cowan, I. M. (1974). Thermal adaptation as a factor affecting colonizing success of introduced Sturnidae (Aves) in North America. *Can. J. Zool.*, **52**, 1559–76.

King, C. (1984). *Immigrant Killers: Introduced Predators and the Conservation of Birds in New Zealand*. Oxford University Press, Auckland.

King, C. M., and Moors, P. J. (1979). On coexistence, foraging strategy, and the biogeography of weasels and stoats (*Mustela nivalis* and *M. erminea*) in Britain. *Oecologia*, **39**, 129–150.

Lack, D. (1947). *Darwin's Finches*. Cambridge University Press, Cambridge.

Long, J. L. (1981). *Introduced Birds of the World*. Universe Books, New York.

Moulton, M. P., and Pimm, S. L. (1983). The introduced Hawaiian avifauna: biogeographic evidence for competition. *Amer. Nat.*, **121**, 669–90.

Moulton, M. P., and Pimm, S. L. (1986). The extent of competition in shaping an introduced avifauna. In: Diamond, J., and Case, T. J. (Eds), *Community Ecology*, pp. 80–97. Harper and Row, New York.

Moyle, P. B. (1985). Exotic fishes and vacant niches. *Environ. Biol. Fishes*, **13**, 315–7.

Myers, K. (1986). Introduced vertebrates in Australia, with emphasis on the mammals. In: Groves, R. H., and Burdon, J. J. (Eds), *Ecology of Biological Invasions*, pp. 120–36. Australian Academy of Science, Canberra.

Newsome, A. E., and Noble, I. R. (1986). Ecological and physiological characters of invading species. In: Groves, R. H., and Burdon, J. J. (Eds), *Ecology of Biological Invasions*, pp. 1–15. Australian Academy of Science, Canberra.

Robbins, C. S. (1973). Introduction, spread, and present abundance of the House Sparrow in North America. *Ornithol. Monogr.*, **14**, 3–9.

Roughgarden, J. (1986). A comparison of food-limited and space-limited animals competition communities. In: Diamond, J., and Case, T. J. (Eds), *Community Ecology*, pp. 492–516. Harper and Row, New York.

Scott, J., Mountainspring, S., Ramsey, F., and Kepler, C. (1986). Forest bird communities of the Hawaiian Islands: their dynamics, ecology, and conservation. *Studies in Avian Biology No. 9*. Cooper Ornithological Society, Lawrence Kansas.

Sumners-Smith, J. D. (1963). *The House Sparrow*. Collins, London.
de Vos, A., Manville, R. H., and van Gelder, R. G. (1956). Introduced mammals and their influence on native biota. *Zoologica*, **41**, 163–94.
Wodzicki, K., and Taylor, R. H. (1984). Distribution and status of the polynesian rat *Rattus execulans Acta Zool. Fennica*, **172**, 99–101.

Biological Invasions: a Global Perspective
Edited by J. A. Drake *et al.*
© 1989 SCOPE. Published by John Wiley & Sons Ltd

CHAPTER 14

Mathematical Models of Invasion

MARK WILLIAMSON

14.1 INTRODUCTION

Models have many uses. The aim of this chapter is to examine some of these uses in relation to the problem of the ecology of biological invasions, and to note what has been said about models elsewhere in the SCOPE programme.

It is possible to examine the use of models for understanding a particular problem either by considering different levels of understanding or by considering different aspects of the problem. For invasions, the latter is more convenient simply because there are many different aspects. So sections below consider separately models of establishment, models of spread and models of community equilibration. While different levels of understanding emerge from the discussion of these stages, I will preface these discussions with a consideration of what models can and can not do.

Amongst the many uses of models there are four I wish to highlight. The first is prediction. It would be highly desirable to be able to predict the success or otherwise of potential invaders. The problem of predicting the consequences of the release of genetically engineered organisms increases the need for successful prediction. But what emerges from the SCOPE programme is that reliable prediction in individual cases is still not possible. What is possible is a statistical indication of which invasion type is more, or less, likely, and *post hoc* explanations, with reference to models, of why a particular invasion has or has not happened.

Models have had appreciable success in two other ways, which are the second and third uses I will mention here. These are to explain what has been observed and to indicate the possible behaviour of ecological systems subject to invasions. Much of this chapter deals with these two uses of models.

The fourth use of models that I wish to mention is to bring out gaps in existing knowledge. In a study of invaders of the British Isles, Williamson and Brown (1986) found, in round terms, that 10% of invaders became established, and 10% of those established became pests. Lawton and Brown (1986) tried to see to what extent establishment could be related to the standard demographic parameters r, R_0 and K, but had to resort to indirect measures of these three. As will be seen. there is not only a dearth of direct measurements of these parameters but also, to some extent, a vagueness about precisely how these and other parameters should be measured.

As Beddington (1983) points out, there is now a great variety of models available in ecology, but the interface between theory and experiment (or observation) is weak. Regrettably few ecologists concern themselves with the problems of testing models and of parameter estimation. So a major theme underlying this chapter is that while models can now offer explanations of invasions in broad terms, detailed understanding and prediction require better and more extensive measurements of population parameters.

The SCOPE programme has included a modelling working party (Drake and Williamson, 1986) and much of this chapter draws on material developed there. All the SCOPE national symposia involve models in many places in passing, but in the South African and Dutch symposia there are no papers explicitly on modelling. In the Australian symposium, Newsome and Noble (1986) set out as a model, if not as a mathematical model, the properties of invaders. In the American symposium, Roughgarden (1986) reviews aspects of establishment, spread and community interactions. However, the major contribution to models is the British symposium, which was deliberately focussed on quantitative aspects of invasions. There are detailed discussions of models in Anderson and May (1986), Crawley (1986), Holdgate (1986), Lawton and Brown (1986), Mollison (1986) and Williamson and Brown (1986), with some discussion in most of the other papers. The joint British Ecological Society and Linnean Society symposium, which was also part of the British contribution to the SCOPE programme, mentions models here and there, particularly in Lawton (1987) and Usher (1987). In this volume, models are also dealt with explicitly by Crawley, by Levin and by Noble.

14.2 ARRIVAL AND ESTABLISHMENT

If an invasive species is going to become a problem, which is what the SCOPE programme is about, it has to arrive and establish itself. There is no doubt that it is in general cheaper, easier and more effective to control an invader as early as possible, preferably at arrival; there is also no doubt that many invaders have evaded and will evade efforts to prevent their arrival. Many arrivals, certainly almost all those at regions remote from their centre of origin, are brought by people, and there is little to be learnt from modelling those processes. Nevertheless, the frequency with which they are brought will affect the probability of establishment, as has often been shown on a more local scale. Grime (1986) gives examples in the British Flora.

Between arrival and establishment, there may be a period of adaptation. It is tempting, despite the lack of evidence, to suppose that this period involves genetic change. Both Orians (1986) and Kruger *et al.* (1986), using data spanning centuries, suggest that establishment is more likely the longer the time since the original introduction. However, the data on weed biocontrol, covering a much shorter period, show such delay in only 7% of releases (Crawley, 1986).

The stage after the arrival, or introduction, of a species I prefer to call establishment. Some authors call it colonization, and certainly historically, as used by the Greeks, a colony is a population that has introduced and established itself away from its home ground. However, a colonizing species is often considered to be one whose life history and habitat requirements lead to most of its populations being new colonies, a species that habitually invades and occupies transient habitats (Gray, 1986; and in the discussion of Grime, 1986). As there is no particular connection between this type of species and those invading species that cause environmental problems, the invaders that are the concern of SCOPE, I shall, as far as possible, avoid the use of the term colonization in this chapter.

14.2.1 Minimum viable population size

Small population sizes or low population densities may result in a species being more vulnerable. This general phenomenon was referred to as undercrowding by Allee *et al.* (1949) and is now often known as the Allee effect. The original Allee effect was the physiological effect on an individual of the presence of other members of the same species; a social effect, such as is seen in flocks and swarms. Other possible effects can be listed. A newly arrived individual may be unable to find a mate, or, more immediately, unable to find the resources to sustain itself. Small populations may be more vulnerable to general predators, and are, by hypothesis, subject to competition from larger populations of other species. A small population may lack genetic variability, and, if it remains small for long, may suffer inbreeding depression. These effects, and others, are reviewed by Roughgarden (1986).

Each of these effects can be put in mathematical form, admittedly quite a complex form for the genetic effects. It is much harder to get quantitative measures of these effects in the field, or to distinguish, say, the difficulty new arrivals have in finding resources from a failure of establishment because of a lack of long-term resources. If the concept of a minimum viable population size is to be useful in controlling and predicting invasions, more measurements need to be made of the relevant parameters of invasion models. Roughly speaking only the epidemiological models discussed below (Section 14.2.3) have been tested and shown to include real effects of small population size.

Nevertheless there is evidence that small population size is critical in determining the success of establishment in some species. Table 14.1 gives data on biological control insects in Canada, based on the numbers released of each species. Ehler and Hall (1982) show similar results with a larger data set, but classified by the number of individuals of all introduced species released against a given pest species. There are explanations other than small population size and Allee effects for the type of data in Table 14.1. For instance: species that are better

Table 14.1. Success rate in biological control of insects by insects in Canada. Data from Beirne (1975)

	Success	Failure	% Success
Total individuals released			
< 5000	9	89	9
5000–31 200	13	20	39
> 31 200	22	6	79
Individuals in a single release			
< 800	—	—	15
> 800	—	—	65
Number of times released			
< 10	—	—	10
> 10	—	—	70

climatically matched may be easier to breed and so release in large numbers; the species that can adapt to the conditions of the breeding cage may also be those that can adapt to the alien environment into which they are being introduced; species with larger r may be easier to breed; an experimenter finding a successful biological control species may go to more trouble to breed large numbers; and so on. Still, the failure of about two-thirds of releases for biological control of insects (Beirne, 1975) and one-third for releases of weed biocontrol agents (Crawley, 1986) can be compared with the failure of around 90% of introductions into Britain (Williamson and Brown, 1986), and so suggest that a large proportion of failures to invade may be ascribed, at least in part, to the effects of small population size.

14.2.2 The parameters of establishment

If the resources are sufficient, and the environmental conditions adequately favourable, can the conditions for the establishment of an invader be summarized in a few parameters? Or, to put it another way, can simple single species or oligo-species models help us understand, explain or predict invasions?

The best known parameters are r and K from the logistic equation

$$d \ln n/dt = r(1 - n/K) \quad (1)$$

where n is the population size (number or density) of a single species, t is time, r is the rate of increase at zero population density, K the equilibrium population size, frequently, if inaccurately, referred to as the carrying capacity. Following the introduction by MacArthur (1962) of the concepts of r and K selection, many authors related life history characteristics, and in particular colonizing ability, to r and K. This idea has fallen into disfavour since the empirical tests of Stearns (1977) and the criticisms of Boyce (1984). In any case, colonizing in the sense used in most r and K studies is, as I have indicated above, not necessarily relevant to invasion.

More elaborate schemes, usually with three categories (Grubb, 1985; Vermeij, 1978) which implies at least three parameters, have been proposed. For instance, Grime's three factors relate to the loss (or failure to grow) from competition (with other species), the constraints on plant production, which he calls stress, and the destruction of biomass, which he calls disturbance. All three imply negative terms, the loss of actual or potential population, in any model (see the discussion after Grime (1986)). In principle, the scheme could be related to invasion into any type of community. Newsome and Noble (1986) find 10 suites of eco-physiological characters associated with success in invasion in Victoria (Australia) and suggest that these can be spanned by three types of invaders: gap-grabbers, competitors and swampers. This is another, verbal, three character model, much like those listed by Grubb and by Vermeij. All these models, which are clearly similar but not identical, are useful in describing and explaining some community patterns, and suggesting parameters that might be measured. Grubb (1985) discusses some of the difficulties of definition. Until measurements have been made, it is impossible to say if the models will be helpful in prediction. The scheme proposed by Newsome and Noble (1986) brings out the point that invaders can be found in all three categories and so, conversely, knowing the category that a species belongs to says nothing about its invasion potential. Noble (this volume) points out further weaknesses in the use of three character schemes in discussing invasion.

Invasion, of the type that is of importance to the SCOPE programme, is perhaps more analogous to succession than to the colonization of transient habitats. So models of succession (Connell and Slatyer, 1977; Lawton, 1987; Usher, 1987) might be helpful in studying invasions, but these models are more closely related to the models of community structure considered in Section 14.4 than to simple few parameter models for the establishment of single species.

Another, possibly important, parameter arises naturally in the mathematics of a population with age structure. This is R_O, the expected number of female offspring produced during a single female's entire life, that is the number of eggs produced by an egg, or the number of seeds produced by a seed. In standard notation, where l_x is the number of survivors to age x, and m_x is the reproductive rate (of females) of age x, then

$$R_O = \int l_x m_x \, dx \qquad (2)$$

and is independent of the age structure of the population. Its value, obviously, depends on the life table and the birth table, and so assumes whatever conditions produce those values. Conversely R_O is dependent on all aspects of the environment and in particular those causing the death rate. Salisbury (1942) seems to have been the first to suggest that the commonness of a plant, that is, both its abundance in one place and its geographical range, reflected the reproductive capacity of the plant. He measured, in many species, seed output in one season. This differs from R_O by ignoring juvenile mortality and by not measuring the reproductive output of one plant over its lifetime (except in annuals), but nevertheless seed output is clearly a major component of R_O.

For an age structured population an important parameter is the intrinsic rate of natural increase, first called *r* and championed by Lotka (see Smith and Keyfitz, 1977), and defined as the dominant root of the equation

$$\int e^{-rx} l_x m_x \, dx = 1. \tag{3}$$

As with R_O, this assumes a particular schedule of death rates, conventionally not dependent on the density of the population. However, while *r* is a rate of increase, it is the rate of increase only in the stable age distribution. Few natural populations are actually in the stable age distribution as that normally requires several generations of density independent increase, so usually *r* has to be calculated rather than observed. In fact the majority of estimates of *r* refer to populations in the laboratory, and are always dependent on such variables as temperature and humidity (Williamson, 1972). What *r* would be for such species under natural conditions is not known, but it would probably be smaller at any given physical conditions because of natural mortality.

Lotka's *r* may be regarded as composed of an intrinsic birth rate *b*, and an intrinsic death rate *d* (Lotka, 1924; Pielou, 1977), giving

$$r = b - d \tag{4}$$

and it is natural to equate it with *r* in the logistic equation, so bringing together *r*, *K* and R_O.

In discussing the immigration of species onto islands MacArthur and Wilson (1967) suggest that b/d is a better measure than *r* for the probability of succesful establishment. Armstrong (1978) shows by an ingenious argument that, for a population with age invariant rates, $b/d = R_O$. Unfortunately this is not true for a general age structured population, as can be seen with a little trouble from (2), (3) and the analytic expression for *b* in the stable age distribution

$$b^{-1} = \int l_x e^{-rx} \, dx. \tag{5}$$

Nevertheless, if invasion success can be predicted to some extent by *r* or R_O, it might be better predicted by *b*, *d*, or some combination of them, or by some other parameters derived from the l_x and m_x distributions. A fundamental point is the definition of conditions assumed for l_x and m_x. Conventionally, the distributions are measured at minimal population density. Even so, which mortality factors are included and which left out is not always clear, but such decisions will have a marked effect on the usefulness of these parameters in understanding and predicting invasions. Lotka (1924), starting from human populations, includes all forms of mortality; Crawley (this volume) follows the modern convention of excluding all mortality appearing elsewhere in his models. For studying invasion, the density independent mortality associated with the site of the invasion would seem to be what is wanted. The discussion of equations (12) and (13) in Section 14.4.3 elaborates on this point, as does Boyce (1984).

These basic models of population dynamics suggest examining r, K and R_0 for their importance in invasion simply because there are no other standard parameters in these models. The difficulty, as already noted for r, is that there are very few measurements of these parameters, and these measurements are not in the natural environment. Lawton and Brown (1986) observe that over groups of animals as a whole, large animals are more successful invaders than small in Britain, but that among insects, small insects are more successful than large. They note that these differences could result from either the mode of introduction, or the biology of the invaders, or both, but nevertheless they attempt to relate these patterns to what has been reported in the literature about the relationships of r, R_0 and K to size. Lawton and Brown conclude 'there are neither theoretical nor empirical grounds for believing that r alone is the principal, or necessarily even an important, arbiter of invasion ability'; but Crawley (1986), using similar arguments with different data, concludes that r is important.

The basis of these contrary conclusions would seem to be both a lack of knowledge of r in real environments, and a well-known generalization which is possibly misleading when comparing invasions by closely related species. The generalization, which dates from Smith (1954), is that larger organisms have smaller r. Smith's diagram, which relates r to R_0 and the generation time T, has been repeated with some additions by several authors; Fenchel (1974) makes the direct comparison of r to size. All these diagrams cover several orders of magnitude. Smith has six orders of magnitude for r and T, Fenchel six for r and 21 for weight.

Fenchel (1974) notes that published estimates of r come mostly from opportunistic, colonizing or pest species, and indicates that the relationship of r to size would be weaker if other types of species had been included. In the study of invasions, and in particular in predicting invasions, it is important to be able to say of closely related species, of species in the same genus or family, which will be the most successful. In contrast to the Smith–Fenchel generalization, it is well known that within a species the largest individuals generally have the highest reproductive rate (Vermeij, 1978). This is particularly striking in angiosperms with their marked phenotypic plasticity (Salisbury, 1942). Maaløe and Kjeldgaard (1966, p. 63) show a four-fold increase in r with a four-fold increase in cell mass for the bacterium *Salmonella typhimurium*. Dobzhansky *et al.* (1964) show a 20–30% variation in r over five genotypes of *Drosophila pseudobscura*. The flies, as far as is known, do not differ significantly in size. In Fenchel's relationship, this difference in r corresponds to a 10-fold difference in weight. So if r decreases with size between organisms in different classes, but increases with size in a species, how does it behave in families and genera?

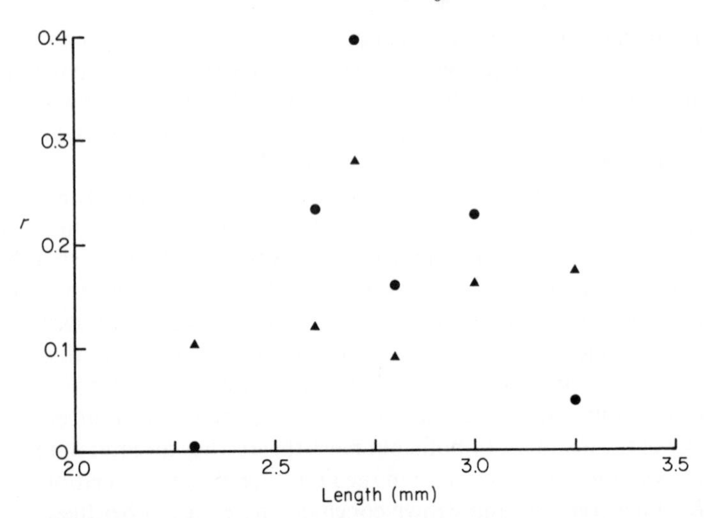

Figure 14.1. The intrinsic rate of increase (r) at two temperatures for six species of ptinid beetles. Lengths from Joy (1976), r from Howe (1953, 1955, 1959) and Howe and Bull (1956). ▲ 20°C, ● 25°C

Table 14.2. Rates of increase of three species of *Daphnia* (Crustacea, Cladocera). Data from Bengtsson (1986)

Species	Adult size (mm)	$r(\mathrm{day}^{-1})$	R_O
D. magna	3.7	0.185	64
D. pulex	2.0	0.215	75
D. longispina	1.8	0.140	38

Adult size is the mean of the size range, r and R_O are averages of the values under four conditions of temperature and food.

In Figure 14.1, I show the best information I can find, which is for spider beetles, which are pests of stored products. The variation in size is not great, and derived from a taxonomic work not from the cultures; overall the two variables are independent. Nevertheless the highest r goes with a middle-sized species, the commonest pest, while the lowest r's are found in the largest and smallest of this set of six species. Bengtsson (1986) measured r (and also R_O) for three species of *Daphnia* under four conditions of temperature and food, and found the highest r in each case in the middle-sized species, *D. pulex*. His data are summarized in Table 14.2. The order of species is the same for r in all four conditions. For R_O the results are more variable. *D. magna* has the lowest R_O of the three at 20 °C and low food, but the highest with both feeding regimes at 15 °C. *D. pulex* is apparently the commonest species of the three. In both ptinids and *Daphnia* middle-sized species have the largest r, and maybe the highest K if commonness is a reliable guide.

For plants, r has scarcely ever been estimated (Crawley, 1983). Salisbury (1942) gives the seed production for adults, which may be related to R_O and hence to r. In his data it seems that the highest seed output is associated variously with the largest or a middle-sized species in each genus. As already noted, seed production is related to commonness; so presumably R_O is related to K measured as biomass. For birds, O'Connor (1986) suggests that reproductive output, again an approximate measure of R_O, is relatively high for successful invaders.

All this work shows that the modelling invasions of macroorganisms there is much to be done in defining the important parameters and measuring them. For pathogenic microbes the situation is somewhat better, as will be seen in the next section.

14.2.3 Models of epidemics

Although the life table and its parameters are a powerful aid in human demography, the logistic curve and similar forms are based more on mathematical convenience than biological reality (Allee *et al.*, 1949). There are, however, two areas of population biology in which simple equations have been developed with parameters with real biological meaning, and in which the equations have been tested by observation and experiment. These are studies on continuous cultures such as chemostats (Williamson, 1972) and the mathematical theory of epidemics (Bailey, 1975).

Conditions for invasions into chemostats have not been studied systematically, except for genetic mutants which are usually the only invaders readily available. Tilman (1977) gives conditions for the stability of a two species, two resource system, and so, by implication, the conditions for one of the species invading a system containing only the other. In simple systems the number of species can not exceed, and may be less than, the number of controlling factors (Williamson, 1957), and Tilman's results are an example of this. It is possible that the study of density-dependent controlling factors might be helpful in understanding invasions.

In contrast to continuous cultures, much is known about invasions by micropathogens, and this shows, as will be seen, how complex the conditions for invasion are even in systems of only three species. The simplest epidemic model has two species and considers a host species with susceptible, infected and recovered individuals, and a parasite (disease) population. Transmission from infected to susceptible is taken to be proportional to random contact. This gives

$$x + y = n = \text{constant} \tag{6}$$

$$\mathrm{d}x/\mathrm{d}t = sxy - (a + b + c)y \tag{7}$$

where x are the susceptibles, y the infecteds, s the transmission rate, a, b and c are loss rates from natural death, disease caused death and recovery. Anderson and May (1983) emphasize the comprehensibility that results from using as few dimensionless parameters as possible, a standard technique in physics. So they rescale the time t on to the natural time scale $(a+b+c)^{-1}$, and call it t', and define a dimensionless parameter R

$$R = sn/(a+b+c) \tag{8}$$

so that (6) becomes

$$\mathrm{d}\ln y/\mathrm{d}t' = (R-1) - Ry. \tag{9}$$

So growth is only possible with $R > 1$. R is the expected number of secondary infections produced within the infectious period of one newly infected host. Anderson and May (1983) say R is precisely analogous to R_O; Anderson and May (1986) call it R_O. Note that R depends on n, which is assumed constant for any one epidemic. If $n < (a+b+c)/s = n_T$, the epidemic can not develop. For an epidemic, the population must be greater than the threshold density n_T. This is the famous threshold theorem of Kermack and McKendrick (1927).

In stochastic models, the threshold becomes blurred but the principle holds (Bailey, 1975). This threshold theorem is yet another reminder that R_O is not constant for a species but depends on conditions. Anderson and May (1986), developing Bartlett's (1957) well known result, show how stochastic modifications to this equation, which have to be solved by simulation, can explain the fluctuations in measles, and how, while a population of 6500 is needed for measles to invade, a population of 250 000 is needed for persistence, perhaps the most interesting result in all models of invasions. The prediction fits data on real island populations well (Williamson, 1981).

When these models are expanded to three species (Anderson and May, 1986), remarkable complexities arise, demonstrating once again that with a little ingenuity a model can be made to produce any result you like. But some results are simple. If a new pathogen invades a host/pathogen system it will displace the previous pathogen if it has a larger R_O, a lower n_T. Others are complex. In some cases a pathogen stabilizes a two species system allowing all three to persist, in others it destabilizes, in others the result depends on the parameter values. Worse, many of the results may depend on the particular form of the equations. There are no critical tests yet for the forms of the equations or the values of the parameters in these three species systems.

Another example of the complexity of three species systems is found in Rejmanek (1984), who developed a model of three competing plant species, together with disturbance and diffusion. Not only did he show that disturbance may mediate permanent coexistence, but he also showed that diffusion increased the zone, in parameter space, of such coexistence. Consideration of diffusion leads to the next topic.

14.3 SPREAD

An invader once established will usually spread out. The speed at which it does this is of some importance to management; it is generally easier to control a slow spreading species than a fast one. However, models of this stage are often of less importance than those of establishment, considered earlier, or those of final equilibrium population density, considered below. It is perhaps, then, not surprising that these models are those that have been best developed and best tested. But even these models can be quite complex and difficult to test.

The most elementary process in population dynamics is the exponential growth of an unconstrained population. In spatial processes, simple diffusion is the natural starting point. So the simplest and most basic model of the spread of an invading species contains exponential growth and random diffusion and nothing else. That is, spread comes from the random movement of individuals which are assumed to be not affected by the behaviour and abundance of other individuals. The model has just two parameters, the intrinsic rate of natural increase, r, and the diffusivity, k. Considering just the radial spread from the starting point gives

$$\partial n/\partial t = rn + k(\partial^2 n/\partial x^2 + x^{-1}\partial n/\partial x) \quad (10)$$

where n is the population size, t time and x the radial distance (Kendall, 1948).

Even this simple equation has surprising complications. Its origin in biology is in the genetical studies of Fisher (1937) and Kolmogoroff *et al.*, (1937), and in that form the solution has an ambiguity in the velocity of the wave of advance, though Moran (1962) shows that there is no ambiguity if the process is started from a minimal population at one point. Murray (1977) discuss many of the complications. The equation was first discussed as an ecological one by Kendall (1948) and Skellam (1951), and they showed that equation (10) leads to an approximate asymptotic relation.

$$\hat{x}/t = 2r^{1/2}k^{1/2} \quad (11)$$

where $\hat{x}$ is the position of the front, as defined by some detection threshold. This represents a wave of advance with constant linear speed, the speed being a function of the two parameters r and k. If equation (11) is a good approximation, invasions will spread in concentric circles with fixed spacing for a fixed time interval. The many maps of the spread of invasion that have been published show that the approximation is a fair fit to much data. Williamson and Brown (1986) have examined in more detail than Skellam (1951) the spread of the North American muskrat, *Ondatra zibethicus*, in central Europe from 1905 to 1927. Skellam used the square root of the area as an average of the radial distance. Williamson and Brown showed that this is indeed the best transformation of the area to produce a linear spread, but even so the rate of spread varies, and the variation in the rate is very large when the spread along individual radii is studied.

If the rate of spread, the intrinsic rate of natural increase and the diffusivity were all known, it would be possible to make more critical tests of equation (11). While

r has seldom been measured, measurements of k are even scarcer (Okubo, 1980). Instead Williamson and Brown estimated r from North American works of reference, and with that and the rate of spread estimated k and thence the mean annual dispersal of muskrats in central Europe. This dispersal came out at 7.65 km per year, a rather high value for random diffusion. However, their estimate of r may have been too low for European populations at low density (Mrs A. Verkaik, personal communication to Dr M. B. Usher); if so the estimate of k, and so also of the mean dispersal, is too high.

In other studies quoted by Williamson and Brown, the rate of spread appears to accelerate, or be more irregular with occasional leaps. Both these phenomena are important for the control of invaders. Mollison (1977) discusses the consequences of models that are more complex then equation (10), stochastic, in discrete time or with different contact distributions, and shown that these can predict occasional 'great leaps forward'. Kot and Schaffer (1986) and Levin (1986) discuss other models that can show complicated behaviour. Problems of estimating parameters in these elaborate models from field data are discussed by Banks *et al.* (1985, 1986). The variability in the rate of spread shown by Williamson and Brown and by others, such as Usher (1986), may well be explicable by some of these models.

The eastern North American grey squirrel, *Sciurus carolinensis*, has been introduced into Britain, California and South Africa. Williamson and Brown (1986) found a modal rate of spread of 5 km yr^{-1} in eastern England. Millar (1980) gives maps and data for the spread near Cape Town from 1900 to 1971, which give rates an order of magnitude or more lower, 0.4 km yr^{-1}. In South Africa the squirrel is only found among introduced trees; in England it is found throughout the countryside. So while the rate of increase and the diffusivity might explain the observed spread in England, for South Africa it is necessary to know why its habitat is restricted. There are no native squirrels there, and the food supply in the natural habitats appears to be sufficient. Possibly there is a community effect of the sort considered in the next section.

14.4 MODELS OF EQUILIBRIUM STATE

Even when an invader has established and spread, it will not usualy be a problem, as the surveys by Simberloff (1981) and Williamson and Brown (1986) show. There will be management problems if the invader reaches high densities, or interacts strongly with the indigenous species, and in some other cases. Can models predict the eventual population density, or the extent of interaction with other species?

14.4.1 The number of parameters

Although some species may limit their own density, by territorial behaviour, or simply by forming a monolayer, the equilibrium of most species will depend on interactions with other species. It is possible that these interactions are diffuse, multifaceted and weak, as seems to be the case for the diatom *Biddulphia sinensis* in the Irish Sea (Williamson and Brown, 1986; Williamson, 1987). In such cases there is little point in trying to model all the interactions; there would be too many of them. Indeed an essential aim in modelling should be to keep the number of parameters small.

Complex models, with many parameters, are difficult to understand, and it is difficult to estimate their parameters with sufficient accuracy to test the models, let alone to use them for prediction. Medawar (1982) contrasts the reliance placed on econometric models with the general scepticism about weather forecasts, built on well founded meteorological models. Montroll and Shuler (1979) point out that the cost, measured in money or time, of getting a satisfactory solution from a complex model increases exponentially with the number of parameters. Usher (1987) makes a similar numerical point: 32 species have 496 possible pairwise interactions and 4×10^9 three way interactions. Mollison (1984, 1986) shows how a few parameters are more intelligible than many. Anderson and May (1983, 1986) recommend the reduction of parameters by finding dimensionless combinations. We should be wary of the claims of any complex ecological model even when we understand ecological systems much more than we do today.

It is now widely recognized that simulation models of real communities, with a large numbers of parameters, such as those used in the IBP, have been generally unsatisfactory (Holdgate, 1986). An example of the futility of this approach is Swartzmann and Zaret (1983). They attempted to build a simulation model of the invasion of Gatun Lake, Panama, by the predatory fish *Cichla ocellaris*, which spread through the lake at about 10 km a year and wrought havoc on the species there. Of eight diurnal fish species (out of about 15 total) four became extinct, two were reduced 90%, one by 50% and one increased 50% (Zaret and Paine, 1973). The simulation model had over 50 parameters, but failed to represent the rate of spread, and in the graphs published all prey species increase in density, though there is a claim of better fit to the data in later results. Watt (1975) has noted the tendency of simulation modellers to claim that their results will be right soon.

Simulation, as such, is often necessary for a solution, as for instance in the stochastic measles model of Section 14.2.3. What is undesirable is the proliferation of parameters. The firm recommendation of the SCOPE working party is that the number of parameters should be kept to a minimum and that the parameters should be both measurable and interpretable.

14.4.2 Simplified models of communities

There are many ways in which the properties of communities may be simplified. Among these are the log-normal and other descriptive distributions, analysis as food webs, and the theory of limiting similarity, all discussed below. Others include Markov compartment models (Usher, 1987), the competitive equivalence model (Goldberg and Werner, 1983), functional groups (Noble, this volume) and the concept of ecological fields (Walkner *et al.*, 1986), which are all designed to avoid discussing individual species, and so are of little help in understanding invasions of particular species. Noble (this volume) emphasizes the need to measure parameters on individual invading species.

Theories of the distribution of the abundance of individual species, such as the log-normal, indicate the pattern of species abundance to be expected, and so might indicate for a given community where there are invasible gaps. However, both the theory and the fit to data are approximate. As populations vary logarithmically (Williamson, 1972), applying the central limit theorem to them will produce a log-normal distribution (Brown and Sanders, 1981). However, as Pielou (1977) points out, the argument only applies to populations of one species. It can be stretched to apply to sets of species which are in some sense of the same sort (Williamson, 1988). For instance, the abundances of the set of copepod species in the mid-Pacific (McGowan and Walker, 1985) are more or less log-normal; those of a wider set of plankton, including copepods, in the North Sea, are clearly not (Williamson, 1972). Gaps in the distribution of abundances cannot be identified, and even if they could would not indicate what sort of species could invade. If invading species come at random in the log-normal distribution, relatively few will be abundant enough to be pests, and, of course, relatively few are.

The theory of limiting similarity, which in one form suggests that species are spread out along resource axes, might indicate what type of gaps exist in real communities. Unfortunately, elaborations of the theory showed, once again, that many other results can be got (Lawton and Hassell, 1984) and this and empirical tests, for instance on size-ratios (Carothers, 1986), led to the conclusion that 'no generalizations have emerged from these studies' (Abrams, 1983), and certainly no results that are useful in studying invasions (Roughgarden, 1986).

For a given quantity of resource, Hanski (1978) develops a model that predicts that the more finely divided the resource the more species can coexist on it, and the higher the invasibility of the guild of consumers. Shorrocks and his group have developed another way of looking at communities on finely divided resources, one that involves relatively few parameters, three per species (Shorrocks and Rosewell, 1986). These are an aggregation measure based on the negative binomial distribution, a rate of increase and a competition coefficient. Green (1986) has criticized the way the negative binomial is used. No doubt there will be some argument about how best to model the aggregation within species of a guild living on a divided and ephemeral resource, but certainly aggregation could reduce competition and make coexistence easier. In this model, invasion and spread are both facilitated by much aggregation or, if there is less aggregation, by low competition rates among the pre-existing species. The invader needs a relatively high rate of increase to enter, and should be competitive and weakly aggregated to spread and become common. Not only is this an interesting theory, but Shorrocks provides measures of his parameters for some *Drosophila*.

14.4.3 Food webs

The study of food webs has become a central part of modern community theory (Pimm, 1982). I will only discuss two aspects here, the mathematics of invasion in food webs, and empirical observations on predator–prey ratios.

The ratio of predators to prey has been studied both empirically and in models. Figure 14.2 shown some data from freshwater systems, derived from taxonomically complete studies. The ratio is more or less invariant with richness, and there are rather less than three prey species to each predator on average. Although the relationship is strong, and possibly useful for explaining success or failure of invaders in some communities, it is not strong enough to produce predictions. Even at the bounding rations given in Figure 14.2, namely 1:1 ad 7:1, it is not

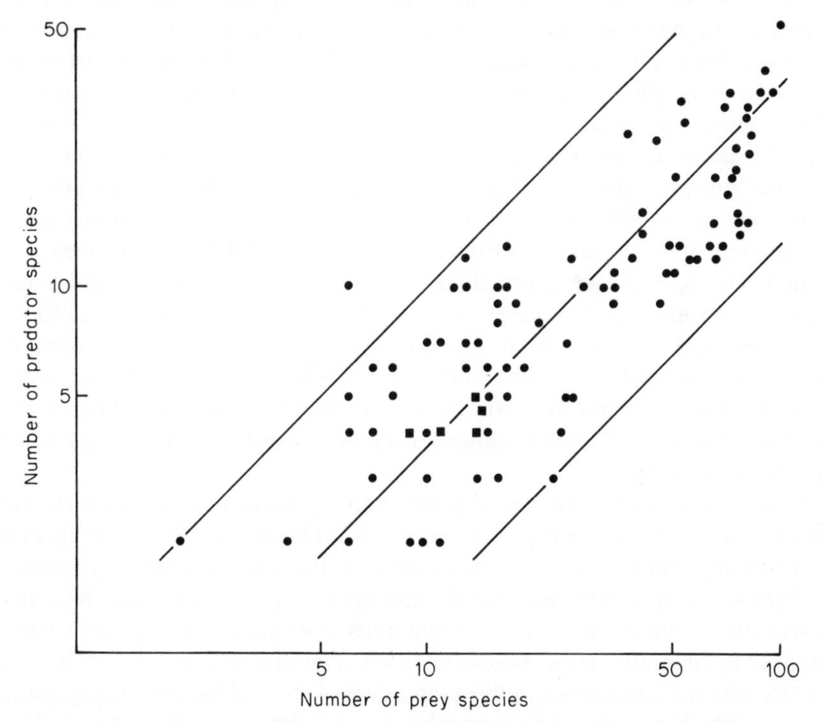

Figure 14.2. Number of predator and prey species in freshwater communities. The central line is 2.7 prey per predator, the mean of the two sets and very close to the reduced major axis. The upper line is the 1:1 ratio, the lower 7.3:1 ($7.3 = 2.7^2$). Square points relate to two identical observations. Modified from Mithen and Lawton (1986), data from Jeffries and Lawton (1985)

possible to be certain what would happen to a new species, though a prey invasion is more likely at the first and a predator one at the second, assuming that predator–prey relationships are important in invasion. In between, more or less anything could happen on the basis of the information in this graph. Even in models that produce qualitatively similar relationships (Mithen and Lawton, 1986), predicting, on the basis of similarity to other members of the web, survival time of species invading the foodweb is impossible.

The predator–prey ratio has also been studied by Cohen and his colleagues, as for instance in Newman and Cohen (1986), marrying ingenious theory with extensive survey. There is a worry, voiced for instance by May (1983), that the studies surveyed are more informative about the way ecological surveys are done than about ecological systems. The predator–prey ratios found in this work differ appreciably from those of Figure 14.2. Until these differences in predator–prey ratios are resolved, there will be even more uncertainty about using them for explanation, let alone prediction.

The other aspect of food webs that I think relevant to the study of invasions is the use of computer models of imaginary species. The most notable of these are by Post and Pimm (1983) and Drake (1985), both using Lotka–Volterra equations, and arbitrarily selected parameters, but giving reasons for their choice of method. The number of parameters is unfortunately large. The possible types of equilibrium were first expounded in an ecological context by Lotka (1924) and are well known. Foa a set of species represented by differential or difference equations, a stability matrix, often called the community matrix, can be formulated, and the stability of the set determined from the eigenvalues of the matrix. The two representations, difference and differential, can be mapped one to one, and so are equally stable as far as this form of stability is concerned (Williamson, 1987). Invasion is only possible if the equilibrium is unstable, but the invader may still go extinct. Many remarkable results can be found in more general models with non-transitive relationships or non-linearities (Gilpin, 1975; May and Leonard, 1975). An example of a non-transitive system is when A beats B beats C beats A.

Different conditions are needed to determine permanent coexistence (Hutson and Moran, 1982; Hutson and Vickers, 1983; Hutson and Law, 1985). These apply to any number of species in a Lotka–Volterra system, but as yet only to systems with three or less species in the more general case. A set of invaders can be represented by an extension to the community matrix, called the invasion matrix (Reed and Stenseth, 1984; Stenseth, 1986). The eigenvalues of this invasion matrix, which are mathematically independent of those of the community matrix, indicate whether an invasion is possible, but not whether it will succeed. These difficulties perhaps indicate some reasons why the computer simulations on the whole did not produce clear-cut results.

Both Post and Pimm (1983) and Drake (1985) found that successful invaders in their models differed only slightly, and on average, from the total set of species considered. The results are in the direction expected. Successful invaders have lower intrinsic mortality, better productivity and are less affected by predators, but, as these are statistical effects, there are plenty of individual exceptions. Note that the models avoided the complications discussed by Hutson and his colleagues by considering only stable and unstable equilibrium points, not trajectories. Bender et al. (1984) and Hastings (1986) using pure competition models, the antithesis of the pure predator–prey models, again found much complexity and surprising results.

One problem with these complex food-chain models is relating them to what is measured in the field. To take the simplest set of modified Lotka–Volterra equations representing a self-limited producer p and a herbivore h (with t time and the other terms all constants),

$$\mathrm{d}\ln p/\mathrm{d}t = r_p - ap - bh$$
$$\mathrm{d}\ln h/\mathrm{d}t = cp - e. \tag{12}$$

In these, r_p is the intrinsic rate of natural increase of the producer, but e is merely a death rate, the rate of decline of the herbivore population under starvation. The value of r for h must be the maximum rate of increase possible in the system, which is at the asymptote for p alone, namely r_p/a so

$$r_h = (cr_p/a) - e. \tag{13}$$

If another producer is now introduced, r_h will increase. The herbivores do better on a mixed diet. It is perhaps not surprising then that polyspecies models have a multitude of possible outcomes.

All these community models are equilibrium models. That is, the starting point as well as the end point is an equilibrium. In contrast, the models of up to three species discussed in Section 14.2 include non-equilibrium behaviour as well, the behaviour of the system in transit between equilibrium points. Real communities are probably never at an equilibrium point, and certainly never stationary (Williamson, 1987a). Disturbance makes invasion easier, and disturbed communities are likely to be further from equilibrium (Rejmanek, 1984). So it is quite possible that non-equilibrium community models, when these are developed, will be more informative than those considered here.

14.5 CONCLUSIONS

There is a very great deal of empirical data about invasions, as the chapters in this book show. There are also very many models that might be of use in understanding invasions. The major deficiency is in connecting the two. Much thought needs to be given to defining parameters, reducing their number where possible, and to testing the validity of equations. A useful way forward would be to get many more measurements in the field of those parameters that have been discussed, such as the intrinsic rate of increase, the reproductive capacity and the diffusivity. Indeed, the models that have been the most successful, for instance those in epidemiology, are those whose development has seen close collaboration between modellers and experimentalists, and this can be expected to be the case in future. Even so, as we are discussing models of real, complex, ecosystems, the best that can be expected for some time will be an improvement in understanding; useful models for prediction will remain rare.

ACKNOWLEDGEMENTS

I am grateful to Mr I. A. W. Macdonald for providing me with the necessary pages of Millar (1980). Dr B. Shorrocks kindly elaborated on the values of the parameters need for invasion in his model. I thank Drs M. J. Crawley, A. H. Fitter, R. Law, J. H. Lawton, S. A. Levin, M. Rejmanek, M. B. Usher and Charlotte Williamson for comment and discussion on many points. I am grateful to the members of the modelling working party, Drs D. A. Andow, J. A. Drake, S. A. Levin, I. Noy-Meir, S. J. Pimm, N. C. Stenseth and M. Rejmanek, for their contributions.

REFERENCES

Abrams, P. (1983). The theory of limiting similarity. *A. Rev. Ecol. Syst.*, **14**, 359–76.

Allee, W. C., Emerson, A. E., Park, O., Park, T., and Schmidt, K. P. (1949). *Principles of Animal Ecology*. Saunders, Philadelphia.

Anderson, R. M., and May, R. M. (1983). The population dynamics of microparasites and their invertebrate hosts. *Phil. Trans. R. Soc. Lond. B*, **291**, 451–524.

Anderson, R. M., and May, R. M. (1986). The invasion, persistence and spread of infectious diseases within animal and plant communities. *Phil. Trans. R. Soc. Lond. B*, **314**, 533–70.

Armstrong, R. A. (1978). A note on the demography of colonization. *Amer. Nat.*, **112**, 243–5.

Bailey, N. T. J. (1975). *The Mathematical Theory of Infectious Diseases and its Application.* Griffin, London.

Banks, H. T., Kareiva, P. M., and Lamb, P. K. (1985). Modeling insect dispersal and estimating parameters when mark-release techniques may cause initial disturbances. *J. Math. Biology*, **22**, 259–77.

Banks, H. T., Kareiva, P. M., and Murphy, K. A. (1986). *Parameter Estimation Techniques for Interaction and Redistribution Models of Species Interactions: a Predator-Prey Example.* Lefschetz Center for Dynamical Systems, Brown University, Providence, RI. USA Report No. 86–29. 26 pp.

Bartlett, M. S. (1957). Measles periodicity and community size. *J. R. Statist. Soc. A*, **120**, 48–70.

Beddington, J. R. (1983). Review of 'The mathematical theory of the dynamics of biological populations II', *Biometrics*, **39**, 536.

Beirne, B. P. (1975). Biological control attempts by introductions against pest insects in the field in Canada. *Can. Ent.*, **107**, 225–36.

Bender, E. A., Case, T. J., and Gilpin, M. E. (1984). Perturbation experiments in

community ecology: theory and practice. *Ecology*, **65**, 1–13.
Bengtsson, J. (1986). Life histories and interspecific competition between three *Daphnia* species in rockpools. *J. Anim. Ecol.*, **55**, 641–55.
Boyce, M. S. (1984). Restitution of r- and K-selection as a model of density-dependent selection. *A. Rev. Ecol. Syst.*, **15**, 427–43.
Brown, G., and Sanders, J. W. (1981). Lognormal genesis. *J. Appl. Prob.*, **18**, 542–7.
Carothers, J. H. (1986). Homage to Huxley: on the conceptual origin of minimum size ratios among competing species. *Amer. Nat.*, **128**, 440–2.
Connell, J. H., and Slatyer, R. O. (1977). Mechanisms of succession in natural communities and their role in community stability and organisation. *Amer. Nat.*, **111**, 1119–44.
Crawley, M. J. (1983). *Herbivory*. Blackwell, Oxford.
Crawley, M. J. (1986). The population biology of invaders. *Phil. Trans. R. Soc. Lond. B*, **314**, 711–31.
Dobzhansky, T., Lewontin, R. C., and Pavlovsky, O. (1964). The capacity for increase in chromosomally polymorphic and monomorphic populations of *Drosophila pseudoobscura*. *Heredity*, **19**, 597–614.
Drake, J. A. (1985). Some theoretical and empirical explorations of structure in food webs. Ph.D. thesis, Purdue University.
Drake, J. A., and Williamson, M. H. (1986). Species invasions of natural communities. *Nature, Lond.*, **319**, 718–19.
Ehler, L. E., and Hall, R. W. (1982). Evidence for competitive exclusion of introduced natural enemies in biological control. *Env. Ent.*, **11**, 1–4.
Fenchel, T. (1974). Intrinsic rate of natural increase: the relationship with body size. *Oecologia*, **14**, 317–26.
Fisher, R. A. (1937). The wave of advance of advantageous genes. *Ann. Eugen.*, **7**, 355–69.
Gilpin, M. E. (1975). Limit cycles in competition communities. *Amer. Nat.*, **109**, 51–60.
Goldberg, D. E., and Werner, P. A. (1983). Equivalence of competitors in plant communities: a null hypothesis and a field experimental approach. *Amer. J. Bot.*, **70**, 1098–104.
Gray, A. J. (1986). Do invading species have definable genetic characteristics? *Phil. Trans. R. Soc. Lond. B*, **314**, 675–93.
Green, R. F. (1986). Does aggregation prevent competitive exclusion? A response to Atkinson and Shorrocks. *Amer. Nat.*, **128**, 301–4.
Grime, J. P. (1986). The circumstances and characteristics of spoil colonization within a local flora. *Phil. Trans. R. Soc. Lond. B*, **314**, 637–54.
Grubb, P. J. (1985). Plant populations and vegetation in relation to habitat disturbance and competition: problems of generalization. In White, J. (Ed.), *The Population Structure of Vegetation*, pp. 595–621. Dr W. Junk Publishers, Dordrecht.
Hanski, I. (1987). Colonization of ephemeral habitats. *Symp. Br. Ecol. Soc.*, **26**, 155–85.
Hastings, A. (1986). The invasion question. *J. Theor. Biol.*, **121**, 211–20.
Holdgate, M. W. (1986). Summary and conclusions: characteristics and consequences of biological invasions. *Phil. Trans. R. Soc. Lond. B*, **314**, 733–42.
Howe, R. W. (1953). Studies on beetles of the family Ptinidae. VIII. The intrinsic rate of increase of some ptinid beetles. *Ann. Appl. Biol.*, **40**, 121–33.
Howe, R. W. (1955). Studies on beetles of the family Ptinidae. 12. The biology of *Tipnus unicolor* Pill. and Mitt. *Ent. Mon. Mag.*, **91**, 253–7.
Howe, R. W. (1959). Studies on beetles of the family Ptinidae. XVII. Conclusions and additional remarks. *Bull. Ent. Res.*, **50**, 287–326.
Howe, R. W., and Bull, J. O. (1956). Studies on beetles of the family Ptinidae. 13. The oviposition rate of *Pseudeurostus hilleri* (Reitt.). *Ent. Mon. Mag.*, **92**, 113–15.
Hutson, V., and Law, R. (1985). Permanent coexistence in general models of three

interacting species. *J. Math. Biol.*, **21**, 285–98.

Hutson, V., and Moran, W. (1982). Persistence of species obeying difference equations. *J. Math. Biol.*, **15**, 203–13.

Hutson, V., and Vickers, G. T. (1983). A criterion for the permanent coexistence of species with an application to a two-prey one-predator system. *Math. Biosci.*, **63**, 253–69.

Jeffries, M. J., and Lawton, J. H. (1985). Predator-prey ratios in communities of freshwater invertebrates: the role of enemy free space. *Freshwater Biology*, **15**, 105–12.

Joy, N. H. (1976). *A Practical Handbook of British Beetles.* Reprinted edition. Classey, Faringdon.

Kendall, D. G. (1948). A form of wave propagation associated with the equation of heat conduction. *Proc. Camb. Phil. Soc.*, **44**, 591–4.

Kermack, W. O., and McKendrick, A. G. (1927). Contributions to the mathematical theory of epidemics. *Proc. R. Soc. Lond. A*, **115**, 700–21.

Kolmogoroff, A. N., Petrovsky, I. G., and Piscounoff, N. S. (1937). A study of diffusion with growth of a quantity of matter and its application to a biological problem. *Bull. de l'Univ. d'Etat a Moscou* (*ser. internat.*), *A*, **1**(6), 1–25. (in Russian and in French).

Kot, M., and Schaffer, W. C. (1986). Discrete-time growth-dispersal models. *Math. Biosci.*, **79**, 1–28.

Kruger, F. J., Richardson, D. M., and van Wilgen, B. W. (1986). Processes of invasion by alien plants. In: MacDonald, I. A. W., Kruger, F. J., and Ferrar, A. A. (Eds), *The Ecology and Management of Biological Invasions in Southern Africa.* Oxford University Press, Cape Town.

Lawton, J. H. (1987). Are there assembly rules for successional communities? *Symp. Br. Ecol. Soc.*, **26**, 225–44.

Lawton, J. H., and Brown, K. C. (1986). The population and community ecology of invading insects. *Phil. Trans. R. Soc. Lond. B*, **314**, 607–17.

Lawton, J. H., and Hassell, M. P. (1984). Interspecific competition in insects. In: Huffaker, C. B., and Rabb, R. L. (Eds), *Ecological Entomology*, pp. 451–95. John Wiley & Sons, New York.

Levin, S. A. (1986). Population models and community structure in heterogeneous environments. In: Hallam, T. G., and Levin, S. A. (Eds), *Mathematical Ecology*, pp. 295–320, Biomathematics 17. Springer-Verlag, Berlin.

Lotka, A. J. (1924). *Elements of Physical Biology.* Williams and Wilkins, Baltimore.

Maaløe, O., and Kjeldgaard, N. O. (1986). *Control of Macromolecular Synthesis.* W. A. Benjamin, New York and Amsterdam.

MacArthur, R. H. (1962). Some generalized theorems of natural selection. *Proc. Nat. Acad. Sci. U. S.*, **51**, 1207–10.

MacArthur, R. H., and Wilson, E. O. (1967). *The Theory of Island Biogeography.* Princeton University Press, Princeton.

McGowan, J. A., and Walker, P. W. (1985). Dominance and diversity maintenance in an oceanic ecosystem. *Ecol. Mon.*, **55**, 103–18.

May, R. M. (1983). The structure of food weds. *Nature Lond.*, **301**, 566–8.

May, R. M., and Leonard, W. J. (1975). Nonlinear aspects of competition between three species. *SIAM J. Appl. Math.*, **29**, 243–53.

Medawar, P. B. (1982). *Pluto's Republic.* Oxford University Press.

Millar, J. C. G. (1980). Aspects of the ecology of the American grey squirrel *Sciurus carolinensis* Gmelin in South Africa. M.Sc. thesis, University of Stellenbosch.

Mithen, S. J., and Lawton, J. H. (1986). Food-web models that generate constant predator-prey ratios. *Oecologia*, **69**, 542–50.

Mollison, D. (1977). Spatial contact models for ecological and epidemic spread. *J. R. Statist. Soc. B*, **39**, 283–326.

Mollison, D. (1984). Simplifying simple epidemic models. *Nature, Lond.*, **310**, 224–5.

Mollison, D. (1986). Modelling biological invasions: chance, explanation, prediction. *Phil. Trans. R. Soc. Lond. B*, **314**, 675–93.
Montroll, E. W., and Shuler, K. E. (1979). Dynamics of technological evolution: random walk model for the research enterprise. *Proc. Nat. Acad. Sci. U. S.*, **76**, 6030–4.
Moran, P. A. P. (1962). *The Statistical Processes of Evolutionary Theory*. Clarendon Press, Oxford.
Murray, J. D. (1977). *Lectures on Non-linear-differential-equation Models in Biology*. Clarendon Press, Oxford.
Newman, C. M., and Cohen, J. E. (1986). A stochastic theory of community food webs. IV. Theory of food chain lengths in large webs. *Proc. R. Soc. Lond. B*, **228**, 355–77.
Newsome, A. E., and Noble, I. R. (1986). Ecological and physiological characters of invading species. In: Groves, R. H., and Burdon, J. J. (Eds), *The Ecology of Biological Invasions: an Australian Perspective*, pp. 1–21. Cambridge University Press, Cambridge.
O'Connor, R. J. (1986). Biological characteristics of invaders among bird species in Britain. *Phil. Trans. R. Soc. Lond. B*, **314**, 583–98.
Okubo, A. (1980). *Diffusion and Ecological Problems: Mathematical Models*, Biomathematics 10. Springer-Verlag, Berlin.
Orians, G. H. (1986). Site characteristics favoring invasions. In: Mooney, H. A., and Drake, J. A. (Eds), *The Ecology of Biological Invasions of North America and Hawaii*, pp. 133–47. Springer-Verlag, New York.
Pielou, E. C. (1977). *Mathematical Ecology*. John Wiley & Sons, New York.
Pimm, S. L. (1982). *Food Webs*. Chapman and Hall, London.
Post, W. M., and Pimm, S. L. (1983). Community assembly and food web stability. *Math. Biosci.*, **64**, 169–92.
Reed, J., and Stenseth, N. C. (1984). On evolutionary stable strategies. *J. Theor. Biol.*, **108**, 491–508.
Rejmanek, M. (1984). Perturbation-dependent coexistence and species diversity in ecosystems. In: Schuster, P. (Ed.), *Stochastic Phenomena and Chaotic Behaviour in Complex Systems*, pp. 220–30. Springer-Verlag, Berlin.
Roughgarden, J. (1986). Predicting invasions and rate of spread. In: Mooney, H. A., and Drake, J. A. (Eds), *The Ecology of Biological Invasions of North America and Hawaii*, pp. 179–81. Springer-Verlag, New York.
Salisbury, E. J. (1942). *The Reproductive Capacity of Plants*. Bell, London.
Shorrocks, B., and Rosewell, J. (1986). Guild size in Drosophilids: a simulation study. *J. Anim. Ecol.*, **55**, 527–41.
Simberloff, D. (1981). Community effects of introduced species. In: Nitecki, M. H. (Ed.), *Biotic Crises in Evolutionary Time*, pp. 53–81. Academic Press, New York.
Skellam, J. G. (1951). Random dispersal in theoretical populations. *Biometrika*, **38**, 196–218.
Smith, D., and Keyfitz, N. (1977). *Mathematical Demography*, Biomathematics 6. Springer-Verlag, Berlin.
Smith, F. E. (1954). Quantitative aspects of population growth. In: Boell, E. (Ed.), *Dynamics of Growth Processes*, pp. 274–94. Princeton University Press, Princeton.
Stearns, S. C. (1977). The evolution of life history traits: a critique of the theory and a review of the data. *A. Rev. Ecol. Syst.*, **8**, 145–71.
Stenseth, N. C. (1986). Darwinian evolution in ecosystems: a survey of some ideas and difficulties together with some possible solutions. In: Casti, J. L., and Karlqvist, A. (Eds), *Complexity, Language and Life: Mathematical Approaches*, pp. 105–45. Biomathematics 16. Springer-Verlag, Berlin.
Swartzmann, G. L., and Zaret, T. M. (1983). Modeling fish species introduction and prey extermination: the invasion of *Cichla ocellaris* to Gatun Lake, Panama. In: Lauenroth,

W. K., Skogerboe, G. V., and Flug, M. (Eds), *Developments in Environmental Modelling 5. Analysis of Ecological Systems: State-of-the-Art In Ecological Modelling.* Proceedings of a symposium held from 24 to 28 May at Colorado State University, Fort Collins. Elsevier, Amsterdam.

Tilman, D. (1977). Resource competition between planktonic algae: an experimental and theoretical approach. *Ecology*, **58**, 338–48.

Usher, M. B. (1986). Invasibility and wildlife conservation: invasive species on nature reserves. *Phil. Trans. R. Soc. Lond. B*, **314**, 695–710.

Usher, M. B. (1987). Modelling successional processes in ecosystems. *Symp. Br. Ecol. Soc.*, **26**, 31–55.

Vermeij, G. J. (1978). *Biogeography and Adaptation.* Harvard University Press, Cambridge, Mass.

Walker, J., Sharpe, P. J., Penridge, L. K., and Wu, H. (1986). Competitive interactions between individuals of different size: the concept of ecological fields. Technical Memorandum 86/11, CSIRO Institute of Biological Resources, Canberra.

Watt, K. E. F. (1975). Critique and comparison of biome ecosystem modeling. In: Patten, B. C. (Ed.), *Systems Analysis and Simulation in Ecology*, **3**, 139–52. Academic Press, New York.

Williamson, M. H. (1957). An elementary theory of interspecific competition. *Nature, Lond.*, **180**, 422–5.

Williamson, M. H. (1972). *The analysis of biological populations.* Edward Arnold, London.

Williamson, M. H. (1981). *Island Populations.* Oxford University Press, Oxford.

Williamson, M. H. (1987). Are communities ever stable? *Symp. Br. Ecol. Soc.*, **26**, 353–71.

Williamson, M. H. (1988). The relationship of species number to area, distance and other variables. In: Myers, A. A., and Giller, P. S. (Eds), *Biogeographic Analysis: Methods, Pattern and Processes.* Chapman and Hall, London.

Williamson, M. H., and Brown, K. C. (1986). The analysis and modelling of British invasions. *Phil. Trans. R. Soc. Lond. B*, **314**, 505–22.

Zaret, T. M., and Paine, R. T. (1973). Species introduction in a tropical lake. *Science*, **182**, 449–55.

Biological Invasions: a Global Perspective
Edited by J. A. Drake *et al.*
© 1989 SCOPE. Published by John Wiley & Sons Ltd

CHAPTER 15

Theories of Predicting Success and Impact of Introduced Species

STUART L. PIMM

15.1 INTRODUCTION

The purpose of this paper is to review some theoretical ideas relating to two questions: can we predict whether an introduced species will successfully invade a community and, if it does, can we predict whether that species will have a disruptive effect on the community? I shall use empirical studies to examine whether the theoretical ideas are reasonable. This demonstration is strictly limited in its scope, as I shall rely heavily on sources which deal with introduced vertebrates. For these taxa, the theories do seem reasonable, but this is not necessarily true of other groups of organisms. I shall try to indicate where the theories will prove to be inadequate for these other organisms.

This paper is in three sections. In the first section, I shall address the problems a species must overcome when it first arrives in a new community—essentially a species-oriented view of the problem of invasion. In the second section, I shall address the resistance to invasion a community may possess, which is the community-oriented view of the problem of invasion. I am not convinced that we can totally separate the prediction of invasion into species and community components, rather it is more likely that we must look to the interaction of species and communities. None the less, the division I shall use is a reasonable way of getting across many of the major ideas. In the third section of the paper, I shall consider the effects of introduced species.

15.2 THE PROBLEMS OF SMALL POPULATIONS

There are many cases of successfully introduced species which started with tiny founder populations. The lower limit in population size for a successful invasion is demonstrated by one pregnant female in the anecdote about the colonization of the Aland archipelago (in the Baltic) by the red squirrel (Jarvinen, personal communication). Some introductions have been made with such large numbers that the problems besetting small populations are irrelevant. But many introductions, particularly accidental ones, have been made with small numbers of individuals. For these, success will not be certain: many more than one pregnant female may be required for success. Small populations are prone to extinction and I shall argue that some species are much more likely to become extinct at a given small population size than others. Clearly, such extinction-prone species are less likely to be successful invaders.

15.2.1 Theory

The theory of extinctions for small populations has usually been presented in the context of extinctions—not introductions. The results, as I have argued, seem readily applicable. Diamond (1984) and Pimm *et al.*, (1988) provide a review of several ideas.

The chance of extinction rapidly increases as population sizes decrease. Even in a perfectly constant environment, small populations face risk of extinction from demographic accidents—the chance fluctuations of deaths and births, and consequent changes in numbers and sex ratios. There are three other factors:

1. Populations are more likely to become extinct if their numbers fluctuate considerably.
2. Populations of long-lived species will have a lower risk of extinction, per year (but not per generation) than short-lived species.
3. Populations with a low intrinsic rate of increase should have an increased risk of extinction because they recover more slowly from reductions in numbers.

These expectations are based on a number of mathematical models: the case of constant environments was considered by MacArthur and Wilson (1967), and Richter-Dyn and Goel (1982): both constant and fluctuating environments were considered by Leigh (1975, 1981). The exact formulations of the models seem less important to me than their general implications, which seem reasonably independent of the exact equations used.

The three factors mentioned above are not independent of each other. Theoretically, there is likely to be a relation between the intrinsic rate of increase, r and a population's long-term variability in density. A high r means a fast recovery from low numbers. Of course, too high an r can mean the population repeatedly overshoots its equilibrium density. So, the sign of the correlation between r and variability is not obvious. For some bird populations the sign is negative; high r does confer some population stability, but the relationship is a relatively weak one (Pimm, 1984). Still, the basic result is that high r and low variability are both advantageous and that the former helps to impose the latter.

There is a much stronger, *negative* correlation between how long organisms live and their r. From protozoa to elephants, and animals in between, there is an order-of-magnitude reduction in r for every order-of-magnitude increase in longevity. Both r and longevity are closely correlated with body size (Peters, 1983).

Thus, we can ask in a given period, is a small, short-lived species with high r more or less likely to become extinct than a large, long-lived species with correspondingly lower r, given that both have the same population size? Sample calculations, using Leigh's equations, show that the combination of small body size and high r is nearly always advantageous. The exception occurs at very low population densities of about half a dozen pairs or less. At these low densities, the extinction rate of large, long-lived species in lower.

15.2.2 Empirical tests 1: island birds

Diamond (1984) has assembled a large number of studies that document the rapidly increased chance of extinction as a population becomes small. But how should we test the additional predictions of the theory? We would need a data base that includes many observations of extinctions under 'natural conditions,' in which species have not been inexorably driven to extinction by man-made changes. Pimm *et al.* (1988) used the data on the repeated, annual, breeding censuses of 16 British islands. These have already been analyzed by several previous studies of species turnover (Lack, 1969; Diamond and May, 1977; Williamson, 1982; Diamond 1984). For each of these islands, the number of breeding pairs of land birds has been determined in consecutive years, for several decades. Some populations became extinct, others were founded, while others underwent successive extinctions and recolonizations. The islands are of different sizes, and the populations exhibit a range of long-term variations in numbers.

These data, too, show the overwhelming effect of population size on extinction rate. The risk of extinction drop steeply with increasing population size. There is, however, a considerable amount of scatter about this relationship. Pimm *et al.* were able to relate this to the three theoretical factors by first partially correcting the data for the dominant effects of population size. They divided the species into two groups: 'large' species were non-passerines plus the Corvidae (crows, jays, etc.) whereas 'small' species were the passerines minus the Corvidae. Large species were expected to have lower r's and longer generation times than small species. The critical results were:

1. Small, short-lived species were less prone to extinction at densities above seven pairs per island, than large, long-lived species.
2. Below seven pairs per island, large, long-lived species were less likely to become extinct than the small, short-lived species.
3. In both large and small species, and at all population densities, species with highly variable numbers were more likely to become extinct than species with relatively constant numbers. Wrens (*Troglodytes troglodytes*), for example, are small insectivores whose numbers are greatly decreased by occasional cold winters. Their numbers are particularly variable, and, for their population sizes, they are particularly prone to extinction.

15.2.3 Empirical tests 2: island insects

Lawton and Brown (1986) have examined the predictors of success among the various animal taxa introduced to Britain. Comparisons of nematodes, spiders, molluscs, insects, flatworms and various vertebrates, show there is a weak, positive correlation between body size and the chance of successful invasion. There is, therefore, an implied negative correlation between invasion success and r. Lawton and Brown argue that the smaller species may fail more often because their small size may make them more vulnerable to the vagaries of the British climate, or because small size often correlates across taxa with the magnitude of population fluctuations. Within the insects, there is a strong, negative correlation of invasion success with body size, implying a positive correlation of invasion success with r. Lawton and Brown write: 'by concentrating on one group of organisms, we may have reduced the range of variation encountered in the amplitude of population fluctuations, making... [intrinsic growth rate and equilibrium density]... rather than... variation in population size the most important determination of population establishment.' There are obvious difficulties with studies as broad ranging as this one, but it tentatively supports the importance of intrinsic growth rate as a predictor of invasion success. Williamson (this volume) also discusses the relationship between r and the chance that a species will successfully invade.

15.2.4 Empirical tests 3: Hawaiian vertebrates

For their size, the Hawaiian Islands have received more vertebrate introductions than anywhere else on earth. These introductions have been the subject of a number of studies by Moulton (Moulton, 1985; Moulton and Pimm, 1983, 1986a, 1986b, 1987). These introductions afford unusual opportunities to test various ideas on species invasions, independent of the ideas on intrinsic growth rate and population variability that I have been discussing.

Moulton and Pimm (1986a) first considered the area of orgin of the bird introductions to see if different faunal regions (palearctic, ethiopian, etc.) were more likely to be sources for successful introductions. They were not. Surprisingly, even if it was known whether the bird introductions came from the topics or temperate regions, invasion success could not be predicted. Of the species that could be strictly attributed to either temperate or tropical areas, 13 to 23 temperate species were successful, but only 14 of 28 tropical species were successful. The Hawaiian Islands, of course, are tropical and the introduced species are largely confined to the lowlands, rather than the more temperate montane areas of the islands.

I take this result to mean that species possess broad, physiological tolerances. We would certainly expect that many introductions will fail because they involve a total mismatch between the introduced species and its environment. None the less, the abundance of temperature forest birds in tropical, wet, lowland forest on

Hawaii is graphic evidence that such 'physiological mismatches' are not the only, or even likely, the most important factor, in determining which species introductions succeed.

Moulton and Pimm did find that the size of a species range predicted invasion success. Species with larger ranges were more likely to be successful than species with smaller ranges. Of course, this might be because species with larger ranges may have been introduced earlier. Species with larger ranges are more likely to be subject to man's attempts at moving them than species with geographically restricted ranges. Date of introduction is an important predictor of success, as species introduced earier faced fewer competitors (see below).

Finally, Moulton and Pimm examined differences between taxa. Bird introductions were more likely to fail than mammal or reptile introductions, for reasons that are not obvious.

Many introductions will succeed only if their numbers can increase quickly, beyond the small population size where extinction is likely. Individual species characteristics, especially the intrinsic growth rate and the propensity for densities to vary, are important predictors of extinction of small populations. These characteristics should also be important predictors of invasion success and some empirical studies suggest that they are. These are not the only population-oriented characteristics that may be useful in predicting success. There are also differences between taxa that do not readily fit into this theoretical framework. For at least some species, the match between the habitats from where the species originated to where it is introduced, does not predict whether the species will succeed.

15.3 COMMUNITY STRUCTURE AND CHANCE OF INVASION

Community characteristics are also likely to modify whether an introduction succeeds. There are many starting points for this discussion. There has been a prolonged debate over community structure and competitive exclusion, which dates from at least 1944 (Ano., 1944) and continues today in the arguments between Florida and California schools (Lewin, 1983a, 1983b). Elton (1946) argued that community patterns could be interpreted to mean that communities were excluding species—i.e. they were probably hard to invade. Others, taking Williams' (1964) lead, have argued that the data imply that many species are absent from a community because they cannot reach that community.

15.3.1 Models of community assembly

I shall take a rather different starting point, and consider models of community assembly (Post and Pimm, 1983). These models use differential-equation systems to model the sequential addition of species to communities. The essential features are illustrated by Figure 15.1. The models start with a given number of plant

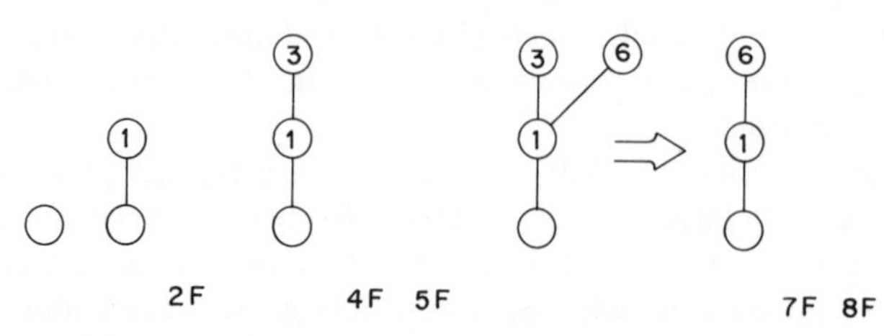

Figure 15.1. Diagram of community assembly process discussed in the text. Circles represent species, lines represent feeding interactions between species. The model starts out with a plant, and then adds successive animal species. Species 1, 3, and 6 succeed, species 2, 4, 5, 7, and 8 fail. Species 6 causes the extinction of species 3

species, then attempt to add animals (and, in some unpublished simulations, both animals and plants). Some attempts fail because the species cannot increase when rare. Other introductions succeed and add to the existing community. Yet other introductions succeed and replace species, or cause widespread extinctions. The principal results (Figure 15.2) involve the number of species in the model communities and the number of attempted introductions it takes before an introduction succeeds. Both these variables are plotted in terms of the total number of attempted invasions. The exact details of the simulations vary slightly with different assumptions. The major results, however, are robust.

Species numbers increase rapidly at first, but then rise to an asymptote. The final level of species numbers depends on the *connectance* of the community. Connectance is the number of actual inter-species interactions divided by the possible number of inter-species interactions. (In an n-species system there are $n(n-1)/2$ possible interactions). Connectance enters as a parameter in the models, by choosing how many species of predator and of prey with which the introduced species is likely to interact. Communities with high connectance have few species compared to those with low connectance. (This is a result well-known from analyses of food web models (Pimm, 1982).)

The difficulty of invasion continuously increases with the accumulated number of attempted invasions. Communities with high connectance are much harder to invade than those with low connectance. In simple terms, high connectance means that the invading species suffers from many competitors and predators.

I interpret these results to show two effects on the difficulty of invading a community. First, a community with relatively few species is likely to be easier to invade than community with many species. This is a familiar result, as it is the same as that suggested first by Elton: the presence of lots of competitors and predators confers on a community a 'biotic resistance' (Simberloff, 1986). Second, once the cummunity has reached its equilibrium number of species, the process of community assembly itself makes the community harder to invade. Simply, each successive, successful invasion makes the community less likely to be invaded.

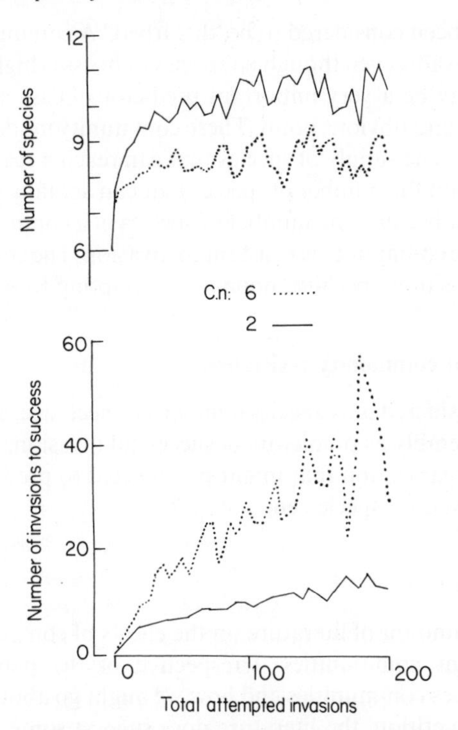

Figure 15.2. Typical results of community assembly models, simplified from Post and Pimm (1983). As assembly proceeds (measured by the total number of attempted invasions), the number of species increases (n) then asymptotes. The number of attempted invasions, before one successfully invades the community, increases continuously, even after the number of species has levelled off. The details of the simulations depend on connectance (C; defined in the text) and n

Ecologists are well aware that 'disturbed' communities seem easy to invade. I suggest that ecologists often use 'disturb' in three separate ways. The first meaning describes communities which are clearly short of individuals. For example, there may be an abundance of bare ground in which plants can germinate. The significance of this is not apparent in the assembly models, because they only consider communities where the individuals have had time to reach their equilibrium densities. The second meaning describes communities that are short of species, i.e. they are 'unsaturated.' The third meaning is more elusive. Communities, whose species compositions are of only recent origins, in the sense that they are not the end point of a long process of community assembly, might also have been considered to be 'disturbed.' A community of recent origin may be easy to invade, even though its species richness is high. Simply, the 'age' of a community may be a very important predictor of ease of invasion.

There is a final and obvious point. These community models assume an infinite supply of species. The supply of species that can reach a real community may be strictly limited, and the number of species that can actually survive there may be even smaller. Just because the number of species in a community is limited, does not mean that the community is resistant to invasion. The community may not be resisting species, simply because none are attempting to invade.

15.3.2 Details of community resistance

So far, I have considered only species number, connectance, and the history, of the community's assembly as predictors of successful invasion. What are the details of community organization that might be expected to predict, more closely, the difficulties an invading species encounters?

15.3.2.1 Competitors

There is a large amount of literature on the effects of competition in structuring animal and plant communities. Irrespective of the debate about whether competition shapes communities and how we might go about demonstrating the existence of competition, the literature does suggest some interesting ideas.

First, there is the effect of morphological similarity. Lack (1947) and later Hutchinson (1959) suggested that morphologically similar species might be unable to coexist. Second, there is the effect of taxonomic similarity. Elton (1946) suggested that there might be limits to the number of species per genus that could coexist.

Both morphology and taxonomy are merely surrogates for ecology. We might use any one of a large number of ecological measurements of similarity and predict that the more similar the invading species to a species already present, the less likely is the success of the invader.

15.3.2.2 Predators

The more predation a species suffers, the less likely it is to successfully invade. But how should we measure this? A rather surprising result about predation is incorporated in the discussions of 'enemy free space,' whose significance has been discussed by Jeffries and Lawton (1984, 1985) and Holt (1977, 1984). Recall an empirical observation: the numbers of species of predator and the number of species of prey are closely correlated over communities that contain various numbers of species (Cohen, 1977; Briand and Cohen, 1984; Jeffries and Lawton,

1984). Jeffries and Lawton argue that we cannot expect such a good correlation on the basis of the number of predatory species being determined solely by the number of prey species. For such a good correlation, the numbers of species of predators may, in addition determine the number of species of prey. For communities where predators exercise a controlling influence on their prey species, one would expect a community with a single predator to be harder to invade than one with several predator species. This leads to the superficially surprising prediction that, for a given number of prey species, communities with the most species of predator may be the easiest for prey species to invade.

15.3.3 Empirical studies 1: the role of competition

15.3.3.1 Hawaiian birds

Many studies of introduced species deal with relatively small numbers of introductions, so they cannot be expected to test the ideas suggested above. The Hawaiian Islands provide an obvious system on which to test ideas of community resistance, because they have received so many introductions. Moreover, the fate of the introductions is relatively well known.

There are virtually no native land birds below 1000 m on the Hawaiian Islands, and the islands are too remote to receive anything other than very infrequent avian visitors from the mainland. There are predators, on the islands: the mongoose (*Herpestes auropunctatus*), the native Hawaiian hawk (*Buteo solitarius*), and two species of owls (the barn owl, *Tyto alba*, is introduced, whereas the shorteared owl, *Asio flammeus*, is native). But, with the exception of the mongoose's absence from Kauai, and the hawks' presence only on Hawaii, each island has the same set of predators. The lowland communities are thus isolated and form close-to-ideal, experimental systems to test ideas on community assembly.

It there any evidence of community-wide effects in determining the success of introduced species? The conclusions of several of Moulton's studies (1985; Moulton and Pimm, 1983; Moulton and Pimm, 1986a, b) are, in brief:

1. Species introductions were generally successful when there were less than 10 introduced species per island. When there were more introduced species, a substantial proportion of the introductions failed.
2. The chance that a species will succeed decreases significantly as the morphological similarity increases between it (the invader) and the nearest congeneric species already present.
3. The surviving species are morphologically more dissimilar than one would expect by chance. Morphologically similar species cannot coexist.
4. The abundances of species present across all the islands are lower on islands

where there are more introduced species, and hence, more potential competitions.

5. Species were *not* less likely to succeed if there were congeners present, than if congeners were absent.

Result 5 rejects Elton's (1946) idea about a limit on species-to-genus ratios imposed by competition. But with this exception, the results demonstrate the dominant effect of competition from existing species is determining which species succeeded on the islands.

15.3.3.2 The application to other taxa

The studies of Hawaiian birds provide large-scale, essentially experimental evidence for the role of community structure in general, and competition in particular, in determining which species can invade a community. The results are almost exactly what one might expect for bird communities shaped by competition. But how applicable are the results to other studies?

To test ideas on how difficult it is for species to invade a community, one needs large numbers of introductions and a knowledge of which species have failed. The successes, of course, are obvious. Not many studies meet these criteria, so it is difficult to evaluate the generality of the Hawaiian results. One comparable study involves insects introduced for biological control.

Early analyses of these data suggested that the patterns where consistent with the dominant role of competition determining which species succeeded (Ehler and Hall, 1982, 1984). Simultaneous introductions of several species at once seemed more likely to fail than when one species was introduced. Simberloff (1986) has argued forcefully that such a result is probably a consequence of the ways in which the introductions were undertaken. Multiple introductions are often undertaken with less care than single species introductions. Moreover, when an introduction is successful, further introductions are not necessary. I agree with Simberloff that the role of competition in determining invasion success for these insects seems minimal.

15.3.4 Empirical studies 2: the role of predation

The role of predation in determining invasion success is difficult to demonstrate. Certainly, introduces species suffer predation, but to what extent to community patterns predict the probability of an invasion failing because of predation?

An intriguing suggestion, based on the ideas of 'enemy free space' discussed earlier, is that communities rich in predatory species might be easier to invade than those with few predatory species. There is some evidence for this seemingly counterintuitive result. Rainbow smelt (*Osmerus mordax*) have been introduced widely into lakes in Canada as a source of food for predatory fish. Evans (1986) has shown that smelt are more likely to be found in lakes with greater numbers of predatory fish species. Initially, I thought that this result might be due to a coincidental correlation with some other variable. That is, predators are found in certain kinds of lakes (large rather than small, for example), and smelt also favor the same feature. But unpublished analyses that I have performed on Evans' data using log-linear models show that irrespective of whether the lakes are large or small, deep or shallow, acid or alkali, clear or turbid, the result still holds. Smelt apparently invade more readily when there are more predatory species.

This result is tantalizing because there is one other explanation. That is, that fisherman introduce smelt differentially to predator-rich lakes, irrespective of the lakes' physical and chemical characteristics. This explanation cannot be ruled out because we do not know in which lakes smelt introductions failed.

15.3.5 Empirical studies 3: community assembly

The final pattern to be discussed involves the difference in ease of invasion between communities with similar numbers of species, but which differ in how long the communities have been accumulating species. Some communities are clearly ancient, while others, such as those defaunated by volcanic eruptions or ice sheets, may be relatively recent. Yet others, like the man-made communities in lowland Hawaii and elsewhere, are only decades old. According to the theory, these age differences should play a major role in determining ease of invasion. Testing this idea is extremely difficult, because communities that differ in age, also differ in many other characteristics.

There is one observation, however, which though non-quantitative and anecdotal, I find interesting. Many introduced species are clearly in man-made and, therefore, only recently assembled communities. Much older, native communities, typically have many fewer introduced species. The upland, Hawaiian forests are good examples of this phenomenon. Of all the introduced bird species, only the Japanese white-eye (*Zosterops japonicus*) occurs in substantial numbers in the upland forests. What is striking is the distribution of habitats and the bird species in them.

The upland forests are cool, often temperature forests. On the leeward sides of the islands, the lowland forests consist of dry woodlands dominated by introduced legumes, for example kiawe (*Prosopis pallidus*). These dry woodlands merge into savanna-like habitats as rainfall decreases. In the lowlands on the windward sides of the islands, wet rain forests of introduced species predominate. In these lowland tropical habitats, the introduced species are temperate-forest bird species, but they do not occur in upland temperature habitats, which seems strange. This observation may be an example of a species-poor, native and thus, relatively old community, being unusually resistant to species invasions.

15.4 WHAT DETERMINES THE IMPACT OF AND INTRODUCED SPECIES?

There are really two aspects to this question: which species are the most damaging, and which communities are the most vulnerable? The data to answer these questions are many, scattered, and of very uneven quality. Their interpretation is often difficult. One of the earliest summaries, by Simberloff (1981), has recently been criticized by Herbold and Moyle (1986) and by a compendium of bird and mammal introductions, assembled by Ebenhard (1988).

Simberloff reviewed 10 papers covering 850 plant and animal introductions. The basic results were:

1. Less than 10% of the introductions caused species extinctions (there were only 71 extinctions). Of the 10 studies, Greenway's book (1967), which was devoted to avian extinction, reported the highest percentage of extinctions (i.e. 30%, which accounted for 55 of the 71 extinctions in Simberloff's analysis). Introductions apparently tend to add species to a community, rather than to cause extinctions.
2. Of Greenway's 55 extinctions, over 90% were on islands.
3. Over all the 10 studies, and within Greenway's study, predation was the principal cause of extinction; this accounted for 51 of the 71 extinctions in Siberloff's analysis, and 42 of the 55 in Greenway's study. Habitat change, which accounted for 11 of 71 extinctions, whereas competition caused only 3 extinctions and so was a distant third.

Simberloff's summary might be interpreted to mean that introductions rarely have much of an impact. But Herbold and Moyle show how difficult it is to interpret his results in this way. Examination of the three papers from which Simberloff tallied 525 of his 854 cases, shows that his determination of 'no' generally conflicts with conclusions of the original papers.

From Elton (1958), Simberloff extracted 241 instances of species introductions and reported only four of them as showing 'any effect at all.' But Elton's accounts included chestnut blight, spruce budworm, gypsy moth, argentine fire ant, Norway rat, black rat, house mouse, starling and house sparrow. Elton, however, found only four species that had no apparent effect and which had 'been able to edge in without producing any noticeable disturbances or making... species extinct.' Atkinson (1985), for example, provides a detailed discussion of the effects of rat introductions world wide, which shows that serious effects and extinctions are legion.

Herbold and Moyle also point out that many of the apparently benigh introductions are into highly man-modified habitats, where species have already been lost for other reasons. Using California fishes as examples, they find that 48 of the 137 species of freshwater fishes come from outside the state. Analysis of the habitat of these introduced species shows that 21 occur mainly in such highly modified habitats as reservoir and farm ponds, whereas another 21 occur mainly in moderately modified habitats, such as streams with altered temperature and flow regimes and six occur in near-pristine habitats. The latter include four species of trout and salmon introduced into lakes without fish, where the introduced fish altered invertebrate and emphibian populations. In all, 24 of 48 introductions have been documented as having a negative impact on the native fauna. The effects of the remaining 22 species occur in limited, artificial habitats of farm ponds and sewage treatments plants.

I have discussed these results with Dr Simberloff. He agrees that his study was a cursory one which probably overlooked effects. But he and I also agree that the impression that there *is* likely to be an effect of an introduced species is not proven. It might be true, but the data are inadequate.

15.5 CAN WE ANTICIPATE THE EFFECT OF AN INTRODUCED SPECIES?

Some useful insights to this question can be obtained from studies of the effect of removing species from communities. This may seem a perverse way of tackling the question. However, an introduction may be a success because its predators or competitors are few or absent in its new surroundings. A successful introduction may then cause extinctions whose effects will cascade through the community. One vehicle for generating insights about introduced species are differential-equation models of food webs (Pimm, 1982). Though the details are complex and tedious, the models' main features are simple and obvious (Figure 15.3).

Removing a plant species from the base of a simple food chain destroys the entire system (Figure 15.3; top left). But the loss of one of the several plant species utilized by a generalized herbivore, in the more complex, system, would have much less of an effect, because the herbivore is not so dependent on one species (Figure 15.3; middle). These effects are obvious and well known. Less obvious are the effects of removing species from the top of food chains. Removing a predator from a monophagous herbivore probably leaves the plant at a lower density, but it is likely to survive (Figure 15.3; top right). Special conditions are required for a predator to eliminate its sole prey, before it, too, becomes extinct. In the more complex system, the predator's absence may lead to the herbivore exterminating all but the one resistant plant species, which then regulates the herbivores' numbers (Figure 15.3; bottom).

There are three 'don'ts' for species introductions. Impacts will likely be severe when:

1. Species are introduced into places where predators are absent: this is equivalent to removing predators from communities. Examples would include the introduction of large herbivores to islands, or predators that feed high in a good chain and which lack predators as a consequence. In a similar way, we might expect introductions to have more severe impacts when competitors are also absent.

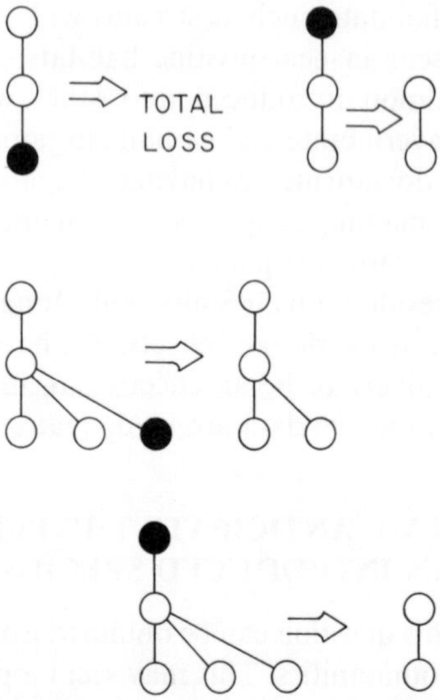

Figure 15.3. Effects of removing species from food webs, simplified from Pimm (1982). Circles represent species, lines represent feeding interactions between species. Removing the plant species at the base of a simple food chain results in the total loss of all the species. Removing a predator from the top of a simple food chain probably leaves the remaining species extant, even if at different densities. These results are reversed in the more complex system. Removing a plant from the base of the more complex system may cause no extinctions. Removing a predator from a polyphagous herbivore may be disastrous, as the herbivore eliminates all but one of the plant species

2. Polyphagous species are introduced: predator removals are likely to be more severe when the herbivores are polyphagous.
3. Species are introduced into relatively simple communities where the removal of a few plant species will cause the collapse of entire food chains.

Ebenhard (1988) provides dramatic evidence of the first effect, suggestive evidence for the second, and there are some compelling case histories to support the third.

By researching over 800 introductions, Ebenhard has classified those species that have an effect in changing abundance, species composition or in causing extinctions. Overall, I suspect effects are likely to be missed (compare Simberloff's and Atkinson's comparisons of the effects of rats). Yet the comparisons of the incomplete data are interesting. Herbivorous mammals introduced to continents were recorded as having effects in 28% of 89 cases. Yet on oceanic islands, which likely lack predators, 50% of 363 introductions show effects.

Almost all of the herbivorous mammals have generalized diets, but among five introductions of specialized herbivores, none were recorded as having an effect.

For the last feature, we can compare the fate of island and continental birds following the destruction of various plant species. In North America, there appear to be no extinctions of vertebrates and few insect extinctions attributable to the loss of chestnut trees following the introduction of chestnut blight (Opler, 1978). Chestnuts were locally one of the commonest trees. Contrast this with Hawaii. Some of the larger nectarivorous birds seem to have been differentially sensitive to extinction, perhaps when a few, important, nectar-producing plants were exterminated by goats and pigs (Pimm and Pimm, 1982). Similarly, the threatened palila depends on the immature seeds of one species of tree, the mamane (*Sophoea chaysophylla*), which does not regenerate when large herbivores are present (van Riper III, 1980), and the akiapolaau, (*Hemignathus munsoi*) an insectivore, depends on the presence of large koa trees (*Acacia koa*) (Ralph and van Riper III, 1985), which are a popular source of wood for objects d'art.

Not surprisingly, the introduction of generalized herbivores to islands is often devastating, because all three 'don'ts' operate. But, equally, some introductions should have little effect on the community to which they are introduced. Ebenhard finds that bird introductions rarely seem to have an effect.

15.6 CONCLUSIONS

There is an inevitable tendency to consider species introductions as isolated events. When one does this, which species succeed or fail and which species, when successful, are benign or damaging, may not be readily appreciated. There do appear to be some general rules about introductions. Some species are much more likely to survive at low population densities than others. Communities with high numbers of species are likely to be resistent to species because the introduction is more likely to encounter a strong competitor. Under some circumstances, communities with large numbers of predatory species might be relatively easy to invade. The age of a community—how long its species have been assembling may be a very important determinant of invasion. Species introduced without their predators are likely to be damaging, especially if the species are polyphagous. Communities with simple food chains are likely to be very vulnerable to the introduction of herbivores.

Details of the ecology of species, and the communities into which they are being introduced are bound to be extremely important in determining whether a particular invasion will succeed. Yet, considering introductions in total, it is clear that there are some simple theoretical expectations and there are some equally simple general patterns.

REFERENCES

Anonymous (1944). British Ecological Society Syposium on 'The ecology of closely allied species.' *J. Anim. Ecol.*, **13**, 176–8.

Atkinson, I. A. E. (1985). The spread of commensal species of *Rattus* to oceanic islands and their effects on island avifaunas. In: Moors, P. J. (Ed.), *Conservation of Island Birds*, pp. 35–81. ICBP Tech. Pub. No. 3.

Briand, F., and Cohen, J. E. (1984). Community food webs have scale-invariant structure. *Nature*, **307**, 254–67.

Cohen, J. E. (1977). Ratio of prey predators in community food webs. *Nature*, **270**, 165–7.

Diamond, J. M. (1984). Historic extinctions: a Rosetta Stone for understanding prehistoric extinctions, In: Martin, P. S., and Klein, R. G. (Eds), *Quaternary Extinctions: a Prehistoric Revolution*, pp. 824–62. University of Arizona Press, Tucson.

Diamond, J. M., and May, R. M. (1977). Island biogeography and the design of natural reserves. In: May, R. M. (Ed.), *Theoretical Ecology*, pp. 228–52. Sinauer, Sunderland, Mass.

Ebenhard, T. (1988). Introduced birds and mammals and their ecological effects. *Swedish Wildlife* Research, **13**, no. 4.

Ehler, L. E., and Hall, R. W. (1982). Evidence for competitive exclusion of introduced natural enemies in biological control. *Environ. Entomol.*, **11**, 1–4.

Ehler, L. E., and Hall, R. W. (1984). Evidence for competitive exclusion of introduced natural enemies in biological control (an addendum). *Environ. Entomol.*, **13**, v-vii.

Elton, C. C. (1946). Competition and the structure of ecological communities. *J. Anim. Ecol.*, **15**, 54–68.

Elton, C. S. (1958). *The Ecology of Invasions by Animals and Plants.* Methuen, London.

Evans, D. O., and Loftus, D. H. (1986). Colonization of inland lakes in the Great Lakes region by Rainbow Smelt: their freshwater niche and effects on indigenous fishes. *Can. J. Fish Aquat. Sci.*, **44**, 249–266.

Greenway, C., Jr. (1967). *Extinct and Vanishing Birds of the World.* Dover, New York.

Herbold, B., and Moyle, P. B. (1986). Introduced species and vacant niches. *Amer. Nat.*, **128**, 751–60.

Holt, R. D. (1977). Predation, apparent competition and the structure of prey communities. *Ther. Pop. Biol.*, **12**, 197–229.

Holt, R. D. (1984). Spatial heterogeneity, indirect interactions, and the coexistence of prey species. *Amer. Nat.*, **122**, 521–41.

Hutchinson, G. E. (1959). Homage to Santa Rosalia, or Why are there so many kinds of animals? *Amer. Nat.*, **93**, 145–59.

Jarvinen, O. (1987). Species introductions to Sweden. *Swedish Wildlife* (in press).

Jeffries, M. J., and Lawton, J. H. (1984). Enemy free species and the structure of ecological communities. *Biol. J. Linn. Soc.*, **23**, 269–86.

Jeffries, M. J., and Lawton, J. H. (1985). Predator-prey ratios in communities of freshwater invertebrates: the role of enemy free space. *Freshwater Biol.*, **15**, 105–12.

Lack, D. (1947). *Darwin's Finiches.* Cambridge University Press, Cambridge.

Lack, D. (1969). The number of bird species on islands. *Bird Study*, **16**, 193–209.

Lawton, J. H., and Brown, K. C. (1986). The population and community ecology of invading insects. *Phil. Trans. R. Soc. B.* (in press).
Leigh, E. G. Jr. (1975). Population fluctuations, community stability, and environmental variability. In: Cody, M. L., and Diamond, J. M. (Eds), *Ecology and Evolution of Communities*, pp. 51–73. Harvard University Press, Cambridge.
Leigh, E.G., Jr. (1981). The average lifetime of a population in a varying environment. *J. Theor, Biol.*, **90**, 213–39.
Lewin, R. (1983a). Santa Rosalia was a goat, *Science*, **221**, 636–9.
Lewin, R. (1983b). Predators and hurricanes change ecology. *Science*, **221**, 737–40.
MacArthur, R. H., and Wilson, E. O. (1967). *The Theory of Island Biogeography*. Princeton University Press, Princeton.
Moulton, M. P. (1985). Morphological similarity and the coexistence of congeners: an experimental test with introduced Hawaiin birds. *Oikos*, **44**, 301–5.
Moulton, M. P., and Pimm, S. L. (1983). The introduced Hawaiian avifauna: biogeographic evidence for competition. *Amer. Nat.*, **121**, 669–90.
Moulton, M. P., and Pimm, S. L. (1986a). Species Introductions to Hawaii. In: Mooney, H. A., and Drake, J. A. (Eds), *Ecology of Biological Invasions of North America and Hawaii*, pp. 231–49. Springer-Verlag, Berlin.
Moulton, M. P., and Pimm, S. L. (1986b). The extent of competition in shaping and introduced Avifauna. In: Diamond, J. M. and Care, T. J. (Eds), *Community Ecology*, pp. 80–97. Haper and Row, New York.
Peter, R. H. (1983). *The Ecological Implications of Body Size*. Cambridge University Press, Cambridge.
Pimm, S. L. (1982). *Food Webs*. Chapman and Hall, London.
Pimm, S. L. (1984). The complexity and stability of ecosystems. *Nature*, **307**, 321–6.
Pimm, S. L., and Pimm, J. W. (1982). Resource use, competition, and resource availability in Hawaiian Honeycreepers. *Ecology*, **63**, 1468–80.
Pimm, S. L., Jones, H. L., and Diamond, J. M. (1988). On the risk of extinction *Amer. Nat.*, (in press).
Post, W. M., and Pimm, S. L. (1983). Community assembly and food web stability. *Math. Biosc.*, **64**, 169–92.
Ralph, C. J., and van Riper III, C. (1985). Historical and current factors affecting Hawaiian native birds. In: Temple, S. (Ed.), *Bird Conservation*, vol. 2, pp. 7–42. University of Wisconsion Press, Madison.
Richter-Dyn, N. and Goel, N. S. (1982). On the extinction of a commonizing species. *Theor. Pap. Biol.*, **3**, 406–33.
Simberloff, D. (1981). Community effects of introduced species. In: Nitecki, M. H. (Ed.). *Biotic Crises in Ecological and Evolutionary Time*, pp. 53–83. Academic Press, New York.
Van Piper III, C. (1980). Observations on the breeding of the palila *Psittrostra baileui* of Hawaii Ibis, **122**, 462–75.
Williams, C. B. (1964). *Patterns in the balance of nature*. Academic Press, London.
Williamson, M. (1981). *Island Populations*. Oxford University Press, Oxford.

Biological Invasions: a Global Perspective
Edited by J. A. Drake *et al.*
© 1989 SCOPE. Published by John Wiley & Sons Ltd

CHAPTER 16

Invasibility of Plant Communities

MARCEL REJMÁNEK

16.1 INTRODUCTION

One of Elton's classical arguments for a positive causal relationship between community species richness and stability was the ease with which alien species can invade species-poor oceanic islands (Elton, 1958, p. 147). This has been often interpreted as 'the more native species present the more invasion resistant the community.' Certain data from small mammal and bird communities support this relationship (Fox and Fox, 1986; Pimm, personal communication). Properly collected data from plant communities are either unavailable; or are confounded by species–area effects or by species richness changes during succession and/or by dependence on environmental gradients. Decline of invasibility (openness to invasions) during the course of successions is generally admitted (Crawley, this volume) but has never been quantified. Similarly, the maximum number of aliens in mesic and riverine environments has been indicated (Fox and Fox 1986; MacDonald *et al.*, 1986) but no generalizations over the whole moisture gradient have been made. Invasions into successionally mature communities with no anthropogenic disturbances are the objects of careful statements and supported by very few persuasive examples. There seems to be no community without some degree of natural disturbance (Pickett and White, 1985; Rejmánek, 1984; Sousa, 1984). What is the role of this disturbance, competitiveness and recolonization ability of native species, and of the amount of imported propagules of aliens in an invasion process? We have only partial answers for a few simple cases but, unfortunately, the existing studies are too few to warrant generalizations.

In this chapter I will first briefly examine a plausible procedure for the collection of data which could tell us something about the species richness–invasibility relationship. Then the available data on invaders in successional series from Europe and North America will be summarized and interpreted. Plant invasions in relation to moisture gradients will be also examined. Finally, some examples of plant invasions into natural communities will be discussed and the role of disturbance will be illustrated using a simple simulation model. All examples and considerations will be limited to vascular plants and their communities.

16.2 SPECIES RICHNESS VERSUS INVASIBILITY

Available data on numbers of native and invading species are not usually comparable, because they are based on samples of different size and different homogeneity. This fact suggests that percentages of invading species, instead of their actual numbers should be used as a measure of invasion vulnerability of sampled communities or areas. A percentage is certainly a useful expression but, because it is in fact the ratio of the number of invader (A) and native (N) species (100A/(N + A)), it can not be used for the study of invasibility dependence on the number of native species. If such a comparison is done, and the percentage of invader species is plotted against the number of native species (Fox and Fox, 1986), a parabolic negative relationship is the inevitable outcome. The resulting relationship is negative even if correlation between absolute numbers of invaded and native species is positive!

It is a commonplace that large samples contain more species than smaller samples. Unfortunately, the rate of species increase with increasing sample area differs from community to community (Moravec, 1973; Williamson, 1988). It is, therefore, virtually impossible to define 'the best' sample size for invasibility comparisons over several communities. One way to approach this problem is (1) take a series of samples of different size from each community under interest, (2) calculate the regression coefficient for the number of invading species on the log of the sample area, b_A, and the regression coefficient for the number of native species on the log of the sample area, b_N, in each community, and (3) calculate the correlation coefficient between the resulting values of b_A and b_N over all communities and test for null hypothesis $r=0$. Unfortunately very few species–area community data sets are available for such an analysis. Species–log area regression coefficients calculated on the basis of check lists or manuals covering large heterogeneous areas (Crawley, 1987) are not helpful in community invasibility investigations. Also, it would be particularly useful to know the values of b_N before invasions started. And finally, even if we have all these data, if there is any dependence of invasibility on species richness, this dependence may in practice be masked by the influence of many other factors. I will deal briefly with three of them: succession, moisture gradient, and natural disturbance.

16.3 INVADERS IN SUCCESSIONAL SERIES

Succession of plant communities from those consisting mainly of ruderals (r-strategists) to communities of competitors and stress-tolerators (*sensu* Grime, 1979), represents a gradient of constraints placed on potential invaders. Because most non-native species are ruderals by their nature (Baker, 1965; Heywood, this volume), it is not surprising that pioneer communities exhibit a higher number and proportion of invaders than successionally more advanced stands. There is a consistent exponential decline of the proportion of invaders in all series where complete lists of species are available (Figures 16.1 and 16.2). The absolute

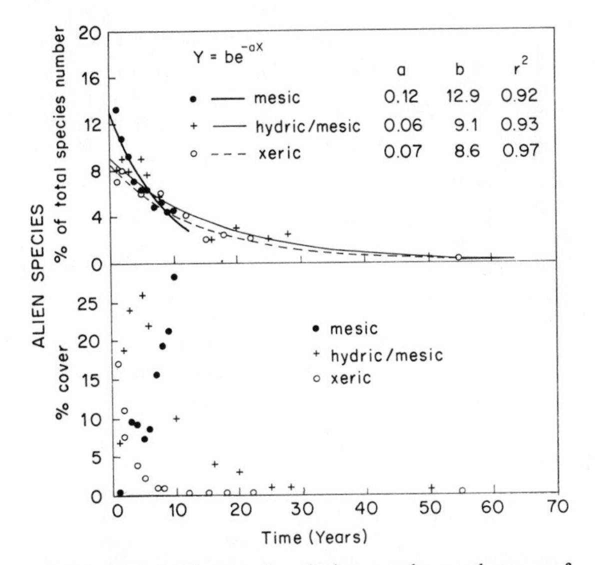

Figure 16.1. Changes in relative number and cover of invading species in three successional series from central Europe. Secondary *mesic* series: old field, Göttingen, West Germany, after Schmidt (1981). Primary *hydric/mesic* series: artificial islands in fishponds, Trebon area, Czechoslovakia, after Rejmánek and Rejmánková (unpublished data). Secondary *xeric* series: old fields in Bohemian Karst Protected Area, Czechoslovakia, after Huml (1977), Baumova (1985) and author (unpublished data)

number of invaded species differs considerably from series to series but also declines exponentially in the course of succession.

How much does this result reflect different exposure of different successional stages to the propagule import? We should expect more intensive human-mediated import of propagules at the beginning of secondary successions. Only one series seems to be free of this confounding effect. All communities along the primary successional gradient in the Atchafalaya delta and Atchafalaya Basin (Rejmánek, *et al.*, 1987) are flooded and exposed to propagule deposition every spring. Possibly, more propagules are trapped in older communities located upstream. However, the trend here is the same as in the remaining series. There are many invaders in initial stages including for example; *Cyperus difformis, Eichhornia crassipes, Sphenoclea zeylandica, Alternathera phyloxeroides*, and *Colocasia esculenta*. These are replaced by only two invaders in stands older than 30 years (*Sapium sebiferum, Vigna unguiculata*); and the Japanese climbing fern (*Lygodium japonicum*) is the only alien species in forests older than 50 years.

Spatial isolation of young successional communities from sources of alien propagules may change the absolute numbers substantially. In the first 3 years of

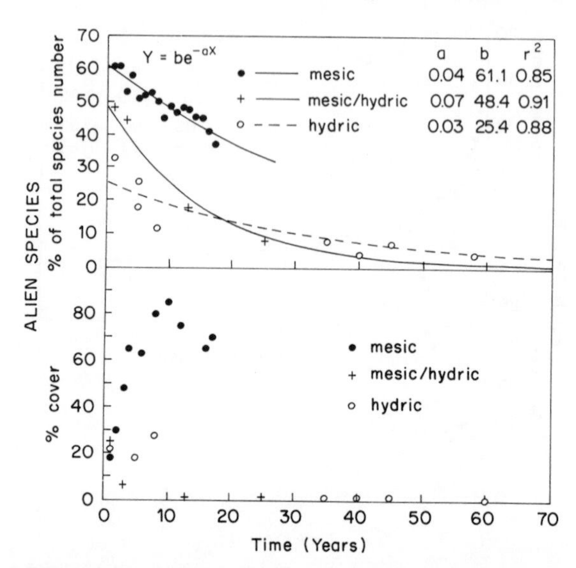

Figure 16.2. Changes in relative number and cover of alien species in three successional series in the USA. Secondary *mesic* series: old field, New Jersey, after Pickett (1982, 1983). Secondary *mesic/hydric* series: old fields in Louisiana, after Bonck and Penfound (1945) and author (unpublished data). Primary *hydric* series, the Atchafalaya Basin and Atchafalaya delta, Louisiana, after Rejmánek and Sasser (unpublished data)

postlogging succession in Oregon only four invading species can be found (less than 3%) (Dyrness, 1973).

The total cover of invading species very often does not have its maximum at the beginning but increases in the first 5 to 10 years and declines later. This trend seems to be characteristic especially for secondary mesic series (Figures 16.1 to 16.3 and Table 16.1, p. 374).

The proportion and absolute number of alien species are higher in American mesic series than in European ones. This is in agreement with general trends in exchange of alien species between North America and Europe (Di Castri, this volume).

It is difficult to find data from other continents for comparison. Nowhere else do studies on succession have such a long tradition as in Europe and the USA. Unfortunately, full species lists are published rather infrequently. Lists of dominant species from different stages of secondary succession in South Africa (Davidson, 1964) indicate the same decline of invaders as in successional series from Europe and North America.

It seems that this successional 'repairing' function of native vegetation is rather universal in continental situations. Of course, the proximity of diaspore sources

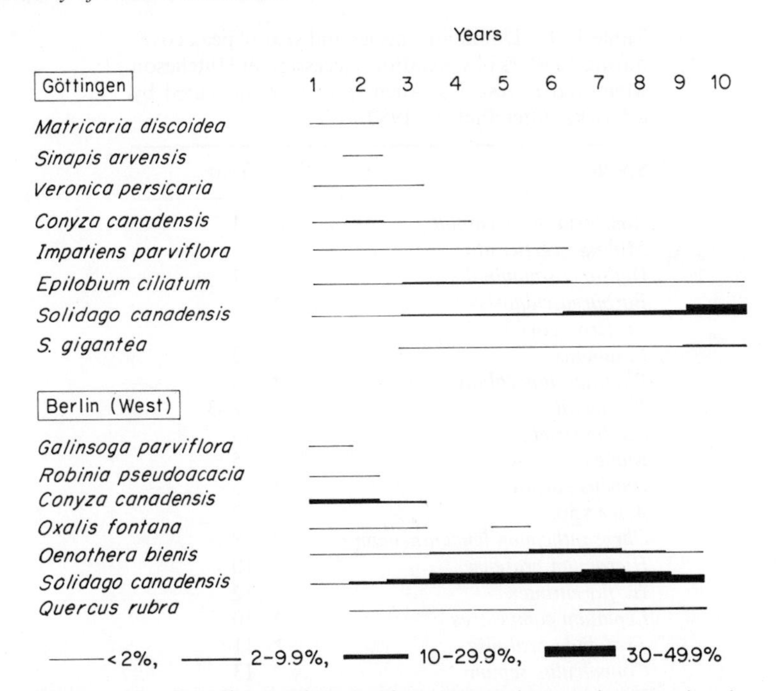

Figure 16.3. Invading species cover in two secondary mesic successional series in Germany (Bornkamm and Hennig, 1982; Schmidt, 1981)

of native plants plays a crucial role in the rate of recovery. Compared with continents, natural recovery of native flora on islands seems to be much slower or incomplete. The reasons for this difference are complex and still not fully understood (Mueller-Dombois, 1981; Loope and Scowcroft, 1985; Loope and Mueller-Dombois, this volume).

16.4 INVADERS ALONG MOISTURE GRADIENTS

In the successional series discussed above, both the maximum cover and the maximum proportion of alien species was found in mesic series, or at least, in inital stages of mesic succession (Figures 16.1 and 16.2). The absolute numbers of species are not directly comparable, because of different sample sizes. But, taking sample sizes into account, it is not only the proportion but also the number of invaded species which is highest in the first years of mesic succession. Invaders seem to be rather rare also in other xeric and hydric series described in literature (Symonides, 1985; Burbanck and Philips, 1983; Carpenter, 1983; Vasek, 1980; Eicke-Jene, 1960). The number of invading species in initial stages of succession is apparently the greatest in mesic environments and decreases at both ends of the

Table 16.1. Dominant species and year of peak cover during 20 years of vegetation succession at Hutcheson Memorial Forest, NJ. Alien species are indicated by asterisks. After Pickett (1982)

Species		Year
Ambrosia artemisifolia		1
Mullugo verticilata	*	1
Digitaria sanguinalis		1
Barbarea vulgaris	*	2
Frigeron cancdensis		2
E. annuus		3
Plantago lanceolata	*	3
P. rugellii		2–3
Oxalis stricta		3
Rumex acetosella	*	5
Daucus carota	*	5
Aster spp.		7
Chrysanthemum leucanthemum	*	8
Hieracium pratense	*	10
H. florentinum	*	12
Lepidium campestres	*	10
Trofolium pratense	*	11
Convolvulus sepium	*	13
Poa pratensis		15
Agrostis alba	*?	12
Rhus glabra		19–20
Lonicera japonica	*	17
Juniperus virginiana		19
Acer rubrum		20
Poa compressa	*	17
Acer negundo		18
Solidago graminifolia		19
Rhus radicans		20
Rosa multiflora	*	20
Solidago juncea		19

moisture gradient. This is in agreement with changes of the total species number in young communities along a moisture gradient (Auclair and Goff, 1971; Peet, 1978).

The only complete list of all invading species in plant communities along the moisture gradient in a rather undisturbed landscape is, as far as I know, the one published by Falinski (1968) from Bialowieza Primeval Forest, Poland. A simplified version of his analysis is summarized in Table 16.2.

Table 16.2. Alien plant species in plant communities of the Bialowieza Primeval Forest in 1968 year (after Falinski, 1968) (+, sporadic; *, common; #, very common)

Species	Moisture Plant community (association or alliance) 1	2	3	4	5	6	7	8	9	10	11	12	13	14	15	16	17	18	19	20
Elodea canadensis	+	+	#	+	#	#	#													
Acorus calamus							*	#	#	*	+	+								
Carex brizoides															*	#			#	
Cytisus scoparius																+	+	*		*
Lupinus polyphyllus															+	#	#	+		
Sambuchus racemosa															+	+		+	+	
Pulmonaria mollissima															+	+	#	+		
Rosa sp. div.																+	+			
Deschampsia flexuosa																	+			+
Luzula luzuloides																		+		+
Juncus macer																		+	+	+
Elsholtzia patrini																+				
Rudbeckia laciniata													+	+	+					
Solidago serotina													+	+	+					
Erigeron ramosus															+	+	+	+		
Viola odorata															+	+				
Myosotis sylvatica																+				
Poa chaixii																+				
Sambucus nigra																+		+		
Quercus rubra															+	+	+			
Acer negundo															+	+				
Reynoutria japonica															+	+				
R. sachalinensis															+	+				

1, *Lemnetum gibbae*; 2, *Lemnetum minoris-trisulcae*; 3, *Potamion*; 4, *Hottonietum palustris*; 5, *Hydrocharo–Stratiotetum*; 6, *Myriophyllo–Nupharetum*; 7, *Spargανio–Sagittarietum*, 8, *Acoretum calami*; 9, *Phragmition*; 10, *Magnocaricion*; 11, *Salicetum pentandro-cinereae*; 12, *Carici elongatae–Alnetum*; 13, *Salicetum triandfro-viminalis*; 14, *Salicetum albo-fragilis*; 15, *Alno–Padion*; 16, *Tilio–Carpinetum*; 17, *Potentillo albae–Quercetum*; 18, *Pino–Quercetum*; 19, *Querco–Pinetum*; 20, *Peucedano–Pinetum.*

Several species from Falinski's list occur only in disturbed stands, such as forest clearings. It is also unclear whether all species included in the list are true aliens (*Carex brizoiodes, Deschampsia flexuosa*). Nevertheless, the maximum number of alien species is found in mesic communities located in the central part of the moisture gradient (*Alno–Padion, Tilio–Carpinetum*). Accordingly, Loope and Mueller-Dombois (this volume) found lack of competitive superiority of alien species in extreme habitats like montane bogs and new lava flows.

At this point we may ask a similar question to the one we asked in connection with successional stages. How much does this result reflect a difference in exposure of mesic and extreme habitats to propagule import? And another question: does this result reflect preadaptation of introduced species (coming from mesic environments) rather than higher vulnerability of mesic communities to invasions? Neither of these possibilities can be completely ruled out. Apparently the number of alien species declines faster from mesic environments towards extreme ends of the moisture gradient than does the number of native species in the same directions. In other words, there seems to be a lower percentage of alien species in extreme environments, at least in young successional stages (Figures 16.1 and 16.2). The above indicates the relevance of both questions. Properly designed experiments in this area are badly needed.

The only experiment in which plants were sown into different communities along a moisture gradient was done by Juhász-Nagy (1964). Unfortunately, he never published the full list of non-native species used, or the examined communities (nine different vegetation types). The most resistant community to the establishment of 11 sown species was, in agreement with our findings, the extremely dry one (*Festucetum pseudodalmaticae*).

Apparently xeric environments are not favorable for germination or the establishment of many non-native species and, on the other hand, wet productive habitats support strong resident competitors (Grime, 1979; del Moral, 1983) which do not leave much space for potential invaders. Both these unfavorable factors for invasion, stress preventing ecesis and competition preventing survival to reproduction stage, seem to be weaker in mesic environments.

In many areas, river banks and floodplains in general often host the highest percentage of alien species (Kopecky, 1967; Crawley, 1987; Nilsson *et al.*, 1988). These habitats represent a transition between mesic and hydric environments. Besides a high frequency of natural disturbance which promotes invasions (see later), riverine habitats have traditionally been disturbed and exposed to propagules of alien plants due to human activities (communications and settlement).

16.5 INVADERS IN 'UNDISTURBED' COMMUNITIES

So far, the list of plant species invading undisturbed (= naturally disturbed) and successionally advanced communities is very short and many questions remain concerning possible direct or indirect effects of human disturbance. The fact that this preliminary list (Table 16.3) is so short is not surprising if we recall what tremendous mortality native and established species have in mature communities. The probability of survival from seed to a reproducing plant is usually in the range 10^{-4} to 10^{-6} for trees (Hett, 1971; Guittet and Laberche, 1974; Van Valen, 1975; Vacek, unpublished data). There is no reason to expect a lower mortality rate for introduced non-native trees in temperate regions. This means that about 100 000 seeds of a non-native tree species would be required to examine invasibility of a relatively undisturbed forest. Nobody has tried to do it, as far as I know.

Table 16.3. Invaders in natural communities

Species (Family) Country of origin	Adventive distribution	Reference
Acacia saligna (Fabaceae) Australia	South African fynbos shrublands	Wilgen and Richardson (1985)
Acaena anserinifolia (Rosaceae) New Zealand	Great Britain, sand dunes	Ratcliffe (1984)
Acer pseudoplatanus (Aceraceae) Continental Europe	British Isles, woodlands	Crawley (1987)
Acorus calamus (Araceae) Himalayas	Europe, reedswamps	Kornas (1988) Wein (1942)
Ailanthus altissima (Simarubiaceae) Asia	California, 'invades areas of' native vegetation	McClintock (unpublished)
Alnus viridis (Betulaceae) Alps, eastern Carpathian Mountains	Sudeten Mts., tall herb sub-alpine communities	Rejmánek *et al.* (1971)
Amaranthus spinosus (Amaranthaceae) North America	Phan Rang, Vietnam, *Barringtonia* floodplain forests	Rejmánek (unpublished)
Ammophila arenaria (Poaceae) Western Europe	California, coastal dunes	Mooney *et al.*, (1986)
Ardisia crenata (Myrsinaceae) East Asia	Mauritius, evergreen wet forest	Lorence and Sussman (1986)
Arundo donax (Poaceae) Southern Europe	South Africa, reedswamps	Wells *et al.* (1980)
Butomus umbellatus (Butomaceae) Europe	Michigan, Ohio, Ontario, marshes along rivers and lakes	Stuckey (1968)
Cakile edentula and *C. maritima* (Cruciferae) Europe	Coastal dunes in western USA and Australia	Barbour and Rodman (1970) Boyd (1986) Rodman (1986)
Centranthus ruber (Valerianaceae) Mediterranean region	England, Wales, Southern Ireland, cliffs, dry banks	Crawley (1987)

(*Contd.*)

Table 16.3. (*Contd.*)

Species (Family) Country of origin	Adventive distribution	Reference
Chysanthemoides molinifera (Asteraceae) South Africa	Coastal dunes in Australia	Weiss and Noble (1984)
Cytisus scoparius and *C. monspesulanus* (Fabaceae) Europe	Western USA, 'invades areas of native vegetation'	McClintock (unpublished)
Elaeagnus angustifolia (Elaeagnaceae) Eurasia	Southwestern USA, arid riparian communities	Knopf and Olson (1984)
Epilobium brunnescens (Onagraceae) New Zealand	British Isles, rocky beds and banks of streams	Crawley (1987)
Epipactis heleborine (Orchideaceae) Europe	Northeastern USA, British Columbia, deciduous and Douglas fir forests	Luer (1975) Antos (unpublished)
Eucalyptus globulus (Myrtaceae) Australia	California, 'spreading from previous plantings, crowding out the native vegetation'	McClintock (unpublished)
Geranium robertianum (Geraniaceae) Europe	British Columbia, along streams in hemlock forests	Antos (unpublished)
Hedera helix (Araliaceae) Europe	California, British Columbia, 'displacing native ground cover plants', Douglas fir forests	McClintock (unpublished) Antosh (unpublished)
Hakea sericea (Proteaceae) Australia	South Africa, fynbos	Macdonald and Richardson (1986)
Hypericum androsaemum (Guttifereae) Southern Europe	New Zealand, native forests	Johnson (1982)
Impatiens glandulifera (Balsaminaceae) Himalaya	Central Europe, England, stream banks	Kopecky (1967) Crawley (1987)
I. parviflora Asia	Central and western Europe, in many types of *Carpinus*, *Fagus*, and *Alnus* forests	Coombe (1959) Trepl (1984) Csontos (1984)
Iris pseudacorus (Iridaceae) Europe	North America, swamps, ponds, stream banks	Raven and Thomas (1970)
Lantana camara (Verbenaceae) Southern America	Southern Africa, riparian vegetation	Macdonald (unpublished)
Ligustrum robustum (Oleaceae) Ceylon	Mauritius, evergreen wet forest	Lorence and Sussman (1986)
Linum austriacum (Linaceae) Southern Europe	Island of Gotland, alvar grasslands	Nilsson (1981)
Litsea glutinosa (Lauraceae) Southeast Asia	Mauritius, evergreen wet forest	Lorence and Sussman (1986)

Table 16.3. (*Contd.*)

Species (Family) Country of origin	Adventive distribution	Reference
Lonicera japonica (Caprifoliaceae) Japan	Eastern USA, deciduous forests	Slezak (1976)
Lygodium japonicum (Schizaceae) Japan	Bottomland hardwood forests from Florida to Eastern Texas	Thieret (1980) Rejmánek (unpublished)
Melaleuca quinquenervia (Myrtaceae)	Southern Florida, dwarf cypress forest	Myers (1983) Ewel (1986)
Melia azedarach (Meliaceae) West Asia	Southern Africa, riparian vegetation along major rivers	Macdonald (unpublished)
Mesembrianthemum spp. (Aizoaceae) South Africa	California, coastal dunes, cliffs, and grasslands	Vivrette and Muller (1977)
Mimulus guttatus (Scrophulariaceae) Western North America	Europe, stream banks	Rothmaler (1976) Crawley (1987)
Mycelis muralis (Asteraceae) Europe	New Zealand, native forests	Johnson (1982)
Myrica faya (Myricaceae) Canary Island	Hawaii, young volcanic areas	Vitousek (1986)
Ossaea marginata (Melastomataceae) Brazil	Mauritius, evergreen wet forest	Lorence and Sussman (1986)
Pinus lutchuensis (Pinaceae) Ryukyu Islands	Bonin Islands, native vegetation lacking tree conifers	Shimuzu and Tabata (1985)
P. pinaster Southern Europe	Southern Africa, Cape fynbos	Kruger (1977)
P. radiata California	Australia, native eucalypt forests, South Africa, fynbos	Burdon and Chilvers (1977)
Pittosporum undulatum (Pittosporaceae) Eastern Australia	Central Victoria, Australia, sclerophyll forests	Gleadow and Ashton (1981)
Psidium cattleianum (Myrsinnaceae) Brazil	Mauritius, evergreen wet forest	Lorence and Sussman (1986)
Pterolepis glomerata (Melastomataceae) Tropical America	Oahu, tropical rain forests	Gerrish and Mueller-Dombois 1980
Rhododendron ponticum (Ericaceae) Turkey	Ireland, semi-natural oakwoods (*Quercus petraea*)	Cross (1981)
Robinia pseudoacacia (Fabaceae) Eastern USA	Central and southern Europe, xerotherm grasslands	Holzner (1982)
Rubus moluccanus (Rosaceae) Southeast Asia	Mauritius, evergreen wet forest	Lorence and Sussman (1986)
Senecio mikanoides (Asteraceae) Southern Africa	California, 'native' plant areas in canyons and gullies'	McClintock (unpublished)

(*Contd.*)

Table 16.3. (*Contd.*)

Species (Family) Country of origin	Adventive distribution	Reference
Smyrnium olusatrum (Umbelliferae) Continental Europe	British Isles, sea-cliffs	Crawley (1987)
Spartina alterniflora (Poaceae) North America	Widely distributed *S. anglica* resulted from hybridization of *S. alterniflora* with *S. maritima*	Ranwell (1972)
S. patens From Newfoundland to Texas	Cox Island, Oregon, low marsh communities	Frenkel and Boss (1982)
Spathoglottis plicata (Orchideaceae) Asia	Oahu, tropical rain forests	Gerrish and Mueller-Dombois (1980)
Tamarix spp. (Tamaricaceae) Eurasia	Southwestern USA, arid riparian communities	Ohmart and Anderson (1982)
Ulex europaeus (Fabaceae) Europe	California, 'invades areas of native vegetation'	McClintock (unpublished)
Viscum album (Loranthaceae) Europe	Sonoma County, California, on 9 native host tree or shrub species	Hawkswort and Scharpf (1986)

The diversity of species invading natural or seminatural communities (about 60 species from 40 families representing all major growth-forms in our table) suggests extreme difficulty or even impossibility of quantifying invasibility of plant communities. It is simply technically impossible to expose an experimental set of some, for example, temperate plant communities to all 50 000 species which could be potential candidates for invasion.

The list in Table 16.3 represents a bizarre collection of extremely diverse adaptations which have been necessary for invasions into a variety of more or less natural communities in different environments. While it seems to be possible to make some generalizations about successful invaders in disturbed and successionally young communities (Baker, 1965; Heywood, this volume), there is apparently nothing unifying for invaders in 'undisturbed' natural communities. Some of them form a new stratum in communities where the tree or tall shrub strata were missing but the environment can support them (*Pinus radiata* and *Hakea sericea* in South African fynbos, see Campbell *et al.*, 1979; *Robinia pseudoacacia* in xerotherm grasslands in central Europe). Some apparently did find an 'open niche' like the climbing fern, *Lygodium japonicum*, in bottomland hardwoods from Louisiana to Florida. No climbing ferns were present in these communities originally. Some are apparently dependent on natural disturbance (*Cakile* spp. and *Chrysanthemoides molinifera* in coastal dunes). Some had been only in disturbed habitats for some time and invaded undisturbed communities only recently (*Impatiens parviflora*, Figure 16.4). What happened with such species?

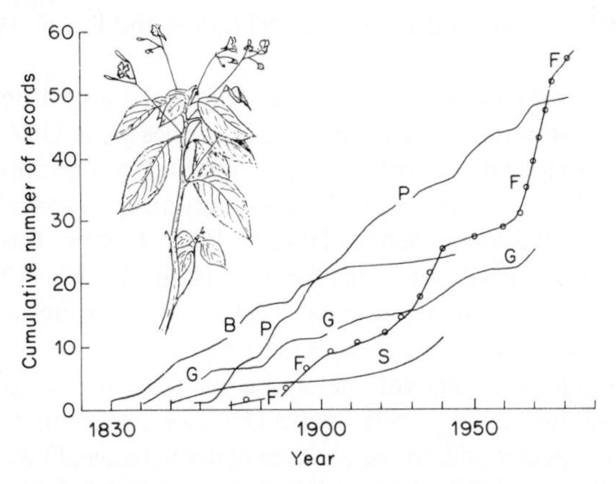

Figure 16.4. Number of *Impatiens parviflora* records from different habitats in central Europe from 1830 to 1971. After Trepl (1984). B, Botanical gardens; G, gardens; P, parks and cemeteries; S, International sowing; F, forests

According to some authors, genetic changes can make some introduced populations better adapted for spread (Baker, 1965). There is no evidence for this in listed species. But very often, it could be simply a matter of the amount of propagules available in disturbed areas and transported into 'undisturbed' ones.

16.6 THE ROLE OF DISTURBANCE

There is overwhelming evidence that several types of disturbance promote biological invasions (e.g. Egler, 1983; Forcella and Harvey, 1983; Pickard, 1984; Ewel, 1986; Hobbs, this volume; Mack, this volume). Basically, the probability of successful invasions seems to be crucially dependent on the extent and type of disturbance, on the number of non-native species propagules deposited in the community per year, and how long the community is exposed to import of propagules. The amount of biomass or cover may be the most efficient indices of community resistance to invasions in some situations.

Peart and Foin (1985) found that invasion success of *Anthoxanthum odoratum* into coastal grasslands in California is a negative exponential function of resident vegetation biomass, rather than dependent on its species composition or species number. Open space, therefore, created by biomass destruction might be the only really general factor responsible for plant invasions.

Some disturbance occurs in all natural communities because of herbivory, rodent digging activity, tree-falls, soil frost disturbance, fires, etc. Senescence in almost all plant species is also responsible for stochastic space openings in plant communities. In spite of the fact that the majority of invasion situations are nonequilibrium ones, almost all mathematical models of invasibility assume an equilibrium (e.g. Robinson and Valentine, 1979; Post and Pimm, 1983; Shigesada *et al.*, 1984).

A simple model may be used to simulate the invasion of a new species (say species D) into a community consisting of three species (A, B, C). Assume that the community is experiencing disturbances that are both temporally and spatially stochastic. Some disturbances could create not only an open space but changes in the amount of available nutrients (Tilman, 1982; Hobbs, this volume). To simplify the problem, I will limit disturbance using Grime's (1979) definition: partial or total destruction of biomass. Technically, the model consists of a system of four Lotka–Volterra competition equations with one dimensional diffusion (see Williamson, this volume) and a spatially and temporally discrete stochastic harvest (Rejmánek, 1984). Even if D is a weaker competitor than any of the three resident species and, in the absence of disturbance, D would be extinct, some realistic regimes of disturbance allow its invasion. Moreover, the invasion might be successful even if an external source of propagules exists only temporarily and D does not disperse faster than B and C (Figures 16.5 and 16.7). The invading species in presented simulations spreads rather slowly like, for example, *Impatiens parviflora* dispersing ballistically only about 2 m per year (Coombe, 1959). Even if in a very crude way, the model shows how a low level of natural disturbance and/or vegetation senescence can allow the spread of a competitively inferior non-native species. On the other hand, from a similar simulation model (Dostalkova *et al.*, 1984) it follows that in the absence of any disturbance, selfinhibition or senescence, even invaders which are stronger

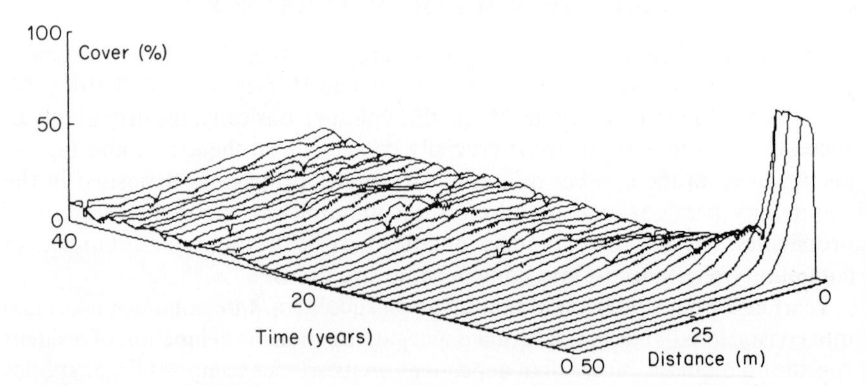

Figure 16.5. A simulation of the invasion of a hypothetical species (D) into a community consisting of three stronger competitors (A, B, C). Propagule diffusion from one end of transect (forest boundary) took place between year 1 and 4. Invading species has a lowest position in a competitive hierarchy (A > B > C > D) and cannot survive in the community without stochastic disturbance. The disturbance in this simulation was realized by a complete harvest of all species in randomly selected 30 (in average) out of 200 intervals forming 50 m long transect every year. Such disturbance corresponds to a complete destruction or death of vegetation in about 15% of an area every year

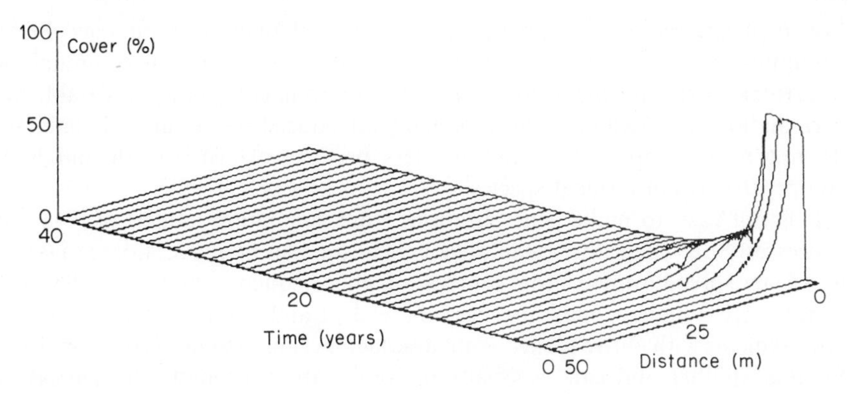

Figure 16.6. The same simulation parameters as in Figure 16.5 but the frequency of disturbance is reduced to two-thirds. The invader (D) is unable to establish itself

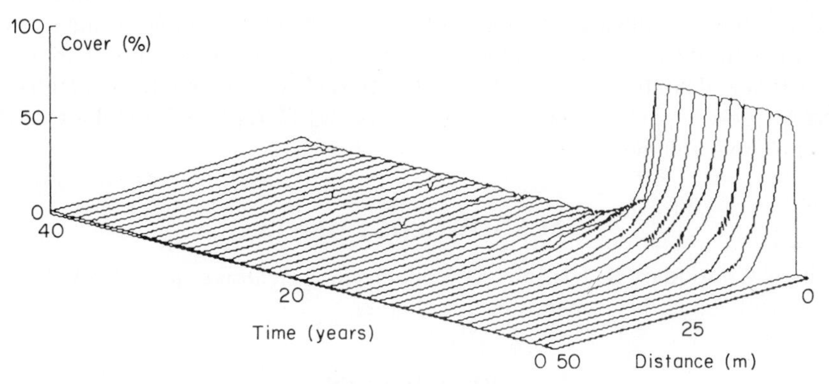

Figure 16.7. The same simulation parameters as in Figure 16.6 but with the propagule diffusion from external source between year 1 and 12. The invader (D) is spreading and survives in the community

competitors than native species can scarcely invade their community. Paleobotanical evidence for communities resistant to invasion of species superior to established ones from the Pleistocene–Holocene period comes from Cole (1985); (see also Levin, 1985; Markgraf, 1986; and Cole, 1986).

16.7 CONCLUSIONS

All analysis of community invasibility based just on field observations (*a posteriori*) are unsatisfying. Usually, it is impossible to separate the resistance of a biotic community from resistance determined by an abiotic environment. Also, in most of the cases we do not know anything about the quality and quantity of imported propagules. Nevertheless, available evidence indicates that only very few alien species invade successionally advanced plant communities. Plant communities in mesic environments seem to be more invasible than communities in extreme environments. Apparently xeric environments are not favorable for germination and seedling survival of many introduced species and wet habitats do not provide open space for invaders because of fast growth and high competitiveness of resident species.

It is not easy to make generalizations from simulation models of invasion processes. Still, one rather qualitative conclusion can be made. Longer lasting and faster spread from external sources; a higher frequency (up to some level) of disturbance; a higher competitiveness of invaders; and lower overlap in resource requirements with native plants;—all these factors can, to some extent, substitute for one another and can, especially in combination, promote the success of invasions in plant communities. This generalization is in agreement or, at least, not in contradiction with existing field data.

While we have identified several factors that are important determinants of plant community invasibility, we still lack experimental data that would permit us to quantify and sort out the combined effects of these traits. Far more experimental studies are needed to add to those few that already exist (Juhasz-Nagy, 1964; Sagar and Harper 1960; Cavers and Harper 1967; Stebbins, 1985; Hobbs, this volume).

ACKNOWLEDGEMENTS

I thank C. Stoddart, R. J. Hobbs, M. Williamson, J. A. Drake and F. J. Kruger for comments on earlier versions of this chapter.

REFERENCES

Auclair, A. N., and Goff, F. G. (1971). Diversity relations of upland forests in the western Great Lakes area. *Amer. Nat.*, **105**, 499–528.

Baker, H. G. (1965). Characteristics and modes of origin of weeds. In: Baker, H. G., and Stebbins, G. L. (Eds), *The Genetics of Colonizing Species*, pp. 147–169. Academic Press, New York.

Barbour, M. G., and Rodman, J. E. (1970). Saga of the west coast sea-rockets: *Cakile edentula* ssp. *Californica* and *C. maritima*. *Rhodora*, **72**, 370–86.

Baumova, H. (1985). Influence of disturbance by mowing on the vegetation of old-fields in the Bohemian Karst. *Folia Geobot. Phytotax.*, **20**, 245–65.

Bonck, J., and Penfound, W. T. (1945). Plant succession on abandoned farm land in the vicinity of New Orleans, Louisiana. *Amer. Mid. Natur.*, **33**, 520–9.

Bornkamm, R., and Henning, U. (1982). Experimentell-ökologische Untersuchungen zur Sukzession von ruderalen Pflanzengesellschaften auf unterschiedlichen Böden. *Flora*, **172**, 267–316.

Boyd, R. S. (1986). Comparative ecology of two west coast *Cakile* species at Point Reyes, California. Ph.D. Thesis, Dept. of Botany, UC Davis. 128 pp.

Burbanck, M. P., and Philips, D. L. (1983). Evidence of plant succession on granite outcrops of the Georgia Piedmont. *Amer. Midl. Nat.*, **109**, 94–104.

Burdon, J. J., and Chilvers, G. A. (1977). Preliminary studies on a native Australian eucalypt forest invaded by exotic pines. *Oecologia*, **31**, 1–12.

Campbell, B. M., McKenzie, B., and Moll, E. J. (1979). Should there be more tree vegetation in the Mediterranean climatic region of South Africa. *J. South Afr. Bot.*, **45**, 453–7.

Carpenter, D. E. (1983). Old-field succession in Mojave desert scrub. MS Thesis, Dept, of Geography, UC Davis. 112 pp.

Cavers, P. B., and Harper, J. L. (1967). Studies in the dynamics of plant populations. 1. The fate of seen and transplants introduced into various habitats. *J. Ecol.*, **55**, 59–71.

Cole, K. (1985). Past rates of change, species richness, and a model of vegetational inertia in the Grand Canyon, Arizona. *Amer. Nat.*, **125**, 289–303.

Cole, K. L. (1986). In defense of inertia. *Amer. Nat.*, **127**, 727–8.

Coombe, D. E. (1959). Biological flora of British Isles: *Impatiens parviflora* DC. *J. Ecol.*, **44**, 701–13.

Crawley, M. J. (1987). What makes community invasible? In: Crawley, M. J., Edwards, P. J., and Gray, A. J. (Eds), *Colonization, Succession and Stability*, pp. 429–53. Blackwell, Oxford.

Cross, J. R. (1981). The establishment of *Rhododendron ponticum* in the Killarney Oakwoods, S. W. Ireland: *J. Ecol.*, **69**, 807–24.

Csontos, P. (1984). Ecological and phytosociological studies in a stand of *Impatiens parviflora* DC, at the Vadallo Rocks, Pilis Mts., Hungary, *Abstracta Botanica*, **8**, 15–34.

Davidson, R. L. (1964). An experimental study of succession in the Transvaal Highveld. In: Davis, D. H. S. (Ed.), *Ecological Studies in Southern Africa*, pp. 113–25. Junk, The Hague.

del Moral, R. (1983). Competition as a control mechanism in subalpine meadows. *Amer. J. Bot.*, **70**, 232–45.

Dostálková, I., Kindlmann, P., and Rejmánek, M. (1984). Simulation of species replacement on environmental gradient in the course of ecological succession. *Ecol. Model.*, **26**, 45–50.

Dyrness, C. T. (1973). Early stages of plant succession following logging and burning in the Western Cascades of Oregon. *Ecology*, **54**, 57–69.

Egler, F. E. (1983). *The Nature of Naturalization II. The Introduced Flora of Aton Forest, Connecticut*. Claude E. Phillips Herbarium Publication No. 6. Delaware State College, Dower. 145 pp.

Eicke-Jene, J. (1960). Sukzessionsstudien in der Vegetation des Ammersees in Oberbayern. *Bot. Jb.*, **79**, 447–520.

Elton, C. S. (1958). *The Ecology of Invasions by Animals and Plants*. Methuen, London. 181 pp.

Ewel, J. J. (1986). Invasibility: lessons from south Florida. In: Mooney, H. A., and Drake, J. A. (Eds), *Ecology of Biological Invasions of North America and Hawaii*. Springer, New York.

Fallinski, J. B. (1968). Stan i prognoza neofityzmu w szacie roslinnej Puszcy Bialowieskej. *Materialy Zakladu Fitosociologii Stosowanej U.W.*, **25**, 175–216.

Forcella, F., and Harvey, S. J. (1983). Eurasian weed infestation in western Montana in relation to vegetation and disturbance. *Madrono*, **30**, 102–9.

Fox, M. D., and Fox, B. J. (1986). The susceptibility of natural communities to invasion. In: Groves, R. H., and Burdon, J. J. (Eds), *Ecology of Biological Invasions: An Australian Perspective*, pp. 57–66. Australian Academy of Science, Canberra.

Frenkel, R. E., and Boss, T. R. (1982). Introduction and establishment of *Spartina patens* in Siuslaw Estuary, Oregon. Paper presented at the Pacific Estuarine Research Society 10th Semiannual Meeting.

Gerrish, G., and Mueller-Dombois, D. (1980). Behavior of native and non-native plants in two tropical rain forests on Oahu, Hawaiian Islands. *Phytocenologia*, **8**, 237–95.

Gleadow, R. M., and Ashton, D. H. (1981). Invasion by *Pittosporum undulatum* of the forests of central Victoria. I Invasion patterns and plant morphology. *Aust. J. Bot.*, **29**, 705–20.

Grime, J. P. (1979). *Plant Strategies and Vegetation Processes.* Wiley, New York.

Guittet, J., and Laberche, J. C. (1974). L'implantation naturelle du pin sylvestre sur pelouse xérophile en forêt de Fontainebleau: II. Démographie des graines et des plantules au voisinage des viex arbres. *Oecol. Plant.*, **9**, 111–30.

Hawksworth, F. G., and Scharpf, R. F. (1986). Spread of European mistletoe (*Viscum album*) in California, U.S.A. *Eur. J. Forest Pat.*, **16**, 1–5.

Hett, J. M. (1971). A dynamic analysis of age in sugar maple seedlings. *Ecology*, **52**, 1071–4.

Holzner, W. (1982). Concepts, categories and characteristics of weeds. In: Holzner, W., and Numata, N. (Eds), *Biology and Ecology of Weeds*, pp. 3–20. Junk, The Hague.

Huml, O. (1977). Konkurencni vztahy dominant a jejich vyznam v iniciálnich sukcesních fázích. MS Thesis, Dept of Botany, Charles University, Prague.

Johnson, P. N. (1982). Naturalised plants in south-west South Island, New Zealand. *New Zealand J. Bot.*, **20**, 131–42.

Juhász-Nagy, P. (1964). Investigations concerning ecological homeostasis. In: *Tenth International Botanical Congress*, p. 240. Abstracts of Papers. Edinburgh.

Kloot, P. M. (1984). The introduced elements of the flora of southern Australia. *J. Biogeogr.*, **11**, 63–78.

Knopf, F. L., and Olson, T. E. (1984). Naturalization of Russian-olive: implications to Rocky Mountain wildlife. *Wildlife Soc. Bull.*, **12**, 289–98.

Kopecky, K. (1967). Die flussbegleitende Neophytengesellschaft Impatienti-Solidaginetum in Mittelmähren. *Presila*, **39**, 151–66.

Kornas, J. (1988). Plant invasions in Central Europe: historical and ecological aspects. In: di Castri, F. (Ed.), *History and Patterns of Biological Invasions in Europe and the Mediterranean Basin* (in press).

Kruger, F. J. (1977). Invasive woody plants in Cape fynbos with special reference to the biology and control of *Pinus pinaster. Proceedings of the Second National Weed Conference of South Africa, Stellenbosch*, pp. 57–74.

Levin, R. (1985). Plant communities resist climatic change. *Science*, **228**, 165–6.

Loope, L. L., and Scowcroft, P. G. (1985). Vegetation response within exclosures in Hawaii: a review. In: Stone, C. P., and Scott, J. M. (Eds), *Hawaii's Terrestrial Ecosystems: Preservation and Management*, pp. 377–402. University of Hawaii, Honolulu.

Lorence, D. H., and Sussman, R. W. (1986). Exotic species invasion into Mauritius wet forest remnants. *J. Trop. Ecol.*, **2**, 147–62.

Luer, C. A. (1975). *The Native Orchids of the United States and Canada.* The New York Botanical Garden, New York.

Macdonald, I. A. W., Powrie, F. J., and Siegfried, W. R. (1986). The differential invasion of southern Africa's biomes and ecosystems by alien plants and animals. In: Macdonald, I. A. W., Kruger, F. J., and Ferrar, A. A. (Eds), *The Ecology and Management of Biological Invasions in Southern Africa*, pp. 209–25. Oxford University Press, Cape Town.

Macdonald, I. A. W., and Richardson, D. M. (1986). Alien species in terrestrial ecosystems of the fynbos biome. In: Macdonald, I. A. W., Kruger, F. J., and Ferrar, A. A. (Eds), *The Ecology and Management of Biological Invasions in South Africa*, pp. 77–91. Oxford University Press, Cape Town.

Markgraf, V. (1986). Plant inertia reassessed. *Amer. Nat.*, **127**, 725–6.

Mooney, H. A., Hamburg, S. P., and Drake, J. A. (1986). The invasions of plants and animals into California. In: Mooney, H. A., and Drake, J. A. (Eds), *The Ecology of Biological Invasions of North America and Hawaii*, pp. 250–72. Springer-Verlag, New York.

Moravec, J. (1973). The determination of minimal area of phytocenoses. *Folia Geobot. Phytotax.*, **8**, 23–47.

Mueller-Dombois, D. (1981). Vegetation dynamics in coastal grassland of Hawaii. *Vegetatio*, **46**, 131–40.

Myers, R. L. (1983). Site susceptibility to invasion by the exotic tree Melaleuca quinquenvervia in southern Florida. *J. Appl. Ecol.*, **20**, 645–58.

Nilsson, C., Grellson, G., Johansson, M., and Sperens, U. (1988). Patterns of species richness along river banks. *Ecology* (in press).

Nilsson, Ö. (1981). Gräsfröinkomlingar och andra kulturspridda växter från Gotland. *Svensk Bot. Tidskr.*, **75**, 65–9.

Ohmart, R. D., and Anderson, B. W. (1982). North American desert riparian ecosystems. In: Bender, G. L. (Ed.), *Reference Handbook on the Deserts of North America*, pp. 433–79. Greenwood Press, Westport.

Peart, D. R., and Foin, T. C. (1985). Analysis and prediction of population and community change: a grassland case study. In: White, J. (Ed.), *The Population Structure of Vegetation*, pp. 313–39. W. Junk, The Hague.

Peet, R. K. (1978). Forest vegetation of the Colorado Front Range: patterns of species diversity. *Vegetatio*, **37**, 65–78.

Pickard, J. (1984). Exotic plants on Lord Howe Island: distribution in space and time, 1853–1981. *J. Biogeogr.*, **11**, 181–208.

Pickett, S. T. A. (1982). Population patterns through twenty years of oldfield succession. *Vegetatio*, **49**, 45–59.

Pickett, S. T. A. (1983). The absence of an Andropogon stage in old-field succession at the Hutcheson Memorial Forest. *Bull. Torey Bot. Club*, **110**, 533–5.

Pickett, S. T. A., and White P. S. (1985). *The Ecology of Natural Disturbance and Patch Dynamics*. Academic Press, Orlando.

Post, W. M., and Pimm, S. L. (1983). Community assembly and food web stability. *Math. Biosci.*, **64**, 169–92.

Ranwell, D. S. (1972). *Ecology of Salt Marshes and Sand Dunes*. Chapman and Hall, London.

Ratcliffe, D. A. (1984). Post-Medieval and recent changes in British vegetation: the culmination of human influence. *New Phytol.*, **98**, 73–100.

Raven, P. H., and Thomas, J. H. (1970). *Iris pseudacorus* in western North America. *Madrono*, **20**, 390–1.

Rejmánek, M. (1984). Perturbation-dependent coexistence and species diversity in ecosystems. In: Schuster, P. (Ed.), *Stochastic Phenomena and Chaotic Behaviour in Complex Systems*, pp. 220–30. Springer, Berlin.

Rejmánek, M., Sasser, C. E., and Gosselink, J. G. (1987). Modelling of vegetation dynamics in the Mississippi River deltaic plain. *Vegetatio*, **69**, 133–40.

Rejmánek, M., Sykora, T., and Stursa, J. (1971). Fytocenologicke poznamky k vegetaci Hrubeho Jeseniku. *Campanula*, **2**, 31–9.

Robinson, J. V., and Valentine, W. D. (1979). The concepts of elasticity, invulnerability and invadability. *J. Theor. Biol.*, **81**, 91–104.

Robinson, T. W. (1965). *Introduction, Spread, and Real Extent of Saltcedar (Tamarix) in the Western States*. Geol. Surv. Prof. Paper 491-A, US Dept of Interior. 12 pp.

Rodman, J. E. (1986). Introduction, establishment, and replacement of sea-rockets (*Cakile*, Cruciferae) in Australia. *J. Biogeog.*, **13**, 159–71.

Rothmaler, W. (1976). *Excursionsflora für die Gebiete der DDR und der BRD*. Volk und Wissen Volkseigener Verlag, Berlin.

Sagar, G. R., and Harper, J. L. (1960). Factors affecting germination and early establishment of Plantains (*Plantago lanceolata, P. media and P. major*). In: Harper, J. L. (Ed.), *The Biology of Weeds*, pp. 236–45. Blackwell, Oxford.

Schmidt, W. (1981). Ungestörte und gelenkte Sukzession auf Brachäckern. *Scripta Geobotanica*, **15**, 1–199.

Shigesada, N., Kawasaki, K., and Teramoto, E. (1984). The effects of interference competition on stability, structure and invasion of multi-species system. *J. Math. Biol*, **21**, 97–113.

Shimizu, Y., and Tabata, H. (1985). Invasion of *Pinus lutchuensis* and its influence on the native forest on Pacific island. *J. Biogeogr.*, **12**, 195–207.

Slezak, W. F. (1976). *Lonicera japonica* Thunb. an aggressive introduced species in a mature forest ecosystem. MS Thesis, Rutgers University, New Brunswick, NJ.

Sousa, P. (1984). The role of disturbance in natural communities. *Ann. Rev. Ecol. Syst.*, **15**, 353–91.

Stebbins, G. L. (1985). Polyploidy, hybridization, and the invasion of new habitats. *Ann. Missouri Bot. Gard.*, **72**, 824–32.

Stuckey, R. L. (1968). Distributional history of *Butomus umbellatus* (flowering-rush) in the western Lake Erie and Lake St. Clair region. *Michigan Bot.*, **7**, 134–42.

Sukopp, H. (1962). Neophyten in natürlichen Pflanzengesellschaften Mitteleuropas. *Ber. Deutsch. Bot. Ges.*, **75**, 193–205.

Symonides, E. (1985). Floristric richness, diversity, dominance and species evenness in old-field successional ecosystems. *Ekol. Polska*, **33**, 61–79.

Thieret, J. W. (1980). *Louisiana Ferns and Fern Allies*. Lafayette Nat. Hist. Mus., Lafayette.

Tilman, D. (1982). *Resource Competition and Community Structure*. Princeton University Press, Princeton, NJ.

Trepl, L. (1984). Über *Impatiens parviflora* DC. als Agriophyt in Mitteleuropa. *Disseratt-iones Botanicae*, **73**, 1–399.

Valen Van, L. (1975). Life, death, and energy of a tree. *Biotropica*, **7**, 260–9.

Vasek, F. C. (1980). Early successional stages in Mojave desert scrub vegetation. *Israel J. Bot.*, **28**, 133–48.

Vitousek, P. M. (1986). Biological invasions and ecosystem properties: can species make a difference? In: Mooney, H. A., and Drake, J. A. (Eds), *Ecology of Biological Invasions of North America and Hawaii*, pp. 163–78. Springer, New York.

Vivrette, N. J., and Muller, C. H. (1977). Mechanism of invasion and dominance of coastal grassland by *Mesembryanthemum crystalinum*. *Ecol. Monogr.*, **47**, 301–18.

Wein, K. (1942). Die älteste Einführungs und Ausgebreitungsgeschichte von *Acorus calamus*. 3. *Hercinia*, **3**, 241–91.

Weiss, P. W., and Noble, I. R. (1984). Status of coastal dune communities invaded by *Chrysanthemoides monilifera*. *Aust. J. Ecol.*, **9**, 93–8.

Wells, M. J., Duggan, K., and Hendersen, L. (1980). Woody plant invaders of the central Transvaal. *Proceedings of the third National Weeds Conference of South Africa, Cape Town*, pp. 11–78.

Wilgen van, B. W., and Richardson, D. M. (1985). The effects of alien shrub invasions on vegetation structure and fire behaviour in South African fynbos shrublands: a simulation study. *J. Appl. Ecol.*, **22**, 955–66.

Williamson, M. (1988). The relationship of species number to area, distance and other variables. In: Giller, P. S. (Ed.), *Biogeographic Analysis: Methods, Patterns and Processes* (in press).

Biological Invasions: a Global Perspective
Edited by J. A. Drake *et al.*
© 1989 SCOPE. Published by John Wiley & Sons Ltd

CHAPTER 17

The Nature and Effects of Disturbance Relative to Invasions

RICHARD J. HOBBS

17.1 INTRODUCTION

Biological invasions are a fact of life today, and many natural communities have been or are being invaded by non-native species which affect natural community and ecosystem process. There is still a lot of debate on the characteristics of invasive species and invasible communities (e.g. Crawley, 1986, 1987; Newsome and Noble, 1986). For plant communities several authors have suggested that some form of disturbance is an important precursor to invasion, especially where this disturbance disrupts strong species interactions and hence reduces competition (e.g. Kruger *et al.*, 1986; Macdonald *et al.*, 1986; Crawley, 1986, 1987; Orians, 1986; Fox and Fox, 1986). Although we have a considerable amount of information on invasions, most of the available data are observational or historical and there has been little attempt to assess experimentally the effects of disturbance on the invasibility of communities.

In this chapter I therefore illustrate the potential for experimentation in the field of biological invasions, especially with respect to disturbance. I will firstly examine the nature of disturbance in natural communities and then discuss in detail a series of experiments on the effects of disturbance on the invasibility of natural plant communities, and finally discuss the relevance of these rather local experiments to the global perspective aimed at in these proceedings.

17.2 DISTURBANCE AND PATCH DYNAMICS

Although disturbance is frequently cited as a factor leading to invasions, the nature of the disturbance is not usually explored. Recent reviews have emphasized the importance of disturbances of various types in the functioning of many natural communities and ecosystems (White, 1979; Sousa, 1984; Pickett and White, 1985; Hobbs, 1987), particularly in relation to patch dynamics. White and Pickett (1985) give a broad definition of disturbance as 'any relatively discrete event in time that disrupts ecosystem, community or population structure and changes resources, substrate availability or the physical environment'. Problems of scale arise when, for instance, events causing disruption are actually part of longer-term environmental fluctuations—a flood or a drought may constitute a disturbance but still be part of the normal hydrologic or climatic cycle. These problems have been addressed elsewhere (e.g. Pickett and White, 1985) and will not be pursued here.

Disturbances are viewed as creating patches of either open ground or increased resource availability, often through the removal of the species present before disturbance. There has been increasing interest in 'patch dynamics', or the processes of patch creation and filling, and many studies have examined the formation and colonization of patches caused by different types of disturbance, e.g. by animal disturbance (Platt, 1975; Collins and Barber, 1985; Hobbs and Mooney, 1985), treefalls (Brokaw, 1985; Collins *et al.*, 1985), or wave action in intertidal and subtidal environments (Sousa, 1985; Connell and Keough, 1985). Although immigration into patches is recognized as an important process, the invasion of pre-existing or newly formed patches by non-native species has been little studied. Indeed, invasive species have often been considered outside normal community processes, but biological invasions are now simply a subset of normal community process (e.g. Diamond and Case, 1986; Chesson, 1986). Human activities have rendered many introduced species part of the 'natural' community. Fragmentation has led to the juxtaposition of natural and human-made systems, and intentional and unintentional human transport of non-native species often provides a pool of such species able to disperse into natural systems.

The invasion of natural plant communities can be characterized by a number of stages. Firstly, seed must disperse into the community and remain there long enough to germinate. Subsequent establishment and growth to seed set, followed by retention of seed in the soil until at least the following season, are then required for persistence in the community. In this chapter I take the view that the invasion process requires not only the availability of an invading species able to disperse into an area but also the formation of a patch suitable for colonization. I will explore the characteristics which make a patch susceptible to invasion and examine whether these vary in different communities. Rather than examining this question on a wide scale, I will first discuss a series of different communities occurring in the same location. I take an experimental approach to these questions to assess whether communities can be made more invasible.

17.3 DISTURBANCE, RESOURCES AND INVASIBILITY

17.3.1 Background to experiments

I carried out a series of experimental manipulations within five different plant communities occurring in close proximity to one another. The studies were carried out on Durokoppin Nature Reserve near Kellerberrin, 200 km east of Perth, Western Australia. The area has a Mediterranean-type climate with a mean annual rainfall of 340 mm, falling predominantly in winter, and a strong summer drought with only occasional cyclonic storms. The study site lies in the Western Australian wheatbelt, an area extensively cleared for agriculture containing a large number of small (mostly < 200 ha) remnants of native vegetation (Kitchener *et al.*, 1980). Durokoppin Reserve (1030 ha) is the largest remnant in the Kellerberrin area where only 2.8% of the land area can be classed as 'natural' vegetation (i.e. uncleared land not grazed by sheep).

The native vegetation of the Western Australian wheatbelt is very diverse, with high species numbers and complex vegetation mosaics, related partially to soil patterns (Beard, 1983; Hopkins and Griffin, 1984). Clearance has converted communities dominated mainly by perennials into pasture and cropland dominated by annual species. Here, as in other parts of Australia, invasion of natural vegetation by non-native annual species is recognized as a major problem in conservation (Bridgewater and Backshall, 1981; Adamson and Fox, 1982; Hopper and Muir, 1984). Invading species include species common in surrounding pasture land, particularly *Arctotheca calendula, Avena fatua*, and *Bromus* spp., and others such as *Ursinia anthemoides* and *Hypochaeris glabra* (nomenclature follows Green (1985) throughout). Soil disturbance is a common phenomenon in the native plant communities, both through the activities of the fauna and by humans. In addition, non-native pasture species are known to respond well to addition of major nutrients, particularly N and P (Rossiter, 1966; McIvor and Smith, 1971). Soil disturbance and nutrient addition were therefore the two principal foci of these experiments.

17.3.2 Communities studied

The vegetation of Durokoppin Reserve consists of a mosaic of many different communities, falling into the main structural categories of woodland, mallee, shrubland and heath. I selected for study examples of the most common communities.

1. Casuarina shrubland dominated by *Allocasuarina campestris*, with a sparse understorey of *Hakea scoparia* and *Verticordia chrysantha* and few annuals.
2. Heathland composed of many species of Proteaceae (including *Grevillea* spp., *Hakea* spp.), Myrtaceae (*Verticordia* spp., *Baeckia* spp.) and Epacridaceae (e.g. *Andersonia lehmanniana*), with scattered annuals including *Blennospora drummondii, Helipterum demissum* and *Trachymene cyanopetala.*
3. 'Edge'; open shrubland with scattered *Hakea recurva, Verticordia crysantha* and *Grevillea paradoxa* on an area cleared approximately 50 years ago for agriculture and subsequently abandoned. Dense patches of native annuals are

present, including *Helipterum hyalospermum*, *Waitzia acuminata* and *Myriocephalus gracilis*.

4. Woodland dominated by jam, *Acacia acuminata*, and York gum, *Eucalyptus loxophleba*, with little understorey and a ground layer of scattered *Stackhousia monogyna*, *Borya nitida*, and abundant native annuals including *Helichrysum lindleyi*, *Waitzia acuminata* and *Trachymene cyanopetala*.
5. Woodland dominated by wandoo (*Eucalyptus wandoo*) with a sparse understorey of *Acacia* spp. and scattered native annuals, particularly *Podolepis lessonii*.

17.3.3 Experimental procedure

All of these communities had large areas of bare ground (i.e. > 30% cover) within them, either between shrubs or underneath tree canopies. The areas chosen for study were relatively free of non-native annual species. I set up 1 m^2 plots in bare areas and carried out the following treatments:

C: control (no manipulation).
D: disturbance of the soil by breaking the soil crust (if any) and turning the soil over with a spade to a depth of 5 cm.
F: fertilized with 50 g slow-release 'Osmocote' fertilizer (18% N as 7.5% NO_3, 10.5% NH_4; 4.8% P; 8.3% K and 3.0% S as K_2SO_4).
DF: Disturbance (as above) combined with fertilizer (as above).

A plot size of 1 m^2 was selected to fit into bare areas between shrubs or herbaceous perennials. Each treatment was replicated four times in each community. Within each set of plots, treatments were allocated randomly to plots, and individual plots were separated by at least 0.5 m. Nitrogen and phosphorus are generally considered the major limiting nutrients in the soils studied (e.g. Groves, 1981; Bettenay, 1984).

The experiment ran from May to September 1985 in the wandoo woodland and May to September 1986 in all other communities. To remove the effect of 'opportunity' (i.e. difference in invasion due to differential availability of propagules), I added seed of introduced species to the plots. I scattered into one half of each plot 100 seeds of *Avena fatua*, a non-native annual grass, and into the other half 100 seeds of *Ursinia anthemoides*, a non-native annual forb. Both are common invasive species which are widespread both around the study area and in southwest Western Australia as a whole, but were absent from the immediate vicinity of the study plots. Seed bank trials indicated that seeds of these species were present in the soil in low numbers; this was not controlled for during the experiments.

I counted numbers of established seedlings in June/July following the start of the experiment and all above-ground material was harvested in September, sorted into species, dried and weighed.

17.3.4 Results and discussion

The experimental treatments could possibly increase the invasibility of the communities in two ways, either by increasing resource availability (adding nutrients, which represents a chemical disturbance) or by producing patches of altered substrate characteristics (representing a physical disturbance). Both manipulations could act by directly enhancing the establishment and survival of non-native species or by altering competition for resources between native and non-native species. The five communities studied had quite different attributes (Table 17.1). The shrub communities (Casuarina and heath) had high densities of shrubs with few native annual species while the woodland communities had few shrubs and more native annuals. The 'edge' or old-field plot was intermediate, with scattered shrubs and relatively high densities of native annuals. Soil nutrient status was roughly similar in all communities except for the jam–York gum woodland which had significantly higher levels of N, P and K (Figure 17.1).

Despite these differences between communities, the effects of experimental manipulation were remarkably constant over all the communities (Figure 17.2). Establishment of *Avena* was greatly enhanced by disturbance, with or without fertilizer addition, while fertilizer on its own had little effect. The same was true for *Ursinia* in the wandoo woodland in 1985, but establishment in 1986 was low in all communities, and there was little treatment effect. This difference in *Ursinia* establishment between years may indicate the importance of climatic variations in determining the success of invasions. We have observed similar links between climatic fluctuations and shrub invasion of grassland in northern California (Williams *et al.*, 1987). There was some establishment of *Avena* and/or *Ursinia* in the control plots in all communities, but the Casuarina and Jam–York gum controls contained the most seedlings.

In all communities, biomass of individual *Avena* and *Ursinia* plants showed little response to disturbance alone (Figure 17.2b). In the Casuarina, edge and wandoo communities there was a small increase in biomass in the fertilized plots and a much larger increase where disturbance was combined with fertilizer addition. In the heath and jam–York gum plots, individual plant biomass in the disturbance + fertilizer plots was not significantly different from that in the fertilized treatment. Although seed output was not measured directly, there was a

Table 17.1. Main biotic characteristics of the communities studied; density and diversity of the shrub layer and of native annuals

	Shrub layer	Native annuals
Casuarina shrubland	High density/diversity	Low
Health	High	Low
Edge	Low	High
Jam–York gum woodland	Very Low	High
Wandoo woodland	Very low	Medium

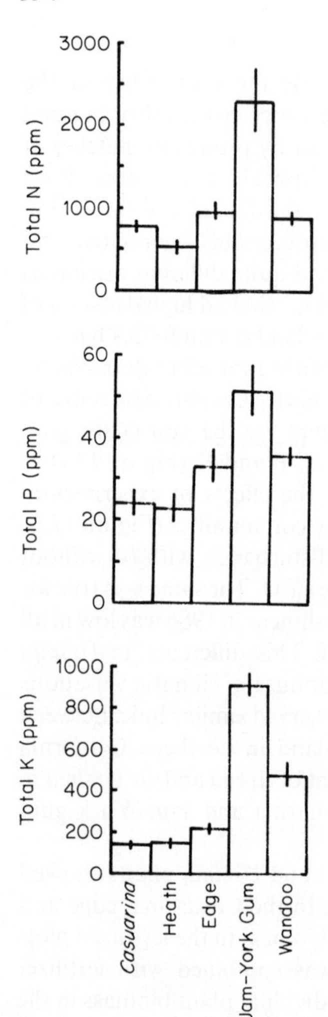

Figure 17.1. Soil nutrient status in the five communities studied. Values for total N (honda digest) and total P and K (nitric acid/sulphuric acid digest), expressed as ppm (mean of 5 samples + 1 SE)

strong correlation between plant biomass and numbers of seed produced for both species. Plant biomass can therefore be equated with reproductive output.

Total biomass of the species added showed the same patterns, with little effect of disturbance or fertilizer alone but a very large effect when they were combined (Figure 17.2c). Only in the jam–York gum community was biomass of *Avena* in the disturbance + fertilizer plots not significantly different from that in the fertilized only plots.

Despite the qualitatively similar treatment responses in different communities, there were some quantitative differences, both in numbers establishing and in biomass. Highest numbers establishing for both species occurred in the woodland sites. The heath site had significantly lower biomass of both species than the other sites.

From these results I conclude that disturbance enhanced the successful establishment of the two added non-native species in all the communities studied, but that these species grow considerably better where disturbance is coupled with

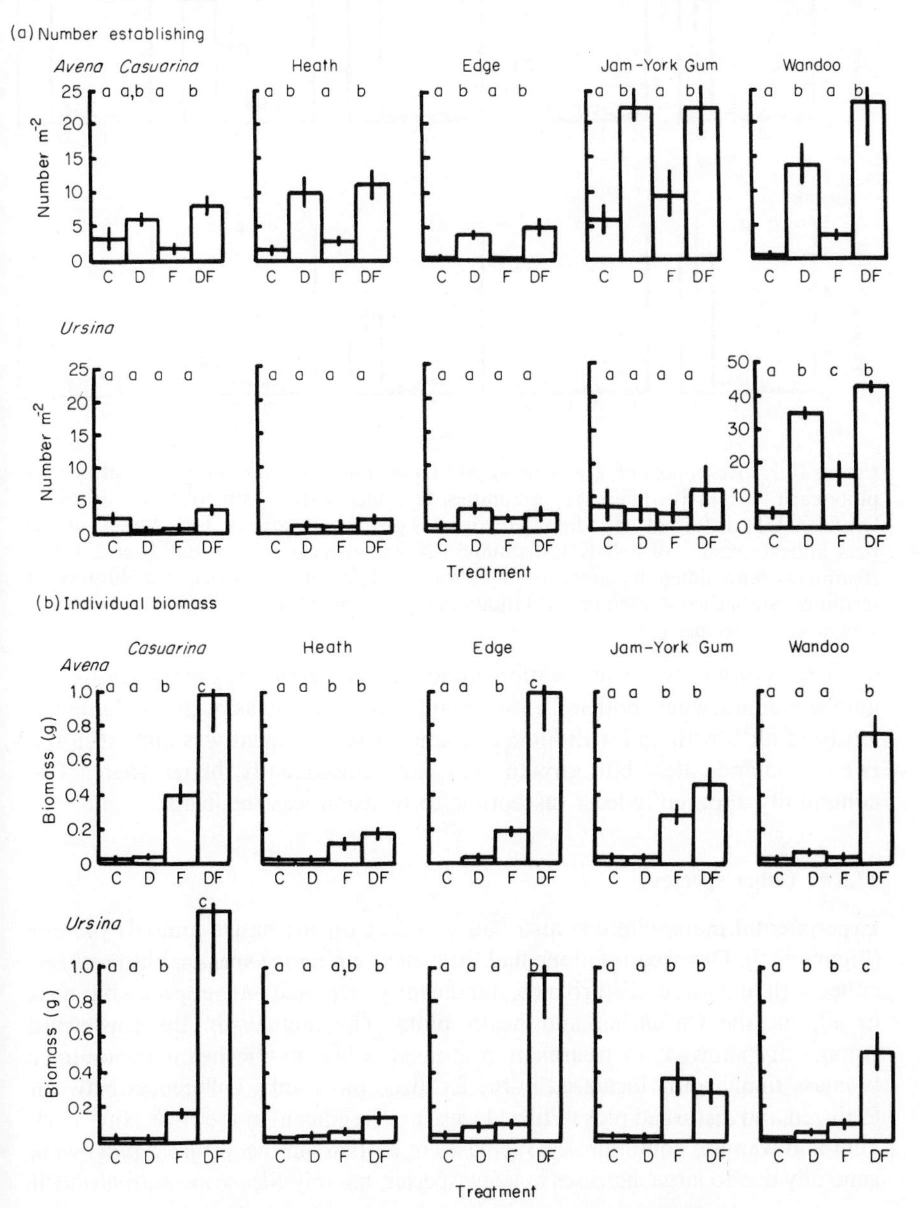

Figure 17.2. (*Caption and continuation overleaf*)

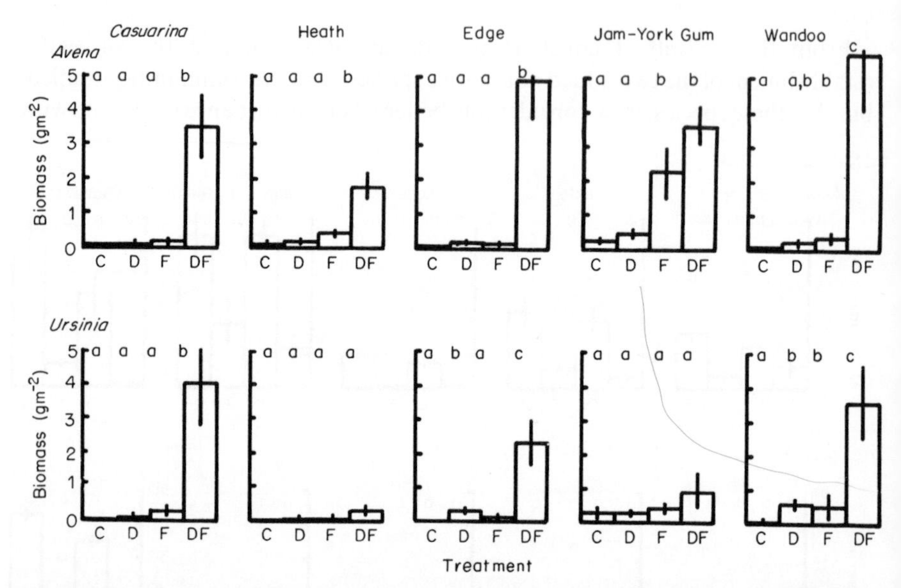

Figure 17.2. Responses of *Avena fatua* and *Ursinia anthemoides* sown in experimental plots (each 1 m^2) in five different communities. Treatments are: C, control; D, disturbed; F, fertilized; DF, disturbed + fertilized. Histograms give mean values + 1 SE. Letters above bars indicate results of S-N-K test (Student-Newman-Keuls) (Steel and Torrie, 1982): treatments with different letters are significantly different at $P < 0.05$. (a) Number of seedlings established in each plot. (b) Individual plant biomass (dry weight). (c) Total dry weight of species per plot

nutrient addition. Nutrient addition alone was important only in the jam–York gum woodland, where both individual plant and total biomass were similar in the fertilized plots with and without disturbance. Establishment was highest in the two woodland sites, but growth was not subsequently better there. The community apparently least susceptible to invasion was the heath.

17.3.5 Other species

Experimental manipulations also had an effect on the native annuals present (Figure 17.3). Disturbance alone had little effect on native species, but fertilizer, either with or without disturbance, significantly increased native species biomass in all but the Casuarina and heath plots. The annuals in the Casuarina community showed no treatment responses, while in the heath community biomass significantly increased in the fertilized plots only. Differences between fertilized and disturbed plus fertilized were not significant in the edge, jam–York gum and wandoo communities. Increases in biomass in the fertilized plots were generally due to large increases in a few species, notably *Blennospora drummondii*

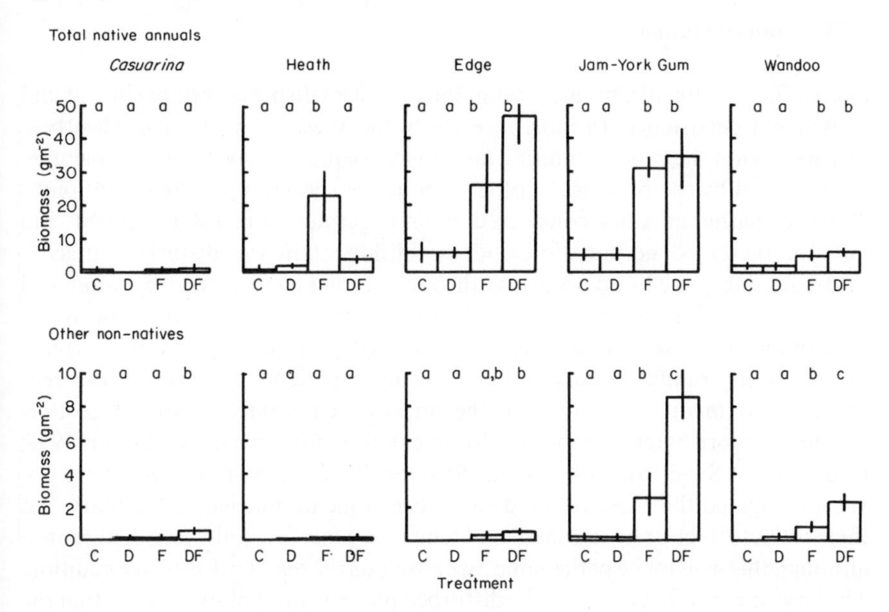

Figure 17.3. Responses of native annuals and non-native species other than *Avena* and *Ursinia* in experimental plots. Treatments and statistical notation as for Figure 17.2.

in the heath, and *Actinobole uliginosum* and *Helipterum hyalospermum* in the edge community. These results are discussed more fully by Hobbs and Atkins (1988).

Non-native species other than those sown (notably *Arctotheca calendula, Hypochaeris glabra* and *Vulpia myuros*) established in control plots only in the Jam–York gum woodland. There and in the edge and wandoo woodland there was a marked increase in other non-natives in the fertilized and disturbed plus fertilized plots. In the Casuarina, wandoo and Jam–York gum communities, biomass of these species was significantly higher in the disturbance plus fertilizer treatment than in the fertilized only plots. Few other non-native species established in the heath plots, again indicating that this community is the least susceptible to invasion.

The result in Figures 17.2 and 17.3 indicate a difference in response between native and non-native annual species, with natives responding to higher nutrients, with or without disturbance, but non-natives responding most to the combination of disturbance and fertilizer addition. Care must be taken, however, when comparing the response of native species already present as seed in the soil with that of the added non-native species, since disturbance would affect placement of the seeds already present.

17.3.6 Interpretation

Successful invasion of a natural community requires dispersal, establishment and subsequent persistence. In most reserves in the Western Australian wheatbelt, dispersal is not the process limiting invasion since most areas receive propagules, as evidenced by the presence of other non-native species in the treatment plots. These experiments thus considered mainly germination, establishment and growth. Increased germination and establishment in the disturbed areas is probably due to reduced seed loss through secondary dispersal by wind and water or predation by ants, combined with increased germination following the disruption of the soil crust. Seeds in disturbed plots were probably partially buried during rainfall events and thus would experience a markedly different microenvironment from those on the undisturbed surface. Disturbance thus provides a more heterogeneous environment and more safe sites (Harper, 1977; Grubb, 1977, Silvertown 1981; Fox, 1985) for the non-native species. I have not yet investigated the effect of soil disturbance alone on nutrient availability, but there is likely to be at least a small and temporary increase following disturbance although this will not be of the same magnitude as the results of fertilizer addition. The large increase in growth in the disturbed plus fertilized plots suggests that the non-native species are usually extremely nutrient limited in the unaltered soil. Disturbance combined with nutrient addition thus not only provides a safe site but increases resource availability to the established plants. Native annual species respond less dramatically to the combination of disturbance and nutrient addition, and observations suggest that they do not germinate preferentially in disturbed areas.

It is usually assumed that disturbance temporarily increases resource availability (Tilman, 1982; Pickett and White, 1985) or causes a resource shift (Fox and Fox, 1986). However, I suggest that certain types of disturbance need not significantly increase resource availability, expecially in community where light is not limiting. In the present case, soil disturbance may alter resource availabilities sufficiently to affect the germination response of the non-native species but does not significantly increase the resources available to the established plant. Such a disturbance will not therefore lead to increased invasion success, whereas a disturbance which increases resource levels is more likely to.

In Western Australian wheatbelt plant communities, many natural small-scale soil disturbances occur through the activities of ants, echidnas (*Tachyglossus aculeatus* Shaw) and other soil fauna. These probably do not lead to large increases in nutrient availability at the same time, although some ant species produce middens composed of discarded seed and other material which probably increases nutrient levels in the immediate vicinity. Rabbit disturbance is also usually accompanied by the deposition of faces, which leads to increased nutrient levels. These areas then provide foci for invasion by non-native annuals. The same result can be produced by fertilizing echidna scrapes. Fires also produce a

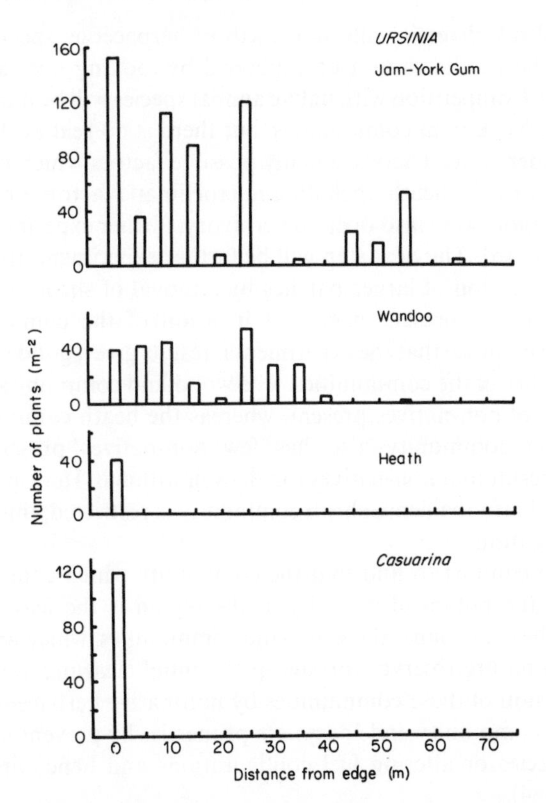

Figure 17.4. Degree of invasion into vegetation remnants from the remnant edge in different communities. Results from transects situated at the edges of Durokoppin Reserve with adjacent road and pasture. 50 × 50 cm quadrats recorded every 5 m in from the edge

temporary increase in nutrient availability and may also lead to increased invasion. Windblown fertilizer, sand or stubble from adjacent farmland into small remnant areas provides another potentially large source of nutrient input. The combination of natural soil disturbance and increased overall nutrient levels is likely to lead to increased invasion by non-native annual species unless transfer of nutrients between farmland and natural vegetation can be minimized.

Although differences between communities were apparent, all those studied were shown to be invasible. Only the heath community showed a significantly greater resistance to invasion. The reasons for this are not immediately obvious. The heath is very species rich, and it could be suggested that the degree of species packing and internal organization influences the susceptibility of the community to invasion. This explanation is not really satisfactory because it is difficult to see how woody shrub diversity affects growth of herbaceous species in the inter-shrub gaps. The results cannot be explained by looking only at the 'guild' of annual species. Competition with native annual species will be most intense in the edge and jam–York gum communities, but there is no real evidence that these 'keep the invaders out'. There are many possible factors which may explain the low invasibility of the heath, including microclimatic factors or differential ant predation. All plots were also open to herbivory; further experiments controlling for this are planned. The next step will be further experimentation in the heath, including the creation of larger patches by removal of shrubs or fire.

If we look at the present degree of invasion of the communities studied (Figure 17.4) we can see that the experimental results give a good indication of the actual invasibility of the communities. The woodland communities already have large numbers of non-natives present, whereas the heath community does not. The Casuarina community also has few non-natives present, despite the experimental result that non-natives can do well within it. Here, perhaps, is a case where dispersal into the dense shrub community is restricted; this again requires further investigation.

It is of some comfort to find that the community which contains the largest proportion of the botanical diversity in the region is the least susceptible to invasion. On the other hand, the woodland communities, which are the least well represented in nature reserves through preferential clearing, are susceptible to invasion. Invasion of these communities by non-native herbaceous species may strongly affect community and ecosystem properties by preventing regeneration of the tree species or altering fuel configurations and hence fire regimes (e.g. Wycherley, 1984).

17.4 COMPARISON WITH OTHER SYSTEMS

I have presented results of an experiment on a series of diverse communities in one geographical location. What relevance do these results have to the broader perspective aimed at in these proceedings? This question is difficult to answer since few data are available which specifically address the problem. Other chapters in this symposium have discussed the importance of disturbance in allowing plant invasion (e.g. Kruger *et al.*, this volume; Rejmanek, this volume). Fox and Fox (1986) concluded that, 'There is no invasion of natural communities without disturbance', while Mack (1985) stated that 'Sufficient examples can now be assembled to indicate that invasion can proceed without continuing disturbance.' Thus, for instance, pines in Australia and South Africa are apparently able to invade native plant communities in the absence of disturbance (Burdon and Chilvers, 1977; Kruger, 1977). This does not preclude the necessity for an initial disturbance to create a focus for establishment, however. There are other examples from Australia which suggest that fertilizer application in the absence of soil disturbance will allow non-native species to invade native communities (e.g. Rossiter, 1964; Armor and Piggin, 1977; Heddle and Specht, 1975). Clearly soil characteristics will be important, and in the Western Australian case reported here, the integrity of the soil surface crust may be an important feature preventing invasion even following nutrient addition. It is clear that a particular disturbance type does not have a 'generic' effect in all situations.

Different responses to disturbance may also arise where the invasive species are predominantly perennial rather than annual. For instance L. Cameron (personal communication) found little interaction between nutrient addition and disturbance when perennials such as *Hypochoaris radicata* and *Lolium perenne* were sown in three different communities in northern New South Wales.

Recent experiments on serpentine grasslands in northern California have shown that it is possible to change the grassland dominated by native forbs into one dominated by non-native grasses by adding nutrients (Hobbs *et al.*, 1988; L. Huenneke and coworkers, personal communication). Hobbs *et al.*, (1988) have shown that physical disturbance of the soil by gophers may reverse the invasion process. After spot applications of fertilizer, they found that non-native grasses, mainly *Bromus mollis* and *B. trinii*, increased greatly in abundance. The mulch formed by the dead grasses led to enhanced germination of *Bromus* in the following year, coupled with reduced survival of the native species. Fertilised plots that were subsequently disturbed by gophers had this mulch removed and native species were able to recolonize. Thus it appears likely that continual soil disturbance by gophers at some sites plays an important part in preventing non-native species from becoming dominant, and in retaining serpentine grasslands as natural 'islands' surrounded by seas of grasslands dominated by non-native Mediterranean grasses. In this case, therefore, although gopher disturbance is important to the internal dynamics of the serpentine grassland (Hobbs and Mooney, 1985), nutrient addition is a much more disruptive 'disturbance' in terms of invasion by non-natives.

17.5 SUMMARY AND CONCLUSIONS

I have outlined a simple experimental approach to the question of what determines the invasibility of natural plant communities. Using the example of a number of different communities in southwest Australia, I investigated the effect of soil disturbance and nutrient addition on invasion by introduced annual species. Soil disturbance enhanced establishment of introduced species, but their subsequent growth was greatly increased when soil disturbance was combined with nutrient addition. Other studies have found similar effects of nutrient adddition alone, and I conclude that the attributes of the system in question determine the influence of disturbances on invasibility. The experiments carried out so far are somewhat exploratory and leave many questions unanswered, but point the way to possible generalizations.

What resources are limiting and the natural disturbance regime will determine which disturbances are likely to have the largest effect. One can predict, for instance, that the creation of gaps might be more important than nutrient increase in light-limited systems. A general statement might be that disturbance will enhance invasibility only if it increases the availability of a limiting resource. This is essentially a truism, since the resource must be limiting to the invading species, but it is nevertheless useful to consider which resources are being significantly altered by disturbance. It is quite possible to have a disturbance which does not significantly increase the availability of a resource which is limiting to an invader, and it is important to be able predict the effects of disturbances on this basis.

While comparisons across systems are important, detailed studies within individual systems are also needed. The long-term conservation of many natural systems now depends on our ability to understand the invasion process and find ways to arrest or prevent it. While we must still rely heavily on historical and observational data on invasions, we must also make use of other tools including modelling and experimentation to help explain some of the questions arising from observational studies. Experimentation has been used relatively little to date and, clearly, some natural systems are more suitable for experimentation than others. There is nevertheless a wealth of information to be obtained from relatively simple and inexpensive experimental techniques.

ACKNOWLEDGEMENTS

I thank L. Atkins, M. Adams and G. Towers for technical support and V. J. Hobbs, G. W. Arnold, P. Vitousek, R. Mack, M. Rejmanek and P. Ramakrishnan for comments on and improvements to the manuscript. I am indebted to Hal Mooney for providing me with the opportunity to attend the symposium and to SCOPE and the National Academy of Sciences for financial support.

REFERENCES

Adamson, D. A., and Fox, M. D. (1982). Change in Australasian vegetation since Europe an settlement. In: Smith, J. M. B. (Ed.), *A History of Australasian Vegetation*, pp. 109–46. McGraw-Hill, Sydney.

Armor, R. L., and Piggin, C. M. (1977). Factors influencing the establishment and success of exotic plants in Australia. In: Anderson, D. (Eds), *Exotic Species in Australia—Their Establishment and Success*. Proc. Ecol. Soc. Aust. 10, pp. 15–26.

Beard, J. S. (1983). Ecological control of the vegetation of Southwestern Australia: moisture versus nutrients. In: Kruger, F. J., Mitchell, D. T., and Jarvis, J. U. M. (Eds), *Mediterranean-type Ecosystems. The Role of Nutrients*, pp. 66–73. Springer, Berlin.

Bettenay, E. (1984). Origin and nature of the sandplains. In: Pate, J. S., and Beard, J. S. (Eds), *Kwongan. Plant Life of the Sandplain*, pp. 51–68. University of Western Australia Press, Nedlands.

Bridgewater, P. B., and Backshall, D. J. (1981). Dynamics of some Western Australian ligneous formations with special reference to the invasion of exotic species. *Vegetatio*, **46**, 141–8.

Brokaw, N. V. L. (1985). Treefalls, regrowth, and community structure in tropical forests.

In: Pickett, S. T. A., and White, P. S. (Eds), *The Ecology of Natural Disturbance and Patch Dynamics*, pp. 53–69. Academic Press, New York.

Burdon, J. J., and Chilvers, G. A. (1977). Preliminary studies on a native Australian eucalypt forest invaded by exotic pines. *Oecologia (Berlin)*, **31**, 1–12.

Chesson, P. L. (1986). Environmental variation and the coexistence of species. In: Diamond, J., and Case, T. J. (Eds), *Community Ecology*, pp. 240–56. Harper and Row, New York.

Collins, B. S., Dunne, K. P., and Pickett, S. T. A. (1985). Responses of forest herbs to canopy gaps. In: Pickett, S. T. A., and White, P. S. (Eds), *The Ecology of Natural Disturbance and Patch Dynamics*, pp. 218–34. Academic Press, New York.

Collins, S. L., and Barber, S. C. (1985). Effects of disturbance on diversity in mixed grass prairie. *Vegetatio*, **64**, 87–94.

Connell, J. H., and Keough, M. J. (1985). Disturbance and patch dynamics of subtidal marine animals on hard substrate. In: Pickett, S. T. A., and White, P. S. (Eds), *The Ecology of Natural Disturbance and Patch Dynamics*, pp. 125–51. Academic Press, New York.

Crawley, M. J. (1986). The population biology of invaders. *Phil. Trans. R. Soc. Lond.* **314**, 711–31.

Crawley, M. J. (1987). What makes a community invasible? In: Crawley, M. J., Edwards, P. J., and Gray, A. J. (Eds), *Colonization, Succession and Stability*. Blackwell, Oxford, pp. 629–54.

Diamond, D., and Case, T. J. (1986). Overview: introductions, extinctions exterminations and invasions. In: Diamond, J., and Case, T. J. (Eds), *Community Ecology*, pp. 65–79. Harper and Row, New York.

Fox, J. F. (1985). Plant diversity in relation to plant production and disturbance by voles in Alaskan tundra communities. *Arctic Alpine Research*, **17**, 199–204.

Fox, M. D. and Fox, B. J. (1986). The susceptibility of natural communities to invasion. In: Groves, R. H., and Burdon, J. J. (Eds), *Ecology of Biological Invasions: an Australian Perspective*, pp. 57–66. Australian Academy of Science, Canberra.

Green, J. W. (1985). *Census of the Vascular Plants of Western Australia.* Western Australian Herbarium, Department of Agriculture, Perth.

Groves, R. H. (1981). Heathland soils and their fertility status. In: Specht, R. L. (Ed.), *Heathlands and Related Shrublands: Analytical Studies*, pp. 143–50. Elsevier, Amsterdam.

Grubb, P. J. (1977). The maintenance of species-richness in plant communities: the importance of the regeneration niche. *Biol. Rev.*, **52**, 107–45.

Harper, J. L. (1977) *Population Biology of Plants.* Academic Press, London.

Heddle, E. M., and Specht, R. L. (1975). Dark Island Heath (Ninety Mile Plain, South Australia) VIII. The effects of fertilisers on composition and growth. *Aust. J. Bot.*, **23**, 151–64.

Hobbs, R. J. (1987). Disturbance regimes in remnants of natural vegetation. In: Saunders, D. A., Arnold, G. W., Hopkins, A. J. M. and Burbidge, A. A. (Eds), *Nature Conservation: the Role of Remnants of Native Vegetation*, pp. 233–40 Surrey-Beattie, Sydney.

Hobbs, R. J., and Atkins, L. (1988). Effect of disturbance and nutrient addition on native and introduced annuals in plant communities in the Western Australian wheatbelt. *Aust. J. Ecol.*, **13**, 43–57.

Hobbs, R. J., and Mooney, H. A. (1985). Community and population dynamics of serpentine grassland annuals in relation to gopher disturbance. *Oecologia* (*Berlin*), **67**, 342–51.

Hobbs, R. J., Gulmon, S. L., Hobbs, V. J., and Mooney, H. A. (1988). Effects of fertiliser addition and subsequent gopher disturbance on a serpentine annual grassland community. *Oecologia* (*Berlin*), **75**, 291–95.

Hopkins, A. J. M., and Griffin, E. A. (1984). Floristic patterns. In: Pate, J. S., and Beard, J. S. (Eds), *Kwongan. Plant Life of the Sandplain*, pp. 69–83. University of Western Australia Press, Nedlands.

Hopper, S. D., and Muir, B. G. (1984). Conservation of the kwongan. In: Pate, J. S., and Beard, J. S. (Eds), *Konwgan. Plant Life of the Sandplain*, pp. 253–66. University of Western Australia Press, Nedlands.

Kitchener, D. J., Chapman, A., Dell, J., Muir, B. G., and Palmer, M. (1980). Lizard assemblage and reserve size in the Western Australian wheatbelt—some implications for conservation. *Biol. Conserv.*, **17**, 25–62.

Kruger, F. J. (1977). Invasive woody plants in the Cape fynbos with special reference to the biology and control of *Pinus pinaster*. *Proceedings of the 2nd National Weeds Conference of South Africa*. A. A. Balkema, Cape Town.

Kruger, F. J., Richardson, D. M., and van Wilgen, B. W. (1986). Processes of invasion by plants. In: Macdonald, I. A. W., Kruger, F. J., and Ferrar, A. A. (Eds), *The Ecology and Management of Biological Invasions in South Africa*, pp. 145–55. Oxford University Press, Cape Town.

Macdonald, I. A. W., Powrie, F. J., and Siegfried, W. R. (1986). The differential invasion of South Africa's biomes and ecosystems by alien plants and animals. In: Macdonald, I. A. W., Kruger, F. J., and Ferrar, A. A. (Eds), *The Ecology and Management of Biological Invasions in South Africa*, pp. 209–25. Oxford University Press, Cape Town.

McIvor, J. G., and Smith, D. F. (1971). The effect of fertiliser application and time of seasonal break on the growth and dominance of capeweed (*Arctotheca calendula*). *Aust. J. Exp. Ag. Anim. Husbandry*, **14**, 553–6.

Mack, R. N. (1985). Invading plants: their potential contribution to population biology. In: White, J. (Ed.), *Studies in Plant Demography: A Festschrift for John L. Harper*, pp. 127–41. Academic Press, London.

Newsome, A. E., and Noble, I. R. (1986). Ecological and physiological characteristics of invading species. In: Groves, R. H., and Burdon, J. J. (Eds), *Ecology of Biological Invasions: an Australian Perspective*, pp. 1–20. Australian Academy of Science, Canberra.

Orians, G. H. (1986). Site characteristics promoting invasions and systems impact of invaders. In: Mooney, H. A., and Drake, J. A. (Eds), *Ecology of Biological Invasions of North America and Hawaii*. Springer, pp. 133–48. New York.

Pickett, S. T. A., and White, P. S. (1985). *The Ecology of Natural Disturbance and Patch Dynamics*. Academic Press, New York.

Platt, W. J. (1975). The colonisation and formation of equilibrium plant species associations on badger disturbances in a tallgrass prairie. *Ecol. Monogr.*, *45*: 285–305.

Rossiter, R. C. (1964). The effect of phosphate supply on the growth and botanical composition of annual-type pasture. *Aust. J. Agric. Res.*, **15**, 61–76.

Rossiter, R. C. (1966). Ecology of the Mediterranean annual-type pasture. *Adv. Agron.*, **18**, 1–56.

Silvertown, J. (1981). Microspatial heterogeneity and seedling demography in species-rich grassland. *New Phytol.*, **88**, 117–28.

Sousa, W. P. (1984). The role of disturbance in natural communities. *Ann. Rev. Ecol. Syst.*, **15**, 353–91.

Sousa, W. P. (1985). Disturbance and patch dynamics on rocky intertidnal shores. In: Pickett, S. T. A., and White, P. S. (Eds), *The Ecology of Natural Disturbance and Patch Dynamics*, pp. 101–24. Academic Press, New York.

Steel, R. G. D., and Torrie, J. H. (1982). *Principles and Procedures of Statistics. A Biometrical Approach*. McGraw Hill, London.

Tilman, D. (1982). *Resource Competition and Community Structure*. Princeton University Press, Princeton, New Jersey.

White, P. S. (1979). Pattern, process and natural disturbance in vegetation. *Bot. Rev.*, **45**, 229–99.

White, P. S., and Pickett, S. T. A. (1985). Natural disturbance and patch dynamics: an introduction. In: Pickett, S. T. A., and White, P. S. (Eds), *The Ecology of Natural Disturbance and Patch Dynamics*, pp. 3–13. Academic Press, New York.

Williams, K., Hobbs, R. J., and Hamburg, S. (1987). Invasion of an annual grassland in Northern California by *Baccharis pilularis* ssp. *consanguinea*. *Oecologia* (*Berlin*), **72**, 461–65.

Wycherley, P. (1984). People, fire and weeds: can the vicious spiral be broken? In: Moore, S. A. (Ed.), *The Management of Small Bush Areas in the Perth Metropolitan Region.* Department of Fisheries and Wildlife, Perth.

Biological Invasions: a Global Perspective
Edited by J. A. Drake *et al.*
© 1989 SCOPE. Published by John Wiley & Sons Ltd

CHAPTER 18

Chance and Timing in Biological Invasions

MICHAEL J. CRAWLEY

18.1 INTRODUCTION

Both chance and timing play such a vital and obvious role in almost all biological invasions that any general analysis of these processes runs the risk of appearing somewhat banal. Despite (or perhaps because of) the importance of chance and timing, however, rather little has been written about them; their importance is taken for granted.

We can assess the relative importance of bad luck and bad timing in failures of establishment of an invading organism (N), by reference to the model presented by Crawley (1986a, 1987):

$$\begin{aligned} dN/dt = {} & \text{intrinsic rate of increase} \\ & - \text{exploitation competition effects} \\ & - \text{interference competition effects} \\ & - \text{natural enemy effects} \\ & - \text{effects of scarce mutualists} \\ & + \text{immigration and refuge effects} \end{aligned} \tag{1}$$

A necessary but not sufficient condition for successful invasion is that dN/dt is greater than 0 when N is small.

18.1.1 Intrinsic rate of increase

While the theoretical notion of maximal intrinsic rate of increase is straightforward, there are formidable practical problems in its estimation. It is easy enough to estimate maximum fecundity, but it is hard to know what constitutes minimal mortality. The best practical solution is to take the maximum recorded value of the slope of a graph of $\log_e$ numbers against time for a population thought to have an approximately stable age distribution, and to use this as the best estimate of r_{max}. By this definition, the measure is unaffected by timing, but the *realized* rate of increase depends upon a wide variety of random variables (e.g. inclement weather conditions) and may be extremely sensitive to the niceties of timing (e.g. synchrony of invasion with resource availability). Other random effects influencing the rate of increase include genetic founder effects, the age and life stage of the immigrant, and the chance of death between arrival and first breeding.

18.1.2 Exploitation competition

Whether or not exploitation competition has any influence on the population dynamics of an invading species is a moot point. In classical Volterra invasion models (Turelli, 1981) it is assumed that interspecific competition is the major structuring force in community dynamics. Invading species, however, may exploit resources that are not limiting (at least at the time of invasion), so that exploitation competition is not necessarily a force resisting invasion (Crawley, 1986a, 1987). Where it is important, then both the timing of resource supply and competitor phenology may be vital, and the luck of the year (whether it is a good year for the invader or a bad one) will influence resource productivity and competitor density.

18.1.3 Interference competition

This is virtually impossible to predict in advance of invasion (see Crawley, 1986a), unless the invading species is known to be exceptionally aggressive (e.g. the Argentine ant, *Iridomyrmex humilis.* However, if there are fierce, resident members of the invader's guild, then the phenology of their fierceness (whether it is continuous or seasonal, for example), could influence the probability of establishment. As with exploiting competitors, the luck of the year will influence competitor density and hence invasibility.

18.1.4 Natural enemies

Chance and timing are both likely to be important here. Timing of enemy abundance and foraging activity, coupled with luck of the year as it affects both enemy density and the abundance of alternative food species, will determine the death rate suffered by the invader. Other random elements affect enemy preferences, the synchrony between enemy and prey age-structures, and the relative logevities of enemies and prey.

18.1.5 Mutualisms

There is a substantial element of luck in having a full set of mutualists available for the invader on arrival. Clearly the most successful invaders are likely to be those with no obligate mutualists, or with only modest needs that can be met by widespread, generalist mutualists. Timing is important in ensuring that the attentions of resident mutualists are not diverted by abundant, more preferred hosts.

18.1.6 Refuges

Both the availability and size of refuges will vary seasonally and from year to year. Luck will be important in determining whether or not there are reinforcing immigrations of the invading species (i.e. natural repeats of the invasion experiment), and this will depend upon the isolation of the habitat and the size of the nearest source of immigrants.

In short, chance and timing have the potential to affect every aspect of population dynamics relating to invasion. However, the relative importance of chance and timing is likely to vary from case to case and from process to process.

18.2 CHANCE

Changes are often said to be random when we do not understand what causes them. On other occasions, we may choose to ignore the complexity of causal factors, and refer to a change as random simply as a matter of convenience (i.e. as unexplained rather than inexplicable). Thus we have events which are, in practice, impossible to predict in anything but probablistic terms (e.g. the precise location of lightning strikes), and events that we cannot (or have not planned to) explain. This latter kind of randomness is the statistical error familiar to experimentalists; it is the variation left over once we have explained all that we can, and, since analysis is retrospective, this residual variation can *never* be explained (even in principle). Thus, while random variation is not always unattributable in principle, it is frequently unattributed in practice.

Mathematical modellers refer to processes involving random elements as stochastic processes, a phrase only vaguely understood by many ecologists. To a mathematician, random events are those with a probability of occurrence that is determined by some probability distribution. A probability distribution is a mathematical relationship giving the probability of a given value, x, as a function of x. It may be derived empirically from a frequency distribution giving the observed frequencies of different values of x, or theoretically from a specified model of the process (like a Poisson or a negative binomial), which relates the probability of x occurring to the value of x. The phrase 'random event' is used to denote events that may or may not happen at a given trial (e.g. a six appears in a throw of a die), or an event that may or may not happen at a given time (e.g. an individual falls prey to a predator). A stochastic process is one where the system incorporates an element of randomness, and it is the opposite of a deterministic process (Kendall and Buckland, 1960).

A great deal has been written on the pros and cons of deterministic versus stochastic modelling (Bailey, 1957; Bartlett, 1966; May, 1973). While both sides have their fervent advocates, there does appear to be a consensus of opinion. Deterministic models of invasion are most likely to yield reliable approximations when the number of arriving individuals is large, or during the later stages of successful invasion, when population densities are high. However, deterministic models may exhibit patterns of dynamics that are qualitatively different from observed patterns, whereas stochastic models may describe observed patterns reasonably well (e.g. extinctions of apparently well-adapted species; see below).

In essence, the two kinds of models are best used to tackle different kinds of questions. Stochastic models are essential if one intends to address problems relating to the probability of extinction of small populations of invaders, to investigate the details of neighbour-dependent contact processes during the spread of invading species, or to estimate the variance of demographic parameters. Deterministic models, on the other hand, will give simpler, and usually clearer, insights into the equilibrium behaviour of large populations of invaders, and into whether or not a particular suite of demographic attributes equips a species for invasion into a specific community. These matters are discussed more fully below.

It is important to distinguish between two different kinds of randomness that are employed in ecological modelling, namely demographic stochasticity and environmental stochasticity (also called environmental noise). Ecological modellers have often referred to these very different kinds of randomness under the umbrella term of stochastic simulation models, frequently without making clear what kind of randomness is being modelled or why that kind of model is the most appropriate. Models containing environmental noise are actually deterministic models of population dynamics, in which the values of the (usually abiotic) driving variables are randomized at each time period, using pseudorandom numbers to select particular values from probability distributions that have specified means and variances. Even when variation in the demographic parameters themselves (e.g. fecundity) is simulated in this way, a given fecundity will uniquely predict the number of progeny born, and in this important sense the dynamics are deterministic.

Demographic stochasticity is quite different. Here, the fate of every individual member of the population is considered separately in every time period. For each individual, questions are asked as to whether it gives birth during the period, whether it emigrates, or whether it dies. Most simulation models of demographic stochasticity address these questions using 'Monte Carlo' techniques. The fate of individuals is resolved by generating a pseudorandom number between 0 and 1 for each demographic process; the random numbers are then compared with probabilities specified within the model, in order to determine the outcome. For example, if the probability of death is 0.1, then a random number between 0 and 1 is generated for every individual: if the number is less than or equal to 0.1, the individual is assumed to have died; if the number is greater than 0.1 the individual lives to fight another day. Analytical models of demographic stochasticity have the advantage of showing unambiguously the consequences of their particular assumptions, an advantage they share with analytical deterministic models. However, as soon as the biological assumptions are made anywhere near realistic, the mathematics quickly become dauntingly complex, then completely insoluble.

Both environmental and demographic stochasticity are likely to be important in biological invasions. Environmental stochasticity encapsulates both bad luck and bad timing; it determines the good years and the bad years for invasion. Demographic stochasticity, however, is the luck of the draw; given that certain individuals will die or fail to reproduce, chance determines which individuals suffer which fates, and ensures that there will be a variable number of births and deaths, even under constant environmental conditions.

Deterministic invasion models like equation (1) define the set of conditions necessary for establishment (i.e. for initial increase following invasion). Whether or not these conditions are sufficient to explain persistence of a population depends upon the context. Crawley and May (1987) have recently proposed a simple stochastic model for competition between an annual and a perennial plant species. Briefly, the annual is assumed to have no effect on the demography of the perennial; the ramets of the perennial always win in competition with seedlings of the annual. The annual, if it persists at all, does so by germinating in the empty spaces that arise from the death of perennial ramets. The deterministic, necessary condition for persistence of the annual is that its fecundity (c) must exceed the reciprocal of the equilibrium proportion of gaps (E^*):

$$c > 1/E^* \qquad (2)$$

Whether or not the annual persists in the face of competition in a stochastic model, however, depends upon the initial conditions. If an equilibrium population of annuals is invaded by the perennial, then stable coexistence is almost certain, because the annual is initially abundant and never becomes really scarce. If, on the other hand, the annual attempts to invade an equilibrium population of perennials, then there is a finite (and often substantial) probability that the invasion will fail, despite the fact that the fecundity is greater than the threshold defined by the deterministic criteria in equation (2). Figure 18.1 shows the probability that the annual will survive invasion by the perennial, or will itself persist as an invader of a perennial community. At a given fecundity the annual may be certain to survive invasion by the perennial, but have only a 25% chance of invading an established population of perennials. The mathematics of this beach-head effect are detailed by Crawley and May (1987). The present point, however, is that we must adopt stochastic models if answers to questions of this kind are required (e.g. how likely is it that an organism with the requisite deterministic demographic parameters will successfully invade a community?).

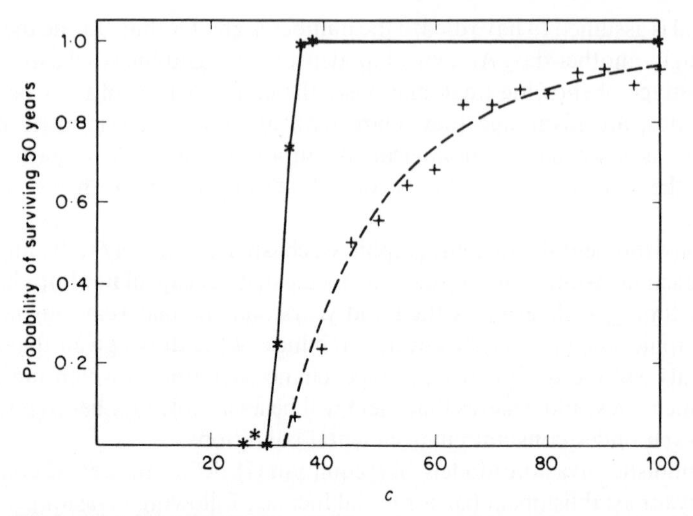

Figure 18.1. The probability that the annual will persist for at least 50 generations as a function of its fecundity, *c*. Solid curve: the annual is invaded by the perennial. Dashed line: the annual invades an established perennial community. Note that persistence is much less likely when the annual invades the perennial (from Crawley and May, 1987)

Other stochastic aspects of invasion are treated by Turelli (1981) and Chesson (1985). Successful invading species of plants and animals tend to exhibit one or more of a variety of mechanisms that buffer them from the full impact of environmental hazards or increase their performance relative to native species, in the aftermath of extreme conditions.

18.2.1 Buffering mechanisms exhibited by plants

Alien plants may be better buffered than natives for a variety of reasons: (1) because they tend to have fewer herbivores and seed predators, their fecundity may be higher; (2) this, in turn, may mean that they form larger seed banks; (3) because of this, there may be a greater likelihood of alien plants dominating gaps that appear in the canopy following disturbances; or (4) the alien's (preadapted) phenology may expose them to less risky periods of the season. This line of argument, of course, is entirely speculative at present.

Alien plants may themselves alter the risks they face. For instance, they may increase the likelihood of exceptionally severe fires occurring, either by accumulating inflammable fuel at a greater rate, or by burning with fiercer heat than native vegetation (G. J. Breytenbach, unpublished). Several alien plants appear to be more fire tolerant than natives (e.g. invaders of the South African fynbos such as *Hakea*), and a number of alien species may be more disturbance tolerant than any of the natives (e.g. *Chamomilla suaveolens*, a virtually indestructible alien weed of gates and roadways).

The buffering ability of aliens in relation to rare events, however, is likely to bc inferior to natives. Most native species of flora and fauna will have been exposed to extreme but rare environmental events (i.e. those occurring less frequently than, say, once per century). Alien species, on the other hand, are unlikely to have been exposed to the full range of different environmental extremes, and unusual weather conditions that might eliminate (or severely set back) an alien, might only cause temporary set-back to the natives. Indeed, it could be argued that if aliens had been around sufficiently long and had been exposed to these rare events, then we would probably not identify them as aliens in the first place. What appear, today, to be well-established, alien members of communities may simply not yet have been exposed to the particular environmental extremes that will eventually prove fatal to them. Unfortunately, this kind of optimism will be of little consolation to those who are charged with managing ecological communities dominated by alien species.

18.2.2 Buffering mechanisms exhibited by animals

I have argued elsewhere that probability of establishment should be positively correlated with the intrinsic rate of increase, and, therefore, negatively correlated with body size (Crawley 1986a, 1987), although the correlation may sometimes be reversed in comparisons within lower taxonomic groups (e.g. within certain genera; Williamson, this volume). Data on introduced species of weed biocontrol insects are consistent with two predicted patterns, with smaller species significantly more likely both to establish and to bring about substantial weed control. Given the central importance of the term for intrinsic rate of increase in invasion models like equation (1), this result is only to be expected.

There is a counterintuitive element to the argument, however. Consider the following correlations of body size: (1) bigger animals are more buffered against climatic extremes (especially for homeotherms); (2) bigger animals are more polyphagous and therefore less likely to starve in a foreign environment: (3) bigger animals have fewer predators; and (4) bigger animals may be able to disperse over greater distances if and when local conditions deteriorate. Weighted against these advantages, however, are all the problems associated with a low rate of increase, and it appears that, for most invaders, the ability to bounce back rapidly following disaster, and the ability to increase rapidly from a rare, initial bridge-head population, are of paramount importance compared with the attributes that buffer individual organisms against chance events. It is clear that there is a continuum of risk adaptation that runs parallel to (if not exactly coincidental with) the r–K continuum, and along which there are a variety of trade-offs between risk spreading (high r) and risk buffering (large size). Many of the successful large invaders are mammals associated with built-up and other human-modified environments (e.g. rats, mice and cats), where the advantage of high buffering capacity is not at all clear. Others, like rabbits and goats on islands, are extremely polyphagous, but, considering their large size, have relatively high rates of increase.

For invading insects, high fecundity tends to be the main means of risk buffering. With species that show alternation of generations, chance may play a vastly different role in the two life stages, as in the two legs of the annual migration pattern of the alien gall wasp, *Andricus quercuscalicis*. Given the enormous difference in the relative abundance in southern England of the two host trees between which it migrates (about 100 *Quercus robur* to one *Q. cerris*), individual sexual females emerging from *Q. cerris* are much more likely to find a suitable *Q. robur*, than are agamic females emerging from *Q. robur* of finding a suitable tree of *Q. cerris*. This asymmetry in host abundance may have been the evolutionary cause of the differential fecundity exhibited by the two generations; each sexual female contains only 10 eggs on emergence, whereas each agamic female has over 1000.

18.3 DATA ON CHANCE IN INVASIONS

I shall illustrate this section on chance using data from the Silwood Project on Weed Biocontrol (see Crawley, 1987, for details), and the section on timing with data from the 130 years of experimental work on the Broadbalk Experiment at Rothamsted (see Crawley, 1986a, for details). I should have liked to draw information on both processes from both data sets, but the phenological data from weed biocontrol are too scanty, and we have no data at all on failures of invasion for the Broadbalk weeds.

18.3.1 Chance in the weed biocontrol data

Several of the variables in the weeds data base allow us to assess the role of stochastic elements in the establishment of insects intentionally released as potential biocontrol agents.

18.3.2 Number of releases

Other things being equal, the more releases that are made, the greater should be the chance of successful establishment. Indeed, the data do show that multiple releases are significantly more likely to establish than are single releases. Multiple releases are no guarantee of success, however, and the same proportion of species fails to establish following 13 years of introduction as fails following 2 years of introduction. The reason for this lies in the extraordinary persistence of the practitioners, who appear to be working on the rationale that 'this species worked somewhat else, so it had better work here, as well'.

If we take only those cases with well-documented releases, and calculate a figure for total release (the product of number of release-sites and the logarithm (base 10) of the number of insects released per site), we find that total release is significantly greater for introductions that succeeded in establishment (87.3, SE 23.1) than for those that failed (9.7, SE 1.9).

Release strategies were also important. Comparing single releases at one site with several releases at one site, single releases at several sites, and many releases at many sites, we obtain the expected increase in probability of establishment (see Table 18.1). Nevertheless, single releases were often effective, and multiple releases often failed. Redistribution of insects within the country of release following successful establishment further reduces the risk of extinction.

18.3.3 Previous experience

Agents are frequently introduced because they have been successful somewhere else against the same species of weed. Given information on prior performance of the agent, it might be expected that predictive ability would be improved. This expectation is only partially fulfilled. If it is known that the agent failed to establish in previous attempts, then it is significantly more likely to fail again. On the other hand, completely untried species are just as likely to establish as species which have successfully established elsewhere. In this context, therefore, the negative data are more valuable than the positive data in contributing to our predictive ability.

18.3.4 Taxonomic mistakes

Bringing together the wrong taxa by mistake introduces a stochastic element into the establishment probability. It might be expected that mismatched organisms

Table 18.1. Release strategy and probability of establishment for agents released against all weeds except *Opuntia* spp. and *Lantana camara*, the category of 'Other Weeds' in Crawley (1986b). Data show the number of releases in each category. Thrity-six species of agents were established accidentally. For 46 species the release strategy is unknown (26 of these failed to establish, 20 established successfully). Chi-squared = 12.15, $v = 4$, $P < 0.05$

Releases	Sites	Failed	Established
1	1	18	14
several	1	9	12
1	several	17	25
several	several	32	8
many	many	3	15

would stand a significantly lower chance of successful establishment, but this is not the case. There is no significant difference in the probability of establishment between those cases where the plant and animal were correctly identified. The reason for this is that the cases where genuinely mismatched species fail immediately just happen to be balanced by cases where the mismatch turns out, quite fortuitously, to benefit the agent. For example, the plant may be a suitable host for the insect even though (or perhaps because) the plant and the insect shared no evolutionary history of association (Crawley, 1986a).

18.3.5 Total effort

Biocontrol effort consists of much more than just the scale of the individual releases. Energy is expended on geographical search for suitable agents, on collection, on taxonomic study, on screening, on specificity testing, and on pre-release rearing (Moran, personal communication). The relationship between total effort (a weighted sum of all these factors, each measured in scientist years) and the probability of establishment, suggests the biocontrol practioners may well be wasting their time. The mean number of scientist years associated with successful establishment is 9.3 as against 6.9 for failures ($t = 0.95$, $v = 214$, NS). This result is coloured, of course, by the fact that repeating successes is relatively cheap in terms of scientist years, an artefact that tends to reduce the mean effort associated with successful establishment. A less biased test would compare only the mean efforts associated with success or failure following the first introduction.

18.3.6 Spatial distribution of eggs

One of the intriguing biological attributes of the agent which influences the probability of establishment is the pattern in which the female distributes her eggs. We compared species where the eggs are laid in batches, with species where the females lay their eggs singly, and found that batch-layers were significantly less likely to establish. This is presumably to do with risk-spreading; a single batch of 100 eggs is more likely to be completely lost to predators or to accidents than are all of 100 spaced-out eggs. Interestingly, once established, a good weed biocontrol is more likely with batch-laying species (Crawley, 1986a), since their impact on plant fitness is substantially greater.

18.3.7 Adult longevity

The only other biological form of risk spreading that emerged from the weed biocontrol data concerned adult longevity. The introduction of species with relatively long-lived adult stages led to significantly increased probabilities of establishment, compared to species that had short-lived adults. The explanation of this involves a combination of good luck and good timing; the longer the adults live, the more likely they are to encounter suitable conditions for successful reproduction.

18.3.8 Failure to establish

All of the factors identified in the data base as likely causes of failure to establish (climate, predators, parasitoids, disease, competition; see Crawley, 1986a) involve an element of chance. However, if they are ranked from most chancy to least chancy, there is a correlation between the importance of the factor (measured by the proportion of cases in which this was the major cause of failure) and the degree of chanciness involved with that factor. Thus bad weather is both the most frequent cause of failure, and the chanciest of the factors; competition is both the least common cause of failure and the most deterministic (it requires relatively high densities of the competing species, and when densities are high, competition is almost inevitable).

18.4 TIMING

In compiling his study on the phenology of British plants and animals, the 19th century naturalist, Leonard Blomefield, had occasion to correspond with Charles Darwin, about contributions that Darwin might make to his *Observations in Natural History*. In his reply, Darwin (see Francis Darwin, 1924) writes 'My work on the species question has impressed me very forcibly with the importance of ... [phenological works], containing what people are pleased generally to call trifling facts. These are the facts which make one understand the working or economy of nature.' Seventy-six years later, Francis Darwin (1924) was to write that phenology is 'a science which has suffered in two ways—viz. from incompleteness in observation, and a too bold style of theorizing'.

18.4.1 Seasonal phenology

Good timing is vital in all invasions. For example, matching an invading plant to its recruitment microsite often requires extremely precise timing (Harper, 1977). Similarly, matching the arrival of an invading animal species with the phenology of its food resources, mating and nesting requirements is crucial to successful establishment (Crawley, 1983). Synchrony of invasion with lows in the abundance of generalist natural enemies is also likely to increase the likelihood of successful establishment (Lawton, 1986).

Many of the best examples of the importance of phenology in animal population dynamics come from studies with herbivorous insects. For instance, a long-term study of English oak showed the key role played by the timing of bud burst in the population dynamics of the two principal herbivores, the winter moth, *Operophtera brumata*, and the green tortrix, *Tortrix viridana*. The tiny hatchling larvae must penetrate between the expanding bud scales, or die. If bud burst is late, then massive insect mortality results (Varley and Gradwell, 1968). Unfortunately, we have little information on the phenology of invading insects, and virtually none on the phenological causes of failure in insect invasions. It may not be without significance, however, that two of the most recent alien additions to the British cynipid gall wasp community have rather extreme phenologies. The marble gall, *Andricus kollari*, lays its eggs in the buds of *Quercus cerris* in the autumn, at a time when most self-respecting gall wasps are already in their overwintering diapause. In contrast, the knopper gall, *A. quercuscalicis*, emerges exceptionally early in the spring (sometimes struggling through snow to get to the surface), and lays its eggs in buds of *Q. cerris* in March, long before the native species that gall male catkins of *Q. robur* (e.g. *Neuroterus quercus-baccarum*) have emerged from their overwintering galls.

With plants, phenological effects are much easier to study. For example, many of the alien herbaceous plants of Californian rangelands remain green longer into the summer than the native plants, a trait that may give them a competitive edge over some natives in years that are not unusually dry, if this increases their relative fecundity (Crawley, 1987). The precise seasonal timing of catastrophes such as fire determines the developmental stage of the plant that is exposed to hazard. A stage (like seedlings) may be killed by a disturbance that would barely set back a different stage (e.g. seeds or thick-barked, mature plants). The same disturbance might even be advantageous if it happened at different time, if, at that time, its effects were relatively more severe on the plant's competitors or natural enemies.

18.4.2 Timing in the Broadbalk experiment

The effect of seasonal phenology on the composition of an invading guild is beautifully illustrated by the classic Broadbalk Experiment at Rothamsted. Every year since 1843, winter wheat has been cultivated in the same field. The experiment consists of 18 manurial treatments each crossed by seven weed control treatments. One block of the experiment has never been treated with chemical weed killers over this entire period. The dynamics of the community of weedy species on this plot provide an unparalleled source of data on plant invasions, as well as on questions relating to why competitive exclusion has failed to occur, even after more than 100 generations (Crawley, 1986a).

Two species of wild oat, *Avena fatua* and *A. ludoviciana*, are found in wheat fields in southern England. Both are aliens introduced from southern Europe, but only *A. ludoviciana* has become a serious pest in the Broadbalk Experiment (and this despite years of laborious, selective, hand-weeding). The reason is phenological. The autumnal cultivation which precedes sowing of winter wheat allows the autumn-germinating *A. ludoviciana* to establish free from competition with other weeds. Spring-germinating *A. fatua* is unable to compete in the shade cast by the well-established canopy of wheat and *A. ludoviciana*. In contrast, *A. ludoviciana* is eliminated from fields where spring cereals are grown, because its autumn-germinated seedlings are killed by spring cultivation.

The Broadbalk data provide other insights on the importance of timing. In the 1920s and 30s, attempts were made to reduce weed problems on the field by fallowing for various lengths of time. This involves sowing no crop, and destroying the weeds as they germinate over the course of the year, timing the cultivations to ensure that none of the weeds set seed before they are cut down. Several successive years of fallow should run down the seed bank of species that possess dormancy, and should virtually eliminate annual weeds that exhibit no seed dormancy. It is interesting to compare the outcome of a 4 year fallow on two of the most important weeds, the poppy *Papaver rhoeas* (a relatively recent invader to this particular field) and the annual black grass *Alopecurus myosuroides*. Poppy has a vast seed bank (20000 seeds m^{-2} in the top 10 cm of soil) and has moderate seed dormancy (it can survive for well over 4 years in the soil). Black grass, on the other hand, exhibits no seed dormancy at all. Given these life history traits, one might predict that fallowing would eliminate black grass, but that it would be ineffective in poppy control. As so often happens, detailed prediction is confounded. A 4 year fallow vastly reduced the poppy problem, but had no effect on black grass control. After fallowing, the wheat crop tends to grow particularly well because there is little competition from weeds, and nutrient levels are relatively high. Those weeds that do germinate after fallowing, therefore, tend to be suppressed by the crop and to set little seed, so the seed bank recovers only very slowly. Poppy is regarded as uncompetitive with winter wheat on Broadbalk soils (Brenchley and Warrington, 1930, 1945), and this is the likely reason that poppy stayed scarce during the 5 years after fallowing.

The case of black grass, however, illustrates the importance of chance and timing. It happened that in the first year of wheat cultivation following the fallow period, the crop was subject to heavy attack by wheat bulb fly, *Leptohylemyia coarctata*. Whether this was entirely coincidental, or whether fallowing in some way increased the probability of bulb fly attack, is unknown. In the large, competition-free gaps in the wheat crop caused by the pest, the few surviving seeds of black grass were able to produce highly fecund individuals, which, in the course of a single year, produced sufficient seed to bring the weed population back to its pre-fallow densities.

18.4.3 Successional timing

The importance of timing in the invasion of successional communities depends upon the kind of succession under consideration. In primary successions, where facilitation is a vital prerequisite for establishment, there are likely to be rather clearly defined windows of time in which invasion is possible; too early and the resources required by the invader may not be established, too late and generalist natural enemies may be too abundant (Crawley, 1986b; Gray *et al.*, 1987).

In secondary successions, where facilitation is usually less important, inhibition of invasion by established, native species may constitute more of a barrier to ill-timed invasions (Connell and Slatyer, 1977). Invasibility may also depend upon the details of the relative abundances of resident species of resources, competitors, mutualists and natural enemies, rather than simply upon which species are resident. In this case, rather detailed models of population dynamics (like equation (1)) for species in the trophic levels both above and below the invader may be necessary to predict success or failure of establishment. This vastly increases the complexity of the task of prediction. Unfortunately, we have no direct data on the importance of successional stage on invasibility, from any of our studies carried out so far. The natural history observation that aliens are more common in early successional communities, while intriguing, may reflect nothing more than the fact that rates of species introductions (and species' pool-sizes) are far higher for early successional communities. Hard data showing that with equal rates of species' immigration, late successional communities are harder to invade, are simply not available (Crawley, 1986b). If there is any evidence, it may actually point the other way; for instance, in southern Britain there are more alien 'late' than 'early' successional tree species.

If, indeed, it is true that late secondary successional communities are more difficult to invade, it is probably because their increased diversity (both structural and biotic) imposes extra constraints on community membership. The more rules that apply, and the more rigorous the selection process, the harder it is for a given species to invade. Because early successional communities are simpler, they tend to be more alike, so a species that is good at invading one kind of disturbed community is likely to be able to invade other communities created by the same agent of disturbance (be it fire, landslip, or urban dereliction). An excellent example is afforded by the African grass *Rhynchelytrum roseum* (= *repens*), that grows as an alien on dry, road-side verges all the way across Mexico from the Gulf to the Pacific, through countless different plant communities, up and down an elevational gradient of over 4000 m. Whether dominance throughout this extraordinary range of environments is due to tremendous phenotypic plasticity on the part of a few genotypes, or to wide genetic polymorphism, remains to be discovered.

18.5 CONCLUSIONS

What emerges most forcibly from this discussion is not that chance or timing are all-important, but that the *interaction* between chance and timing is the vital ingredient. Rare chance events that occur at just the right time may well be the cause of major, long-term structural and dynamic changes in ecological communities. However, exactly the same extreme event might lead to the development of a radically different community, if it occurred at a slightly different time (e.g. depending upon what propagules were abundant at the time of the disturbance). Because of this, existing communities bear the marks of countless past contingencies. A good example is seen in oak woodlands in Silwood Park in southern England. This naturally regenerated forest owes its origin to the introduction of the myxoma virus to Britain in the mid-1950s. This brought about the complete collapse of the resident rabbit population and allowed a cohort of oaks to become established. Before the introduction of myxomatosis, the rabbits had maintained these sites as grasslands, selectively predating the seeds and seedlings of the trees. In recent years, rabbit numbers have risen again, as their resistance to myxoma has increased and the virulence of the virus has declined (Fenner, 1983). At present there is virtually no natural regeneration of oaks in rabbit-grazed areas. This dense oak woodland owes its existence to the chance event of the introduction of the virus. Had the timing been different, the low in rabbit numbers might not have coincided with any years of high acorn production, in which case the oak's many other seed predators might never have been satiated, and an entirely different king of woodland would have developed (e.g. of *Betula, Acer* or *Tilia*; Crawley, 1983). Similarly, dramatic consequences of rare, chance events are described by Noble and Slatyer (1980) in relation to unusually severe fires, and Gleason (1927) in relation to the vagaries of plant immigration to newly created freshwater ponds.

The relative importance of deterministic elements of invasion, like those listed in equation (1), and the stochastic elements involved in good luck and good timing, are rather difficult to assess in practice. In principle, the deterministic model allows us to establish necessary conditions for invasion; values of the demographic parameters that ensure that dN/dt is greater than 0 when N is small. The stochastic elements should enable us to obtain estimates of how likely it is that a given set of (apparently suitable) demographic attributes will actually lead to establishement. Unfortunately, for untried species in new environments, we have virtually no idea of how to estimate these probabilities. Even for the deterministic models, many of the parameters are unknown, and some of them are even unknowable (e.g. the coefficients of interspecific interference between resident and invading species). In this sense, the condition that dN/dt be greater than 0 is not especially helpful, and, for the time being, we may have to accept that quantitative predictions on the probability of successful invasion may be impossible for most systems. This is not to say that deterministic criteria of invasion are inevitably unhelpful. Great steps have been made in quantitative epidemiology, for example, by analysing simple models for establishment (see the work of Anderson and May (1986) on R_0, the basic reproductive rate; see also Williamson, this volume). Can we learn anything about the importance of chance and timing from an examination of those cases where chance and timing appear to be relatively unimportant in determining the course of invasions? A list of the great invaders (cats and goats on oceanic islands, rats and mice in human settlements, human diseases in previously isolated populations, birds such as starling and partridge, mammals like rabbits and foxes from Europe, grey squirrels and muskrats from North America, mongoose and axis deer from the Orient, and so on) is a list of species that can be confidently expected to invade successfully. There is a certain uniformity to the habitats involved (most are human-made or highly altered by people), to the physiological tolerances and buffering of the individual organisms (most are warm-blooded vertebrates) and to the population-level traits associated with high intrinsic rate of increase (relatively small size, rapid development and high fecundity). For the cold-blooded species, and for the insects in particular, it appears that great successes are less repeatable. The combination of higher climate-sensitivity, coupled with the small absolute size and low buffering capacity of the individuals, means that invasions are always going to be chancy. We know the attributes likely to increase the probability of establishment (Crawley 1986a, 1987), but we are not in a position to make accurate predictions about individual cases.

REFERENCES

Anderson, R. M., and May, R. M. (1986). The invasion, persistence and spread of infectious diseases within animal and plant communities. *Phil. Trans. R. Soc. Lond., B*, **314**, 533–70.

Bailey, N. T. J. (1987). *The Mathematical Theory of Epidemics.* Griffin, London.

Bartlett, M. S. (1966). *An Introduction to Stochastic Processes with Special Reference to Methods and Applications.* Cambridge University Press, Cambridge.

Brenchley, W. E., and Warrington, K. (1930). The weed seed population of arable soil. I. Numerical estimation of viable seeds and observations on their natural dormancy. *J. Ecol.*, **18**, 235–72.

Brenchley, W. E., and Warrington, K. (1945). The influence of periodic fallowing on the prevalence of viable weed seeds in arable soil. *Ann. Appl. Biol.*, **32**, 285–96.

Chesson, P. L. (1985). Coexistence of competitors in spatially and temporally varying environments: a look at the combined effects of different sorts of variability. *Theor. Pop. Biol.*, **28**, 263–87.

Connell, J. H., and Slatyer, R. O. (1977). Mechanisms of succession in natural community stability and organization. *Amer. Nat.*, **111**, 1119–44.

Crawley, M. J. (1983). *Hervivory. The Dynamics of Animal Plant Interactions* Blackwell Scientific Publications, Oxford.

Crawley, M. J. (1986a). The population biology of invaders. *Phil. Trans. R. Soc. Lond. B*, **314**, 711–31.

Crawley, M. J. (1986b). The structure of plant communities. In: Crawley, M. J. (Ed.), *Plant Ecology.* Blackwell Scientific Publications, Oxford.

Crawley, M. J. (1987). What makes a community invasible? In: Gray, A. J., Crawley, M. J., and Edwards, P. J. (Eds), *Colonization, Succession and Stability*, Blackwell Scientific Publications, Oxford.

Crawley, M. J., and May, R. M. (1987). Population dynamics and plant community structure: competition between annuals and perennials. *J. Theor. Biol.*, **125**, 475–89.

Darwin, F. (1924). *Darwin. Life and Letters.* John Murray. London.

Fenner, F. (1983). Biological control as exemplified by smallpox eradication and myxomatosis. *Proc. R. Soc. Lond. B*, **218**, 259–85.

Gleason, H. A. (1927). Further views on the succession concept. *Ecology*, **8**, 299–326.

Gray, J., Crawley, M. J., and Edwards, P. J. (1987). *Colonization, Succession and Stability.* Blackwell Scientific Publications, Oxford.
Harper, J. L. (1977). *The Population Biology of Plants.* Academic Press, London.
Kendall, M. G., and Buckland, W. R. (1960). *A Dictionary of Statistical Terms.* Oliver and Boyd, Edinburgh.
Lawton, L. H. (1986). The effect of parasitoids of phytophagous insect communities. In: Waage, J. K., and Greathead, D. (Ed.), *Insect Parasitoids.* Academic Press, London.
May, R. M. (1973). *Stability and Complexity in Model Ecosystems.* Princeton University Press, Princeton, New Jersey.
Noble, I. R., and Slatyer, R. O. (1980). The use of vital attributes to predict successional changes in plant communities subject to recurrent disturbances. *Vegetatio*, **43**, 5–21.
Turelli, M. (1981). Niche overlap and invasion of competitors in random environments. I. Models without demographic stochasticity. *Theor. Pop. Biol.*, **20**, 1–56.
Varley, G. C., and Gradwell, G. R. (1968). Population models for the winter moth. *Symposium of the Royal Entomological Society of London*, **9**, 132–42.

Biological Invasions: a Global Perspective
Edited by J. A. Drake *et al.*
© 1989 SCOPE. Published by John Wiley & Sons Ltd

CHAPTER 19

Analysis of Risk for Invasions and Control Programs

SIMON A. LEVIN

19.1 INTRODUCTION

Dispersal, invasion, and range expansion are manifestations of some of the most fundamental aspects of life history evolution. The changing nature of the environment spells doom for the species that does not find some way to temper the mercy of a fluctuating local resource base, or soften the impact of a sudden increase in its natural enemies. Dispersal is a basic evolutionary response to such harsh or unpredictable aspects of environment; it allows bet-hedging in the absence of information concerning environmental change, or cued responses to direct or indirect hints of impending catastrophe. Even in a completely predictable and stable environment, competitive pressures may select for dispersal, because populations that do not at least occasionally send out propagules must ultimately be displaced by those that do (Hamilton and May, 1977).

Thus, dispersal and associated competitive displacement are inescapable features of the natural world, and powerful fundamental processes in the shaping of community patterns. The nature of the biosphere reflects a balance between the forces of local disturbance, reinvasion, and successional development; human activities have upset the balance, increasing the frequency of disturbance and invasion, and reducing biotic diversity.

Nearly 30 years ago, in his treatise on invasions, Elton (1958) wrote 'We must make no mistake: we are seeing one of the great historical convulsions in the world's fauna and flora.' The causes primarily are anthropogenic, as the activities of societies disturb natural areas, increasing their vulnerability to colonization by invaders, and as humans deliberately and accidentally serve as agents for transporting propagules from one habitat to another. What often is missed in much of the recent debate regarding the deliberate release of genetically engineered organisms is that it is the issue of the frequency and scale of introductions, as much as the nature of any particular introduction, that impels the concerns of ecologists. Our regulatory systems are not designed to deal with the cumulative effects of multiple ' ·w probability events, and yet this is a primary challenge facing us in the case of introductions.

19.2 RISK ASSESSMENT AND RISK MANAGEMENT

Many of the most destructive biological invasions involved accidental introductions: epidemics of human diseases, plant pathogens such as asiatic chestnut blight (*Cryptonectria parasitica*), or disease vectors such as the malaria mosquito (*Anopheles gambiae*). A second major category, including such examples as the Africanized bee (*Apis mellifera scutellata*), the muskrat (*Ondatra zibethica*), and the gypsy moth (*Lymantria dispar*), involved species that were imported for very limited purposes, but escaped from inadequate biological or physical containment.

The third category of introductions includes those for which environmental release is deliberate. These sometimes can be carried out relatively safely provided the ecological characteristics receive adequate attention prior to the release. However, this is not to suggest that such deliberate introductions are somehow risk-free, since there have been numerous incidents arising from faulty assessments; rather, it says that care to ecological detail can minimize risk. Years of experience with plant breeding has led to a wealth of information concerning the hazards associated with introductions, and the ability to control risks. Biological control has had a number of spectacular successes with pests such as the cactus weed *Opuntia*, the European rabbit (*Oryctolagus cuniculus*) in Australia, the citrus mealy bug (*Cryptolaemus montrovzier*), cottony cushiony scale (*Icerya purchasi*), and Klamath weed (*Hypericum perforatum*). In contrast, introduced fish species have driven many indigenous freshwater fish populations in the western United States and elsewhere to the brink of extinction; and more generally, huge shifts in species composition have occurred throughout subtropical areas of North America, where introduced cichlids largely have displaced centrarchids (Moyle, 1986). Thus, where ecological information has been lacking or faulty, deliberate introductions have led to major and sometimes catastrophic ecological occurrences.

The differences among these categories demonstrate the need to separate the various stages in the invasion process: introduction; escape from containment; the dispersion of individual propagules around the original source; the process of establishment, population growth and spread; and the manifestation of effects on other populations and on system processes. These will be discussed individually in the sections that follow. Depending on the nature of the proposed introduction, one or several of these aspects may assume paramount importance. For those species intended for contained uses, as is the case for many products of biotechnology, it is the escape from containment and survival in the external environment that is the key consideration. For deliberate releases, attention shifts to growth, spread, and ecological effects.

For practical purposes, it often is convenient to distinguish between the processes of risk assessment and risk management. In general, however, these are not cleanly separable. Levin and Harwell (1986) suggest that a regulatory program for the deliberate release of genetically engineered organisms must have four components.

1. Pre-release assessment of the likely fate and effects of the novel material.
2. Monitoring after release.
3. A plan for biological or physical containment.
4. A plan for mitigation of undesired side effects.

These elements are inextricably intertwined: the confidence that one can place upon prediction must influence the design of monitoring and containment programs, and in some instances the need for the latter will be minimal. Moreover, containment is not an absolute, and its evaluation is part of the initial assessment. Thus, although some parts of the above scheme assume primary importance in particular situations, the four are interrelated and essential elements in any generic management system.

Consideration of any of the above four points is influenced fundamentally by the category of introduction, in the sense of Levin and Harwell (1986). If the introduction involves a species that exists elsewhere in the natural environment, then it is critical to determine its characteristics in its natural environment, and how they might pertain to the target environment. Is the species absent from the target environment because it never previously has reached there in sufficient proportions to become established, or because it is incapable of competing against the native biota? Are there thresholds associated with local establishment? What are the natural mechanisms of population control in its native environment, and can they be transferred to the new environment if needed?

19.3 SPREAD OF PROPAGULES AND INDIVIDUALS

The first step in an assessment of the potential for spread of an introduced species is an identification of the primary modes of movement. It is essential to determine the relative importance of physical factors such as winds and currents, of biological vectors, and of active mechanisms of locomotion. The scales of movement differ by many orders of magnitude for different species, making it impossible to discuss risk assessment and management without a primary classification based on the scales of movement. For example, influenza virus and airborne fungal spores can traverse huge distances in short periods of time, while individual *Rhizobia* can move only very short distances, on the order of meters (Andow, 1986). The nature of risk and the potential for containment thus differ qualitatively among organisms with such different characteristics.

The basis for classification and extrapolation rests upon a clarification of the relationship between easily measurable parameters, such as those relating to individual movements, and the patterns of spread of populations. The beginnings of appropriate theories exist, but are incomplete. For vertebrates, there have been isolated studies of individual movements, derived, for example, from radio-telemetry data; moreover, there is a vast and fascinating literature on orientation and migration. However, no general body of principles exists that allows one to proceed from the individual to the population. Similarly, for insects, information has been obtained regarding the responses of individuals to chemical cues, or regarding swarming behavior; but much of the most useful information has been derived from population level studies, from which inferences concerning the statistical distribution of the movements of individuals sometimes can be drawn. Most studies of this kind have been long-term, integrating over a large number of individual steps (e.g. Dobzhansky and Wright, 1943); but a number of recent studies (Kareiva, 1983) provide more detailed short-term information.

Phenomenological approaches to describing the distribution of distances traversed by individual spores are a well accepted part of the methodology of plant disease epidemiology (Gregory and Read, 1949; Gregory, 1968; Frampton *et al.*, 1942; Minogue, 1986; Fitt and McCartney, 1986), and can be extended to the distribution of seeds and pollen (Harper, 1977; Okubo and Levin, in press). The simplest models, the power law (Gregory, 1968):

$$y = ax^{-b}, \tag{1}$$

and the negative exponential (Kiyosawa and Shiyomi, 1972; Frampton *et al.*, 1942):

$$y = ae^{-bx}, \tag{2}$$

provide excellent agreement with observed data in a wide variety of situations. In these expressions, x is the distance from the release point, and y is the probability density.

Models of this form have the advantages of simplicity; they are linear respectively on logarithmic or semilogarithmic paper, facilitating parameter estimation. However, they do not explain the sometimes-observed displacement of the dispersal peak from the source. More importantly, because the parameters are derived from curve fits rather than from consideration of underlying physical properties such as wind velocity or height of release, there is no basis for extrapolation to new situations.

Gaussian plume models (Pasquill, 1962), which are the simplest of atmospheric dispersion models, relate the distribution of spores downwind from a source of a known height. These models are used widely in the air pollution literature to describe particulate distributions, but are valid only for very light particles. For heavier propagules, one must employ either a 'tilted plume' model, in which gravitational settling is included, or consider the full diffusion–advection–settling model with all forces considered explicitly, and with uptake kinetics at the surface of the earth also treated explicitly. The latter approach holds great potential for the prediction of spread. Our investigations to date (Okubo and Levin, in press) indicate that such models can provide the basis for a useful classification system for seed and spore dispersal, one in which the critical parameters—settling velocity, wind velocity, and release height—are measurable directly.

Vectored organisms introduce another level of complexity, since one must consider both the movements of the vectors and the transfer kinetics by which organisms board and deplane. Treatment of such phenomena is central to the development of epidemiological models; usually the treatment is phenomenological, restricted to the calibration of standard models based on observed dynamics of outbreaks rather than on more mechanistic observations.

In the case of the release of genetically engineered organisms, an analogous set of considerations regarding infectious transfer via plasmids or viruses is recognized to be of major importance. Such transfers have been observed under laboratory conditions, and are known to occur in the field; but data on the frequency of occurrence under field conditions are in short supply.

19.4 INTEGRATIVE MODELS OF POPULATION GROWTH AND SPREAD

The classical mathematical models for the spread of invading species were developed to represent the central aspects of spread in terms of a few measurable and biologically meaningful parameters, and to provide a basis for prediction. The most widely used models are based on a modified random walk or diffusion model, in which advective forces and growth terms are added. The diffusion term, as well as the other parameters, may be allowed to vary in space and time, or to depend upon population density. Depending on the context, diffusion may occur in one, two, or three dimensions, and not necessarily at the same rates; in two dimensions, the basic form (see Okubo, 1980) is

$$\partial n/\partial t = \partial^2(Dn)/\partial x^2 + \partial^2(Dn)/\partial y^2 - u\,\partial n/\partial x - v\,\partial n/\partial y + F(n, x, y, t) \qquad (3)$$

where F represents the local growth term and $n = n(x, y, t)$ is the population density. Here, D is the diffusion coefficient, which could be allowed to vary in the x and y directions, and u and v denote mean advective effects.

In this model, the means and variances of the lengths of individual steps are represented explicitly in the advection and diffusion coefficients, respectively; higher order moments are ignored. More general models incorporating higher order moments have been considered by Mollison (1977); these have the advantage of being able to account for a wider range of patterns of spread, but the disadvantage of being much more difficult to parameterize. As with any mathematical modeling, one must reach a compromise between detail and utility, and in general the resolution is to settle for the simplest possible model that can account for the major observed patterns.

A large literature, discussed in part by Williamson in this volume, concerns itself with the diffusive spread of species introduced into new environments, and interprets patterns of spread on the basis of equation (3). For the case of an invading organism, the form of the density dependence in equation (3) is of little importance at the front, where the population density is low. It is not surprising, therefore, that the asymptotic rate of advance of population fronts depends only upon the intrinsic rate of natural increase (r), the diffusion coefficient, and advective effects. In the absence of advection, the speed of that front is $2(rD)^{1/2}$. This result was proposed originally by Fisher (1937) for the spread of advantageous alleles, and was demonstrated rigorously by Kolmogorov *et al.*, (1937) for a variety of forms F and initial distributions. Numerous authors since have perfected these results, and have applied them to particular case studies. Certainly the most important paper in this regard was that of Skellam (1951), who analyzed the case of the muskrat (*Ondatra zibethica*) that had been documented by Ulbrich (1930), and the spread of oaks. By relating observed patterns of spread to the distances covered by individual acorns, he concluded that the simple model was inadequate: active dispersal agents had to be involved in the spread of the oaks. On the other hand, the case of the muskrat remains of interest to theoreticians to this day (Williamson, this volume; Andow *et al.*, in press), and provides an excellent example of the usefulness of the basic approches.

One can test such predictions at the coarsest level simply by seeing whether observed rates of advance are linear; but this is not a compelling test because the predicted constant rate does not apply for the early stages of spread, and it is impossible to specify how long that transient period should be. The stronger test of the model, and the situation in which useful predictions might be possible, occurs when independent estimates of r and D are available. We (Lubina and Levin, 1988) have done this for the linear spread of the California sea otter, and found excellent agreement with the theory. In related work (Andow *et al.*, in press), we have carried out similar studies for the muskrat, the cereal leaf beetle (*Oulema melanopus*), the rice water weevil (*Lissorhoptrus oryzophilus*), and the small cabbage white butterfly (*Artogeia rapae*), with mixed success. It is clear that the framework provided by equation (3) is an excellent one for understanding observed rates of spread. It is equally clear that a more refined theory and much better data sets are needed before we have available a truly predictive theory. At the very least, however, the approach points the way to the essential data, and provides a basis for identifying the controlling factors.

For the cabbage white butterfly, the simple picture presented above is complicated by the fact that multiple invasions occurred, involving separate introductions and saltational movements. Such long jumps are demonstrated by Mollison (1977) for his more general model, and are treated in a somewhat different way by Moody and Mack (personal communication). When leaps of this kind can occur, one really must recognize that there are two scales of spread: the longer one on which new centers of spread are established, and the shorter one on which diffusive spread from centers occurs. Patterns of this kind must be of fundamental concern for control programs, for it is the great leaps that are the least predictable and that can cause the major difficulties. One of the most interesting of such situations occurs for the great pandemics of influenza. For these, a most impressive model for the establishment of new centers has been put forth by Rvachev (see Rvachev and Longini, 1985), and is based largely on scheduled airline travel between major world cities. In general, this entire area of research is an extremely active one, and we should expect to see the development of more reliable empirically-based spread models for many species of interest. The major difficulty is that it is the tail of the distribution of dispersal distances that is most important for the questions of spread, but most difficult to deal with statistically.

19.5 ECOLOGICAL EFFECTS

Parallels exist between the evaluation of the consequences of species invasions, and the assessment of the effects of chemical releases. In the case of the environmental release of chemicals, the first step in assessment is the determination of the fate and transport of the material, as it is for the case of biological invasions. The determination of effects is much more problematical, for although it is possible to use laboratory bioassays to evaluate direct toxic effects on organisms, problems begin to multiply when one tries to extrapolate to field conditions, or to the population level. The reductionistic approach does not suffice for the evaluation of community and ecosystem effects, because of the multiplicity of pathways through which effects can be propagated. One cannot expect to gain very much information about indirect effects on community structure and system process without carrying out higher level studies in microcosms or in field situations, and the importance of such higher level effects has received increasing recognition in the ecotoxicological literature (Levin and Kimball, 1984; Kimball and Levin, 1985; Cairns, 1983, 1986).

Similar considerations apply to the introduction of biological material. The anecdotal literature is inconclusive on the effects of species introductions. As discussed earlier in this paper, and more completely in a variety of reviews (see, for example, the various papers in Mooney and Drake, 1986), the effects of a species introduction may range from benign to the elimination of major indigenous taxa. System processes in general are more buffered than are individual species, but even this generalization is debatable. Vitousek (1986) has documented the effects that invading species can have upon such system properties as productivity, soil structure, and nutrient cycling. Thus the need is clear for case by case consideration of introductions, using information derived from the experimental manipulation of whole systems where that is practicable.

Mathematical models may provide some guide to the assessment of effects on community structure and ecosystem processes, but these never will provide very detailed information. Highly detailed and specific models cannot be adapted to contexts very different from those in which they were developed. Furthermore, they are limited in their predictive power because of the difficulty in obtaining reliable estimates of parameters, and because of the multiplicity of ways that error can arise and propagate. For environmental management in general, a compromise must be reached. No single model suffices, and one needs a combination of models at different levels of detail, and techniques for simplification and aggregation of the most complicated models.

One promising approach, still in its infancy, is the development of integrated models of disturbance, spread, colonization, and community interactions. We (Levin and Buttel, 1987) have begun the development of computer models of mosaic systems, in which recurrent disturbance is an essential feature and continually creates new opportunities for invaders. Our interests have been in gap phase systems in forests, grasslands, and the intertidal; but these models also can incorporate man-made disturbances such as logging or road construction. In such models, as patterns of disturbance change, competitive release may allow invasions by species that previously were suppressed. By coupling such models for community structure with those for dispersal and colonization, we hope to develop a method for relating disturbance to invasion (see also Crawley and May, 1987).

A fundamental difficulty regarding the assessment of ecological effects is the determination of which system level measures are of interest. Any ecosystem is valued differently by different people, and the multiple uses to which systems are put place different management objectives in conflict. A related but distinct issue, treated in much greater detail elsewhere (Levin, 1987, 1988; O'Neill *et al.*, 1986), is that of scale, neglected for too long in ecological studies. An ecosystem is not something that can be crisply defined, like the objects of study in molecular biology; rather, its characteristic patterns and responses to stress vary with the scale of investigation. For example, questions relating to how diverse and stable ecosystems are make no sense without reference to the scale of interest, and environmental management must not oversimplify this complexity and ambiguity in objectives.

19.6 CONCLUSIONS

Risk assessment and risk management are interrelated activities; the capability to do one eases the task in approaching the other. Because uncertainty is an ineluctable aspect of prediction, some potential for monitoring and adaptive management is essential.

Species invasions can result from deliberate or accidental introductions, or from disturbance of established biotic communities. In the case of the former perturbation, the considerations bear rough similarity to those that must be addressed in connection with the release of chemicals into the environment.

Principally, these involve the fate and transport of the released material, and their possible ecological effects. For the movement of organisms and propagules, a sound basis exists for the description of spread, and substantial theoretical advances are likely within the next few years. The issue of effects is much more problematical, and emphasizes the need for a general theory of the responses of ecosystems to stress. Such a theory can draw inspiration and generalization from mathematical models, but cannot rely exclusively upon them. To develop the understanding of process that is essential to a capability to extrapolate from one experience to another, we desperately need more information derived from the experimental manipulation of whole ecosystems.

Finally, as in any case of environmental management, we must deal explicitly with the recognition that ecosystems have vital characteristics on a multiplicity of spatial, temporal, and structural scales. The multiple uses that may pertain to a single ecosystem argue against any attempt to reduce all complexity to a minimum number of descriptors that somehow balance all societal interests. We must not cloak what should be societal decisions in the mantle of scientific objectivity when the determinations are not purely scientific. The scientist must consider the problem on multiple scales, making clear the conflicts and the levels of uncertainty, facilitating but not usurping the societal role in managing risks.

ACKNOWLEDGEMENTS

This work was supported in part by NSF DMS-8406472, by US Environmental Protection Agency Cooperative Agreement CR 812865, and by McIntire-Stennis Project NYC-183568 to Simon A. Levin. This paper is also Ecosystems Research Center Report No. ERC-137. The views expressed are those of the author, and do not necessarily reflect those of the sponsoring agencies. Mick Crawley, Jim Drake, and Mark Williamson provided very helpful comments.

REFERENCES

Andow, D. A. (1986). Dispersal of microorganisms, with emphasis on bacteria. pp. 470–87 In: Gillett, J. W. *et al.*, Potential impacts of environmental release of biotechnology products: assessment, regulation, and research needs. *Environm. Manage.*, **10**(4), 433–563.

Andow, D. A., Kareiva, P. M., Levin, S. A., and Okubo, A. Spread of invading organisms: patterns of spread. In: Kim, K. C. (Ed.) *Evolution of Insect Pests: The Pattern of Variations.* John Wiley, New York (in press).

Cairns, J., Jr. (1983). Are single species toxicity tests alone adequate for estimating hazard? *Hydrobiologia*, **100**, 47–57.

Cairns, J., Jr. (1986). The myth of the most sensitive species. *Bio Science,* **36**(10), 670–2.

Crawley, M. J., and May R. M. (1987). Population dynamics and plant community structure: competition between annuals and perennials. *J. Theor. Biol.*, **125**, 475–89.

Dobzhansky, T., and Wright, S. (1943). Genetics of natural populations. X. Dispersion rates of *Drosophila pseudoobscura. Genetics*, **28**, 304–40.

Elton, C. S. (1958). *The Ecology of Invasions by Animals and Plants.* Methuen, London.

Fisher, R. A. (1937). The wave of advance of advantageous genes. *Ann. Eugen. London*, **7**, 355–69.

Fitt, B. D. L., and McCartney, H. A. (1986). Spore dispersal in relation to epidemic models. In: Leonard, K. J., and Fry, W. E. (Eds), *Plant Disease Epidemiology: Population Dynamics and Management*, Vol. 1. Macmillan, New York.

Frampton, V. L., Linn, M. B., and Hansing, E. D. (1942). The spread of virus diseases of the yellow type under field conditions. *Phytopathology*, **32**, 799–808.

Gregory, P. H. (1968). Interpreting plant disease dispersal gradients. *Annu. Rev. Phytopathol.*, **6**, 189–212.

Gregory, P. H., and Read, D. R. (1949). The spatial distribution of insect-borne plant-virus diseases. *Ann. Appl. Biol.*, 36, 475–82.

Hamilton, W. D., and May, R. M. (1977). Dispersal in stable habitats. *Nature*, **269**, 578–81.

Harper, J. L. (1977). *Population Biology of Plants*. Academic Press, London.

Kareiva, P. M. (1983). Local movement in herbivorous insects: applying a passive diffusion model to mark-recapture field experiments. *Oecologia*, **57**, 322–7.

Kimball, K. D. and Levin, S. A. (1985). Limitations of laboratory bioassays and the need for ecosystem level testing. *BioScience*, **35**, 165–71.

Kiyosawa, S., and Shiyomi, M. (1972). A theoretical evaluation of the effect of mixing resistant variety with susceptible variety for controlling plant diseases. *Ann. Phytopathol. Soc. Jap.*, **38**, 41–51.

Kolmogorov, A., Petrovsky, I., and Piscunov, N. (1937). Etude de l'Equation de la diffusion avec croissance de la quantite de la matiere et son application a un probleme biologique. *Bull. Univ. Moscou Ser. Internation., Sec. A*, **1**(6), 1–25.

Levin, S. A. (1987). Scale and predictability in ecological modeling. In: Vincent, T. L., Cohen, Y., Grantham, W. J., Kirkwood, G. P., and Skowronski, J. M. (Eds), Modeling and Management of Resources Under Uncertainty, Proc., Honolulu 1985. *Lecture Notes in Biomathematics* 72, Springer-Verlag, Heidelberg.

Levin, S. A. (1988). Pattern, scale, and variability: an ecological perspective. In: Hastings, A. (Eds), *Community Ecology. Lecture Notes in Biomathematics* 77, Springer-Verlag, Heidelberg.

Levin, S. A., and Buttel, L. (1987). Measures of patchiness in ecological systems. Report No. ERC-130, Ecosystems Research Center, Cornell University, Ithaca, New York.

Levin, S. A., and Harwell, M. A. (1986). Potential ecological consequences of genetically engineered organisms. pp. 495–513. In: Gillett, J. W. *et al.*, Potential impacts of environmental release of biotechnology products: assessment, regulation, and research needs. *Environm. Manage.* **10**(4), 433–563.

Levin, S. A., and Kimball, K. D. (1984). New perspectives in ecotoxicology. *Environm. Manage.*, **8**(5), 375–442.

Lubina, J. A., and Levin, S. A. (1988). The spread of a reinvading species: range expansion in the California sea otter, *Amer. Nat.*, **131**(4), 526–43.

Minogue, K. P. (1986). Disease gradients and the spread of disease. In: Leonard, K. J., and Fry, W. E. (Eds), *Plant Disease Epidemiology: Population Dynamics and Management*, pp. 285–310, Vol. 1. Macmillan, New York.

Mollison, D. (1977). Spatial contact model for ecological and epidemic spread. *J. Roy. Statist. Soc. B.*, **39**, 283–326.

Mooney, H. A., and Drake, J. A. (Eds) (1986). *Ecology of Biological Invasions of North America and Hawaii*. Springer-Verlag, New York.

Moyle, P. B. (1986). Fish introduction into North America: Patterns and ecological impact. In: Mooney, H. A., and Drake, J. A. (Eds), *Ecology of Biological Invasions of North America and Hawaii*, pp. 27–43. Springer-Verlag, New York.

Okubo, A. (1980). *Diffusion and Ecological Problems: Mathematical Models*. Biomathematics, Vol. 10. Springer-Verlag, Berlin, Heidelberg, New York.

Okubo, A., and Levin, S. A. (1988). A theoretical framework for the analysis of date on the wind dispersal of seeds and pollen (*manuscript*) (Accepted *Ecology*).
O'Neill, R. V., DeAngelis, D. L., Waide, J. B., and Allen, T. F. H. (1986). *A. Hierarchical Concept of Ecosystems. Monographs in Population Biology 23.*, Princeton University Press, Princeton, New Jersey.
Pasquill, F. (1962). *Atmospheric Diffusion*, 1st Edn. Van Nostrand, London.
Rvachev, L. A., and Longini, I. M. (1985). A mathematical model for the global spread of influenza. *Math. Biosci.*, **75**, 3–22.
Skellam, J. G. (1951). Random dispersal in theoretical populations. *Biometrika*, **38**, 196–218.
Ulbrich, J. (1930). *Die Bisamratte*. Heinrich, Dresden.
Vitousek, P. M. (1986). Biological invasions and ecosystem properties: Can species make a difference. In: Mooney, H. A., and Drake, J. A. (Eds), *Ecology of Biological Invasions of North America and Hawaii*, pp. 163–78. Springer-Verlag, New York.

Biological Invasions: a Global Perspective
Edited by J. A. Drake *et al.*
© 1989 SCOPE. Published by John Wiley & Sons Ltd

CHAPTER 20

Ecological Control of Invasive Terrestrial Plants

RICHARD H. GROVES

20.1 INTRODUCTION

Methods of controlling terrestrial invasive plants have evolved as land use systems changed and diversified over time. But the aim of control has always remained the same—namely, to limit the number of plant propagules in the long term to a level tolerable to human activities. Rarely has eradication been a management aim. In this chapter I shall review some of the ways by which populations of invasive plants have been deliberately limited. I define ecological control of an invasive plant as the planned use of one or several methods of control when integrated with an understanding of the dynamics of the ecosystem in which the plant occurs. Control methods used in agricultural ecosystems usually simplify the system. Ecological control methods in natural systems, on the other hand, aim to maintain or even enhance biological diversity in the longer term.

In this chapter I shall discuss, using examples from invaded natural vegetation wherever possible, the different methods of control in terms of their relative importance in reducing plant populations. Case histories of three groups of invasive plants will be presented in relation to ecological aspects of their control: firstly, *Hypericum perforatum* in temperate grasslands; secondly, a group of *Eupatorium* (*Chromolaena*) species which invade areas of subtropical forests; and thirdly, some invasive tall shrubs in mediterranean-climate shrublands. From these and other examples I shall present some general principles for more effective control of invasive terrestrial plants in natural ecosystems.

20.2 METHODS OF CONTROL

In this section I discuss the different methods of control of invasive plants as they apply to natural areas. Wherever possible, my examples will be drawn from those identified by the contributors to national symposia (Groves and Burdon, 1986; Kornberg and Williamson, 1986; Macdonald *et al.*, 1986; Mooney and Drake, 1986).

One method to control invasive terrestrial plants is to prevent their entry to a country or region. Such a method is a policy of exclusion enacted by parliamentary legislation (Navaratnam and Catley, 1986). The method is presumably effective in limiting the number of invasive plants entering a country, although I have been unable to obtain figures on the numbers and identities of invasive plants which reach an entry point but are then detected and thereby excluded. This method does not usually keep out 'new' species of known taxonomic identity unless they are especially troublesome in another country. Even then, legislation may not prevent their entry, as in the case of various *Hieracium* species native to Europe and known to be invasive in Canada and in New Zealand but which have yet to reach Australia (Groves, 1986).

An allied method of legislative control for invasive plants is to declare them 'noxious' once they have entered a country. Whilst in theory this method gives management authorities the legal power to control growth and reproduction of such plants, in practice it seems neither to reduce the rate of spread of an invasive plant nor to lead to more effective control (Moore, 1971; Amor and Twentyman, 1974). A. M. Gill (personal communication) showed how ineffective this control method has been for the European plant *Hypericum perforatum* over the timespan of its invasion of southeastern Australia. Legislation concerning noxious plants is being revised currently in several regions to take this point into account, especially for plants invading non-agricultural land. How effective the revised legislation will be remains to be assessed.

Both these legislative methods of control are usually retrospective but I believe the emphasis to be changing gradually to become more forward-looking as more attempts are made to predict the potential of a plant to spread and be invasive (see, for example, Medd and Smith, 1978; Williams and Groves, 1980; Gunn *et al.*, 1981; Patterson, 1983). A change has also occurred as knowledge of the biology of invasive plants moves from a regional to an international perspective, as has occurred as a result of this SCOPE programme.

Physical methods of control include the planned mechanical or manual cultivation of invaded land or the manipulation of fire regimes to benefit indigenous species at the expense of invasive species. Cultivation of land to control invasive species mechanically is usually inappropriate to natural areas, although it has been used very effectively for millennia in agricultural areas. Hand-pulling of invasive plants is often practised and has the potential to be effective—as, for instance, in the control of the South African shrub *Chrysanthemoides monilifera* in urban parks in southern Australia (see later). Manual slashing of the Australian shrub *Hakea sericea* has been effective as a physical method of control in conjunction with burning of South African mountain fynbos. Hand-pulling and slashing have been completely ineffective, however, in controlling *Rhododendron* in British nature reserves (Usher, 1986).

The planned use of fire is usually a preferred method to conserve natural plant, communities and to control the growth of undesirable plants invading those communities (Christensen and Burrows, 1986). For these two groups of plants alterations in the different components of the fire regime (Gill, 1975) will have different consequences. Fires at too frequent an interval often favour plant invasion. Generally, frequent fires favour resprouting perennials over non-sprouting species, disadvantage plants which rely solely on seed stored on the plant, promote grasses and forbs over dicotyledonous plants and may reduce species diversity (Vogl, 1977). Baird (1977) attributed the spread of the South African grass *Ehrharta calycina* in eucalypt woodland at King's Park, Perth, in part to fires every one or two years.

Watsonia species are iridaceous components of the South African fynbos which have invaded parts of southern Australia and New Zealand (Parsons, 1973), especially the species *W. bulbillifera*. Another species, *W. pyramidata*, was shown to be more prominent in autumn-burned vegetation in South Africa (Kruger, 1977). On the basis of this observation, burning of areas invaded by *W. bulbillifera* in Australia and New Zealand in seasons other than autumn should limit its invasiveness, especially if season of burning can be combined with other control methods.

Manipulation of fire intensities can also be used to control invasions. For instance, in areas of South African fynbos invaded by woody Australian acacias, fires of high intensity will increase the invasiveness of these shrubs because they possess a high proportion of soil-stored hard seeds which are stimulated to germinate by the rupturing of the testa induced by the high fire temperatures. Fires of lower intensities may benefit the indigenous component of the vegetation, especially the proteaceous element, and thereby reduce the invasiveness of the introduced leguminous element in the flora.

The most effective method of control of invasions by fire can often be to try and greatly reduce its frequency. In the example discussed earlier on the effects of fire frequency, Baird (1977) also found that in one area unburnt for 15 years the number of clumps of the invasive *Ehrharta calycina* gradually decreased from 115 to six. This result thus has the potential to change the almost self-perpetuating cycle of increased fire frequency and decreased fire intensity leading to increased colonization by *E. calycina* to one of a greatly reduced fire frequency and decreased colonization by the invasive species. In general, hard-seeded legumes, species with wind-dispersed seeds and bulb- and corm-producing plants will be at a disadvantage on infrequently burnt reserves. The effectiveness of this control method depends on the interaction between time since the last fire, the lifespan of the invasive plant and its propagules and the successional status of the invaded community. The South African plant *Senecio pterophorus* has short-lived wind-dispersed seeds. When previously grazed land in the Adelaide Hills, South Australia, was reserved for nature conservation, *S. pterophorus* rapidly dominated open areas no longer grazed by sheep. But after 10–15 years without fire a tree cover established and *S. pterophorus* plants no longer dominated the more shaded understorey. Whilst the plant is still present in the area, it is no longer as invasive as it once was because of the changed environmental conditions induced by less frequent fire (P. M. Kloot, personal communication). Generally, frequent fires keep a natural community in an early successional stage; infrequent fire may enable a community to change and this change may be unsuited to the growth of invasive species such as *S. pterophorus*.

Fire remains the cheapest form of management available to conserve and perpetuate natural plant communities. The different components of a fire regime—frequency, season and intensity—may be used effectively to retain the natural element and control the invasive element in the flora of a nature reserve.

Chemical methods of control are used widely to control the growth and development of invasive plants, although their use in areas set aside for natural values is less extensive than in agriculture, and even less desirable. Control of invasive plants by herbicide application is usually short term and directed at individual 'target' species. For example, *Chrysanthemoides monilifera* is sprayed regularly with herbicide in a hill reserve in southern Victoria whilst no control is directed at plants of the invasive grass genus *Ehrharta*. Spraying of one invader may well be leading to its replacement by a second plant also capable of outcompeting the indigenous species. Regular spraying with herbicides seems to have kept the invasive European shrub *Cytisus scoparius* from spreading further in Barringon Tops National Park in eastern Australia until reductions in funding halted the spraying programme. The subsequent extensive spread of the invader through the eucalypt woodland can be dated from this temporary cessation in chemical control (see also Macdonald *et al*., this volume). For this leguminous species a control programme which does not lead to a reduction in the input of new seed into the store of dormant but viable seed in the soil will be ineffective in the long term. To reduce the level of *C. scoparius* cover by spraying with a herbicide will be effective if it allows for regeneration of native tussock grass (in the short term) or of native trees (in the long term), but it may not necessarily be effective in reducing the number of long-lived seeds in the soil.

Some invasive plants such as blackberry (*Rubus fruticosus* sp. agg.) provide food and refuge for invasive animals, such as foxes and rabbits in Australia and New Zealand. Spraying of blackberry with herbicides may reduce the suitability of the habitat for these introduced animals in the short term but the high and increasing costs of herbicides and the cost of application in difficult terrain usually means that continuity of spraying is interrupted and the invasion returns. Application of herbicides for blackberry control has the added disadvantage that because blackberries often occur densely along watercourses the chances of chemical contamination of waterways is thereby increased. I conclude that chemical control may be effective in limiting newly discovered infestations which have yet to spread, e.g. of *Onopordum tauricum* in Victoria (W. T. Parsons, personal communication). On the other hand, chemical control of already widespread invasive plants in nature reserves is often expensive, usually ecologically undesirable and rarely, if ever, effective in the long term, unless integrated with other methods of control.

The deliberate promotion of growth of indigenous plants to compete with and thereby control invasive plants is inadequately researched. It is a control method widely used in pasture research to promote the growth of desirable species which then better compete with the undesirable species. This imbalance in research effort can lead to the situation where a plant such as *Hypericum perforatum*, when invasive in pastures, can be controlled by competition from desirable pasture species such as *Trifolium subterraneum* and/or *Phalaris aquatica* (see later); *H. perforatum* can, however, remain dominant in more natural vegetation in reserves adjoining pasture lands because little is known of the characteristics of indigenous species which may compete effectively with the invader. Some *Eupatorium* spp. may be controlled by shading from indigenous tree species if the shading effect continues beyond the active reproductive output of the invader, as can occur in northeastern India (see later).

Because of the agronomic bias in most previous research on this control method, the index of competition is usually measured as the enhanced yield of plant or livestock product, whereas in the context of invasions in nature reserves, numbers of propagules per unit area of land or volume of soil may be of greater ecological significance. The method has the advantage that the controlling effects are expressed over a much longer time period than, say, are the effects of chemical control. It is an ecological aspect of control of invasive plants which is in need of much more research effort. If this enhanced research can include a study of rooting characteristics of the competing species, the results may be even more applicable to the management of biological invasions.

Biological methods to control invasive terrestrial plants have sometimes had a spectacular success in the long term. The control of *Opuntia* spp. by *Cactoblastis cactorum* in various countries (e.g. Mann, 1970, for Australia; Zimmermann *et al.*, 1986, for South Africa) is probably the best known example of success. Zimmermann *et al.* (1986) even considered that the present distribution, as well as abundance, of three invasive *Opuntia* species in South Africa was determined to a large extent by pressure from imported insect herbivores. The method is not without risks, however (see, for example, Howarth, 1983), although these risks are minimized by careful specificity testing before release in regions such as Australia and California. I know of no documented cases where a natural enemy after its deliberate introduction to control an invasive plant has caused a reduction in the population of a native congener of that plant, although such cases are known for the biological control of insects (Howarth, 1983). Programmes for biological control of genera such as *Convolvulus, Rubus, Rumex* and *Solanum* will need to carefully assess economic benefits against biological risk.

Results of a recent survey of biological control of invasive plants, both terrestrial and aquatic, show that only between 25 and 40% of programmes could be considered effective (Julien, 1982). When the method has been successful, density or cover measurements of the invasive plant have decreased as a result of the planned release of introduced or native arthropods or fungi or both.

Sometimes this reduction has been quantified, sometimes not. Success is usually attained when the effects of the invasive plant no longer exceed a 'threshold' level which may be economically, agronomically or, more rarely ecologically based. Research programmes on biological control commonly ignore increases in cover or numbers of the species replacing the invasive one (but see, as an exception, Huffaker and Kennett, 1959, for *Hypericum perforatum* control in California). This deficiency has the potential to lead to one invasive plant replacing another, as with several other methods of control (see earlier).

For plants invading nature reserves where low-cost control and minimal disturbance are important considerations, biological control methods have a considerable and continuing role to play. When effective, biological control is the ideal method, but it is not always effective and rarely is it predictable. A greater effort at evaluation, in ecological terms, of the successes and failures in the biological control of a range of invasive plants may help to overcome this deficiency from which formulation of a theoretical basis for the method can commence (see e.g. Crawley, 1986 and this volume).

The theme developed in this section has been that although individual control methods are sometimes effective at controlling invasive plants, a combination of control methods carefully timed to coincide with critical stages in the reproductive cycle of the plant will be even more effective. Such control is increasingly being termed 'integrated' control (see, for example, Kluge *et al.*, 1986), following the terminology developed for systems for invertebrate control. For my purposes, I prefer to use the term 'ecological' control for those methods not only attuned to the plant's life cycle but also to the dynamics of the ecosystem in which it occurs.

20.3 SOME EXAMPLES OF ECOLOGICAL CONTROL

In the three case histories which follow I shall develop the concept of ecological control further and endeavour to show that control methods which result in a more diverse ecosystem, such as the planned use of fire, competing plants and/or biological control, may make control of terrestrial invasive plants more effective in the long term and more ecological. Mechanical or chemical control methods seem to have the opposite effect of making the invaded ecosystem simpler. Whilst the latter may be desirable in an agroecosystem it is less desirable in the natural ecosystems which are the subject of this volume.

20.3.1 Control of *Hypericum perforatum*

The 200 or more species of *Hypericum* (family Clusiaceae) generally are distributed world-wide in temperate and subtropical regions. Several species with large, bright yellow flowers are valued as garden plants and some for their herbal properties. Because of these two sets of characteristics, *Hypericum* species have been introduced deliberately to regions where several have become naturalized and invasive. Of this latter group, *H. perforatum*, native to a large area of Europe, western Asia and northern Africa, is a particularly invasive perennial species of temperate grasslands and woodlands. It reproduces vegetatively, both from crowns and rhizomes, and sexually from seeds (Campbell and Delfosse, 1984); each plant of *H. perforatum* produces an average of about 30000 seeds (Salisbury, 1942; Tisdale *et al*., 1959; Parsons, 1973), which are small, sticky and dormant. Genetically, *H. perforatum* is variable, with several hybrids of different ploidy levels known (Robson, 1968). Its breeding system is almost entirely apomictic (Robson, 1968). The species is very variable (Robson, 1968), especially in leaf width.

Hypericum perforatum was brought to the east coast of North America in 1793, to California 100 years later (Tisdale *et al*., 1959), and to British Columbia soon after (Harris *et al*., 1969). It is known to have been introduced deliberately to Australia in the 1880s (Parsons, 1973), although A. M. Gill (personal communication) showed that multiple introductions of the plant were highly probable; certainly it was cultivated in Melbourne in 1858 and in Adelaide in 1859. The plant is widespread on both islands of New Zealand (Healy, 1972, Campbell and Delfosse, 1984). *H. perforatum* was introduced to South Africa as a contaminant in seed from Australia in 1942 (Stirton, 1983) and at the end of the 19th century to Chile from Argentina (Villanueva and Fauré, 1959). The genetic identity of the material introduced to these different regions is unknown.

H. perforatum presence reduces the capacity of grasslands to provide grazing for livestock and it alters grassland composition. The plant has been of considerable economic importance to the western USA and southeastern Australia, where formerly productive areas have been abandoned and land use has sometimes been changed radically.

To be effective in the long term, control programmes for *Hypericum perforatum* need to reduce seed production to close to zero; in the short term, reduction in growth of *H. perforatum* and thereby reductions in replenishment of root reserves are essential for control. Some control methods seemingly have no effect on *H. perforatum* populations. As most of the introductions of *H. perforatum* predate the implementation of quarantine procedures, legislation enacted subsequently seeks only to prevent the importation of new and possibly different genetic material to a region. *H. perforatum* is a declared noxious plant in several countries, and in Australia at least this form of legislative recognition seems to have done very little if anything to slow its rate of spread or to bring about a more effective level of control (A. M. Gill, personal communication). Cultivation of invaded land is not an effective control method on its own, even in arable areas (Davey, 1917). Burning increases, rather than decreases, the density of the plant (Dodd, 1920; Moore and Cashmore, 1942), although it may temporarily reduce growth and destroy some seeds (Campbell and Delfosse, 1984). These various methods, considered either alone or together, thus appear to have no effect on limiting the numbers of seeds produced by *H. perforatum* per unit area or on reducing reserves in the root system.

Application of herbicides to stands of *Hypericum perforatum* may reduce growth and seeding of the plant, depending on time of application and level of active ingredient (a.i.) in the herbicide mixture. The present recommendation, based on results from pasture research, is to use either 2, 4-D ester applied at the rate of 3.36 kg a.i./ha at early flowering (late spring) or glyphosate (1.68 kg a.i./ha) applied in summer or early autumn before annual pasture species germinate (Campbell and Delfosse, 1984). This recommendation may affect associated plants to varying extents and may not be appropriate to situations where values other than agronomic ones are important, as in nature reserves. The effectiveness of these recommendations is based on reductions in percentage ground cover of *H. perforatum* after 2 years (Campbell *et al.*, 1979); I can find no results for concomitant reductions in the level of seeding of *H. perforatum* in response to herbicide application or any results for a period longer than 2 years.

The growth of *Hypericum perforatum* may be controlled in pasture by competition from other plants, especially from a mixture of subterranean clover (*Trifolium subterraneum*) and perennial grasses (Moore and Cashmore, 1942). Four and a half years after sowing various pasture species into land heavily infested with *H. perforatum*, Moore and Cashmore (1942) showed that the number of *H. perforatum* plants was reduced by 96% on plots containing the winter-growing *T. subterraneum* and by 64% of the level on an unsown 'control' on plots containing the summer-growing *T. repens*. A perennial grass such as *Phalaris aquatica* was more effective in reducing yield of *H. perforatum* than was the annual grass *Lolium rigidum* over 4 years of measurements in New South Wales. Moore and Cashmore attributed control by this means to shading of the procumbent shoots of *H. perforatum* by the dense canopy of *T. subterraneum* produced in winter. More probably, as Clark (1953) has suggested, the mature plants of *H. perforatum* are being controlled by perennial grasses in summer when competition for moisture is severe, and seedlings of *H. perforatum* are being shaded in winter by a dense canopy of *T. subterraneum*.

A more extreme form of competition imposed on *H. perforatum* plants is to radically change land use of the invaded area from either grassland or woodland to a plantation of *Pinus radiata* which, when canopy closure is reached in 10–12 years, completely shades *H. perforatum*. As a plantation of *P. radiata* lasts about 40 years, it is an effective long-term control method used in several regions, especially northeastern Victoria (Parsons, 1973). Planting of *P. radiata* has the obvious disadvantage, however, that the natural ecosystem is obliterated for ever—one invader is replaced by another with economic value, as in the deliberate promotion of growth of *T. subterraneum*. Use of indigenous species to control *H. perforatum* populations has not been evaluated experimentally, although in southern Australia the indigenous perennial grass *Themeda australis* is able to suppress *H. perforatum* growth (Davey, 1919).

Attempts to control *H. perforatum* by the introduction of insects from the plant's region of origin date back to 1919 when a search began in England for potential biological control agents. Ten insect species were subsequently introduced to Australia from both England and southern France and six are known to have been released in Australia between 1930 and 1940 (Campbell and Delfosse, 1984). Of these insects, only the chrysomelid *Chrysolina quadrigemina* from southern France survived in sufficient numbers to cause significant damage to *H. perforatum* in grassland areas. The adult *C. quadrigemina* exerts its controlling effect by completely defoliating the plant in spring, whilst in late autumn and winter its larvae feed on the young buds and leaves. Various of the other insects introduced attack different parts of the plant but in Australia they have relatively minor effects on plant density at other than a local level. The same insect has been the most successful biocontrol agent for *H. perforatum* subsequently in the western USA (Huffaker, 1967; Dahlsten, 1986), Canada (Harris *et al*., 1969), Chile (Villanueva and Fauré, 1959) and South Africa (Stirton, 1983), whilst in New Zealand, *C. hyperici* has survived better than *C. quadrigemina* (Harris *et al*., 1969) and in Hawaii the gall midge *Zeuxidiplosis giardi* seems to be the main controlling agent (Davis and Krauss, 1967; Julien, 1982).

The long-term effects of release of *C. quadrigemina* in Californian rangelands containing *H. perforatum* were followed for up to 10 years by Huffaker and Kennett (1959). These rangelands were probably composed originally of perennial grasses and forbs which were replaced largely by annual plants as a result of overgrazing by domestic livestock (Clements and Shelford, 1939). At the time of insect release (early 1946; Huffaker, 1967) the percentage cover of *H. perforatum* plants varied between 26 and 51% depending on site (Figure 20.1), with other plant cover being from other invasive plants, some legumes and a group of annual forage grasses. As a result primarily of the winter feeding of the larvae of *C. quadrigemina* (cf. adult beetles feeding for a shorter period in spring–summer), Huffaker and Kennett (1959) measured a substantial reduction in *H. perforatum* cover to almost zero in a period of 4–5 years and a concomitant increase in cover of all other plant groups, especially of annual forage grasses. At one site in Humboldt County the native perennial grass *Danthonia californica* increased in cover. At no site did other noxious plants show any consistent increases as *H. perforatum* cover decreased. The larvae of *C. quadrigemina* kept the plants of *H. perforatum* defoliated from midwinter through to early spring of each year and hence root reserves were progressively depleted. With further time from release, the seed crop (number and yield?) was also depleted, although Huffaker and Kennett (1959) give no quantitative data on this aspect. We may conclude that entomological control of *H. perforatum* has been highly successful in California. In fact, St John's wort has been removed from the state's list of primary noxious plants.

In regions other than California the success rate of entomological control has not been as satisfactory, however (Huffaker, 1967). At most Californian sites

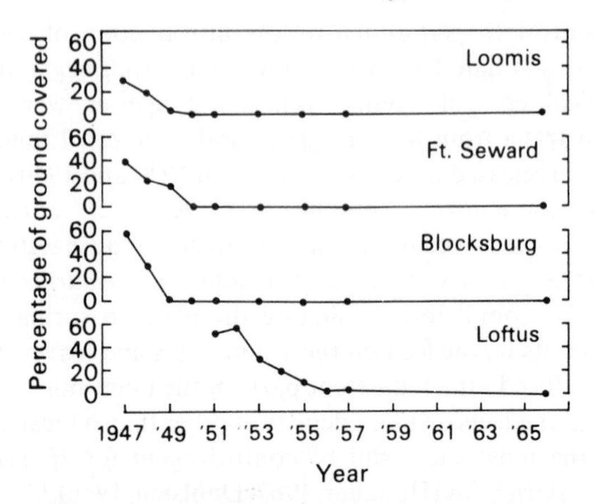

Figure 20.1. Reduction in ground cover of *Hypericum perforatum* by *Chrysolina quadrigemina* at four sites in California for the period 1947 to 1966 (redrawn from Huffaker, 1967)

nearly all plants died as a result of a single severe attack by *C. quadrigemina* larvae and a very high degree of control of *H. perforatum* (namely, greater than 99%) was maintained over 16 years of measurement (1950–1966) at three sites (Figure 20.1) (Huffaker, 1967). Huffaker claimed an 80–90% level of control to be satisfactory because domestic animals can feed on light infestations without ill effect and thereby help to maintain a low plant density. In southeastern Australia the level of mortality of *H. perforatum* was much less (about 54% averaged over 16 sites) and Huffaker attributed this significant difference to the differing incidence of summer rainfall in the two regions (Figure 20.2). Absence of summer rain in many areas of California, as represented by Loomis, Alderpoint and Redding (Figure 20.2), kills *H. perforatum* plants already defoliated by *C. quadrigemina* larvae. Summer rainfall in southeastern Australia, as represented by Myrtleford, Benalla and Mudgee (Figure 20.2), promotes regrowth of the defoliated plants and enables them to survive. The insects were imported originally from Mediterranean France to southeastern Australia and then were sent to California after at least 10 years' acclimatization in Australia. Obviously, the insect's phenology was still attuned to a typical summer-dry mediterranean-type climate. Huffaker observed a greater level of success for *C. quadrigemina* in South and Western Australia, as represented by Clare and Dwellingup (Figure 20.2), because these regions are much more summer-dry than are northeastern Victoria and southern New South Wales. In these latter regions the plant continues to be a major invader of nature reserves and national parks, such as Kosciusko.

H. perforatum increased initially in grasslands in several regions of the world from the sites to which it was introduced deliberately because it was able to invade ground made bare as a result of overgrazing by domestic stock. The

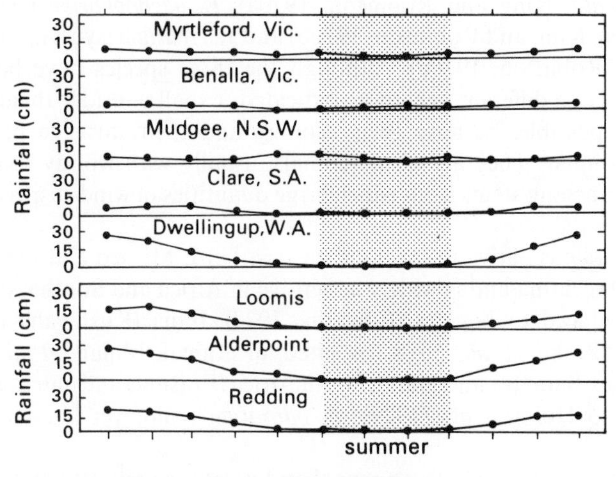

Figure 20.2. Patterns of annual rainfall, as long-term averages at five locations in southern Australia and three in California, all representative of regions invaded by *Hypericum perforatum*. Periods of summer rainfall have been synchronized and centred to show contrasts (redrawn from Huffaker, 1967). Entomological control has been highly successful at Clare, Dwellingup, Loomis, Alderpoint and Redding (see text)

perenniality, profuse seeding and deep root system of *H. perforatum* enabled it to compete successfully with annual grasses and forbs and to become dominant within about 20 to 30 years from its time of naturalization. Because seeds of *H. perforatum* may remain viable in soil for at least 20 years or longer (A. M. Gill, personal communication), *H. perforatum* has the capacity to germinate and reinvade an area should there be significant reductions in vegetative cover at any time over this long period. Clark (1953) commented appropriately concerning *H. perforatum* in plantations of *P. radiata* that '*Hypericum* is generally the last plant to be excluded by the pines and the first to reappear' (p. 98). In the temperate-climate regions which *H. perforatum* has invaded, I conclude that control will always be more effective if it arises from the long-term interaction between the effects of one or several natural enemies and of competing perennial plants as modified by appropriate grazing regimes and by the incidence of summer rainfall.

20.3.2 Control of *Eupatorium* spp.

Three species of the genus *Eupatorium* (family Asteraceae), all originating in Central or South America, are common invaders of formerly forested land in subtropical regions. The three species are *Eupatorium odoratum* (syn. *Chromolaena odorata*, King and Robinson, 1970a), *E. adenophorum* (syn. *Ageratina adenophora*, King and Robinson, 1970b) and *E. riparium* (syn. *Ageratina riparia*, King and Robinson, 1970b). Although the three species have been assigned recently to two different genera as indicated, I shall consider them collectively whenever possible, because their ecology and their invasive properties are basically similar. They are all perennials, usually with woody rootstocks and upright branching stems from which large quantities of wind-dispersed seeds are produced.

E. odoratum is native to the southeastern USA, Mexico and the West Indies south to Argentina and is widely adventive in Africa and Southeast Asia. It has been introduced to Nigeria (Edwards, 1977), India (Kushwaha *et al.*, 1981), Thailand (Zinke *et al.*, 1978, as cited in Ramakrishnan *et al.*, 1981), the Phillippines, Sumatra and Natal, South Africa (Erasmus and van Staden, 1986b). Contrary to Holm *et al.*, (1979), *E. odoratum* is not yet known to occur in Australia.

The species depends both on sexual and vegetative reproduction for increase. In north eastern India it can produce as many as 48000 seeds per plant (Kushwaha *et al.*, 1981). Seeds of *E. odoratum* apparently do not survive in soil for long although there is some evidence for seed dormancy (Erasmus and van Staden, 1986b). *E. odoratum* is variable (Edwards, 1977; Edwards and Stephenson, 1974).

E. adenophorum was introduced deliberately from Mexico to Maui in about 1864 as an ornamental and it now occurs on all the other major Hawaiian islands except Kaui (Bess and Haramoto, 1972). It also occurs in Australia on the northern coast of New South Wales (Auld and Martin, 1975) and in Queensland (Dodd, 1961). For this region, Auld (1969) showed that there was about a 75% chance of the occurrence of a dense population of *E. adenophorum* in areas which had a combination of steep land (> 20°), no tree cover and an annual rainfall greater than 1900 mm. *E. adenophorum* also occurs in India (Ramakrishnan and Misra, 1981), the northern region of New Zealand, Nigeria, the Philippines, Thailand, Trinidad (Holm *et al.*, 1979) and California (Auld, 1972).

Population increase in *E. adenophorum* occurs by both sexual and vegetative reproduction. *E. adenophorum* is an apomictic triploid which forms seeds by agamospermy (Holmgren, 1919, as cited by Auld and Martin, 1975). Seeds (cypselas) germinate in late summer in New South Wales (Figure 20.3) (Auld and Martin, 1975) and they have an absolute requirement for light (Auld and Martin, 1975), so that germination is effectively limited to sites free from plant competition. Seedlings have a high relative growth rate and are fully established within 8 weeks of germination. New vegetative growth begins as resprouts from the crown of the plant with the first sustained rains in summer (Auld and Martin, 1975). In India invasion occurs in early successional vegetation up to 6 years from clearing of forest; mortality of seedlings reached 100%

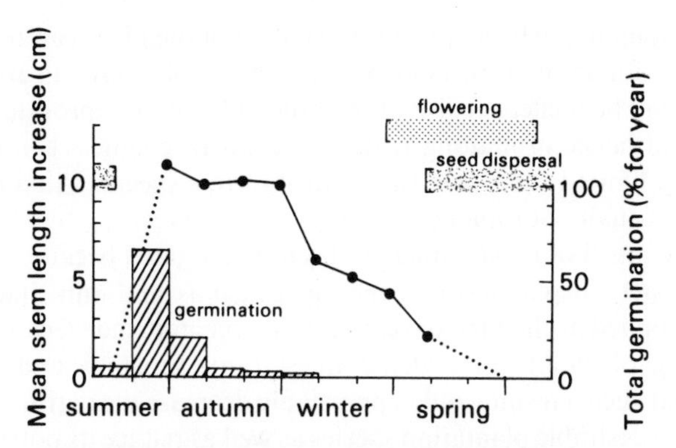

Figure 20.3. Seasonal growth and germination of *Eupatorium adenophorum* in eastern coastal Australia. Monthly increase in stem length, monthly germination and length of flowering and seed dispersal periods are shown (redrawn from Auld and Martin, 1975)

in vegetation older than this and mortality of vegetative resprouts followed a similar pattern (Ramakrishnan *et al.*, 1981). As for *E. odoratum* (see earlier), *E. adenophorum* is also variable (Ramakrishnan *et al.*, 1981), albeit by different environmental and genotypic means.

E. riparium, another invasive species in the genus co-occurs with *E. adenophorum* in eastern Australia (Auld and Martin, 1975) and northeastern India (Ramakrishnan *et al.*, 1981). It is an herbaceous perennial with a scrambling habit. It is rarely as troublesome as *E. adenophorum* and, at least in Australia, has a more restricted distribution although it was introduced much earlier.

In most countries in which *Eupatorium* species occur, their presence reduces stocking capacity of grazing land and restricts the movement of stock and machinery. They reduce carrying capacity because they compete with more desirable pasture plants and they contain aromatic chemicals which make the foliage unpalatable to cattle (Ramakrishnan and Misra, 1981). In south east Asia generally, the presence of *Eupatorium* spp. can arrest secondary succession of former forested land which was cleared for short-term cropping; only by lengthening the slash-and-burn cycle can their deleterious effects be overcome (Ramakrishnan *et al.*, 1981). *E. odoratum* is highly flammable and thus its presence increases the chance of fire which may further retard secondary succession. *Eupatorium* spp. can limit early growth in plantations of forest tree species, of coconuts in Sri Lanka and of other tropical tree crops (Anon., 1977).

Control of *E. adenophorum* by cultivation is feasible in some arable situations because its root system is usually confined to the top 40 cm of soil (Auld and Martin, 1975). These authors assessed the potential for root segments cut from different positions on the plant to regrow. They showed that only cuttings which included part of the crown regrew and that regrowth took place only from crown tissue. Thus ploughing which'uproots' the crown of the plant can be an effective control method for agricultural land, but slashing of shoot growth above the level of the crown will be ineffective. Because of the difficulty of uprooting crowns on steep land, and hence of limiting regrowth, most mechanical control methods have had only limited success in the control of this species (Auld and Martin, 1975) and are usually not applicable to nature reserves.

Herbicidal control of *E. odoratum* has been investigated because of the plant's economic impact on plantation cropping (Erasmus and van Staden, 1986a). Triclopyr, if applied at the time of year and rate recommended, caused at least a 90% mortality of shoots in a dense infestation. Erasmus and van Staden considered that such a result would appreciably decrease the detrimental effect of *E. odoratum* on desirable plantation species as well as reduce its potential as a fire hazard. The level of shoot mortality obtained would undoubtedly cause a major decrease in seed production, but the extent of that reduction was not measured.

Many chemicals have been used to control dense populations of *E. adenophorum* in pasture (Auld, 1972). Auld and Martin (1975) concluded that progress on an effective chemical control programme for *E. adenophorum* was restricted by the requirement for high volume application of herbicides and by seasonal variability in results: their conclusion probably holds for *Eupatorium* spp. generally and especially when considered in the ecological context of this review.

Competing plants have been shown to control the growth and reproduction of *Eupatorium* spp. in two ecosystems. On land cleared permanently for grazing *Eupatorium* invades poor quality pasture dominated by species such as *Axonopus affinis* (carpet grass). More productive grasses such as *Pennisetum clandestinum* (kikuyu) need to be established during spring when the probability of effective rainfall is low and before *Eupatorium* seeds germinate in late summer (Auld and Martin, 1975). As *Eupatorium* seeds require light to germinate (Erasmus and van Staden, 1986b) and seedlings can tolerate a shading level of as much as 10% of full sunlight (Auld and Martin, 1975), the growth of pasture grasses must be early and substantial to limit growth of *Eupatorium* shoots in this way. Growth of *Eupatorium* spp. may also be controlled in the long term by shading by the canopy of tree species either regenerating naturally or planted as tree crops. This is an effective 'natural' control method provided that the duration of the forested stage is longer than the viability of seeds in the soil or, in the case of *E. adenophorum*, root crowns. In northeastern India the period of tree cover required is at least 20 years (Ramakrishnan *et al.*, 1981).

Biological control of *E. adenophorum* has been tried with some success in Hawaii (Bess and Haramoto, 1959, 1972,), Australia (Dodd, 1961) and India (Rao *et al.*, 1981) using primarily a tephritid gall fly *Procecidochares utilis*. This fly was introduced from Mexico to Hawaii in 1945 and its progressive effects in reducing the abundance of *E. adenophorum* on Maui especially have been studied over a 22-year period (Bess and Haramoto, 1959, 1972). The same insect was introduced from Hawaii to eastern Australia in 1952 (Dodd, 1961) and to India in 1963 (Rao *et al.*, 1981). On Maui the release of *P. utilis* has led to substantial reductions in the invasiveness of *E. adenophorum* in the long term. No regrowth of the plant has occurred on areas from which it was freed between 1950 and 1957 (Bess and Haramoto, 1972). In general, the degree of control achieved is related to rainfall—control on Maui has been good in low rainfall areas but negligible in higher rainfall areas, where mowing and herbicide applications are still necessary.

In both eastern Australia and India *P. utilis* seems to have been less effective at limiting *Eupatorium* numbers than in Hawaii because, when introduced to these former regions, *P. utilis* has been parasitized by several native hymenoptera (Dodd, 1961). Apparently concurrently with the establishment of *P. utilis* in Australia, the fungus *Cercospora eupatorii* appeared, possibly arriving as a contaminant of a consignment of gall flies from Hawaii (Dodd, 1961). *C. eupatorii* is specific to *Eupatorium* and is damaging to seedlings especially; it is native to America (Dodd, 1961). Attack of seedlings by this fungus is probably a factor limiting further spread of the plant in southern Queensland. The fungus is now being evaluated as a candidate organism for introducion to South Africa to control *E. odoratum* (M. J. Morris, personal communication). In Australia *Eupatorium* is also attacked by a native crown-boring beetle of the genus *Dihammus*, which can weaken plants to the point where pasture species can compete successfully in most open, dry situations. Biological control of *Eupatorium* has been more effective in situations such as on Maui where growth of pasture grasses (e.g. *Pennisetum clandestinum*) has also exerted a controlling effect.

Control of *Eupatorium* spp. by any means depends on achieving substantial reductions in seed numbers and in the amount of regenerative tissue in root crowns. Despite the considerable literature on *Eupatorium* control (Anon., 1977), I can find no data on these critical measures. Bess and Haramoto (1959) presented data on the comparative growth in height of *E. adenophorum* plants infested with *P. utilis* in relation to the developmental stage of the fly; they showed that infestations of shoots by *P. utilis* led to a reduction in height growth (Figure 20.4) but the relationship of this reduction to seed output is not clarified. Erasmus and van Staden (1986b) showed that application of some herbicides could cause 100% mortality in *E. odoratum* shoots in the short term. Where control was less than 100%, the relationship between incomplete shoot mortality and seed production is not clear, nor are the longer term consequences of the treatment on growth and reproduction of the species apparent. Only in the documented case of *E. odoratum* and *E. adenophorum* in different-aged communities in northeastern India have these data been collected (Table 20.1). These results show a complete reduction in seed production after more than 5 years for *E. odoratum* and after 10 years for *E. adenophorum* as the forest reverts to its former state. Substantial reductions in plant density took longer to occur in these systems, from which we may conclude that established plants of each species may be long-lived.

I conclude that the most effective ecological control for *Eupatorium* spp., at least in eastern Australia, as in India, may be to allow the invaded land to revert to

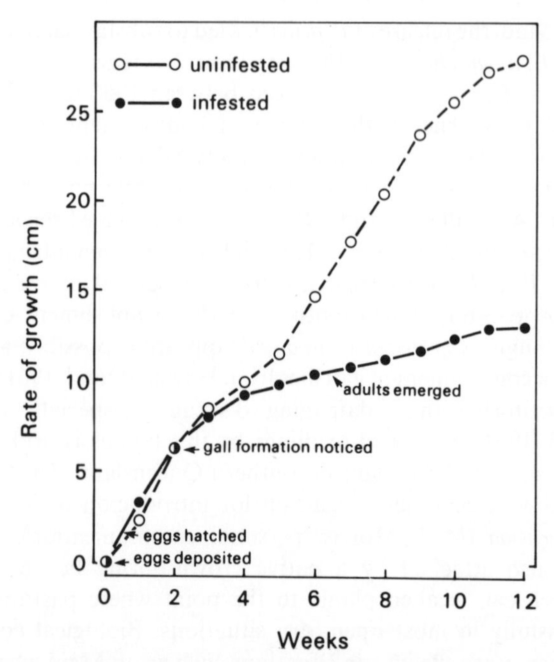

Figure 20.4. Comparative height growth of plants of *Eupatorium adenophorum* infested or not infested with the fly *Procecidochares utilis* grown in a glasshouse. The duration of the developmental period of the fly is shown (redrawn from Bess and Haramoto, 1959)

Table 20.1. Reproductive characteristics of *Eupatorium adenophorum* (*a*) and *E. odoratum* (*o*) in 'fallows' of different ages in northeastern India (from Table 3 of Ramakrishnan and Misra (1981) and Table 3 of Kushwaha *et al.* (1981) respectively)

	Age of 'fallow' (years)								
	1		3		6	5	10		15^+
	a	*o*	*a*	*o*	*a*	*o*	*a*	*o*	*a*
Plants/m²*	8	2	28	5	24	35	2	3	3
Capitula/m²	536	428	1988	14635	1287	910	62	—	—
Seeds/m²	33768	12818	123256	436185	79794	13975	3782	—	—

*Excluding seedlings
⁺20-year fallow measured and similar to 15 years for *a*

subtropical forest and/or to deliberately plant in chosen indigenous tree species to compete with other invasive plants such as *Cinnamomum camphora* or *Baccharis halimifolia.*

20.3.3 Control of *Acacia longifolia* and *Chrysanthemoides monilifera*

The two previous case histories have involved essentially one-way movements of invasive plants; my next case history is more a two-way exchange. In this section I shall review the ecology and control of several woody shrubs native to two regions of mediterranean-type climate. Some southeastern Australian species in the widespread genus *Acacia* (Mimosaceae) are invasive in South Africa, whilst two subspecies of the South African species *Chrysanthemoides monilifera* (Asteraceae) are invasive in southeastern Australia. The main coastal species of *Acacia* being displaced by *C. monilifera* ssp. *rotundata* is the structurally similar *Acacia longifolia* (Weiss and Noble, 1984a), which is itself strongly invasive in South African shrubland (Stirton, 1983), which, in turn, may contain *C. monilifera* ssp. *monilifera* as a common component.

Acacia longifolia varies in eastern Australia, especially in phyllode width and growth habit. Two distinct varieties are recognized, viz. *A. longifolia* var. *longifolia* (Sydney golden wattle), a tall shrub of coastal forests, and *A. longifolia* var. *sophorae* (coastal wattle), a low, bushy spreading shrub of coastal sand dunes. There is some doubt as to the taxon present in South Africa (Weiss, 1983) as var. *sophorae* is itself variable. The var. *sophorae*, although originally introduced deliberately to South Africa in the 1820s and 1830s (Shaughnessy, 1980) was first identified as an invasive plant only as recently as 1945 from a riverine occurrence (Boucher and Stirton, 1983), possibly because it can be confused in the vegetative state with *A. cyclops* and/or *A. saligna*. Despite the long time since its introduction, the distribution of *A. longifolia* is still 'relatively restricted' (Boucher and Stirton, 1983, p. 47). *A. longifolia* has also been introduced to several regions of the Americas and is naturalized in Uruguay, Argentina and California, but apparently not invasive there (Boucher and Stirton, 1983).

Two (ssp. *monilifera* and ssp. *rotundata*) of the six subspecies of *Chrysanthemoides monilifera* described from southern African material (Norlindh, 1943), have been introduced to Australia at various times. *C. monilifera* seems to be invasive only in Australia—I can find no reference to its occurrence elsewhere. *C. monilifera* ssp. *monilifera* occurs in coastal areas of Australia from about Sydney, New South Wales, to Streaky Bay in western South Australia (H. McBeth, personal communication), as well as in Tasmania and as isolated occurrences in southwest Western Australia and at inland sites in northern Victoria and southwestern New South Wales (Lane, 1984).

C. monilifera ssp. *rotundata* was introduced to Stockton, near Newcastle, in 1908 (Gray, 1976) whence it has both spread naturally and been planted deliberately (to stabilize sand dunes and to revegetate areas following mining) northwards and southwards of the original site along the eastern coastline of New South Wales and Queensland. The subspecies is now an invasive plant of major importance to nature conservation in this region, where it may co-occur with *C.*

monilifera ssp. *monilifera* on coastal dune systems, but usually seaward of the latter. Some hybridization may occur between the two subspecies (Weiss, 1983), but is of apparently limited occurrence both in South Africa and Australia.

Both species invade disturbed areas of natural vegetation where they may displace indigenous plants. Both species produce large numbers of long-lived dormant (hard) seeds which may be stored in the soil. Wildfires in both countries occur periodically and are usually of a sufficiently high intensity to stimulate the soil-stored seeds to overcome dormancy, to germinate prolifically and thereby to ensure seedling establishment. Alternatively, in the absence of fire, germination may occur after weathering of the fruits. Seeds of both species are also dispersed naturally by birds in both countries. It seems then as though *Acacia longifolia* and *Chrysanthemoides monilifera* are almost ecological analogues of one another in South Africa and Australia respectively, the only really significant difference for their control being that *C. monilifera* spp. *rotundata*, and to a lesser extent ssp. *monilifera*, are also able to regenerate by resprouting after fire.

Both species may be controlled mechanically—slashing of mature *Acacia longifolia* shrubs and some hand-pulling of young plants in South Africa, and hand-pulling, especially of seedlings, of *Chrysanthemoides monilifera* in south-eastern Australia, as mentioned earlier. These methods reduce the numbers of plants and if hand-pulling follows a fire, the method is probably effective at reducing the numbers of soil-stored seeds as well.

An appropriate fire regime can control *C. monilifera* effectively. For instance, Lane and Shaw (1978) showed that if a prescriptive fire of low intensity followed a wildfire of higher intensity, the plant density of *C. monilifera* ssp. *monilifera* was greatly reduced. Weiss (1983) advocated a similar fire regime to control *C. monilifera* in coastal New South Wales, provided the second fire could be timed before the commencement of flowering of the seedlings induced to germinate by the first fire. In this way, the numbers of soil-stored seeds could be considerably reduced. Alternatively, Weiss (1983) suggested an application of herbicide, either applied broadly over an invaded area or as a 'spot' spray, to replace the first fire. The use of a high-intensity fire followed by a second has yet to be tried for *A. longifolia* in South Africa.

As at March 1978 no herbicide was registered for chemical control of *A. longifolia* in the Cape Province (Stirton, 1983). Chemical control of *C. monilifera* is registered and recommended in Australia and involves a range of herbicides, dilution rates and methods of application (Cooney *et al.*, 1982; Love, 1984) to either mature plants or to seedlings. Profuse germination may follow such treatments and seedlings establish densely. Continued control measures are then needed for the next 3 or 4 years to reduce the seed supply in the soil if the method is to be effective on its own.

C. monilifera seedlings appear to establish only poorly on areas where competition from dense grasses and herbs is present, such as *Lomandra longifolia* (Weiss and Noble, 1984a); conversely, they are very vigorous in the absence of competition, e.g. on rocky outcrops, cliffs and pathways (Lane, 1984) and as initial colonizers after disturbance. When seedlings of both *A. longifolia* and *C. monilifera* were grown together in pots, *C. monilifera* was found to be more competitive than *A. longifolia* and this competitive advantage was lessened, but not reversed, under conditions of water stress (Weiss and Noble, 1984b). Apart from these results little is known of the characteristics of indigenous plants which may be successful competitors against either species.

A biological control programme for *Acacia longifolia* has been commenced by South Africa (Neser and Kluge, 1986) and a programme on *Chrysanthemoides monilifera* will commence shortly. In each case there is optimism about the chances for success (see, for example, Neser and Morris, 1984, for *C. monilifera*). One of the most hopeful candidates for *C. monilifera* control is a group of tephritid flies in the genus *Mesoclanis* (Munro, 1950) which render the seeds inviable while they are still on the mature shrubs.

Weiss and Milton (1984) provided an excellent quantitative basis to assess the effectiveness of any control programme (see Table 20.2). They tested the thesis that the reproductive output of the invader is higher than the indigene, using *A. longifolia* and *C. monilifera* in both Australia and South Africa. *A. longifolia* had about a thousand-fold fewer viable seeds in Australian soil than in South Africa and conversely, *C. monilifera* had about fifty-fold less viable seeds in South African soil than in Australia, although the number of whole seeds of *C. monilifera* in soil was similar in the two countries. Thus the level of predation of buried seeds was greater for *C. monilifera* in South Africa (see also Noble, this volume). For *A. longifolia* in Australia fewer seeds were incorporated into the soil seed pool because they had been preyed upon earlier in the life cycle. If control programmes for these two groups of plants, especially those integrating different control methods (Groves, 1984), could all be assessed in terms of the differential reproductive outputs presented in Table 20.2, then a truly ecological basis for control of invasive plants could be formulated.

Table 20.2. Reproductive characteristics of *Acacia longifolia* and *Chrysanthemoides monilifera* ssp. in Australia and South Africa (modified from Table 1 of Weiss and Milton, 1984)

	A. longifolia		*C. monilifera*	
	Australia	South Africa	Australia	South Africa
Main flowering time	Aug.–Oct.	July–Sept.	Apr.–Aug.	June–Sept.
Ripe seeds/m^2	364	2923	4450	2160
Soil seeds/m^2				
fragmented	25	—	6380	2352
whole	7.5	7600	2475	2320
viable	5.6	7370	2030	46

20.4 SUMMARY AND CONCLUSIONS

Of the invasive plants considered in the preceding sections some common attributes emerge.

1. They have nearly all been introduced deliberately and usually because of their perceived value to horticulture. The only exceptions seem to be the accidental introductions of *Hypericum perforatum* to South Africa and of *Chrysanthemoides monilifera* ssp. *rotundata* to Australia, although the latter was then spread deliberately because of its perceived value for sand stabilization.
2. Because they have usually been introduced many times, some genetic variation occurs, which may be expressed both morphologically and physiologically.
3. All species are early colonizers of disturbed sites in their countries of origin.
4. Their seedlings have a high growth rate which can be reduced substantially by shading.
5. Plants produce large numbers of seeds early in their life cycles and a proportion of seed is usually dormant.
6. They can reproduce vegetatively from perennial rootstocks once they are established, with the apparent exception of *Acacia longifolia.*
7. The leaves of *H. perforatum* and *E. adenophorum* can be toxic to domestic herbivores.

These plants thus have many of the characteristics of invasive plants generally (see Noble, this volume). Given these attributes in common, what aspects of their control may also be general? I shall present four principles of control which seem to be general for invasive plants in nature reserves.

1. One aspect of control, and a recurring theme throughout this review, is that only rarely is one control method effective in limiting the numbers of propagules of invasive plants per unit area. A rare exception seems to be the entomological control of *Opuntia* spp. by *Cactoblastis* and this successful case history has been so widely quoted as to give an incorrect interpretation for the success of entomological control methods generally. More often, as the case histories presented previously show, it has been a combination of methods which has led to effective control of terrestrial invasive plants. Examples of a conscious integration of methods for control of invasive plants in natural systems are few (see Kluge *et al.*, 1986) and even more rarely is such an integrative approach itself integrated with the dynamics of the natural ecosystem being invaded.
2. Control methods which simplify the ecosystem and reverse the trend towards diversification of the system seem to be more prone to subsequent invasion by other groups of invasive plants, the end result of such actions being to replace one invasion with another. On this basis control methods which add to diversity, e.g. a marked reduction in fire frequency, deliberate promotion of

competing plants, use of arthropods and/or fungi, have the potential to produce an ecosystem which may be better able to resist further invasion.

3. Once initiated, control methods have to be maintained (see also Macdonald *et al.*, this volume). This need for continued action has not always been recognized and a short-term interruption to a control programme can have disastrous consequences in the longer term, e.g. as with chemical control of *Cytisus scoparius* in a national park in eastern Australia (see earlier).
4. An aspect of previous research on control of invasive plants is that the monitoring of control is sometimes inadequate, either because it is not done at all or else it is done for too short a time. A further deficiency is that often the index measured may not always be the critical one by which to assess the effectiveness of control in ecological terms. As a previous section showed, a control programme which measures reduction in yield of an invasive plant such as *Hypericum perforatum* may be appropriate in pasture research but one which measures the number and viability of seeds per unit volume of soil and the change in cover of associated species (see Huffaker and Kennett, 1959) may be more appropriate for *H. perforatum* control in a nature reserve.

Invasions have been occurring naturally for millennia. This chapter has been concerned with a few more recent invasions of plants not indigenous to the region being invaded. As reserves of natural ecosystems become increasingly the only remnants of vegetation types formerly widespread, a study of invasive plant control becomes more urgent if those ecosystems are to be retained and conserved for the future. In this contribution I have assessed internationally the present status of research on some examples of invasive plant control in three regions—temperate grassland/woodland, subtropical forests and mediterranean-climate shrublands. If other plant invasions in other regions are to have a similar potential for successful control a major requirement is to know more of the ecology of the species or species aggregate in its country of origin. For obvious reasons researchers in biological control have been better able to contribute to such knowledge. But legislators, appliers of herbicides and manipulators of vegetation dynamics also need to be more international in their approach to a control programme for terrestrial invasive plants. I hope this review may be catalytic in bringing about such a widening of outlook and approach to all controllers of invasive plants.

ACKNOWLEDGMENTS

I thank Bruce Auld, David Briese, Professor P. S. Ramakrishnan and Paul Weiss for comments on a first draft of parts of this chapter and Tricia Kaye for expert help in producing the final copy against a tight deadline.

REFERENCES

Amor, R. L., and Twentyman, J. D. (1974). Objectives of, and objections to, Australian noxious weed legislation. *J. Aust. Inst. Agric. Sci.*, **40**, 194–203.

Anonymous (1977). *Selected References to the Biology and Control of Eupatorium spp.* Agric. Res. Council, Weed Res. Org., Annot. Bibliog. No. 109.

Auld, B. A. (1969). The distribution of *Eupatorium adenophorum* Spreng. on the far north coast of New South Wales. *J. Proc. R. Soc. N. S. W.*, **102**, 159–61.

Auld, B. A. (1972). Chemical control of *Eupatorium adenophorum*, crofton weed. *Trop. Grassl.*, **6**, 55–60.

Auld, B. A., and Martin, P. M. (1975). The autecology of *Eupatorium adenophorum* Spreng. in Australia. *Weed Res.*, **15**, 27–31.

Baird, A. M. (1977). Regeneration after fire in King's Park, Perth, Western Australia. *J. R. Soc. W. Aust.*, **60**, 1–22.

Bess, H. A., and Haramoto, F. H. (1959). Biological control of pamakani, *Eupatorium adenophorum*, in Hawaii by a tephritid gall fly, *Procecidochares utilis*. 2. Population studies of the weed, the fly and the parasites of the fly. *Ecology*, **40**, 244–9.

Bess, H. A., and Haramoto, F. H. (1972). Biological control of pamakani, *Eupatorium adenophorum*, in Hawaii by a tephritid gall fly *Procecidochares utilis*. 3. Status of the weed, fly and parasites of the fly in 1966–71 versus 1950–57. *Proc. Hawaii Entomol Soc.*, **21**, 165–78.

Boucher, C., and Stirton, C. H. (1983). Long-leaved wattle. In: Stirton, C. H. (Ed.), *Plant Invaders. Beautiful but Dangerous*, pp. 44–7. 3rd Edn. Dept Nature Env. Conserv., Cape Prov. Admin., Cape Town.

Campbell, M. H., and Delfosse, E. S. (1984). The biology of Australian weeds. 13. *Hypericum perforatum* L. *J. Aust. Inst. Agric. Sci.*, **50**, 63–73.

Campbell, M. H., Dellow J. J., and Gilmour, A. R. (1979). Effect of time of application of herbicides on the long-term control of St. John's wort (*Hypericum perforatum* var. *angustifolium*). *Aust. J. Exp. Agric. Anim. Husb.*, **19**, 746–8.

Christensen, P. E., and Burrows, N. D. (1986). Fire: an old tool with a new use. In: Groves, R. H. and Burdon, J. J. (Eds.), *Ecology of Biological Invasions: An Australian Perspective*, pp. 97–105. Australian Academy of Science, Canberra.

Clark, N. (1953). The biology of *Hypericum perforatum* L. var. *angustifolium* DC (St John's wort) in the Ovens Valley, Victoria, with particular reference to entomological control. *Aust. J. Bot.*, **1**, 95–120.

Clements, F. E., and Shelford, V. E. (1939). *Bio-ecology*, Wiley, New York. 425 pp.

Cooney, P. A., Gibbs, D. G., and Golinski, K. D. (1982). Evaluation of the herbicide "Roundup" for control of bitou bush (*Chrysanthemoides monilifera*) *J. Soil Conserv. Serv. N. S. W.*, **38**, 6–12.

Crawley, M. J. (1986). The population biology of successful invaders. In: Kornberg, H. and Williamson, M. H. (Eds), *Quantitative Aspects of the Ecology of Biological Invasions*, pp. 209–29. Royal Society, London.

Dahlsten, D. L. (1986). Control of invaders. In: Mooney, H. A. and Drake, J. A. (Eds), *The Ecology of Biological Invasions of North America and Hawaii*, pp. 275–302. Springer-Verlag, Berlin and New York.

Davey, H. W. (1917). Weeds. St. John's wort. *J. Agric. Vic.*, **15**, 427–34.

Davey, H. W. (1919). Experiments in the control of St. John's wort. *J. Agric. Vic.*, **17**, 378–9.

Davis, C. J., and Krauss, N. L. H. (1967). Recent introductions for biological control in Hawaii. XII. *Proc. Hawaii Entomol. Soc.*, **20**, 375–80.

Dodd, A. P. (1961). Biological control of *Eupatorium adenophorum* in Queensland. *Aust. J. Sci.*, **23**, 356–66.

Dodd, S. (1920). St. John's wort and its effects on livestock. *Agric. Gaz. N. S. W.*, **31**, 265–72.

Edwards, A. W. A. (1977). The ecology of *Eupatorium odoratum* L. VI. Effects of habitats on growth. *Int. J. Ecol. Environ. Sci.*, **3**, 17–22.

Edwards, A. W. A., and Stephenson, S. N. (1974). The ecology of *Eupatorium odoratum* L. II. Physiological studies of *E. odoratum* from the new and old world tropics. *Int. J. Ecol. Environ. Sci.*, **1**, 97–105.

Erasmus, D. J., and van Staden, J. (1986a). Germination of *Chromolaena odorata* (L.) K. fR. achenes: effect of temperature, imbibition and light. *Weed Res.*, **26**, 75–81.

Erasmus, D. J., and van Staden, J. (1986b). Screening of candidate herbicides in field trials for chemical control of *Chromolaena odorata. S. Afr. J. Plant Soil*, **3**, 66–70.

Gill, A. M. (1975). Fire and the Australian flora: a review. *Aust. Forestry*, **38**, 4–25.

Gray, M. (1976). Miscellaneous notes on Australian plants. 2. *Chrysanthemoides* (Compositae). *Contr. Herb. Aust.*, **16**, 1–15.

Groves, R. H. (1984). Ecological consequences of biological control of *Chrysanthemoides monilifera*. In: Love, A. and Dyason, R. (Eds), *Bitou Bush and Boneseed: A National Conference on Chrysanthemoides monilifera*, pp. 111–16. NSW Natl Parks Wildlife Service and Dept of Agric., Sydney.

Groves, R. H. (1986). Plant invasions of Australia: an overview. In: Groves R. H., and Burdon, J. J. (Eds), *Ecology of Biological Invasions: An Australian Perspective*, pp. 137–49. Australian Academy of Science, Canberra.

Groves, R. H., and Burdon, J. J. (Eds), (1986). *Ecology of Biological Invasions: An Australian Perspective.* Australian Academy of Science, Canberra. 166 pp.

Gunn, C. R., Ritchie, C. A., and Poole, L. (1981). *Exotic Weed Alert. Noxious Weed Printout*. USDA, Beltsville.

Harris, P., Peschken, D. P., and Milroy, J. (1969). The status of biological control of the weed *Hypericum perforatum* in British Columbia. *Can. Ent.*, **101**, 1–15.

Healy, A. J. (1972). Weedy St. John's worts (*Hypericum* spp.) in New Zealand. *Proceedings 25th N. Z. Weed Pest Control Conference*, pp. 180–90.

Holm, L., Pancho, J. V., Herberger, J. P., and Plucknett, D. L. (1979). *A Geographical Altas of World Weeds.* Wiley, New York, 391 pp.

Holmgren, I. (1919). Zytologische Studien uber die Fortpflanzung bei den Gattungen *Erigeron* und *Eupatorium. Kgl Svenska Veternskapskad. Handl.*, **59**, 1–117.

Howarth, F. G. (1983). Classical biocontrol: panacea or Pandora's box. *Proc. Hawaii Entomol. Soc.* **24**, 239-244.

Huffaker, C. B. (1967). A comparison of the status of biological control of St. John's wort in California and Australia. *Mushi*, **39**, (Suppl.), 51–73.

Huffaker, C. B., and Kennett, C. E. (1959). A ten-year study of vegetational changes associated with biological control of Klamath weed. *J. Range Manage.*, **12**, 69–82.

Julien, M. H. (1982). *Biological Control of Weeds. A World Catalogue of Agents and Their Target Weeds.* Commonw. Agric. Bur., Slough, UK. 108 pp.

King, R. M., and Robinson, H. (1970a). Studies in the Eupatorieae (Compositae). XXIX. The genus *Chromolaena. Phytologia*, **20**, 196–209.

King, K. M., and Robinson, H. (1970b). Studies in the Eupatorieae (Compositae). XIX. New combinations in *Ageratina. Phytologia*, **19**, 208–29.

Kluge, R. L., Zimmermann, H. G., Cilliers, C. J., and Harding, G. B. (1986). Integrated control for invasive alien weeds. In: Macdonald, I. A. W., Kruger, F. J., and Ferrar, A. A. (Eds), *Ecology and Management of Biological Invasions in Southern Africa*, pp. 294–302. Oxford University Press, Cape Town.

Kornberg, H., and Williamson, M. H. (Eds), (1986). *Quantitative Aspects of the Ecology of Biological Invasions.* Royal Society, London, 240 pp.

Kruger, F. J. (1977). Ecology of Cape fynbos in relation to fire. In: Mooney, H. A., and Conrad, C. E. (Eds), *Symposium on the Environmental Consequences of Fire and Fuel*

Management in Mediterranean Ecosystems, pp. 230–44. USDA Forest Serv. Gen. Tech. Rept WO-3.

Kushwaha, S. P. S., Ramakrishnan, P. S., and Tripathi, R. S. (1981). Population dynamics of *Eupatorium odoratum* in successional environments following slash and burn agriculture. *J. Appl. Ecol.*, **18**, 529–35.

Lane, D. W. (1984). The current status of boneseed within Australia. In: Love, A., and Dyason, R. (Eds), *Bitou Bush and Boneseed: A National Conference on Chrysanthemoides monilifera*, pp. 65–8. NSW Natl Parks Wildlife Service and Dept of Agric., Sydney.

Lane, D. W., and Shaw, K. (1978). The role of fire in boneseed (*Chrysanthemoides monilifera* (L.) Norl.) control in bushland. *Proceedings of 1st Conference of the Council of Australian Weed Science Society*, pp. 333–5. Melbourne, Victoria.

Love, A. (1984). Bitou bush and boneseed current control techniques and programmes. In: Love, A. and Dyason, R. (Eds), *Bitou Bush and Boneseed: A National Conference on Chrysanthemoides monilifera*, pp. 95–104. NSW Natl Parks Wildlife Service and Dept of Agric., Sydney.

Macdonald, I. A. W., Kruger, F. J., and Ferrar, A. A. (Eds) (1986). *Ecology and Management of Biological Invasions in South Africa.* Oxford University Press, Cape Town. 304 pp.

Mann, J. (1970). *Cacti Naturalised in Australia and Their Control.* Govt Printer, Brisbane. 128 pp.

Medd, R. W., and Smith, R. C. G. (1978). Prediction of potential distribution of *Carduus nutans* (nodding thistle) in Australia. *J. Appl. Ecol.*, **15**, 603–12.

Mooney, H. A., and Drake, J. A. (Eds.) (1986). *Ecology of Biological Invasions of North America and Hawaii.* Springer-Verlag, Berlin and New York, 321 pp.

Moore, R. M. (1971). Weeds and weed control in Australia. *J. Aust. Inst. Agric. Sci.*, **37**, 181–91.

Moore, R. M., and Cashmore, A. B. (1942). The control of St. John's wort (*Hypericum perforatum* L. var. *angustifolium* D. C.) by competing pasture plants. *Coun. Sci. Industr. Res. Aust. Bull.*, No. 151.

Munro, H. K. (1950). Trypetid flies (Diptera) associated with the Calendulae plants of the family Compositae in South Africa. 1. A bio-taxonomic study of the genus *Mesoclanis*. *J. Ent. Soc. Sthn Africa*, **13**, 37–52.

Navaratnam, S., and Catley, A. (1986). Quarantine measures to exclude plant pests. In: Groves, R. H., and Burdon, J. J. (Eds), *Ecology of Biological Invasions: An Australian Perspective*, pp. 106–12. Australian Academy of Science, Canberra.

Neser, S., and Kluge, R. L. (1986). The importance of seed-attacking agents in the biological control of invasive alien plants. In: Macdonald, I. A. W., Kruger, F. J., and Ferrar, A. A., (Eds), *Ecology and Management of Biological Invasions in Southern Africa*, pp. 285–93. Oxford University Press, Cape Town.

Neser, S., and Morris, M. J. (1984). Preliminary observations on natural enemies of *Chrysanthemoides monilifera* in South Africa. In: Love, A., and Dyason, R. (Eds), *Bitou Bush and Boneseed: A National Conference on Chrysanthemoides monilifera*, pp. 105–9. NSW Natl Parks Wildlife Service, and Dept of Agric., Sydney.

Norlindh, T. (1943). *Studies in the Calendulae. 1. Monograph of the genera Dimorphotheca. Castalis. Osteospermum. Gibbaria and Chrysanthemoides*, pp. 374–99. Gleerup, Lund.

Parsons. W. T. (1973). *Noxious Weeds of Victoria.* Inkata Press, Melbourne. 300 pp.

Patterson, D. T. (1983). Research on exotic weeds. In: Wilson, C. L., and Graham, C. L. (Eds), *Exotic Plant Pests and Northern American Agriculture*, pp. 381–93. Academic Press, New York.

Ramakrishnan, P. S., and Misra, B. K. (1981). Population dynamics of *Eupatorium adenophorum* Spreng. during secondary succession after slash and burn agriculture (jhum) in north eastern India. *Weed Res.*, **22**, 77–84.

Ramakrishnan, P. S., Toky, O. P., Misra, B. K., and Saxena, K. G. (1981). Slash and burn agriculture in northeastern India. In: Mooney, H. A., Bonnicksen, T. M., Christensen, N. L., Lotan, J. E., and Reiners, W. A. (Eds), *Proceedings of Conference on Fire Regimes and Ecosystem Properties*, pp. 570–86. Gen. Tech. Rept WO-26, USDA, Washington DC.

Rao, V. P., Ghani, M. A., Sankaran, T., and Mathur, K. C. (1971). *A Review of the Biological Control of Insects and Other Pests in South-East Asia and the Pacific Region.* Comm. Inst. Biol. Contr. Tech. Commun. No. 6.

Robson, N. K. B. (1968). Guttiferales. CIX. Guttifereae (Clusiaceae). In: *Flora Europaea 2*, pp. 261–9. Cambridge University Press, Cambridge.

Salisbury, E. J. (1942). *The Reproductive Capacity of Plants.* G. Bell and Sons, London. 244 pp.

Shaughnessy, G. L. (1980). Historical ecology of alien woody plants in the vicinity of Cape Town, South Africa. Ph.D. thesis, University of Cape Town.

Stirton, C. H. (Ed.) (1983). *Plant Invaders. Beautiful, but Dangerous*, 3rd Edn. pp. 84–7. Dept Nature Env. Conserv., Cape Prov. Admin., Cape Town.

Tisdale, E. W., Hironaka, M., and Pringle, W. L. (1959). Observations on the autecology of *Hypericum perforatum. Ecology*, **40**, 54–62.

Usher, M. B. (1986). Invasibility and wildlife conservation: invasive species on nature reserves. In: Kornberg, H., and Williamson, M. H. (Eds.), *Quantitative Aspects of the Ecology of Biological Invasions*, Royal Society, London pp. 193–208.

Villanueva H. L., and Fauré, G. O. (1959). Biological control of St. Johns-wort in Chile. *FAO Plant Prot. Bull.*, **7**, 144–6.

Vogl, R. J. (1977). Fire frequency and site degradation. In: Mooney, H. A., and Conrad, C. E. (Eds), *Symposium on the Environmental Consequences of Fire and Fuel Management in Mediterranean Ecosystems*, pp. 193–201. USDA Forest Serv. Gen. Tech. Rept WO-3.

Weiss, P. W. (1983). Invasion of coastal *Acacia* communities by *Chrysanthemoides*. Ph.D. thesis, Australian National University.

Weiss, P. W., and Milton, S. J. (1984). *Chrysanthemoides monilifera* and *Acacia longifolia* in Australia and South Africa. In: Dell, B. (Ed.), *Proceedings of 4th International Conference on Mediterranean Ecosystems*, pp. 159–60. University of Western Australia Botany Dept, Perth.

Weiss, P. W., and Noble, I. R. (1984a). Status of coastal dune communities invaded by *Chrysanthemoides monilifera. Aust. J. Ecol.*, **9**, 93–8.

Weiss, P. W., and Noble, I. R. (1984b). Interactions between seedlings of *Chrysanthemoides monilifera* and *Acacia longifolia. Aust. J. Ecol.*, **9**, 107–15.

Williams, J. D., and Groves, R. H. (1980). The influence of temperature and photoperiod on growth and development of *Parthenium hysterophorus* L. *Weed Res.*, **20**, 47–52.

Zimmermann, H. G., Moran, V. C., and Hoffmann, J. H. (1986). Insect herbivores as determinants of the present distribution and abundance of invasive cacti in South Africa. In: Macdonald, I. A. W., Kruger, F. J., and Ferrar, A. A. (Eds), *Ecology and Management of Biological Invasions in Southern Africa*, pp. 269–74. Oxford University Press, Cape Town.

Zinke, P. J., Sabhasri, S., and Kundstadter, P. (1978). Soil fertility aspects of the Lua forest fallow system of shifting cultivation. In: Kundstadter, P., Chapman, E. C., and Sabhasri, S. (Eds), *Farmers in the Forest.* East-West Center, Honolulu.

Biological Invasions: a Global Perspective
Edited by J. A. Drake *et al.*
© 1989 SCOPE. Published by John Wiley & Sons Ltd

CHAPTER 21

Ecological Effects of Controlling Invasive Terrestrial Vertebrates

MICHAEL B. USHER

21.1 INTRODUCTION

21.1.1 The perception of the problem

In concluding the British meeting of the SCOPE programme, Holdgate (1986) posed three questions, one of which was 'what management strategies are appropriate to control invading species?' In reply, he said of Great Britain 'very little consideration is given to management questions'. In reviewing the worldwide literature on introduced vertebrates, it is apparent that there is a wealth of data documenting introduced and invasive species, far fewer data recording accurately the habitats that are invaded, very few data on the subject of controlling the introduced species, and extreme scarcity of data on the effects of control on the biota of the invaded environment. Both Brown (this volume) and Ehrlich (1986, this volume) have documented invasive vertebrate species, and hence this paper will concentrate on their management.

Although there are relatively few published data on management, conservationists, in particular, are aware of the potential effects of invasive species and of the formidable management problems that they can pose. An example of this concern is Henderson Island, an uninhabited island of about 4000 ha in the Crown Colony of the Pitcairn Islands (South Pacific; 24° 22′ S, 128° 18′ E). Four of the 17 species of birds breeding on the island are endemic (Bourne and David, 1983, 1985, 1986), as are 10 of the 63 species of ferns and flowering plants. In April and May 1983 many conservation organizations lobbied the British Government to prevent the development of an airstrip and private recreational facilities on the island. The case was made on the grounds of the effects of invasive species (especially black or brown rats) coming with human visitors and of the management impracticability of zoning the island into separate developed and conserved sections. Eventually, Her Majesty's Government's reply (*House of Lords Hansard*, 15 December 1983, col. 385) rejected the development proposals 'on administrative and environmental grounds'.

Another example of the perception of the problem of controlling invasive mammals is the management policy of Barrow Island (Western Australia) during the development, since 1966, of a commercial oil field (Butler, 1987). This island has a varied community of native marsupial mammals (Butler, 1970), and stringent, almost draconian, powers are available to the oil industry's management to conserve the flora and fauna and prevent the colonization of the island by non-native species.

Both the Henderson Island and Barrow Island examples show that the perception is that it is extremely difficult, or impossible, to control invasive vertebrates before they have damaged, often irreparably, the native biota of islands. Perception can, however, be dangerous; the recent discussions of *Rattus* species have indicated that colonization by *R. exulans* (Polynesian rat) may not necessarily lead to the loss of endemic island species (*R. exulans* exists on Henderson Island, for example), but that colonization by either *R. rattus* (black rat) or *R. norvegicus* (brown rat) is more likely to lead to species losses (Atkinson, 1985). However, Diamond (1985) considered that it is not inevitable that these *Rattus* species would arrive on all rat-free islands. Despite these comments, case studies from around the world, and especially from oceanic islands, have shown that native populations can be greatly reduced or exterminated by invasive mammals, but equally Holdgate's (1986) remarks indicate that control measures have often not been either attempted or completely documented. The aims of this chapter are to investigate how frequently control measures have been attempted and to ask how frequently control might be successful.

21.1.2 The scope of this paper

The word 'control' is not necessarily used in the sense of 'eradication'; once an invasive species is established, eradication may well be impossible. Control will, however, refer to the management of that species so that its effect on other biota, or on the environment, are acceptable. Such a definition begs one question; to whom are the effects 'acceptable'? Acceptability has to be defined in relation to the aims of managing the piece of land that carries the invasive population. The situation is analogous to pest management; the manager has to decide how much inconvenience is acceptable, and once that has been decided a management strategy can be devised. Hence, in the remainder of this paper, control will relate to the management of the population of the introduced species, not necessarily to its eradication.

Management can have three different kinds of effects. First, there are the effects on the alien species itself. There are very few studies that relate to this effect, especially for vertebrates, but they are reviewed in Section 21.2. Second, there are the effects on the other biota in the community that has been invaded. The majority of studies address these effects, and these are reviewed in Section 21.3. Third, there are the effects on the abiotic environment. There are very few data that could be used to investigate these effects. Indeed Verkaik (1987) specifically stated that the muskrat control programme in the Netherlands had failed to gather reliable data on the effects of control on the 'burrow-digging time'. In Australia, Backhouse (1987) indicated that invasion of nature reserves by plant and animal pests led to 'habitat degradation'. However, due to the lack of precise data, this third set of effects will not be addressed further.

After reviewing the effects of management on both the invader and the invaded, the final Sections, 21.4 and 21.5, aim to make some prognoses and draw some conclusions. The reasons for wishing to control invasive species are frequently related to conservation activities, and hence the prognoses will be based not only on the review material in Sections 21.2 and 21.3, but also on the data compiled by the Working Group studying Invasions into Nature Reserves (see Macdonald *et al.*, this volume, and Usher (1988)).

21.1.3 The frequency of control measures

If the terrestrial vertebrates are divided into three groups—mammals, birds, herpetofauna—one finds some interesting differences in the approaches of land managers to the control of invasive species. All three groups feature prominently in lists of introduced species, e.g. in Moulton and Pimm's (1986) study of introduced species in Hawaii, they recorded that 22 out of 53 passerine bird species introductions on individual islands persisted, as did 68 out of 74 mammal species introductions and 48 out of 53 reptile species introductions. Mooney *et al.* (1986) recorded 17, nine and five species of invading mammals, birds and herpetofauna respectively in California. In southern Africa, the situation is similar, with species of these three terrestrial vertebrate groups being introduced (Brown, 1985; Griffin and Panagis, 1985; Brooke *et al.*, 1986). However, although all three groups have been introduced, the approach to control differs.

There seems to be almost no control of invasive amphibians and reptiles, although control measures may need to be adopted in the future for some species. *Bufo marinus* (giant toad) was introduced into Queensland from South America in 1935 to control coleopteran pests of sugar cane, but it is now posing a threat to native fauna. Floyd and Easteal (1986) considered two strategies for its control; first, a campaign to prevent its further spread by prohibiting transportation of produce and agricultural machinery from toad-inhabited to toad-free areas, and, second, by reducing the population density in toad-inhabited areas. The former is still being discussed, and methods for the latter have not yet been investigated in Australia.

There are rather more data for the control of birds, some of which can be considerable pests, e.g. the control of *Quelea quelea* (red-billed quelea), a native but invasive pest of grain crops in Africa (Ward, 1979). In some instances an active form of control is not wanted since birds provide a recreational activity for a portion of the population. This is the case, in Sweden, of *Branta c. canadensis* (Canada goose) which provides sport for the wildfowlers (Fabricius, 1983). Casual control of birds is more frequent. In Australia, for example, farmers in particular have shot many of the 27 species that have become established (Newsome and Noble, 1986). Active control operations of introduced bird species are rare; Brooke *et al.* (1986) recorded the unsuccessful attempt to eradicate *Corvus splendens* (Indian house crow) from Zanzibar, whilst Falla *et al.* (1966) reported the deliberate extirpation of *Pycnonotus cafer* (red-vented bulbul) from New Zealand. In the Hawaiian Islands, Berger (1981) recorded the attempted extermination of *Urocissa erythrohyncha* (red-billed blue magpie); S. L. Pimm (personal communication) is reasonably certain that the campaign was successful as there are no recent sightings of the species. Several single individuals of *Sturnus vulgaris* (starling) have been seen in the Hawaiian Islands, and all have been shot (S. L. Pimm, personal communication) so that the species has never become established. Despite these examples, the usual attitude towards invasive birds tends to be *laissez-faire*, and can be summed up by quoting a discussion of *Passer domesticus* (house sparrow) in Africa (Brown, 1985): 'It would be almost impossible to eliminate this species from SWA/Namibia, as recolonization from adjacent areas would probably take place as quickly as areas could be cleared. No control of this species has been attempted, and none is recommended.'

There are many more examples of the control of invasive mammals. Considerable experience has been gained in New Zealand on their control (Lockley, 1970; King, 1984) and the most successful campaigns have been on islands where the aim was total eradication. Although New Zealand scientists may have pioneered invasive mammal control methods (see, for example, the review by Wodzicki and Wright (1984)), examples of such control can be found from all continents except Antarctica, which has yet to be invaded successfully by species such as *Rattus norvegicus* (brown rat) or *Mus musculus* (house mouse). A number of these species has, however, become established on subantarctic islands, and on these control methods appear to have been reasonably successful. An example is the control of *Felis catus* (feral cat) on Marion Island by the feline panleucopaenia disease (Aarde and Skinner, 1981; Aarde, 1984; Howell, 1984; Rensburg, 1985a).

Although there are now a considerable number of examples of the control of invasive mammals, there are few examples of the control of invasive birds, reptiles or amphibians. The remainder of this paper will, therefore, concentrate on the control of mammals.

21.2 THE INVASIVE SPECIES

In general, management tends to concentrate on the ecosystem or on the rare and endangered species which it is hoped to help, and only to think of the invasive species as something to eliminate. One could view this as the 'only good rat is a dead rat' mentality. However, before analysing the effects of control on species and ecosystems, it is as well to ask what effects the control measures will have on the invasive species itself. A perception of the possible effects may be important in determining on appropriate management strategy. In this section two possible effects will be considered—ecological and genetical—though the concept of the two often overlap.

21.2.1 How is the invasive species' ecology affected by control?

Although this question has not been studied in detail, it is apparent that control measures can result in the realization of a larger than normal reproductive capacity. *Myocastor coypus* (coypu) populations in England have a large proportion of young animals, which Gosling *et al.* (1981) attributed to the intensive and prolonged trapping. Verkaik (1987) found that *Ondatra zibethicus* (muskrat) in the Netherlands had a relatively large reproductive capacity, had a large proportion of the females reproducing, and had a high juvenile survival rate. She also attributed these findings to the control campaign. The data for both the coypu and muskrat indicate that, at the relatively low densities maintained by management, the populations are not being regulated by density dependent effects which are likely to operate in higher density, natural populations.

Another interesting effect could be thought of as a very short-term co-evolutionary process. Gerell (1985) recorded the behaviour of *Mustela vison* (mink) and the breeding success of *Somateria mollissima* (common eider) on islands in southern Sweden. Invasive mink caused widespread destruction of colonies of breeding birds, which were subsequently only able to nest successfully on the outer, mink-free islands (R. Gerell, personal communication). Later, the birds have started to nest again on islands occupied by a stationary population of mink, and it thus appears that co-existence of the birds and mink is possible. R. Gerell (personal communication) considers that an ecological effect of controlling mink would be a delay in the adaptation of the sea birds to the new predator.

Although there are relatively few data, control of an invasive mammal may mean that, although many adults are removed, the remaining individuals realize a large reproductive rate and have a greater juvenile survival rate. As demonstrated for the mink in Scandinavia (R. Gerell, personal communication), this large reproductive rate means that the effects of a campaign to control an invasive mammal can be of short duration. Resurgence of *Oryctolagus cuniculus* (rabbit) after myxomatosis, or of coypu after a succession of cold winters, indicates that the ecological effects of control on the invasive species itself need further study. In the terms of the models discussed by Williamson (this volume), what is the effect of control on the life history and demography, represented either r or by R_0, of the population?

21.2.2 Are there genetical effects?

Control changes the environment of the invasive species such that the great majority of the population is at risk and only a few individuals survive. If these few individuals are a truly random sample of the whole population, there may be no genetical influence of the control other than a possible loss of genetic variability. If, on the other hand, the few individuals surviving have special characteristics that increased their probability of survival, then one has a selective pressure and the possibility of genetic change conferring increased resistance in the invasive species.

An example of such adaptive change is the control of *Rattus norvegicus* (brown rat) by the anticoagulant poison warfarin. The use of warfarin began in 1950, and the first case of resistance was reported in Scotland in 1958. By 1972, there were three well-established resistant populations in Great Britain (Berry, 1977). Resistance is determined by a single dominant gene, and homozygous animals with this gene require about 20 times more vitamin K, which is an essential part of the blood-clotting mechanism, than ordinary rats. Berry (1977) discussed the engima of why, in the resistance area, the incidence of resistant rats stabilized at about 50% of the population, and he argued that this was a trade-off between the benefits of resistance and the disadvantage of large vitamin K requirements. Genetical aspects of warfarin resistance are discussed by Bishop *et al.* (1977) who showed that the heterozygote for the resistant allele, Rw_2, may be at a selective disadvantage in the absence of warfarin.

Another example of genetic change of an invasive species relates to the control of *Oryctolagus cuniculus* (rabbit) by the myxomatosis virus. When this virus was first spread in Britain in 1953, rabbit mortality was about 99%; the survivors had contracted the disease but recovered (Fenner, 1983). After six ensuing epidemics, mortality rates dropeed to about 40%. At first sight this may appear as a genetic change in susceptibility in the rabbit population, but such a change is confounded by decreasing viral virulence (Berry, 1977) and by a behavioural change in the rabbits (L. M. Cook, personal communication). In Britain, unlike Australia (Ross and Tittensor, 1986), myxomatosis is transmitted by *Spilopsyllus cuniculi* (rabbit flea) in burrows, but as the myxomatosis epidemics occurred rabbits tended to den less frequently in burrows and more frequently in open habitats where there was less contact with fleas. There is, however, still controversy (discussion of Ross and Tittensor, 1986) as to whether the behavioural change affected the rate of myxomatosis transmission.

These two examples, of control by chemical and disease agents, indicate that there may be genetical consequences which will, in time, make control successively more difficult. Although the whole subject of resistance to poisons and diseases is well documented in the literature on the control of invertebrate pests and microbes, it is less well known for invasive vertebrates. However, scenarios can be envisaged for selection pressures resulting from many forms of control. Consider, for example, a population of trap-happy and trap-shy animals. If all of the former are trapped and killed, the proportion of trap-shy animals in the population would increase with the subsequent difficulty in controlling that population (assuming, of course, that such a behavioural trait can be inherited).

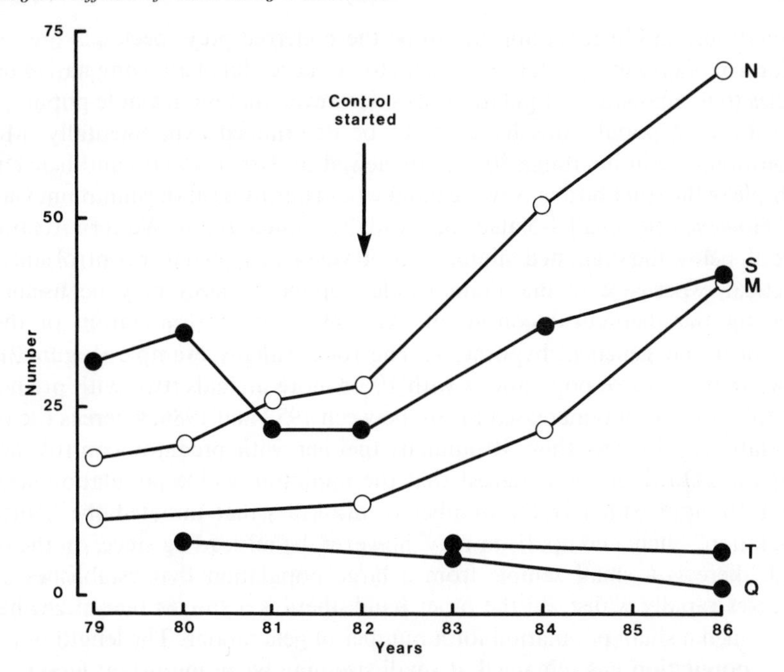

Figure 21.1. Population estimates of *Petrogale lateralis* (rock wallaby) on five rocky hills in the wheatbelt of Western Australia. The fox and feral cat populations were controlled on two of the hills (M and N) between 1982 and 1986, whereas the populations were uncontrolled on hills Q, S, and T. Between 1982 and 1986, rock wallaby populations increased by an average of 10% on hills where predators were controlled and by an average of 43% on hills without control (unpublished data provided by J. E. Kinnear)

cats. Extermination of the cats resulted in a rapid six-fold increase in the stitchbird population, which was then used to re-introduce the species to islands where it had previously occurred. There appears to be no suggestion that genetical variation in the stitchbird population had been lost.

On subantarctic Marion Island the feral cat population has been reduced by the spread of a feline parvovirus. Between 1974 and 1976, when cats were abundant, van Aarde (1980) estimated that a single cat would kill an average of 213 petrels (*Halobaena, Pachyptila, Procellaria* and *Pterodroma* species) per year. However, after control of the cats, field data collected between 1981 and 1983 (van Rensburg, 1958b) indicated that the residual cats were replacing petrels in their diet with *Mus musculus* (house mouse) and that there was now no evidence of predation on *Pterodroma brevirostris* (Kerguelen petrel). The interesting feature is that the introduced cat, at low population density, was feeding mainly on the introduced mouse, and was having a smaller effect on the native nesting bird species. The total consumption of mice was similar in high and low density cat populations, and hence it appears to be the preferred prey species.

Control of invasive predators appears to be successful in allowing native prey species to increase their population sizes. However, minimum viable population sizes for field populations have yet to be determined experimentally. Most suggestions are in the range 50–500 (reviewed by Usher, 1987b) and hence the example of the stitchbird in New Zealand would hardly be at the minimum viable size. However, the small, isolated rock wallaby populations in Western Australia were all below the suggested minimum of 50 when alien predator control started. Practical experience of minimum viable population sizes may be useful in assessing the theoretical considerations that form the foundation of these ecological and genetical hypotheses. The rock wallaby example (Figure 21.1) shows that the three populations with 10 or more animals (two with predator control, one without) increased in size between 1982 and 1986, whereas the two populations with less than 10 animals (neither with predator control) both decreased. Do these data suggest that the minimum viable population size is about 10, approximately the number of muskrats that initiated the Scottish population? Such a comparison may, however, be misleading since, on the one hand, there is a small sample from a large population that establishes and increases rapidly whilst, on the other hand, there is a species that might have maintained a small population for a number of generations. The length of time that a population has remained at small size may be an important factor.

21.3.2 Does the community recover?

Amongst the earliest experiments investigating the effect of herbivore control on vegetation were the rabbit-proof exclosures on the Breckland grasslands (Watt, 1957). The data, collected over a 21-year period beginning in 1936, are amongst the most detailed, long-term data available for ecological analysis and for modelling (Usher, 1987a). Although these data related to small exclosures, more recent data on rabbits have investigated their effect on larger areas. North and Bullock (1986) indicated that a reduction of rabbit numbers on Round Island (Mauritius) had allowed the ground vegetation to recover and some palms to regenerate (discussed in Section 21.3.1). However, the new community, although visually similar to the pre-rabbit community, was less diverse since some of the rarest native species failed to regenerate. Myers (1986) adapted a 'picture model' of the vegetation/rabbit/predator system in Australia indicating that an invasive predator, controlling the invasive herbivore, is likely to result in an improved vegetation cover (though Myers avoided a discussion of whether it would be native or introduced plants that benefit).

A number of studies in New Zealand, and oceanic islands administered by New Zealand, show the effects of controlling large herbivorous mammals (Wodzicki and Wright, 1984). The eradication of *Ovis aries* (feral sheep) on the northern half of the subantarctic Campbell Island (separated from the uncontrolled southern half by a trans-island fence) led to a spectacular regeneration of the native endemic flora, especially the herbfield, turf meadow and upper alpine tall rush associations (Meurk, 1982). On two of the smaller Chatham Islands, eradication of sheep in the 1960s led to a change in the vegetation back to native forest (Ritchie, 1970) so successfully that these islands are now essential for the conservation of two endangered species, *Cyanoramphus auriceps forbesi* (Forbes parakeet) and *Petroica traversi* (Chatham Island robin). An 80–90% reduction of *Cervus elaphus* (red deer) in the Fiordland National Park resulted in the alpine and subalpine vegetation recovering spectacularly (Evans *et al.*, 1976), and control of both red deer and goats in the forests of the Ruahine Range (Cunningham, 1979) changed the forest, which had been modified 'to the point of collapse', to a forest that showed signs of recovery. The examples from New Zealand could be extended, but they demonstrate two important points. First, control of invasive herbivores leads to the recovery of the plant community, at least structurally. The speed of recovery is obviously dependent on the composition of the community, being fastest for grasslands and herbfields and slowest for forests. Second, complete studies of recovering ecosystems are very rare, so it is generally not possible to assess whether recovery results either in a less diverse community (e.g. the Round Island study) or in a community as diverse as that which existed pre-colonization by invasive herbivores.

The same general remarks can be applied to the control or eradication of predators. In southern Africa, Cooper *et al.* (1985) reviewed the effects of introduced mammals on island biota. At Marcus Island (South Africa), serious damage to breeding birds followed constructions of a causeway linking the island to the mainland in 1976. Subsequent construction of a predator-proof wall greatly increased the breeding success of *Haematopus moquini* (black oystercatcher). Similarly, in New Zealand, Veitch (1983) and King (1984) recorded increases in native bird populations following eradication of island populations of feral cats (see also Section 21.3.1). During the period 1984 to 1986 about 20000 *Trichosurus vulpecula* (Australian brush-tailed possum) were killed on Kapiti Island, almost eradicating the species. Lovegrove (1986) recorded that, as the number of possums decreased, the number of birds increased (about two-fold during the first 2 years of the campaign).

In Australia, comparisons of kangaroo population densities both historically and on either side of the NSW/Queensland dingo fence have led Caughley *et al.* (1980) to hypothesize that the past and present densities are directly attributable to predation by dingoes. Dingoes, an introduced species, can keep kangaroos and emus at low population densities provided that there is an abundant alternative prey species. Although direct predation on kangaroos is important in keeping their numbers low, Robertshaw and Harden (1986) also showed that the predator can disrupt the breeding cycle as a result of the loss of young in the pouch due to harassment of the mothers.

Once again these studies of predator control indicate that the herbivore community can recover after the predation pressure, exerted either by killing or by harassment, is either relieved or removed. However, the assessments are usually not quantified, although large increases in bird or kangaroo abundance have been reported. The question that is never addressed is whether the post-control community is the same as, or similar to, the pre-invasion community. One reason why this question is so difficult to answer is that there are often no accurate records of the pre-invasion communities. However, the question 'Does control lead either to a full or to a partial recovery of the ecosystem?' is an important one to be able to answer. The answer must be in terms of the ecological and genetical components of that ecosystem (e.g. diversity and genetical variability) and not only in terms of its landscape contribution.

21.3.3 Are there relationships with other introduced species?

There are a number of possible relationships that could be considered, but only three will be discussed in this review, namely

Alien 1 depends on Alien 2 entirely (Case 1),
Alien 1 eats Alien 2 (but not exclusively) (Case 2), and
Alien 1 facilitates Alien 2 (Case 3).

Case 1 is trivial since Alien 1 is unlikely to be thought of either as invasive or as damaging native species or ecosystems. Elimination of Alien 2 will automatically lead to eradication of Alien 1. This is often a goal of biological control, since Alien 1 is introduced either to control or eradicate Alien 2.

Case 2 is of more interest, since control of Alien 1 may allow Alien 2 to become far more abundant and, in its turn, require control. An example of this is the control of rabbits on Motunau Island (New Zealand); they were finally eradicated in 1963 (Taylor, 1967; Mason, 1967), after which a complete vegetation cover was re-created, mainly by *Disphyrna australe* (ice plant) and *Hordeum musinum* (barely grass). *Dactylis glomerata* (cocksfoot) and *Festuca arundinacea* (tall fescue), both non-native grasses, became established for the first time, and *Carduus tenuiflorus* (winged thistle) grew in great abundance. Subsequently *Lycium ferocissimum* (boxthorn) has become a major weed problem on the island. The rabbit clearly had had an impact on the non-native vegetation. This kind of argument has also been used in considering whether the control or eradication of feral goats is desirable (Daly and Goriup, 1986). Although on balance the decision is often to reduce, exclude or eradicate the goats, it must be remembered that this can have an important impact on releasing potentially invasive plants from grazing pressure.

As well as herbivore–plant relationships, predator–prey relationships should also be considered. In the Kinchega National Park (New South Wales, Australia) the rabbit was the main food of the fox in all seasons of the year, except the autumn (when insects, and especially centipedes, were eaten in greater quantities) (Ryan and Croft, 1974). A study of 899 foxes collected throughout New South

Wales (Croft and Hone, 1978) found that the major food items, in terms of both percentage occurrence and percentage volume, were rabbit, sheep and house mouse, all non-native species. Consumption of native mammal and bird species was small, varying with season and location. What neither of these studies is able to answer is the question 'what would happen to the rabbit (and house mouse) population if foxes were reduced in number or eradicated?' An interesting answer comes from the study of the control of the feral cat on Marion Island. As the number of cats decreased, their diet changed from one where native breeding birds predominated (van Aarde, 1980) to one in which the introduced house mouse predominated (van Rensburg, 1985b), the total mouse consumption not increasing. It would be an ideal world if the reduction in the numbers of an invasive predator always led to its feeding on other invasive species and not killing the native species!

Case 2, like Case 1, may arise from a biological control situation, especially where a predatory vertebrate (Alien 1) is introduced to control some other species (Alien 2); examples are the introduction of the giant toad to Queensland (see Section 21.1.3) to control arthropod pests of sugar cane (Floyd and Esteal, 1986) or the introduction of *Herpestes auropunctatus* (small Indian mongoose) to Maui, Hawaiian Islands, in the 1880s to control rats in sugar cane fields (Brockie *et al.*, 1988). Both are examples of biological control that has gone wrong; the control agent has become an invasive species in its own right, preying both on native species and on the introduced species that they were planned to have controlled. Attempts to control vertebrates with vertebrate predators have generally been unsucessful (J.H. Brown, personal communication), but pathogens have been much more successful biological control agents of vertebrates due to their greater specificity (Case 1 rather than Case 2 relationships).

Case 3 involves the possibility of Alien 1 facilitating Alien 2. There is relatively little information on this possibility in more or less natural ecosystems, but one example is of *Cervus nippon* (sika deer) introduced into the Irish oakwoods (Cross, 1981). Overgrazing and disturbance is thought to make the environment more suitable for the regeneration and establishment of seedlings of *Rhododendron ponticum* (rhododendron). There are a number of examples of feral stock (sheep, goats, cattle) facilitating introduced grasses; examples on island nature reserves are reviewed by Brockie *et al.* (1988). Unlike the feral cat/house mouse and fox/rabbit examples, where the introduced species have come from the same part of the world, the rhododendron/sika deer example involves an invasive plant from southern Europe and an invasive mammal from Japan. Although this is an isolated example, will there be many species pairs from different parts of the world where one introduction will facilitate another? For management purposes, control of the facilitator will have an effect on the facilitated species, though the magnitude of this effect will depend on the nature of the facilitation. If sika deer were to be controlled in Ireland, the oakwoods would become less accessible for *Rhododendron* colonization and establishment, though those already colonized would continue to have an invasive *Rhododendron* population. Facilitation at the colonization or establishment stage is probably most difficult for management, since control of both introduced species, by different methods, will be required.

21.3.4 Are there undesirable side-effects?

Any method of controlling one vertebrate species may lead to the accidental death of other species. But, to what extent does this actually happen in practice?

The Scottish muskrat population originated from nine animals (including five females) that escaped from captivity in Perthshire in 1927. A compaign to eradicate the species from Scotland started in 1932, and Munro (1935) recorded the number of trap deaths between the inception of the campagin and September 1934 (see Table 21.1). Although public opinion would not tolerate jaw traps in the 1980s as it did in the 1930s, it is nevertheless interesting to ask what were the side-effects of the successful muskrat eradication campaign. Besides the muskrat, the jaw traps caught at least 26 other species of vertebrates, and there was a death rate of 6.97 other vertebrates for every muskrat killed. Although the deaths of rats and rabbits (both invasive, non-native species) would probably be acceptable, the loss of otters and many of the species of birds would not now be acceptable on conservation grounds.

The Scottish muskrat example used traps which could have been predicted to kill many other species of wildlife. In Australia, there has been experimentation with appropriate traps for *Canis familiaris dingo* (dingo). Newsome *et al.* (1983) indicated that at least 20 species of protected wildlife were trapped at a rate of two to three individuals per dingo trapped. Their study showed that a smaller trap

Table 21.1. Number of animals caught during the programme to eradicate the muskrat in the Forth and Earn Valleys, Scotland (data from Munro, 1935)

Group and species trapped	Number killed
Ondatra zibethica (muskrat)	945
Mammals	
Arvicola terrestris (water vole)	2305
Rattus norvegicus (brown rat)	1745
Mustela nivalis (weasel)	57
Mustela erminea (stoat)	36
4 other species (hare, mole, otter, rabbit)	8
Birds	
Gallinula chloropus (moorhen)	2178
Duck (unspecified species)	101
Gallinago gallinago (snipe)	28
Seagull (unspecified species)	23
Ardea cinerea (heron)	18
12 other species	87
Fish	
Anguilla anguilla (eel)	1

caught three times less protected wildlife (and 15 times less large marsupials) than a larger trap, but that the efficiency of trapping dingoes was similar. Both trap design and trap placement are considerations in the reduction of undesirable side-effects.

So far the discussion has been on trapping, but poisoning can also have undesirable side-effects. Spurr (1979) assessed the accidental deaths following the use of 1080 (sodium monofluoroacetate), incorporated in chipped carrots, oats, etc., to control the invasive *Trichosurus vulpecula* (brush-tailed possum) in New Zealand. Native birds which were killed included *Gallirallus australis* (weka), *Nestor meridionalis* (kaka), *Nestor notabilis* (kea), *Mohoua albicilla* (whitehead) and *Petroica australis* (robin), as well as several introduced species.

It is probably true that no method of control using traps or poisons can be aimed solely at the target species. Live-trapping, as with the coypus in England (see Section 21.4.1), with subsequent shooting of trapped animals, offers a reasonable degree of specificity, reducing the chances of other species being killed in the traps. However, awareness not only of accidental deaths but of the effects of human disturbance whilst trapping was evident in the evidence presented to the Coypu Strategy Group by wildlife conservations (Morton *et al.*, 1978). The coypu lives in wetland habitats, and the conservationists were particularly concerned at possible disturbance, during the breeding season, to two rare bird species, *Botaurus stellaris* (bittern) and *Circus aeruginosus* (marsh harrier); and they were also concerned about three other species, *Panurus biarmicus* (bearded tit), *Cettia cetti* (Cetti's warbler), and *Locustella luscinioides* (Savi's warbler). Many of these difficulties have been overcome by close cooperation between the conservation organizations and the trappers.

All of the examples indicate that there are a few general principles in relation to the control of invasive vertebrates. First, a few accidental deaths of other species, or limited disturbance, cannot be avoided. Second, awareness of such undesirable side-effects before a control campaign begins allows the campaign to be planned so that these effects are minimized. Third, although attempts at biological control by the use of predatory vertebrates have usually been unsuccessful, the use of pathogens, which are much more specific, is likely to be far more successful. Fourth, before the campaign proceeds, an assessment must be made both of the long-term benefits of control and of the short-term disadvantages of the side-effects (unless, of course, the side-effects are so large that their effects will also be long-term). Fifth, when all parties are satisfied that control is beneficial, then it should proceed.

21.4 PROGNOSES FOR THE FUTURE

21.4.1 Is control possible?

There appear to be three primary factors that influence the answer to this question: the dispersal ability of the invasive species, the extent of the area over which control is required, and the stage in the establishment/invasion process when control begins.

Reviewing the effects of invasive alien species on nature reserves in Great Britain, Usher (1986a) compared the campaigns to control *Mustela vison* (mink) and *Myocastor coypus* (coypu). The campaign to control the former was largely unsuccessful, whereas that to eradicate coypu by 1990 appears to be heading for success. Mink have a lower rate of increase than coypu, but a greater dispersal ability. Although it is said that mink are relatively easy to trap (Anon., 1981), Gerell (1971) showed that it was the juvenile males that were particularly trappable. Where local campaigns against mink have stopped, R. Gerell (personal communication) has shown that the population rapidly increases to its pre-campaign level, and he concluded that the effects were only of short duration. Chanin (1981) concluded that eradication of mink from Britain would be extremely difficult and expensive. The comparison between mink and coypu is interesting since coypu does not disperse so far, can also be relatively easily trapped (Gosling, 1981a), and is being effectively controlled. Trapping studies with this species indicate many subtle effects, such as the fact that traps on rafts are more effective than traps on the banks of water bodies (Anon., 1984) and that weather has a major effect on trapping success (Gosling, 1981b). Comparing these two species with the successful eradication of the muskrat in Britain in the 1930s led Usher (1986a) to suggest that successful eradication was related to small dispersal rates rather than to small rates of increase.

Studies on other alien mammals point to the same conclusion. Rabbits are difficult to control and have one of the largest mean annual dispersal distances. Auld and Tisdell (1986) discussed the control of the *Sus scrofa* (feral pig) in Australia, where individuals are known to have moved 20 km in 48 hours. They argued that the only method of control that is likely to be successful is cooperative since, if an individual farmer controls pigs on his property, other farmers benefit as the destroyed pigs are replaced by other pigs moving from elsewhere. The dearth of examples of the control of invasive birds is due to their high mobility; this is admitted by Brown (1985) for the invasive bird species in South West Africa/Namibia (see Section 21.1.3) and is admitted by Ward (1979) when discussing the control of the native *Quelea quelea* (red-billed quelea) in the dryer regions of Africa. There seems to be increasing evidence to suggest that the probability that control will be successful is inversely related to the dispersal ability of the invasive species (see Figure 21.2a).

The examples quoted in the previous sections suggest a second factor in determining success of a campaign to control an invasive species: action early in the invasion process. The control of bulbuls in New Zealand or of the red-billed blue magpie in Hawaii (see Section 21.1.3) are both examples of species that can disperse widely, have become invasive in some areas of the world, but where eradication on individual islands was successful because it started early in the invasion process.

The third important factor relates to the isolation of the area in which the invasive species is to be managed. The success stories listed in Section 21.3 are generally located on islands; the control of feral cats on Marion Island (van Rensburg, 1985b), the control of sheep on Campbell Island (Meurk, 1982) and the series of successes on New Zealand's offshore islands (King, 1984) are examples. Not all of New Zealand's offshore islands have had introduced mammals successfully controlled or had a successful recovery of a damaged ecosystem; Mark and Baylis (1982) quoted the case of Secretary Island (8000 ha) on which the continuing effects of red deer impact on the vegetation could be observed even after an almost successful control campaign. The experience in South Africa is interesting since the natural ecosystems seem to be less prone to invasion by introduced mammals than ecosystems elsewhere in the world. Macdonald and Richardson (1986) indicated that attempts to control *Sciurus carolinensis* (grey squirrel) were futile, but that *Hemitragus jemlahicus* (Himalayan tahr) is contained on Table Mountain by shooting. In Europe, muskrat was eradicated in Scotland (cf. Munro, 1935), whereas in continental Europe it has spread and control campaigns have been unsuccessful of (cf. Verkaik, 1987). These various studies all indicate that the control of an invasive species on islands is likely to be more successful than control on continental areas.

However, isolation can also be achieved in an artificial manner. Cronk (1986b) discussed the decline of *Commidendrum robustum* (St Helena gumwood) on St Helena due to cutting for firewood, grazing of stock in the wooded areas, and, once again, the depredations of feral goat. The species will, however, regenerate freely, and new woodlands can be created in areas which are fenced against goats. The construction of a predator-proof wall (Cooper *et al.*, 1985), or of the dingo fence (Shepherd, 1981), or of fences in the Hawaiian national parks for goat control (Brockie *et al.*, 1988), effectively increases the isolation of an area within which the invasive species can be controlled or eradicated. Although the barriers quoted above are physical barriers that prevent entry of widely dispersing predators or herbivores, less tangible barriers can also be effective. Quarantine barriers are used in some places with reasonable success, whilst control of the giant toad in Australia (Floyd and Easteal, 1986) could be referred to as a legal barrier if ever legislation prohibited the movement of produce from toad-inhabited to toad-free areas.

Isolation of an area of land is difficult to define precisely, but it is nevertheless an intuitive concept that is useful when considering the probability that management to control or eradicate an invasive species will be successful. Isolation is, of course, not completely unrelated to dispersal ability, since a short distance of sea or a fence would present no barrier to a bird but may be an impermeable barrier to a mammal. The concept of isolation is shown in Figure 21.2b.

Two of these three factors, dispersal of alien species and isolation of management area, can be combined as in Figure 21.2c. In this diagram only three contours have been drawn and, because of the difficulty of giving precise definitions to dispersal and isolation, no points have been plotted. However,

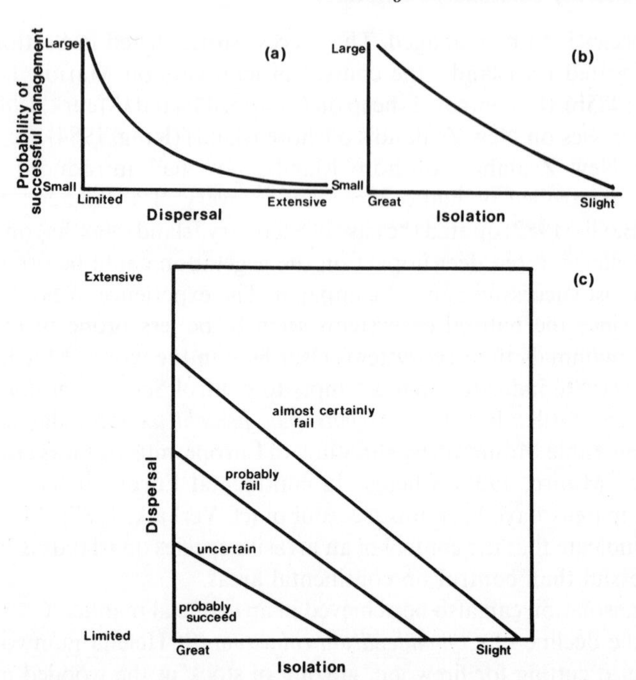

Figure 21.2. Diagrammatic representations of the possible relationships between the probability of successfully managing an invasive species and (a) its dispersal ability and (b) the isolation of the site being managed. These two factors are brought together in (c) to provide a 'picture model' to predict the probability that management will succeed. The real shapes of the curves in (a) and (b) are unknown, but both would be monotonically decreasing functions. The contours in (c) are drawn as straight lines; their precise shape would depend on the shape of the curves in (a) and (b). There is a discussion of the effects of the timing of control operations on the location of the contours in (c) in the text

within the very rough framework of these two factors, Figure 21.2c provides a picture model that could be used to predict how successful management to control an invasive species might be. If control begins early in the invasion process, then the contours in Figure 21.2c can be moved away from the origin towards the upper right-hand corner of the square. As the invasion process proceeds, the contours move closer towards the origin.

21.4.2 Is modelling helpful?

It could, perhaps, be argued that models are not yet able to predict sufficiently accurately the establishment and spread stages of a biological invasion

(Williamson, this volume), and hence that they are unlikely to be useful in designing control strategies.

The position with models is, however, not quite so bleak. In order to eradicate the coypu in Great Britain, the population not only needed to be censussed but estimates needed to be made of the population size in previous years. A simple, statistical modelling approach, estimating monthly frequencies of recruitment, was used to solve the problem of estimating retrospectively the census counts (Gosling *et al.*, 1981). Simple regression models were used by Gosling (1981c) to relate the proportions of female coypus littering in spring to the severity of the winter, and to relate the trap efficiency (number of adults killed per night of trapping) to the number of adults in the population. These simple, statistical models only address specific aspects of the overall control operation.

More sophisticated models have been developed by Anderson (1982) in relation to the fox and rabies interaction. Although his model of foxes under control by shooting and disease (rabies in this case) is applied in an area where the fox is native, the modelling approach adopted could well be used for non-native invasive species. In the fox/rabies model, the management objective was to reduce the size of the fox population below the threshold needed for rabies transmission; for introduced species the management objective would be to reduce the size of the population so that its effects on the native fauna and flora were perceived to be negligible. Anderson's (1982) model was deterministic; Mollison (1986) argued that for policy makers a stochastic modelling approach would be preferable. Whatever these advantages and disadvantages, there is clearly considerable scope for the development of suitable models for vertebrate control.

21.4.3 How is success measured?

The words success and failure have been used frequently both in the foregoing accounts and in the references cited. However, brief consideration should be given to the assessment of success. Holdgate (1986) said 'examples of successful elimination of an invader (like that of muskrat in Britain) will remain limited...success in most cases will be reflected in the reduction of the invader's population size to a level that is compatible with the perpetuation of the invaded ecosystem in only slightly altered form'. Note that, in this quotation, 'slightly altered form' is not defined and it could form the basis of an argument about how much constitutes 'slightly'. It highlights the need for an objective of control to be specified precisely before the control begins. The results of that control can then be compared with the specification; if they match the specification or exceed it then control is successful, whilst if they fall short of the *a priori* specification then management has been unsuccessful.

It may, however, be difficult to monitor the management sufficiently precisely to match results with the specification. As an example, the campaign against the coypu in Britain aimed to eradicate the species from the country (Morton *et al.*, 1978). However, as the population density has decreased, it is more and more difficult to estimate how many remain. This problem is discussed by Gosling (1985), who recently (in August 1986) expressed the opinion that there may only be 10 feral coypu females remaining in Britain (L. M. Gosling to S. Reeves, personal communication). Mark and Baylis (1982) highlighted this problem on Secretary Island (New Zealand), where, during a visit, they failed to sight red deer and only found two groups of pellets. If the aim of management is eradication, it is extremely difficult to be absolutely certain that eradication has been achieved; conversely, sighting an individual proves conclusively that eradication has not been achieved.

21.4.4 Are control and conservation compatible?

Perhaps the overwhelming reason for controlling an invasive vertebrate species is its effect on the native species and communities of its new environment. Macdonald *et al.* (this volume) have concentrated on aliens and conservation, and Loope and Mueller-Dombois (this volume) have stressed the effects of introduced species on the endemic species of the Hawaiian Islands. Frequently there can be different views as to the value of an alien species in conservation management; this was highlighted by Daly and Goriup (1986) who stated that control or eradication of feral goats was not always welcomed by conservations.

From a wildlife conservation point of view, many of the examples quoted in this paper are on the negative side of the non-native/native species interaction. Goats have grazed the native flora till only a few individuals of a native species remain (the St Helena ebony, for example) or foxes have killed the native marsupials so that only a few small, isolated populations remain (the rock wallaby, for example). Control of the non-native species is overtly beneficial to the native flora or fauna.

However, there are also examples that are on the positive side of the non-native/native species interaction. In Sweden, Danell (1979) reported that, if not too numerous, the presence of muskrats increased the waterfowl populations of well-vegetated lakes by expanding the area of open water and hence the feeding areas available to both adult and juvenile birds. Also in Sweden, the spread of the Canada goose has given wildfowlers another species to shoot, reducing the pressure on native species (Fabricius, 1983). In Western Australia the diet of the *Aquila audax* (wedge-tailed eagle) contains a large proportion of introduced species. In the dry Nullarbor Plain, Brooker and Ridpath (1980) recorded that 'rabbits were by far the major food item', whereas in other areas of Western Australia rabbits, sheep, foxes and cats all formed substantial proportions of the eagle's diet.

Even more complex interactions can occur. Christensen and Burrows (1986) explained the complicated management strategy to control foxes in Western

Australia, to encourage *Macropus eugenii* (tammar wallaby) and to burn the thickets of *Casuarina*, *Melaleuca* and *Gastrolobium*. To protect the tammar from excessive predation, the thickets require a fire of high intensity every 25 to 30 years. This leads to mass germination of the thicket species, but there are many management problems associated with small area fires, such as control of the fire, prevention of undue kangaroo and wallaby grazing during regeneration, etc. Management of an appropriate habitat for the tammer needs to consider the introduced species, balancing losses due to predation with losses due to habitat change if an appropriate form of management was not adopted.

21.4.5 Does control increase or decrease genetic diversity?

One of the main purposes of wildlife conservation is to retain as diverse a genetic resource as possible. If control of invasive species prevents the extinction of any native species, it could be argued that genetic diversity has been retained.

However, if a native species has been reduced to a very small population size, has the amount that it contributes to global genetic diversity already been irretrievably impaired? Miller (1979) catalogued some of the genetical problems that have to be thought about by conservationists; one of these is the 'bottleneck', which is particularly relevant in a study of invasive species. As an example, Cronk (1986a) documented the re-finding of *Trochetiopsis melanoxylon* (St Helena ebony). After being considered as extinct for more than a century, two individuals were recently found (see Section 21.3.1), both low bushes rather than shrubs or small trees as would have covered St Helena prior to the arrival of people and goats. Although genetical studies on this species have yet to be reported, Cronk (1986a) considered that these two remaining individuals represent a cliff ecotype and that the 'normal' ecotype is extinct.

Changes in genetic diversity can occur through interbreeding between native and feral populations. This has occurred around the world with the cats. In Britain, some rather bizarre animals, even recently suspected by amateurs and 'The Press' of being a new and undescribed cat species, have resulted from crosses between *Felis catus* (feral cat) and *Felis sylvestris* (wild cat) (D.D. French, personal communication). In southern Africa, the feral cat has hybridized with *Felis lybica* (the indigenous African wild cat) to such an extent that Griffin and Panagis (1985) considered that 'pure' *F. lybica* no longer existed. This raises the taxonomic problem of whether all *Felis* species are distinct, or whether they are essentially subspecies. With the feral cat in localities where there are wild cats it may not now be possible to eliminate the former, and indeed the genetic diversity of the wild population may have been increased with feral cat genes.

General features of the genetic aspects of the invasion process are considered by both Barrett and Richardson (1986) and Gray (1986). What appear not to have been studied are the genetic aspects of the control process. Studies of this are clearly needed.

21.5 CONCLUSION

There are three general points that emerge from a discussion of the control of invasive vertebrate species.

First, there is now much more awareness of the problems and of the need for management policies. The discussion of the need for control is moving from the scientific to more general literature (e.g. the book by Fitter, 1986), and there are lively debates when proposals are brought forward to introduce mammals, e.g. the discussions reported in Pinder (1981) concerning the re-introduction of *Castor fiber* (European beaver) into Britain. More awareness of the problems caused by invasive vertebrates will probably help in the solution of those problems.

Second, conservationists are beginning to recognize the importance of thinking about invasive species. In reviewing the criteria used for selecting nature reserves, Usher (1986b) listed 24 criteria that have been used in 17 evaluation studies around the world. Not one of these studies considered the presence or management consequences of non-native species. Usher (in press) has suggested that, in Africa, the potential problem posed by invasive species should be a criterion in the selection of nature reserves, national parks, etc. The search for the naturalness of reserves (see Brooke *et al.*, 1986) should be as much related to the impacts of invasive species as to the impact of human societies.

Third, to end on an optimistic note, control of invasive vertebrate species can frequently be achieved, as shown in the discussion by Wodzicki and Wright (1984) of the many invasive vertebrate species in New Zealand; the only species that they say is still unmanageable is *Trichosurus vulpecula* (brush-tail possum). Control, or eradication, is likely to be relatively easy for species/site pairs that are located towards the bottom left corner of the diagram (Figure 21.2c). Moving either to the right or upwards, control becomes more difficult. This, of course, is tempered by the timing of control operations; wherever one is in Figure 21.2c, the operation is likely to be easier the earlier control starts in the invasion process. Difficulty of control does not generally mean impossibility; it usually means that more effort is required, and this in turn requires more resources (people and money) and operations over a longer period of time or over a larger territorial area. If the management objectives are defined, and if the resources are available, invasive vertebrate species control or eradication has usually proved possible. Scientists have developed the methods; what is needed is the political will to use them.

21.6 SUMMARY

The chapter aims to investigate how frequently control measures have been attempted for invasive vertebrate species and to explore the factors that lead to successful control (or management). The review starts by asking how the control operations affect the invasive species itself; genetical aspects are little studied but they may be tremendously important, as, for example, the development of warfarin resistance in rats.

Control is usually commenced due to damage either to individual native or endemic species, or to the structure of communities. It is seen that control operations are often successful when individual species are to be protected from invasive species, and that communities can often recover at least structurally if not to their full pre-invasion diversity. For future management, control is usually possible, there being relatively few examples where invasive vertebrates are considered to be unmanageable. Three factors are seen to predispose any campaign to success. First, campaigns that start at the establishment phase of an invasion, or very soon after establishment, have a much greater probability of success. Second, management success is inversely proportional to the dispersal ability of the species to be controlled. Third, success and isolation of the invaded habitat are correlated; thus campaigns to control an invasive vertebrate species are more likely to be successful on an island than on a part of a continent, and barriers (such as fences or even quarantine enforcement zones) increase isolation of continental areas, making control that much more possible.

ACKNOWLEDGMENTS

In preparing this review a number of people around the world have provided information and case histories. I should like to thank Dr R. H. Groves and Dr J. E. Kinnear (Australia); Dr V. Landa, Dr J. Lepš and Dr J. Rusek (Czechoslovakia); Professor R. H. Drent, Dr H. van Emden Dr W. Joenje and Mr. A. J. Verkaik (Netherlands); Dr R. E. Brockie and Professor A. F. Mark (New Zealand); Mr I. A. W. Macdonald (South Africa); Dr K. Danell, Professor E. Fabricius, Dr R. Gerell, Dr G. Görannson, Dr O. Liberg and Dr S. Ulfstrand (Sweden); Miss J. Barber, Dr L. M. Cook, Dr L. M. Gosling and Dr M. Rands (UK); and Dr J. H. Brown, Dr J. A. Drake, Professor P. R. Ehrlich and Dr S. L. Pimm (USA). I should like to thank Professor H. A. Mooney and SCOPE for financial assistance and Professor M. H. Williamson for reading through the manuscript.

REFERENCES

van Aarde, R. J. (1980). The diet and feeding behaviour of feral cats, *Felis catus*, at Marion Island. *S. Afr. J. Wildl. Res.*, **10**, 123–8.

van Aarde, R. J. (1984). Population biology and the control of feral cats on Marion Island. *Acta Zool. Fennica*, **172**, 107–10.

van Aarde, R. J., and Skinner, J. D. (1981). The feral cat population at Marion Island: characteristics, colonization and control. *Com. Nat. fr. Rech. Antarctiques*, **51**, 281–8.

Anderson, R. M. (1982). Fox rabies. In: Anderson, R. M. (Ed.), *Population Dynamics of Infectious Diseases: Theory and Applications*, pp. 242–61. Chapman & Hall, London, New York.

Anonymous (1981). *Feral mink*. Ministry of Agriculture, Fisheries and Food Leaflet No. 794. HMSO, London.

Anonymous (1984). Coypu Research Laboratory. *Agr. Dev. & Advisory Serv. Ann. Rep., 1983*, pp. 96–9. HMSO, London.

Atkinson, I. A. E. (1985). The spread of commensal species of *Rattus* to oceanic islands and their effects on island avifaunas. In: Moore, P. J. (Ed.), *Conservation of Island Birds*, pp. 35–81. ICBP Technical Publication No. 3.

Auld, B. A., and Tisdell, C. A. (1986). Impact assessment of biological invasions. In: Groves, R. H., and Burdon, J. J. (Eds), *Ecology of Biological Invasions: an Australian Perspective*, pp. 79–88. Australian Academy of Science, Canberra.

Backhouse, G. N. (1987). Management of remnant habitat for conservation of the helmeted, *Lichenostomus melanops cassidix*. In: Saunders, D. A., Arnold, G. W., Burbidge, A. A., and Hopkins, A. J. M. (Eds), *Nature Conservation: the Role of Remnants of Native Vegetation*, pp. 287–94. Surrey-Beatty, Sydney.

Barrett, S. C. H., and Richardson, B. J. (1986). Genetic attributes of invading species. In: Groves, R. H., and Burdon, J. J. (Eds), *Ecology of Biological Invasions: an Australian Perspective*, pp. 21–33. Australian Academy of Science, Canberra.

Berger, A. J. (1981). *Hawaiian Bird Life*. University of Hawaii Press, Honolulu.

Berry, R. J. (1977). *Inheritance and Natural History*. Collins, London.

Bishop, J. A., Hartley, D. J., and Partridge, G. G. (1977). The population dynamics of genetically determined resistance to warfarin in *Rattus norvegicus* from mid Wales. *Heredity*, **39**, 389–98.

Bourne, W. R. P., and David, A. C. F. (1983). Henderson Island, central South Pacific, and its birds, *Notornis*, **30**, 233–52.

Bourne, W. R. P., and David, A. C. F. (1985). Henderson Island. *Notornis*, **32**, 83.

Bourne, W. R. P., and David, A. C. F. (1986). Henderson Island. *Nature*, **322**, 302.

Brockie, R. E., Loope, L. L., Usher, M. B., and Hamman, O. (1988). Biological invasions of island nature reserves. *Biological Conservation*, **44**, 9–36.

Brooke, R. K., Lloyd, P. H., and Villiers, A. L. de (1986). Alien and translocated terrestrial vertebrates in South Africa. In: Macdonald, I. A. W., Kruger, F. J., and Ferrar, A. A. (Eds), *The Ecology and Management of Biological Invasions in Southern Africa*, pp. 63–74. Oxford University Press, Cape Town.

Brooker, M. G., and Ridpath, M. G. (1980). The diet of the wedge-tailed eagle, *Aquila audax*, in Western Australia. *Aust. Wildl. Res.*, **7**, 433–52.

Brown, C. J. (1985). Invasive alien birds in South West Africa/Namibia. In: Brown, C. J., Macdonald, I. A. W., and Brown, S. E. (Eds), *Invasive Alien Organisms in South West Africa/Namibia*, pp. 41–3. South African National Scientific Programmes Report, 119.

Butler, W. H. (1970). A summary of the vertebrate fauna of Barrow Island, W. A. *West. Aust. Nat.*, **11**, 149–60.

Butler, W. H. (1987). Management of disturbance in an arid remnant: the Barrow Island experience. In: Saunders, D. A., Arnold, G. W., Burbidge, A. A., and Hopkins, A. J. M. (Eds), *Nature Conservation: the Role of Remnants of Native Vegetation*, pp. 279–85. Surrey-Beatty, Sydney.

Caughley, G., Grigg, G. C., Caughley, J., and Hill, G. J. E. (1980). Does dingo predation control the densities of kangaroos and emu? *Aust. Wildl. Res.*, **7**, 1–12.

Chanin, P. (1981). The feral mink—natural history, movements and control. *Nature in Devon*, **2**, 33–54.

Christensen, P. E., and Burrowns, N. D. (1986). Fire: an old tool with a new use. In: Groves, R. H., and Burdon, J. J. (Eds), *Ecology of Biological Invasions: an Australian Perspective*, pp. 97–105. Australian Academy of Science, Canberra.

Cooper, J., Hockey, P. A. R., and Brooke, R. K. (1985). Introduced mammals on South and

South West African islands: history, effects on birds and control. *Proceedings of Symposium on Birds and Man, Johannesburg, 1983*, pp. 179–203.

Croft, J. D., and Hone, L. J. (1978). The stomach contents of foxes, *Vulpes vulpes*, collected in New South Wales. *Aust. Wildl. Res.*, **5**, 85–92.

Cronk, Q. C. B. (1986a). The decline of the St. Helena ebony *Trochetiopsis melanoxylon*. *Biol. Conserv.*, **35**, 159–72.

Cronk, Q. C. B. (1986b). The decline of the St Helena gumwood *Commidendrum robustum*. *Biol. Conserv.*, **35**, 173–86.

Cross, J. R. (1981). The establishment of *Rhododendron ponticum* in the Killarney oakwoods, S. W. Ireland. *J. Ecol.*, **69**, 807–24.

Cunningham, A. (1979). A century of change in the forests of the Ruahine Range, North Island, New Zealand, 1870–1970. *New Zealand J. Ecol.*, **2**, 11–21.

Daly, K., and Goriup, P. (1986). *Eradication of Feral Goats from Small Islands*. Report of the International Council for Bird Preservation and of the Fauna and Flora Preservation Society. 44 pp.

Danell, K. (1979). Reduction of aquatic vegetation following the colonization of a northern Swedish lake by the muskrat, *Ondatra zibethica*. *Oecologia*, **38**, 101–6.

Diamond, J. (1985). Rats as agents of extermination. *Nature*, **318**, 602–3.

Ehrlich, P. R. (1986). Which animal will invade? In: Mooney, H. A., and Drake, J. A. (Eds), *The Ecology of Biological Invasions of North America and Hawaii*, pp. 79–95. Springer-Verlag, New York.

Evans, G. R., Payton, I. J., Burrows, L. E., Parkes, J. P., and Batchelor, G. L. (1976). *Summary Report on a Vegetation Resurvey of Part of Fiordland National Park*. New Zealand For. Serv. Prot. For. Rep., No. 141.

Fabricius, E. (1983). *Kanadagasen i Sverige*. Naturvardsverket Rapport, 1678.

Falla, R. A., Sibson, R. B., and Turbott, E. G. (1966). *A Field Guide to the Birds of New Zealand and Outlying Islands*. Collins, London.

Fenner, F. (1983). The Florey Lecture, 1983. Biological control as exemplified by smallpox eradication and myxomatosis. *Proc. R. Soc. Lond. B*, **218**, 259–85.

Fitter, R. (1986). *Wildlife for Man: How and Why we should Conserve our Species*. Collins, London.

Floyd, R. B., and Easteal, S. (1986). The giant toad (*Bufo marinus*): introduction and spread in Australia. In: Groves, R. H., and Burdon, J. J. (Eds.), *Ecology of Biological Invasions: an Australian Perspective*, p. 151. Australian Academy of Science, Canberra.

Gerell, R. (1971). Population studies on mink, *Mustela vison* Schreber, in southern Sweden. *Viltrevy*, **8**, 83–114.

Gerell, R. (1985). Habitat selection and nest predation in a common eider population in southern Sweden. *Ornis Scand.*, **16**, 129–39.

Gosling, L. M. (1981a). The coypu. In: Boyle, C. L. (Ed.), *RSPCA Book of British Mammals*, pp. 129–35. Collins, London.

Gosling, L. M. (1981b). The effect of cold weather on success in trapping feral coypus (*Myocastor coypus*). *J. Appl. Ecol.*, **18**, 467–70.

Gosling, L. M. (1981c). The dynamics and control of a feral coypu population. In: Chapman, J. A., and Pursley, D. (Eds), *Proceedings of Worldwide Furbearer Conference*, pp. 1806–25.

Gosling, L. M. (1985). Coypus in East Anglia (1970 to 1984). *Trans. Norfolk & Norwich Nat. Soc.*, **27**, 151–3.

Gosling, L. M., Watt, A. D., and Baker, S. J. (1981). Continuous retrospective census of the East Anglian coypu population between 1970 and 1979. *J. Anim. Ecol.*, **50**, 885–901.

Gray, A. J. (1986). Do invading species have definable genetic characteristics? *Phil. Trans. R. Soc. Lond. B*, **314**, 655–74.

Griffin, M., and Panagis, K. (1985). Invasive alien mammals, reptiles and amphibians in South West Africa/Namibia. In: Brown, C. J., Macdonald, I. A. W., and Brown, S. E. (Eds), *Invasive Alien Organisms in South West Africa/Namibia*, pp. 44–7. South African National Scientific Programmes 119.

Holdgate, M. W. (1986). Summary and conclusions: characteristics and consequences of biological invasions. *Phil. Trans. R. Soc. Lond. B*, **314**, 733–42.

Howell, P. G. (1984). An evaluation of the biological control of the feral cat *Felis catus* (Linnaeus, 1758). *Acta Zool. Fennica*, **172**, 111–3.

King, C. (1984). *Immigrant Killers: Introduced Predators and the Conservation of Birds in New Zealand.* Oxford University Press, Auckland.

Kinnear, J., Onus, M., and Bromilow, B. (1984). Foxes, feral cats and rock wallabys. *Swans*, **14**, 3–8.

Lockley, R. M. (1970). *Man Against Nature.* Andre Deutsch, London.

Lovegrove, T. (1986). *Counts of Forest Birds on Three Transects on Kapiti Island 1982–1986.* New Zealand For. Serv. Rep. 17 pp.

Macdonald, I. A. W., and Richardson, D. M. (1986). Alien species in terrestrial ecosystems of the fynbos biome. In Macdonald, I. A. W., Kruger, F. J., and Ferrar, A. A. (Eds), *The Ecology and Management of Biological Invasions in Southern Africa*, pp. 77–91. Oxford University Press, Cape Town.

Mark, A. F., and Baylis, G. T. S. (1982). Further studies on the impact of deer on Secretary Island, Fiordland, New Zealand. *New Zealand J. Ecol.*, **5**, 67–75.

Mason, R. (1967). Motunau Island, Canterbury, New Zealand: vegetation. *New Zealand Dep. Sci. Ind. Res. Bull.*, **178**, 68–92.

Meurk, C. D. (1982). Regeneration of subantarctic plants on Campbell Island following exclusion of sheep. *New Zealand J. Ecol.*, **5**, 57–8.

Miller, R. I. (1979). Conserving the genetic integrity of faunal populations and communities. *Environm. Conserv.*, **6**, 297–304.

Mollison, D. (1986). Modelling biological invasions: chance, explanation, prediction. *Phil. Trans. R. Soc. Lond. B*, **314**, 675–93.

Mooney, H. A., Hamburg, S. P., and Drake, J. A. (1986). The invasions of plants and animals into California. In: Mooney, H. A., and Drake, J. A. (Eds), *The Ecology of Biological Invasions of North America and Hawaii*, pp. 250–72. Springer-Verlag, New York.

Morton, J., Calver, J., Jefferies, D. J., Norris, J. H. M., Roberts, K. E., and Southern, H. N. (1978). *Coypu: Report of the Coypu Strategy Group.* HMSO (Ministry of Agriculture, Fisheries and Food), London.

Moulton, M. P., and Pimm, S. L. (1986). Species introductions to Hawaii. In: Mooney, H. A., and Drake, J. A. (Eds), *The Ecology of Biological Invasions of North America and Hawaii*, pp. 231–49. Springer-Verlag, New York.

Munro, T. (1935). Note on musk-rats and other animals killed since the inception of the campaign against musk-rats in October 1932. *Scot. Nat.*, (1935), 11–6.

Myers, K. (1986). Introduced vertebrates in Australia, with emphasis on the mammals. In: Groves, R. H., and Burdon, J. J. (Eds), *Ecology of Biological Invasions: an Australian Perspective*, pp. 120–36. Australian Academy of Science, Canberra.

Newsome, A. E., and Noble, I. R. (1986). Ecological and physiological characters of invading species. In: Groves, R. H., and Burdon, J. J. (Eds), *Ecology of Biological Invasions: an Australian Perspective*, pp. 1–20. Australian Academy of Science, Canberra.

Newsome, A. E., Corbett, L. K., Catling, P. C., and Burt, R. J. (1983). The feeding ecology of the dingo. I. Stomach contents from trapping in south-eastern Australia, and the non-target wildlife also caught in dingo traps. *Aust. Wildl. Res.*, **10**, 477–86.

North, S. G., and Bullock, D. J. (1986). Changes in the vegetation and populations of introduced mammals of Round Island and Gunner's Quoin, Mauritius. *Biol. Conserv.*, **37**, 99–117.

Parkes, J. P. (1984). Feral goats on Raoul Island, II. Diet and notes on the flora. *New Zealand J. Ecol.*, **7**, 95–101.
Pinder, N. (Ed.) (1981). *Conservation and Introduced Species.* University College London Discussion Papers in Conservation No. 30.
Rensburg, P. J. J. van (1985a). Feral cats and sub-Antarctic skuas on Marion Island: competition or co-existence? *S. Afr. J. Sciences*, **81**, 691.
Rensburg, P. J. J. van (1985b). The feeding ecology of a decreasing feral house cat, *Felis catus*, population at Marion Island. In: Siegfried, W. R., Condy, P. R., and Laws, R. M. (Eds), *Antarctic Nutrient Cycles and Food Webs*, pp. 620–4. Springer-Verlag, Berlin.
Ritchie, I. M. (1970). A preliminary report on a recent botanical survey of the Chatham Islands. *Proc. New Zealand Ecol. Soc.*, **17**, 52–6.
Robertshaw, J. D., and Harden, R. H. (1986). The ecology of the dingo in north-eastern New South Wales, IV. Prey selection by dingoes, and its effect on the major prey species, the swamp wallaby, *Wallabia bicolor* (Desmarest). *Aust. Wildl. Res.*, **13**, 141–63.
Ross, J., and Tittensor, A. M. (1986). The establishment and spread of myxomatosis and its effect on rabbit populations. *Phil. Trans. R. Soc. Lond. B*, **314**, 599–606.
Ryan, G. E., and Croft, J. D. (1974). Observations on the food of the fox, *Vulpes vulpes* (L.), in Kinchega National Park, Menindee, N. S. W. *Aust. Wildl. Res.*, **1**, 89–94.
Shepherd, N. C. (1981). Predation of red kangaroos, *Macropus rufus*, by the dingo, *Canis familiaris dingo* (Blumenbach), in north-western New South Wales. *Aust. Wild. Res.*, **8**, 255–62.
Spurr, E. B. (1979). A theoretical assessment of the ability of bird species to recover from an imposed reduction in numbers, with particular reference to 1080 poisoning. *New Zealand J. Ecol.*, **2**, 46–63.
Taylor, R. H. (1967). Motunau Island, Canterbury, New Zealand: an ecological survey. *New Zealand Dep. Sci. Ind. Res. Bull.*, **178**, 42–67.
Usher, M. B. (1986a). Invasibility and wildlife conservation: invasive species on nature reserves. *Phil. Trans. R. Soc. Lond. B*, **314**, 695–710.
Usher, M. B. (1986b). Wildlife conservation evaluation: attributes, criteria and values. In: Usher, M. B. (Ed.), *Wildlife Conservation Evaluation*, pp. 3–44. Chapman & Hall, London, New York.
Usher, M. B. (1987a). Modelling successional processes in ecosystems. In: Gray, A. J., Edwards, P. J., and Crawley, M. J. (Eds), *Colonization, Succession and Stability*, pp. 31–55. Blackwell, Oxford.
Usher, M. B. (1987b). Effects of fragmentation on communities and populations: a review with applications to wildlife conservation. In: Saunders, D. A., Arnold, G. W., Burbidge, A. A., and Hopkins, A. J. M. (Eds), *Nature Conservation: the Role of Remnants of Native Vegetation*, pp. 103–21. Surrey-Beatty, Sydney.
Usher, M. B. (1988). Biological invasions of nature reserves. *Biol. Conserv.*, **44**, 1–35.
Usher, M. B. (1987). The evaluation of potential conservation locations in Africa. *Proc. Internat. Symp. African Wildlife, Mweya, Uganda, December 1986.*
Veitch, C. R. (1983). A cat problem removed. *Wildlife: a Review* (*New Zealand Wildl. Serv.*), **12**, 47–9.
Verkaik, A. J. (in press). The muskrat in the Netherlands. *Proc. Kon. Nederl. Akad. Wetens.* C, **90**, 67–72.
Ward, P. (1979). Rational strategies for the control of queleas and other migrant bird pests in Africa. *Phil. Trans. R. Soc. Lond. B*, **287**, 289–300.
Watt, A. S. (1957). The effects of excluding rabbits from grassland B (Mesobrometum) in Breckland. *J. Ecol.*, **45**, 861–78.
Wodzicki, K., and Wright, S. (1984). Introduced birds and mammals in New Zealand and their effect on the environment. *Tuatara*, **27**, 77–104.

Biological Invasions: a Global Perspective
Edited by J. A. Drake *et al.*
© 1989 SCOPE. Published by John Wiley & Sons Ltd

CHAPTER 22

Biological Invasions: a SCOPE Program Overview

H. A. Mooney and J. A. Drake

22.1 INTRODUCTION

Here we give a general overview of the results of the SCOPE program on the Ecology of Biological Invasions. We concentrate our attention on species invasions as a global phenomenon rather then a local event. The biological invasions program provided a global appraisal of the phenomenon of species invasions, and the effect such invaders have on the ecosystems they colonize. We define a species as being an invader when it colonizes and persists in an ecosystem in which it has never been before. Essentially we are viewing invasions in ecological rather then evolutionary time. Those wishing an in-depth treatment of the dynamics of biological invasions are referred to Elton's (1958) classic work, and to the publications which arose out of this program (Appendix 1).

22.2 HOW MANY INVADERS ARE THERE?

One of the primary outcomes of the program was to provide new documentation on the extent of invasions of different types of organisms in various parts of the world. For example, there are between 1500 and 2000 introduced species of plants in Australia, and nearly half of the flora of New Zealand (1570 invading species) is composed of invaders (Heywood, this volume). Over 1300 species of insects have successfully invaded the United States, and we do not doubt that the catalogue of invaders worldwide greatly underestimates the extent of species invasions.

Of the many data sets available, we highlight just two—that of invasions into nature reserves around the world and invasions onto islands, with Hawaii as our central example. These contrasts illustrate two extremes in ecological systems, unmanaged island systems versus carefully managed reserves which are primarily continental. Islands have incurred some of the greatest numbers of invaders, whereas reserves are managed to prevent invasions or to eliminate them, if possible, when they do occur. These examples also serve to illustrate that certain types of organisms have been more successful invaders than others and certain regions have been comparatively more resistant as well as prone to invasion.

22.2.1 Invasions into nature reserves

Twenty-four nature reserves distributed throughout the world were examined for the numbers of invading organisms, with an emphasis on higher plants and vertebrates (Usher *et al.*, 1988). Usher (1988) comments that 'No nature reserve included in the case studies is without at least one species of invasive vertebrate and at least several species of invasive vascular plants.' In the case of island reserves about 30% of the vascular plant species, and almost 18% of the terrestrial vertebrates are invaders (Usher, 1988). Interestingly, reserves on continental land masses are somewhat less heavily impacted. Further, certain kinds of ecosystems like savannas and dry woodlands appear less vulnerable to invasion. This was a particularly important evaluation since it yielded information about invasions into systems that are being biologically protected and are subjected to minimal disturbance. Despite this fact, many invaders have been successful in these systems.

22.2.2 Invasions on islands

All else being equal, ecosystems subject to disturbance and those not directly protected from invasion appear more likely to be successfully colonized by an invading species. As our example of the impact of invaders on a non-protected island system we point to the data provided by Vitousek *et al.* (1987) for the Hawaiian Islands. In Hawaii nearly half of the approximately 2000 species of flowering plants are invaders. These species have radically altered the character of the flora. Before the influence of humans over 90% of the flowering plants were totally endemic to the islands. Similar impacts are seen on other indigenous groups. In the case of reptiles and amphibians the invaders represent the only members of this group. Essentially all of the mammals are likewise established introductions. Around 100 species of birds have invaded or have been introduced to the Hawaiian Islands. A direct result of these invasions has been the extinction of numerous native bird species (Moulton and Pimm, 1986).

Thus, these data from both protected and non-protected regimes indicate that invasive species have substantially altered the nature and patterning of the earth's biota. Essentially, after an invading species becomes established the ecological 'rules' or ecological processes which operated in a given system change. In some cases this change is negligible and the invading species simply adds to the species richness of the system. In other cases the 'rules' or processes which operated, permitting coexistence of a species complex, are drastically altered leading to the extinction of native species. Concomitant with changes in the flora and fauna of an ecological system are changes in that system's resistance and vulnerability to subsequent species invasions. A species which formerly was excluded from a community may now be a successful colonist.

22.3 WHO ARE THE INVADERS?

22.3.1 Functional groups

In considerations of invading species accounting is usually made in terms of the taxonomic groups to which they belong. An alternative approach is to categorize invaders by the role they play in community and ecosystem processes. For example, the invader might be a specialist or generalist consumer, it may function as an herbivore, woody plant, predator, pathogen, or nitrogen fixer. To date few analyses have searched for trends in invasion success as a function of the role species play. However, there are some robust theoretical predictions about what kinds of species are better colonists in differing ecological situations (Drake, 1983; Sugihara, 1983). Using radically different models Drake (1983) and Sugihara (1983) have found that invasion success is a function of both community complexity (number of species) and the degree of trophic specialization. As community complexity increases species which are more specialized in diet tend to be more successful invaders. An empirical evaluation of this trend is forthcoming.

Here we explore invasions in a single functional group—that of plant pathogens. Von Broembsen (this volume) gives an assessment of the extent and effect of invading plant pathogens into natural ecosystems. She noted that there are only five examples of highly successful and destructive invading pathogens into natural forested systems: chestnut blight, white pine blister rust, dutch elm disease, pine wilt disease, and phytophthora root disease. The pathogenic agents in these cases are varied, and include nematodes and fungi but not bacterial or viral phytopathogens. The types of forests attacked have been in temperate regions and have been characterized by uniformly distributed dominant trees of one or a few closely related species. Some of the invading pathogens have relatively complex life cycles, dependent on insects or alternate hosts for portions of their life cycles. All are the result of introductions of diseased plant materials prior to strict quarantine procedures.

Although there have been only a few cases of successful invading pathogens, it should be noted that these cases have had impressive impacts. The chestnut blight effectively eliminated the American chestnut throughout its range. Similarly, large areas of jarrah eucalyptus forests have been affected by phytophthora with impacts on the water balance of these regions. There are many additional examples of invading pathogens which have had only minor effects or no discernible effect at all; however, many of these have been into agroecosystem and are not considered by Von Broembsen.

Why have there been so few invading plant pathogens natural ecosystems? Von Broembsen suggests that they often have specialized hosts (although this is not the case for phytophthora) and are thus not good invaders into new systems. Intact ecosystems appear particularly resistant to invasions; however, it may be that invasions are often unnoted or there have been limited opportunities for invaders. It is instructive to note that of the successful examples given above none are important pathogens at their origins. The picture is substantially altered if one considers the successful invading pathogens of crops, where a given crop species may have large numbers of potential pathogens.

What of other functional groups? Unfortunately, much of the data on invaders does not suggest which functional group species belong to. Hence, an analysis of trends is not currently possible. Nevertheless, we offer a few thoughts based on theoretical considerations. Given a species' ability to tolerate a new environment, a species which represents a new functional role, or perhaps fills a vacent role, might be expected to be a better colonist then a species which invades a system containing a species which already fills that role. If this is the case, we may be unable to find general trends across communities because there is no reason to expect that all communities will be missing the same functional groups. We suggest that at the level of functional groups trends are not likely to exist.

22.3.2 Taxonomic groups

As noted above, invading species may be found among all types of organisms, microorganisms, plants, and animals. However, certain taxa appear to be more successful invaders than others. Unfortunately, it is not possible to evaluate all groups uniformly because of an inadequate data base. Heywood (this volume) has, however, been able to assess terrestrial plants. He notes that of the 250 000 species of higher plants only about 250, or 0.1%, are serious agricultural pests although a much larger number, 8000, are considered weeds of agriculture. Of the world's 18 worst weeds, over half are grasses and all of these are of tropical origins. Heywood notes that the most common plant families from which weedy species originate are Compositae, Gramineae, and Leguminosae. This is not surprising since these represent the largest families of flowering plants, with approximately 20 000, 13 000, and 8 000 species respectively, occupying a tremendous range of habitats throughout the world. In the case of insects, Simberloff (1986) notes though that the invading species into North America are not representative of the taxonomic pool; some taxa are significantly more successful then others. this may be function of differences in opportunity for successful transport.

22.4 HOW DO THEY GET THERE?

The increasing breakdown of the biogeographic barriers that once limited the interchange of the world's biota has been accomplished through the intercontinental transport of humans and their goods. The biological invaders have either been purposely carried with humans or have inadvertently accompanied them. The history of these movements has been chronicled by Crosby (1986) and di

Castri (this volume). Although quarantine efforts have altered the rate of invasions of accidental introductions they have not totally controlled them. Further, as the means and frequency of human transport has changed so have the kinds of organisms accompanying them (Sailer, 1983). Unfortunately, although in many countries there are strict quarantine programs directed toward agricultural pests there is often free interchange and continued introductions of organisms not associated with agriculture that have the potential to alter the properties of native ecosystems.

22.5 WHERE DO THEY COME FROM?

Plant invading species of the New World have their principal origins in 'the eastern and southern fringes of the Mediterranean Basin and the adjacent Mediterraneo-Irano-Turanian steppes of the Middle East' (Heywood, this volume). It is these regions that have had a long history of human modification and where agriculture originated. It is not surprising that plants that co-evolved with agricultural practices have attributes that make them successful invaders. These regions have also served as centers of world commerce and thus as sources of plant materials.

Insect invaders of the New World also are predominately from the western palearctic region (Sailer, 1983). Simberloff (this volume) argues, however, that these origins reflect principally pathways of commerce, and thus opportunities for invading species, rather than any intrinsically aggressive properties they may possess. This trend is likely to be found with plants as well as animals.

22.6 HOW DO THEY GET ESTABLISHED?

Once an invading species arrives in a new environment its successful establishment is contingent on a variety of environmental and biological factors. Successful established represents a significant barrier in the invasion process. It is difficult to present quantitative estimates of the proportion of unsuccessful invasions; one must of course witness the invasion to know it even occurred. Nevertheless, some information on the proportion of invaders that become established is available in the biological control literature (DeBach, 1964, 1974). Simberloff (1981) summarized the attempted introductions reported by DeBach (1974), and found that for over 11 of 30 control efforts one or more of the species introductions failed (100 attempted introductions). Hypothetically, the proportion of unsuccessful invaders should be considerably lower then in natural invasions, because in biological control efforts an attempt is often made to match the ecological and environmental requirements of the introduced species to the target environment.

Given that environmental conditions are sufficiently benign to permit an invading species to colonize, what factors are likely to influence whether or not that population will persist? A variety of factors ranging from the genetics of small populations to community level interactions have been implicated in affecting invasion success. We will begin by considering the process of successful invasion as a sequence of barriers that must be overcome for successful establishment.

22.6.1 The minimum viable population

The concept of the minimum viable population is central to any consideration of species establishment after invasion. Simply stated, a minimum viable population represents some population abundance that is immune to local extinction due to genetic, demographic and environmental stochasticity (Shaffer, 1981; Gilpin and Soule, 1981; Richter-Dyn and Goel, 1972; Leigh, 1975; Frankel and Soule, 1981; Schoenwald-Cox *et al.*, 1983; MacArthur and Wilson, 1967). Given the magnitude of population fluctuations caused by such factors, a population might have a minimum size below which extinction is likely and above which extinction is unlikely.

Richter-Dyn and Goel (1972) modelled the population dynamics of a hypothetical invading species that was resource-limited. They considered the effect chance fluctuations had on invader population dynamics by adding a simple stochastic term to the population growth equation. Although their conclusions are model-dependent, Richter-Dyn and Goel found that a population of about 20 individuals could increase in abundance and persist in the face of stochastic fluctuations in abundance. Conversely, a population below 20 individuals either became extinct at a rate equivalent to the reciprocal of the per-capita birth rate, or, due to random fluctuations the population eventually exceeds the threshold above which extinction is unlikely.

Clearly, factors other than those subject to stochastic variation in population size also influence successful invasion and persistence. These factors effectively increase the minimum viable population size, that is they increase the likelihood of extinction. Because invading species tend to be rare upon colonization, the dynamics and genetic structure of small populations are likely to be important. For example, the sex ratio, breeding system and genetic diversity of an invading population can be a critical determinant of success as an invader. A sex ratio skewed towards females will have less of an effect in some mating systems (say polygyny) than others. Additional complications such as genetic bottlenecks, which arise from the often severe reduction in genetic diversity of the founder population, may have serious consequences as well. These are but a few of the factors which can influence a species' minimum viable population size.

22.6.2 Higher-order effects

If all communities had the same structure and organization one might expect that a successful invader in one system would be equally successful in another.

Communities, however, are not all the same—indeed they posses such a diversity of structures and patterns that a community-level 'phenomenology' does not yet exist (see, however, Drake, 1989). What makes one community vulnerable to invasion while another is resistant to invasions? Clearly, disturbed systems tend to be more susceptible to invaders than intact systems, although we note cases where disturbance prohibitis invasion by many species (Hobbs, this volume; Mooney and Drake, 1986). But what of two undisturbed systems or two equally disturbed systems; what factors are responsible for differences in vulnerability to invasion? Should species-poor communities be more susceptible to invasion than species-rich systems, all else being equal? These are difficult questions to answer and represent a potentially rich avenue of future research.

At least part of the answer will come from integrated community-level studies. Resistance and susceptibility to invasion is a community-level property, not a property of guilds, trophic levels or plant, bird, fish or insect ensembles. Demonstrations of unlikely interactions across disparate trophic levels or levels of scale are common. Such interactions define the boundary of the system that must be explored if progress is to be made in understanding vulnerability to invasion.

22.7 HOW FAST DO THEY MOVE?

After an invading species becomes established in a new environment it is unlikely that it will remain localized. The final stage of a species invasion is usually the spread or dispersal of the invading species into the surrounding environment (Williamson, this volume; Elton, 1958; Levin, this volume; Okubo, 1980, Skellam, 1951; Williamson and Brown, 1986; Anderson and May, 1986). Clearly, a species that invades but does not spread is unlikely to become as serious a problem as an invader that rapidly expands its range. For example, both the gypsy moth (*Prostoia dispar*) and the Africanized honey bee have rapidly expanded their ranges across many habitat types. However, many invaders remain localized, dispersing only short distances from the original introduction.

An important component of spread is the rate at which the invading species colonizes new territory. Absolute rate of spread depends on two population parameters, namely the population's rate of growth and dispersal ability (Levin, this volume; Williamson and Brown, 1986; Williamson, this volume). A species with a low rate of increase but substantial dispersal ability may spread at a rate roughly equal to a prolific breeder but poor disperser. Williamson and Brown show that the spread of the muskrat (*Ondatra zibethica*) and grey squirrel (*Sciurus carolinesnsis*) appears as a '...generally steady advance, which could be ascribed to random dispersal with occasional major advances.'

Rate of spread is not only a function of species characteristics but the ecosystems through which the species spreads. Mack (1986) has suggested that the successful invasion of some non-native grasses into the Intermountain West occurred only when disturbance from large mammals was removed. The spread of zander (*Stizostedios lucipperca*), a piscivorous fish now invading England from eastern Europe, is strongly limited by characteristics of the lakes and rivers themselves (Hickley, 1986). Hence zander spreads more rapidly in slow moving riverine systems then in rapidly moving waters.

Generally, the rate of spread of an invader is a complex function of a species' population ecology and the amount of consequential habitat heterogeneity. Spread through a patchy environment is likely to depend on the degree of habitat heterogeneity, size and distribution of patches, distance between suitable patches, and population characteristics such as growth rate and vagility or dispersal ability.

22.8 WHAT IS THE NATURE OF THE SYSTEMS THEY INVADE?

A primary goal of the invasions program was to assess the significance of invasions of organisms into natural systems. Elton (1958) had argued that the lack of invaders into natural systems was due to some resistance attributes they had that were determined by their complement of competitors, predators, parasites, and diseases. Successful invasions occurred when this resistance system was in some way broken down by disturbance. Simberloff (this volume) argues that the greater apparent vulnerability of native systems is not necessarily related to such system properties but simply to the greater frequency of introductions into the human-dominated disturbed systems. Crawley (1987) makes similar arguments. The data do not yet exist on successes and failures into matched natural and disturbed systems to resolve this issue.

22.9 HOW DO THEY IMPACT THE SYSTEMS THEY INVADE?

Since the time of Elton's book there has been a significant development in the study of the properties of ecosystems. It is thus not surprising then that one of the most important contributions of the SCOPE program on the ecology of biological invasions has been an analysis of the impact of invaders on ecosystem properties. The most telling examples are again from nature reserves—systems that are managed for the maintenance of natural processes. Macdonald *et al.*, (this volume), in their analysis of nature reserves around the world, found examples of ecosystem disruption of the following types:

1. Accleration of soil erosion rates (feral mammals)
2. Alteration of biogeochemical cycling (feral pigs, invasive nitrogen fixers, salt accumulators)
3. Alteration of geomorphological processes (dune and marsh grasses)
4. Alteration of hydrological cycles (phreatophytes, *Phytophthora*, invasive trees)

5. Alteration of fire regimes (invasive grasses and shrubs)
6. Prevention of recruitment of native species (alien plants, mammlas, and ants)

The important point of the above list is that virtually all ecosystem functional (water and mineral cycling, productivity) and structural properties (community structure and succession) can be influenced strongly by an invading species. Further, invaders from all taxonomic categories may play this disruptive role (microbes, plants, and animals).

The impact of invading species on non-protected, yet non-agricultural ecosystems can be considerable. Mack (this volume) notes that 'in less than 300 years...much of the temperate grassland outside Eurasia...has been irreparably transformed by human settlement and the concomitant introduction of alien plants. Few other changes in the distribution of the earth's biota since the end of the Pleistocene have been as radical...or as swift...'. In these instances there has been a complete transformation of the properties of these systems. Mooney, *et al.*, (1986) note, for example, how the perennial grasslands of California have been totally transformed by the apparently irrevocable conversion to a grassland dominated by annuals originating in the mediterranean Basin.

The degree of specific impact an invader can have on an ecosystem will vary depending on its function. Vitousek (1986) has noted that large animal invaders can alter such system properties as productivity, soil structure, and nutrient cycling. Plants, particularly those that differ in functional properties from the native species, can have similarly dramatic impacts.

Pimm (1987), (this volume), on the basis of theory and natural history observations, provides the following guidelines for predicting the potential impact of introductions on biotic communities:

1. Species introductions where predators and competitors of the introduced species are absent are likely to have severe impacts.
2. Introductions into relatively simple communities (few species) are also likely to have large impacts.
3. Finally, animal introduction that utilize multiple food plants are likely to have large effects especially if predators are removed.

22.10 WHAT CAN WE PREDICT AT PRESENT?

In reviewing the information available on biological invasions with the goal of building a predictive base the SCOPE program has been hindered by two fundamental data lacks:

1. Most of the evidence comes from qualitative natural history observations. There has been very little experimentation utilized in the study of biological invasions.
2. Although we have abundant information on the number of successful invading

species for most groups we have essentially no information on the numbers of failures nor an analysis of why they failed. Thus we cannot easily give adequate probabilities of establishment of a given organism into a new system.

Although, as noted above, there are some general rules relating to invasibility based on the properties of systems, there are fewer general predictive guidelines relating to properties of the organism. There have been, however, repeated attempts to make predictions of the potential invasibility success of a given organism based on their traits. These attempts (Baker, 1986, for plants; Ehrlich, this volume, for invertebrates) have generally taken the form of lists of attributes, genetic, physiological, and ecological, that are most often associated with successful invaders. Crawley (1987) presents a somewhat pessimistic assessment of these analyses, though, stating that 'reluctantly, it has to be admitted that at the moment we have virtually no predictive ability in relation to the species characteristics of plants associated with invasibility in a particular kind of community.' He does note, however, that the intrinsic rate of increase, for insects at least, is a predictor of probability of successful establishment as is the widespread nature of their distribution.

Increasingly we are beginning to understand the significance of episodic events in structuring biological communities. Rare events entrain a whole series of responses that may totally change the dynamics of a systerm. An invading species may be successful during these brief intersections of the chance of being there just at the right time. These intersections may be rare indeed and very difficult to predict for an untested species (Crawley, this volume).

Simberloff (this volume) nevertheless believes that detailed studies of the natural history of a particular organism and of the physical and biotic characteristics of the host environment should yield the information necessary to make sound predictions of potential invasion success. In spite of this lack of predictability for a specific case, knowledge of traits of organisms and of the target environment gives important generic guidelines of the probability of success that can be utilized for planning.

The emphasis in the SCOPE program has been on the numbers and impacts of the invading species and very little on the failed introductions because of the lack of data on the latter. There are some studies on certain groups of organisms that do, however, indicate the difficulty of establishing an alien organism in spite of a concerted effort and considerable knowledge of the biology of the organism. Only about a quarter of the attempts to introduce exotic birds have established breeding populations. In some cases (Japanese quail into the United States) no breeding populations have been established in spite of the release of hundreds of thousands of individuals (Long, 1981). In cases of carefully planned introductions, where there was great care to match host and target habitats in biocontrol efforts, high establishment rates have been achieved.

22.11 GOOD INTENTIONS ARE NOT ENOUGH

Of course all purposeful introductions have been with good intentions—increasing agricultural productivity, providing erosion control, enhancing recreation and so forth. In most cases these objectives have been achieved, but in some cases at costs not originally envisioned. As our understanding of biological systems becomes more complete we should be able to reduce the probability that an intentional introduction will have an adverse effect. We should be beyond making such mistakes as we did in purposefully introducing plants such as crabgrass and Johnsongrass, two of North America's most noxious weeds, and animals such as carp. Other purposeful introductions into North America include water hyacinth, dandelion, (Foy *et al.*, 1983) and kudzu vine. Yet even in fairly recent times there have been examples of purposeful introductions that have had fairly drastic ecological and economic consequences. The point of the following example is that great care must be taken to consider all aspects of the consequences of an introduction, not just an immediate one of interest, such as increasing productivity.

In 1957 the Nile perch *Lates niloticus* was introduced into Lake Victoria in East Africa because it was large and relatively easily caught. The consequences of this introduction, as chronicled below, are given by Barel *et al.*, (1985). The intent of the introduction was to increase the availability of food for human populations. This act was taken in spite of concerns over the consequences of introducing a large carnivorous fish weighing over 200 kg into a system where few other fishes exceed 1 kg. These concerns were validated as subsequent events proved. The introduction was very successful in terms of establishment but had many detrimental effects including driving to extinction many endemic fishes. It is claimed that 'the potential loss of vertebrate genetic diversity as a result of this single ill-advised step is probably unparalleled in the history of man's manipulation of ecosystems.' Other undesirable consequences of this introduction have been considerable. The larger fish proved to be a less favored food besides being difficult to preserve. The oily perch have to be smoked rather than sun dried as was done with the native smaller fish. Wood to support the drying has led to deforestation of a number of islands (Coulter *et al.*, unpublished manuscript). Establishment of the Nile perch is now so entrenched that there appears little possibility of removing them. Thus a lake ecosystem, the size of Switzerland, has been irrevocably modified in a relatively short time. The changes, in spite of good intent, have been mostly detrimental. Barel *et al.*, (1985) note that except in the cases of fishless lakes and man-made reservoirs, all examples of movement of fish from one system to another with the intent of increasing yields of human food have been either unsuccessful or simple disasterous. They note that although the basis for these failures could be predicted in some cases often they cannot since information from controlled experiments has not been available.

The lessons from the above example are many and include having a control program in place before a release and utilizing experimental data to base predictions of possible system impact. Importantly the total impact of the introduction should be carefully considered before release.

22.12 THE COSTS AND BENEFITS OF CONTROL

Once an invading species becomes established eradication may become virtually impossible, although there are some success stories, more so for animals than for plants (Macdonald *et al.*, this volume). Attempts to completely eradicate pest insects have generally been costly exercise that in time fail and occasionally cause environmental damage (Dahlsten, 1986). Greater emphasis is generally placed on keeping populations of invading organisms down to an acceptable level. Usher (1988) has proposed that the most likely cases of successful eradication of invading mammals will be those which are naturally poor dispersers and occupy a circumscribed geographical locality.

Since most generally there are no target-specific eradication methods for most kinds of organisms there will be unavoidable elimination of individuals of other species in addition to the adverse environmental effects. The question of whether control efforts are positive depends on the effect the invader is having on the system in relation to the possible effects of the controlling methods utilized. Whether control is actually attempted depends on the economics of the effort as well as the public perception of the problem.

Comprehensive approaches to pest control (integrated pest management) that carefully consider the control options (mechanical, chemical, and biological); the details of the biology of the target organisms; and the economic, social, and ecological impacts of control efforts have unfortunately not been utilized extensively against invasive organisms in rangelands or natural ecosystems (Kluge *et al.*, 1986). The needs are there and the techniques are available for these approaches, but evidently economics has limited their application.

22.13 THE NEW ORDER

The characteristics of the earth we inhabit are being transformed at an ever accelerating rate principally due to the burgeoning human population and accopanying industrialization. Our atmosphere is being transformed as are our waterways. The terrestrial landscapes are being modified, in many instances, inalterably. Eight of the 30 major global vegetation types have been reduced to between 25 and 45% of their original area. Thirteen percent of the total ice-free area has been totally transformed by cultivation (Macdonald *et al.*, this volume). In specific regions the impact of agriculture has been profound. In Great Britain agriculture affects 80% of the land surface and the native vegetation was mostly destroyed several hundred of years ago (Heywood, this volume).

All of these alterations are providing a new landscape with an abundance of disturbed habitats favoring organisms with certain traits. This massive alteration of the biosphere has occurred in conjunction with the disintegration of the great barriers to migration and interchange of biota between continents due to the development by human of long-distance mass transport systems. The introduction of a propagule of an organisms from one region to a distant one has changed from a highly unlikely event to a certainty. The establishment and spread of certain kinds of organisms in these modified habitats, wherever they may occur, is enhanced. The net result of these events is a new biological order. Favored organisms are now found throughout the world and in ever increasing numbers. It is evident that these changes have not yet totally stabilized either in the Old or New World. In the former the success of invading species has changed through time with differing cultural practices and new directions and modes of transport. Old invaders are being replaced by new ones (Heywood, this volume). In the New World additional invading species are still being added.

The kinds of disruptions that non-intentionally introduced invading species can play in natural systems have been outlined above and have been the focus of the SCOPE study. These disruptions may in time stabilize on the basis of a new system equilibrium. Usher (1988) gives an example of the development of a co-existence between invading mink and native ducks in Sweden. Where at first there was widespread destruction of ducks by the mink within a couple of decades an apparent equilibrium has been reached between a stable mink population and nesting ducks. Usher comments that recent invaders may have a greater impact on a new community than long-standing invading species.

Not much attention has been centered on the ecosystem role that these invading species may serve in repairing disrupted systems. As one example, weedy herbaceous invaders may be more effective in sequestering the nitrate that is released from disturbed soils than the natives. Many of these invading species may provide fuel, fiber, and wood for certain societies although in others they may interfere with human activities and goals, e.g. papyrus, cat-tail, common reed (National Academy of Sciences, 1977).

Another viewpoint on the new order is provided by Wells *et al.* (1986), who noted that in South Africa 'most of our alien invader plants were intentional introductions and it is extremely likely that many of the rest were transported in association with them. Since these imports were made to establish industries and gardens, the problems associated with today's plant invaders can be seen as a tax being paid by the community for the prosperity and emenities that have been and are still being enjoyed'.

22.14 INVASIONS BY GENETICALLY DESIGNED ORGANISMS

An issue with which the scientific community at large is debating is the intentional introduction of genetically engineered species. What have we learned from the

SCOPE invasions program that may be of use in considering the release of genetically engineered organisms (GEO) into the environment?

First, there are many components of the issue that must be considered. These include those factors that are involved in the success of establishment and maintenance of the populations of interest. The lessons from the SCOPE program indicate that there are *general* rules for predicting success for a given organism type. However, it is clear that a detailed knowledge of the natural history of the GEO and of the characteristics of the target environment will be required if we are to make accurate predictions about organism performance in the open environment. There is no escaping the fact, however, that careful experimentation will be required to provide the information necessary to make such releases 'safe'.

Then there is the issue of spread out of the target area. If the factors for success of the GEO are the same outside of the target area as within, the organism will likely spread depending primarily on the dispersal properties of the organisms. As has been shown repeatedly, organisms of all types are accidently transferred from one region to another dependent on the degree of and direction of movement of vectors especially humans. It has further been shown that once established it may become extremely difficult to completely eradicate an invader. The lesson is that in many cases there will be no recall once a species becomes established and begins to spread.

There is abundant evidence of the considerable impact certain invading species have on the systems they colonize. Again general predictions can be made based on the properties of the invader. Introductions that perform a system function considerably different from that of the resident species can have a large impact. This may be of particular concern with GEOs since in many cases unusual functional features in relation to those of other members of the systems would generally be desired. Further introductions that assume a 'keystone' role in the organization of the total community can also have a potentially large impact. These are general guidelines that would need to be tested experimentally for any particular case.

Many of the cases of good intent in purposeful introductions that ultimately have bad consequences stem from a narrow view of the potential good that can come from the introduction. Often the desired feature of the organisms is considered in isolation from the total impact that the organism will have on the target system as well as on those who depend on that system for a variety of purposes. The potential effects of release must be considered in a total system context.

REFERENCES

Anderson, R. M., and May, R. M. (1986). The invasion, persistance and spread of infectious diseases within animals and plant communities. In: Kornberg, H., and

Williamson, M. (Eds), *Quantitative Aspects of the Ecology of Biological Invasions*. The Royal Society of London.
Baker, H. G. (1986). Patterns of plant invasions in North America. In: Mooney, H. A., and Drake, J. A. (Eds.), *Ecology of Biological Invasions of North America and Hawaii*, pp. 44–57. Springer-Verlag, New York.
Barel, C. D. N., Dorit, R., Greenwood, P. H., Fryer, G., Hughes, N., Jackson, P. B. N., Kawanabe, H., Lowe-McConnell, R. H., Nagoshi, M., Ribbink, A. J., Trewavas, E., Witte, F., and Yamaoka, K. (1985). Destruction of fisheries in Africa's lakes. *Nature*, **315**, 19–20.
Coulter, G. W., Allanson, B. R., Bruton, M. N., Greenwood, P. H., Hart, R. C., Jackson, P. B. N., and Ribbink, A. J. Unique qualities and special problems of the African great lakes. Unpublished manuscript.
Crawley, M. J. (1987). What makes a community invasible? In: Gray, A. J., Crawley, M. J., and Edwards, P. J. (Eds), *Colonization, Succession and Stability*, pp. 424–53. Blackwell Scientific Publications, Oxford.
Crosby, A. W. (1986). *Ecological Imperialism: The Biological Expansion of Europe, 900–1900*. Cambridge University Press.
Dahlsten, D. L. (1986). Control of invaders. In: *Ecology of Biological Invasions of North America and Hawaii*, pp. 275–302. Springer-Verlag, New York.
DeBach, P. (1964). *Biological Control of Insect Pests and Weeds*. Reinhold, New York.
DeBach, P. (1974). *Natural Control by Natural Enemies*. Cambridge University Press, London.
Drake, J. A. (1983). Invasibility in Lotka-Volterra interaction webs. In: DeAngelis, D., Post, W. M., and Sugihara, G. (Eds), *Current Trends in Food Web Theory*. Oak Ridge National Lab.
Drake, J. A. (Eds) (1986). *Ecology of Biological Invasions of North America and Hawaii*, Springer-Verlag, New York.
Drake, J. A. (1988). Community assembly and structure of ecological landscapes. In: Hallam, T., Gross, L., and Levin, S. (Eds), *Mathematical Ecology*, World Press, New York.
Elton, C. S. (1958). *The Ecology of Invasions by Animals and Plants*. Methuen, London.
Foy, C. L., Forney, D. R., and Cooley, W. W. (1983). History of weed introductions. In: *Exotic Plant Pests and North American Agriculture*, pp. 65–92. Academic Press, New York.
Frankel, O. H., and Soule, M. E. (1981). *Conservation and Evolution*. Cambridge University Press, London.
Gilpin. M., and Soule, M. Minimum viable populations: the process of species extinction, (in press).
Hickley, P. (1986). Invasion by zander and the management of fish stocks. In: Kornberg, H., and Williamson, M. (Eds), *Quantitative Aspects of the Ecology of Biological Invasions*. The Royal Society of London.
Kluge, R. L., Zimmerman, H. G., Culliers C. J., and Harding, G. B. (1986). Integrated control of invasive alien weeds. In: Macdonald, I. A. W., Kruger, F. J., and Ferrar, A. A. (Eds), *The Ecology and Management of Biological Invasions in Southern Africa*, pp. 295–303. Oxford Press, Cape Town.
Leigh, E. G. (1981). The average lifetime of a population in a varying environment. *J. Theor. Biol.*, **90**, 213–39.
Long, J. L. (1981). *Introduced Birds of the World*. Universe Books, New York.
MacArthur, R., and Wilson, E. (1967). *The Theory of Island Biogeography*. Princeton University Press, Princetion.
Mack, R. (1986). Alien plant invasion into the intermountain west: a case history. In:

Mooney, H., and Drake, J. A. (Eds). *Ecology of Biological Invasions of North America and Hawaii.* Spring Verlag, New York.

Mooney, H. A., and Drake, J. A. (Eds) (1986). *Ecology of Biological Invasions of North America and Hawaii.* Springer-Verlag, New York.

Mooney, H. A., Hamburg, S. P., and Drake J. A. (1986). The invasions of plants and animals into California. In: Mooney, H. A., and Drake, J. A. (Eds), *Ecology of Biological Invasions of North America and Hawaii*, pp. 250–72. Springer-Verlag, New York.

Moulton, M. P., and Pimm, S. L. (1986). Species introductions to Hawaii. In: Mooney, H. and Drake, J. (Eds.), *The Ecology of Biological Invasions of North America and Hawaii*, Springer-Verlag, New York.

National Academy of Sciences (1977). *Making Aquatic Weeds Useful: Some Perspectives for Developing Countries.* 174 pp.

Okubo, A. (1980). *Diffusion and Ecological Problems: Mathematical Models.* Springer-Verlag, Berlin.

Pimm, S. L. (1987). Determining the effects of introduced species. *Trends in Ecology and Evolution*, **2**, 106–8.

Richter-Dyn, N., and Goel, N. S. (1982). On the extinction of a colonizing species. *Theor. Pop. Biol.*, **3**, 406–33.

Sailer, R. I. (1983). History of insect introductions. In: *Exotic Plant Pests and North American Agriculture*, pp. 15–38. Academic Press, New York.

Schoenwald-Cox, C. M., Chambers, S. M., MacBryde, B., and Thomas, L. (Eds) (1983). *Genetics and Conservation.* Benjamin/Cummings, Menlo Park, CA.

Shaffer, M. L. (1981). Minimum population sizes for speices conservation. *BioScience*, **31**, 131–4.

Simberloff, D. (1981). Community effects of introduced species. In: Nitecki, M. H. (Ed.), *Biotic Crises in Ecological and Evolutionary Time.* Academic Press, New York.

Simberloff, D. (1986). Introduced species: a biogeographic and systematic perspective. In: Mooney, H., and Drake, J. (Eds), *The Ecology of Biological Invasions of North America and Hawaii*, Springer-Verlag, New York.

Skellam, J. G. (1951). Random dispersal in theoretical population. *Biometrike*, **38**, 196–218.

Sugihara, G. (1983). Holes in niche sapce: a derived assembly rule and its relation to intervality. In: DeAngelis, D., Post, W. M., and Sugihara, G. (Eds), *Current Trends in Food Web Theory.* Oak Ridge National Lab.

Usher, M. B. (1988). Biological invasions of nature reserves: a search for generalizations. *Biol. Conserv.*, **44**, 119–35.

Usher, M. B., Kruger, F. J., Macdonland, I. A. W., Loope, L. L., and Brockie, R. E. (1988). The ecology of biological invasions into nature reserves: an introduction. *Biol. Conserv.* (in press).

Vitousek, P. M. (1986). Biological invasions and ecosystem properties: can species make a difference? In: Mooney, H. A., and Drake, J. A. (Eds), *Ecology of Biological Invasions of North America and Hawaii*, pp. 163–76. Springer-Verlag, New York.

Vitousek P. M., Loope, L. L., and Stone, C. P. (1987). Introduced species in Hawaii: biological effects and opportunities for ecological research. *Trends in Ecology and Evolution* **2**, 224–7.

Wells, M. J., Poynton, R. J., Balsinhas, A. A., Musil, K. J., Joffe, H., Van Hoepen E., and Abbott, S. K. (1986). In Macdonald, I. A. W., Kruger, F. H., and Ferrar A. A. (Eds), *The Ecology and Management of Biological Invasions in Southern Africa*, pp. 21–35. Oxford University Press, Cape Town.

Williamson, M. H., and Brown, K. C. (1986). The analysis and modelling of British invasions. In: Kornberg, H., and Williamson, M. (Eds), *Quantitative Aspects of the Ecology of Biological Invasions*, The Royal Society of London.

APPENDIX. PUBLICATIONS RESULTING FROM THE SCOPE PROGRAM ON THE ECOLOGY OF BIOLOGICAL INVASIONS

I. National Workshops, Projects and Syntheses

A. South Africa

Brooke, R. K. *Bibliography of Alien Birds in Southern and South-central Africa.* South African National Scientific Programmes series. CSIR, Pretoria. (In Prep.)

Brown, C. J., Macdonald, I. A. W., and Brown, S. E. (1985). *Invasive Alien Organisms in South West Africa/Namibia.* South African National Scientific Programmes Report No. 119. CSIR, Pretoria.

Bruton, M. N., and Merron, S. V. (1985). *Alien and Translocated Aquatic Organisms in Southern Africa: A General Introduction, Checklist and Bibliography.* South African National Scientific Programmes Report No. 113, CSIR, Pretoria.

Macdonald, I. A. W., and Jarman, M. L. (Eds) (1984). *Invasive Alien Organisms in the Terrestrial Ecosystems of the Fynbos Biome, South Africa.* South African National Scientific Programmes Report No. 85, CSIR, Pretoria.

Macdonald, I. A. W., and Jarman, M. L. (Eds) (1985). *Invasive Alien Plants in the Terrestrial Ecosystems of Natal, South Africa.* South African National Scientific Programmes Report No. 111, CSIR, Pretoria.

Macdonald, I. A. W., Jarman, M. L., and Beeston, P. (Eds) (1985). *Management of Invasive and Alien Plants in the Fynbos.* South African National Scientific Programmes Report No. 111, CSIR, Pretoria.

Moran, V. C., and Moran, P. M. (1982). *Alien Invasive Vascular Plants in South African Natural and Semi-natural Environments: Bibliography from 1830.* South African National Scientific Programmes Report No. 65, CSIR, Pretoria.

Macdonald, I. A. W., Kruger, F. J., and Ferrar, A. A. (1986). *The Ecology and Management of Biological Invasions in Southern Africa.* Oxford University Press, Cape Town. 324 pp.

B. United Kingdom

Kornberg, H., and Williamson, M. H. (1986). Quantitative aspects of the ecology of biological invasions. *Phil. Trans. R. Soc. Lond. B.*, Vol **314**.

C. Australia

Groves, R. H., and Burdon, J. J. (1986). *Ecology of Biological Invasions.* Australian Academy of Science, Canberra.

D. United States

Mooney, H. A., and Drake, J. A. (Eds) (1986). *Ecology of Biological Invasions of North America and Hawaii.* Springer-Verlag, New York.

Mooney H. A., and Drake, J. A. (1987). The Ecology of Biological invasions. *Environment*, **29**, 10–15; 34–7.

Drake, J. A. (1988). Biological invasions into nature reserves. *Trends in Ecol. and Evol.* **3**, 186–87.

E. The Netherlands

Joenje, W., Bakker, K., and Vlijm, L. (Eds) (1987). The Ecology of Biological Invasions. *Proc. Koninkliijke Netherlandse Akad. van Wetenschappen.* **90**, 1–80.

II. Working Groups

A. Invasions into Nature Reserves

Macdonald, I. A. W., Loope, L. L. Usher, M. B., and Hamann, O. *Wildlife conservation and the invasion of nature reserves by alien species: A Global Perspective.* Submitted to Biological Conservation

B. Invasions in Mediterranean-climate regions

di Castri, F., and Groves, R. H. (Eds). *Biogeography of Biological Invasions in Mediterranean Climate Regions.* (In prep.)

C. Modelling the Invasion Process

Drake, J. A., and Williamson, M. H. (1986). Invasions of natural communities. *Nature*, **319**, 718–19.

D. History of Invasions

di Castri, F. (Ed.). *History and Patterns of Biological Invasions in Europe and the Mediterranean Basin.* (In prep.)

Species Index

Topical and Correlative Index

This index contains cross references to critical books on biological invasions.